Geotechnical Earthquake Engineering and Soil Dynamics

Geotechnical Earthquake Engineering and Soil Dynamics

Edited by **Agnes Nolan**

NY Research Press
New York

Published by NY Research Press,
23 West, 55th Street, Suite 816,
New York, NY 10019, USA
www.nyresearchpress.com

Geotechnical Earthquake Engineering and Soil Dynamics
Edited by Agnes Nolan

International Standard Book Number: 978-1-63238-214-6 (Hardback)

Printed in the United States of America.

Contents

Preface

This book was inspired by the evolution of our times; to answer the curiosity of inquisitive minds. Many developments have occurred across the globe in the recent past which has transformed the progress in the field.

Various aspects of geotechnical earthquake engineering and soil dynamics are highlighted in this all-inclusive book. The current progress in the field of earthquake engineering has been discussed with primary focus on the seismic safety of dams and underground monuments, Bryan's effect, and the mitigation plans against landslide and fire whirlwind. The book discusses various interesting researches that have been contributed by researchers and experts from many countries. The researches presented in this book will be helpful for graduates, researchers and scientists working in these areas of structural and earthquake engineering. It will also be of significance to civil engineers working on building and reconstruction of structures such as dams, buildings, roads and others.

This book was developed from a mere concept to drafts to chapters and finally compiled together as a complete text to benefit the readers across all nations. To ensure the quality of the content we instilled two significant steps in our procedure. The first was to appoint an editorial team that would verify the data and statistics provided in the book and also select the most appropriate and valuable contributions from the plentiful contributions we received from authors worldwide. The next step was to appoint an expert of the topic as the Editor-in-Chief, who would head the project and finally make the necessary amendments and modifications to make the text reader-friendly. I was then commissioned to examine all the material to present the topics in the most comprehensible and productive format.

I would like to take this opportunity to thank all the contributing authors who were supportive enough to contribute their time and knowledge to this project. I also wish to convey my regards to my family who have been extremely supportive during the entire project.

Editor

Review on Liquefaction Hazard Assessment

Neelima Satyam
Earthquake Engineering Research Centre
International Institute of Information Technology Hyderabad
India

1. Introduction

Experience from past earthquakes has demonstrated the vulnerability of structures to seismically induced ground deformation. During earthquake, soil can fail due to liquefaction with devastating effect such as land sliding, lateral spreading, or large ground settlement. The phenomenon of liquefaction of soil had been observed for many years, but was brought to the attention of engineers after Niigata (1964) Alaska earthquakes (1964). Liquefaction is a phenomenon in which the strength and stiffness of a soil is reduced by earthquake shaking or other rapid loading. Liquefaction and related phenomena have been responsible for tremendous amounts of damage in historical earthquakes around the world (Borchardt, 1991). During the Bhuj earthquake on 26th January 2001 (M=7.7) lot of damages had been occurred due to liquefaction and other ground failures (Rao and Mohanty, 2001a). From these investigations it was observed that a vast majority of liquefaction occurrences were associated with sandy soils and silty sands of low plasticity.

For the last four decades, investigations on understanding the liquefaction phenomena were carried out and have resulted in several different perspectives in describing various liquefaction-related phenomena. Liquefaction is looked upon as the condition at which the effective stress reaches (temporarily) a value of zero by few, while others consider liquefaction to have occurred when the soil deforms to large strains under constant shearing resistance. The first phenomenon is referred to as cyclic mobility and the second as flow liquefaction which may result in significant lateral deformations by either of them. To date, most research into liquefaction hazards has concentrated on the question of liquefaction potential, i.e., whether or not liquefaction will occur. The influence of liquefaction on the performance of structures, however, depends on the *effects* of liquefaction. While estimation of liquefaction effects has been improved by development of empirical procedures, the uncertainty involved in predicting these effects is still extremely high. More reliable prediction of structural performance requires more accurate prediction of liquefaction effects (Steven L Kramer, et.al; 2001).

2. Mechanism of soil liquefaction

It is necessary to understand the mechanism of soil liquefaction, where it occurs and why it occurs so often during earthquakes. Figure 1 clearly depicts the mechanism of soil liquefaction. Liquefaction of soil is a process by which sediments below water table temporarily lose shear strength and behave more as a viscous liquid than as a solid. The water in the soil voids exerts pressure upon the soil particles. If the pressure is low enough, the soil stays stable. However, once the water pressure exceeds a certain level, it forces the soil particles to move relative to each other, thus causing the strength of the soil to decrease and failure of the soil follows. During earthquake when the shear wave passes through saturated soil layers, it causes the granular soil structure to deform and the weak part of the soil begins to collapse.

The collapsed soil fills the lower layer and forces the pore water pressure in this layer to increase. If increased water pressure cannot be released, it will continue to build up and after a certain limit effective stress of the soil becomes zero. If this situation occurs then the soil layer losses its shear strength and it can not certain the total weight of the soil layer above, thus the upper layer soils are ready to move down and behave as a viscous liquid. It then is said that soil liquefaction has occurred.

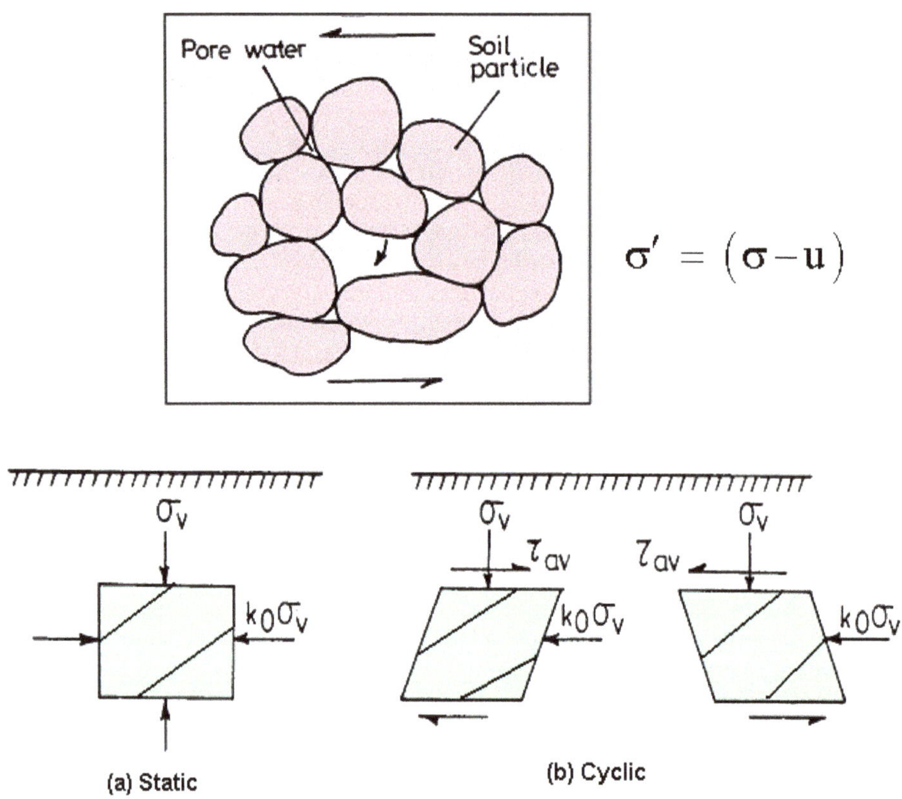

Fig. 1. Mechanism of Soil Liquefaction

2.1 Stress condition at liquefaction

The basic difference between solid state and liquid state of a substance is that the substance in its solid state shows resistance to deformation when subjected to external forces, whereas the substance in its liquid state does not have this property. Therefore, the process of transformation of any substance from solid state into a liquid state is substantially a process of diminishing in shear resistance of the material. The shear resistance of cohesion less soil is mainly proportional to the intergranular pressure and the co-efficient of friction between solid particles which is usually given by the following relationship

$$\tau_s = \sigma' \tan\phi \quad (1)$$

Since $\sigma' = (\sigma - u)$

$$\tau_s = (\sigma - u)\tan\phi \quad (2)$$

where, τ_s is shear resistance, σ' is the effective normal stress, σ is the total normal stress, u is the pore water pressure and ϕ' is the angle of internal friction in terms of effective stress.

The condition for liquefaction is $\sigma' = 0$ then u will be equal to σ. This can be defined in terms of Lambe parameters as

$$q = ½(\sigma_1 - \sigma_3) = ½(\sigma_1' - \sigma_3') \quad (3)$$

$$p = ½(\sigma_1 + \sigma_3) \quad (4)$$

$$p' = 1/2(\sigma_1' + \sigma_3') \quad (5)$$

where, σ_1 and σ_3 are maximum and minimum total principal stress, σ_1' and σ_3' maximum and minimum effective principal stress respectively . Then the condition for liquefied soil will be $q = 0$, $p' = 0$ and $p = u$.

Although the condition of soil liquefaction obeys the same condition as explained above, the mechanism of liquefaction process is different. Three typical mechanisms of soil liquefaction are identified as explained below:

2.2 Liquefaction caused by seepage pressure only: Sand boils

If the pore water pressure in a saturated sand deposit reaches and excesses the overburden pressure, the sand deposits will float or boil and lose entirely its bearing capacity. This process is nothing to do with the density and volumetric contraction of sand. Therefore, it has been usually considered as a phenomenon of seepage instability. However, according to the mechanism behavior of the material, it also belongs to the category of soil liquefaction.

2.3 Liquefaction caused by monotonous loading or shearing: Flow slide

The concept of critical void ratio has been suggested by Casagrande. The skeleton of loose saturated sand exhibits irreversible contraction in bulk volume under the action of monotonous loading or shearing, which cause increase of pure water pressure and decrease of effective stress and finally brings about an unlimited flow deformation.

2.4 Liquefaction caused by cyclic loading or shearing: Cyclic mobility

With various experimental techniques and testing apparatus it has been revealed that cohesion less soil always show volumetric contraction at low shear strain level, but may dilate at higher shear strain level depending upon the relative density of soil. Therefore, under the action of cyclic shearing a saturated cohesion less soil could show liquefaction at time intervals when shear strain is low, but may regain shear resistance in time intervals when the shear strain level is higher. A sequence of such sort of intermittent liquefaction would bring about the phenomenon of cyclic mobility with limited flow deformation. If the saturated cohesion less soil was loose enough to keep contraction at high shear strain level, then it also could come out to be an unlimited flow deformation.

3. Evaluation of liquefaction potential

The liquefaction potential of any given soil deposit is determined by a combination of the soil properties, environmental factors and characteristics of the earthquake to which it may be subjected. Specific factors that any liquefaction evaluation desirably takes into account include the following:

SOIL PROPERTIES

- Dynamic shear modulus
- Damping characteristics
- Unit weight
- Grain size characteristics
- Relative density
- Soil structure

ENVIRONMENTAL FACTORS

- Method of soil formation
- Seismic history
- Geologic history (aging, cementation)
- Lateral earth pressure coefficient
- Depth of water table
- Effective confining pressure

EARTHQUAKE CHARACTERISTICS

- Intensity of ground shaking
- Duration of ground shaking

Some of these factors cannot be determined directly, but their effects can be included in the evaluation procedure by performing cyclic loading tests on undisturbed samples or by measuring the liquefaction characteristics of the soil by means of some in-situ tests. The evaluation of liquefaction potential is based on two approaches (i) macroscopic evaluation and (ii) microscopic evaluation. Evaluation of liquefaction potential can be done using field tests or laboratory tests. The preliminary assessment of the liquefaction potential of a soil deposit over a large area in a seismically active region can be done using the following indices, which are characteristics of liquefiable soils:

- Mean grain size, D_{50} = 0.02- 1.0 mm
- Fines content < 10%
- Uniformity coefficient < 10

- Relative density < 75%
- Plasticity index < 10
- Intensity of an earthquake > VI
- Depth < 15m

There are three different ways to predict liquefaction susceptibility of a soil deposit in a particular region. They are (a) Historical criteria, (b) Geological and Geomorphological criteria and (c) Compositional criteria (Kramer, 2000). According to historical criteria soils that have liquefied in past can liquefy in future also. With the help of past earthquake records one can predict the liquefaction in future.

The type of geological processes that created a soil deposit has strong influence on its liquefaction susceptibility. Deposits formed by rivers, lakes, and wind and by man made deposits particularly those created by the process of hydraulic filling are highly susceptible to liquefaction. It also depends on soil type. Uniform graded soils are highly susceptible than well-graded soil deposits also, soils with angular particles are less susceptible than soils with rounded particles.

Cohesive soils with the following properties are vulnerable to significant strength loss under relatively minor strains (Seed et al.1983) i.e. if percent finer than (0.002 mm) is less than 30 percent, liquid limit less than 35 percent and if the moisture content of the insitu soil is greater than 0.9 times the liquid limit (i.e., sensitive clays).

In addition to sandy and silty soils, some gravely soils are potentially vulnerable to liquefaction. Most gravely soils drain relatively well, however gravelly soils are also liquefiable when the voids are filled with fine particles and if it is surrounded by less pervious soils, drainage can be impended and may be vulnerable to cyclic pore pressure generation and liquefaction.

Gravels tend to be deposited in a more turbulent depositional environment than sands or silts, tend to be fairly dense, and so generally resist liquefaction. Accordingly, conservative preliminary methods may often suffice for evaluation of their liquefaction potential. For example, gravely deposits that can be shown to be pre-Holocene in age (older than about 11,000 years) are generally not considered susceptible to liquefaction. Andrus and Stokoe (2000) compiled 225 liquefaction case histories from the United States, Taiwan, Japan and China. Among these case history sites 90% of the liquefied soils had a critical layer thickness of less than 7 m, an average depth below land surface of less than 8 m, and water table depth is at less than 4 m below ground surface.

3.1 Field methods

The use of insitu testing is the dominant approach in common engineering practice for quantitative assessment of liquefaction potential. Calculation of two variables is required for evaluation of liquefaction resistance of soils. They are as follows:

1. The seismic demand on a soil layer, expressed in terms of CSR and
2. The capacity of the soil to resist liquefaction, expressed in terms of *CRR*.

The models proposed by Seed and Idriss (1971), Seed and Peacock (1971), Iwasaki (1978) and Robertson and Wride (1998) methods are extensively used for predicting liquefaction potential using field data. Youd et al. (2001) reviewed in detail the available field methods available for the evaluation of liquefaction potential of soils.

3.1.1 SPT based methods

Standard penetration test is widely used as an economical, quick and convenient method for investigating the penetration resistance of non-cohesive soils. This test is an indirect means to obtain important design parameters for non-cohesive soils. The use of SPT as a tool for evaluation of liquefaction potential began to evolve in the wake of a pair of devastating earthquakes that occurred in 1964; the 1964 Great Alaskan Earthquake (M=9.2) and 1964 Niigata Earthquake (M=7.5), both of which produced significant liquefaction related damage.

It should be ensured that the energy of the falling weight is not reduced by friction between the drive weight and the guides or between rope and winch drum. The rods to which the sampler is attached for driving should be straight, tightly coupled and straight in alignment. For driving the casing, a hammer heavier than 63.5 kg may be used. Standard Penetration Test set up and accessories are Standard split spoon sampler, 65 kg hammer, guide pipe assembly, anvil and drill rod.

In the standard penetration test, a standard split spoon sampler is driven into the soil at the bottom of a borehole by giving repeated blows (30-40 blows per minute), using a 65 kg hammer released from a height of 75 cm. The blow count is found for every 150 mm penetration. If full penetration is obtained, the blows for the first 150 mm are ignored as those required for the seating drive. The number of blows for the next 300 mm of penetration is recorded as the Standard Penetration Resistance, called the 'N' value.

- If number of blows to drive 15 centimeters exceeds 50, the test has to be repeated.
- If the stratification is homogeneous and denseness is not very erratic, the spacing for the test depth can be increased suitably beyond a depth of 6 meters or so.
- Wide variations in N-value at given depth along the section would show heterogeneity of the subsoil and denseness.

3.1.1.1 Factors affecting test results

i. Effective over burden pressure: Effective over burden pressure affects results considerably.

- Desai (1968) reported that N value at shallow depths underestimates relative density. N-value corrected for the surcharge effects represent normal pressure is changed.
- It was proved that positive or negative pore pressure developed in fine sands and silty sands depend upon the denseness of the sub-soil and thus the effective normal pressure against split spoon sampler is altered.
- It was observed that the removal of 5 meters of soil affected N value considerably. In cohesive soils N-value do not reflect the precompression load and shear strengths if soils are partly saturated.

ii. Grain size and shape

- Gravels have reduced friction and penetration resistance, which will block the SPT sampler and gives erratic results.

- The particle size effect on N value is prominent if 30 % soil particles are less than 0.1 mm and soils is saturated or dry.
- In case of silty fine sands and very fine sand, positive or negative pore pressure can be generated depending on the state of compactness and N values will change according

iii. Degree of saturation

- N values will be reduced by 15 % due to saturation and this is more pronounced in case of loose soils.
- Penetration resistance increased due to saturation in case of denser soils while in case of loose, fine and silty sand N value is considerably reduced.

3.1.1.2 Corrections applied in standard penetration test

i. Corrections due to Overburden

It is an established fact that SPT blows are greatly affected by the overburden pressure at the test point. The effect of length or weight of the driving rods is not so pronounced and may be neglected. According to their investigations, Terzaghi-Peck correlation between SPT blows and density index is valid under an overburden pressure of approximately 280 kPa.

The curves are based on results for air dry and partially wetted, cohesion less sands and are considered conservatively reliable in all sands, saturated or unsaturated. But it is generally felt that the corrections provide over-estimate of density index.

For interpretation and correlations of SPT results the current thinking is to adopt 100kPa ($1 kg/cm^2$) as the reference overburden pressure and the N blows corrected for this pressure are called the normalized or corrected values, Nc.

$$Nc = Cn * Nr \tag{6}$$

$$Cn = 0.77\log_{10}\left(\frac{20}{\sigma'}\right), \text{ where } \sigma' > 0.25 \text{ kg/cm}^2 \tag{7}$$

where, Nr= Observed N value in the field and Cn = Correction factor

The another simple relation for the correction factor Cn that greatly cover more research work on the correction factor carried in the USA was given in Eqn 8 as below.

$$Cn = \sqrt{100/\sigma'} \tag{8}$$

where, σ′ in kPa

ii. Corrections due to Dilatancy

In submerged very fine or silty sands below the water table, the observed value of N may be too great (compared to the penetration resistance of permeable submerged soils of equal density index) if the void ratio is below the critical voids ratio which corresponds approximately to N=15. Submerged fine sands and silty sands offer increased resistance due to excess pore water set up during driving and unable to dissipate immediately (dilatancy effect). The corrected value of N is defined in IS: 2131(1981) is as follows

$$N'=15+\frac{1}{2}(Nc-15) \tag{9}$$

where, Nc is corrected value after over burden correction

Wherever both the overburden and submerged corrections are necessary, the overburden correction is applied first.

3.1.1.3 Seed and Idriss (1971) method

The initial approach for evaluating behavior of soils in the field during dynamic loading was developed by Seed and Idriss (1971). The procedure is referred to as the simplified procedure, and involves the comparison of the seismic stresses imparted onto a soil mass during an earthquake (Cyclic Stress Ratio, CSR) to the resistance of the soil to large magnitude strain and strength loss (Cyclic Resistance Ratio, CRR). The CSR estimation is based on the estimated ground accelerations generated by an earthquake, the stress conditions present in the soil, and correction factors accounting for the flexibility of the soil mass (Youd and Idriss 1997). Seed and Idriss developed this empirical method by combining the data on earthquake characteristics and in-situ properties of soil deposits, which is widely used all over the world for the assessment of liquefaction hazard. For earthquakes of other magnitudes, the appropriate cyclic strength is obtained by multiplying with a factor called magnitude scaling factor MSF. The factor of safety against liquefaction, F_L can then be estimated as the ratio of CSR and CRR.

3.1.1.4 Seed and Peacock (1971) method

In the Seed and Peacock (1971) method, the average shear stress τ_{av} will be computed same as in Seed and Idriss method. Using corrected SPT 'N' value and the proposed chart by Seed and Peacock, τ_Z can be calculated at the desired depth of the soil strata. If $\tau_{av} > \tau_Z$ then soil will liquefy at that zone.

3.1.1.5 Iwasaki et al. (1982) method

Iwasaki et al. (1982) proposed a simple geotechnical method as outlined in the Japanese Bridge Code (1991). In this method, soil liquefaction capacity factor R, is calculated along with a dynamic load L, induced in a soil element by the seismic motion. The ratio of both is defined as 'liquefaction resistance'. The soil liquefaction capacity is calculated by the three factors, which take into account the overburden pressure, grain size and fine content. In this method it is assumed that the severity of liquefaction should be proportional to the thickness of the liquefied layer, proximity of the liquefied layer to the surface, and the factor of safety of the liquefied layer.

The prediction by the liquefaction potential index is different than that made by the simplified procedure of Seed and Idriss (1971). According to Toprak and Holzer (2003), the simplified procedure predicts what will happen to a soil element whereas the index predicts the performance of the whole soil column and the consequences of liquefaction at the ground surface. Sonmez (2003) modified this method by accepting the threshold value of 1.2 of factor of safety as the limiting value between the categories of marginally liquefiable to non-liquefiable soil.

The NCEER workshops in 1996 and 1998 resulted in a number of suggested revisions to the SPT based procedure. Cetin et al. (2000) reexamined and expanded the SPT case history database. The data set by Seed et al. (1984) had 125 cases of liquefaction/ no liquefaction in 19 earthquakes, of which 65 cases pertain to sands with fines content ≤ 5%, 46 cases had fines content between 6 and 34% and 14 cases had ≥ 35%. Cetin et al. (2000) used their expanded data set and site response calculations for estimating CSR to develop revised relationships. Idriss and Boulanger (2004) presented a revised curve between CSR and modified SPT value based on the reexamination of the available field data.

3.1.2 CPT based method

The CPT test has become one of the most common and economical methods of subsurface exploration. The cone penetrometer is pushed into the ground at a standard velocity of 2 cm/sec and data is recorded at regular intervals (typically 2 or 5 cm) during penetration. The results provide excellent stratigraphic detail and repeatability provided proper care has been taken in calibration of the equipment (transducers and electronics). The cone penetrometer is instrumented to record a number of different parameters, with the most common being the force of the tip, the force of the sleeve, and the pore pressure behind the tip. Cone penetrometers have also been used to provide or measure electrical properties, shear wave velocities, visual images of the soil, acoustic emissions, temperature and water samples.

The CPT is a versatile sounding method that can be used to determine the materials in a soil profile and their engineering properties. The equipment consists of a 60^0 cone, with 10 cm^2 base area and a 150 cm^2 friction sleeves located above the cone. A sensor is attached for measuring tip resistance, pore pressure and sleeve resistance. To evaluate the potential for soil liquefaction it is important to determine soil stratification and in-situ soil state. The CPT is an ideal in-situ test to evaluate the potential for soil liquefaction because of its repeatability, reliability, continuous data and cost effectiveness.

3.1.2.1 Robertson and Wride (1998) method

A simplified method to estimate cyclic shear resistance (CSR) was developed by Seed and Idriss (1971) based on maximum ground acceleration at the site as under:

$$CSR = \tau_{av} / \sigma_0' = 0.65 \, (MWF) \, (\sigma_0 / \sigma_0') \, (a_{max} / g) \, r_d \tag{10}$$

$$MWF = (M)^{2.56} / 173 \tag{11}$$

where, MWF is the magnitude weighting factor and M is the earthquake magnitude, commonly M = 7.5

Seed et al. (1985) also developed a method to estimate the cyclic resistance ration (CRR) for clean sands and silty sands based on the CPT using normalized penetration resistance.

The cone penetration resistance q_c can be normalized as

$$q_{c1N} = C_Q (q_c / p_a) \quad (12)$$

$$C_Q = (P_a / \sigma_0')^n \quad (13)$$

where, C_Q is normalized factor for cone penetration resistance, P_a is the atmosphere of pressure in the same units as σ_0'and n ia an exponent that varies with soil type (= 0.5 for sands and 1 for clays) and q_c is the field cone penetration resistance at tip. The normalized penetration resistance (q_{c1N}) for silty sands is corrected to an equivalent clean sand value $(q_{c1N})_{CS}$ as

$$(q_{c1N})_{CS} = K_C\, q_{c1N} \quad (14)$$

where, K_C is the correction factor for grain characteristics and is defined as below by Robertson and Wride (1998).

$$K_C = 1.0 \qquad \text{for } I_C \leq 1.64 \quad (15)$$

$$K_C = -0.403\, I_C^4 + 5.581\, I_C^3 - 21.63\, I_C^2 + 33.75\, I_C - 17.88 \quad \text{for } I_C > 1.64 \quad (16)$$

If $I_C > 2.6$, the soil in this range are likely to clay rich or plastic to liquefy. I_C is the soil behavior type index and is calculated as

$$I_C = [(3.47 - \log Q)^2 + (1.22 + \log F)^2]^{0.5} \quad (17)$$

where Q is normalized penetration resistance

$$= [(q_c - \sigma_0)/P_a][P_a/\sigma_0']^n \quad (18)$$

$$F = [f_s / (q_c - \sigma_0)] * 100\% \quad (19)$$

where f_s being the sleeve friction stress

$$CRR_{7.5} = 0.833\left[\frac{(q_{c1N})_{cs}}{1000}\right] + 0.05 \qquad \text{if } (q_{c1N})_{cs} < 50 \quad (20)$$

$$CRR_{7.5} = 93\left[\frac{(q_{c1N})_{cs}}{1000}\right]^3 + 0.08 \qquad \text{if } 50 \leq (q_{c1N})_{cs} < 160 \quad (21)$$

where, $(q_{c1N})_{cs}$ is clean sand cone penetration resistance normalized to approximately 100 kPa (1atm). Then, using the equivalent clean sand normalized penetration resistance $(q_{c1N})_{cs}$, CRR can be estimated from the Fig. 2.

The CPT based liquefcation correlation was reevaluated by Idriss and Boulanger (2004) using case history data compiled by Shibata and Teparaksa (1988), Kayen et al. (1992), Boulanger (2003) and Moss (2003).

Moss (2003) has provided a most comprehensive compilation of field data and associated interpretations. He used friction ratio R_f instead of the parameter I I_C, soil behavior type index and examined for the cohesion less soils with fines content greater than or equal to 35%.

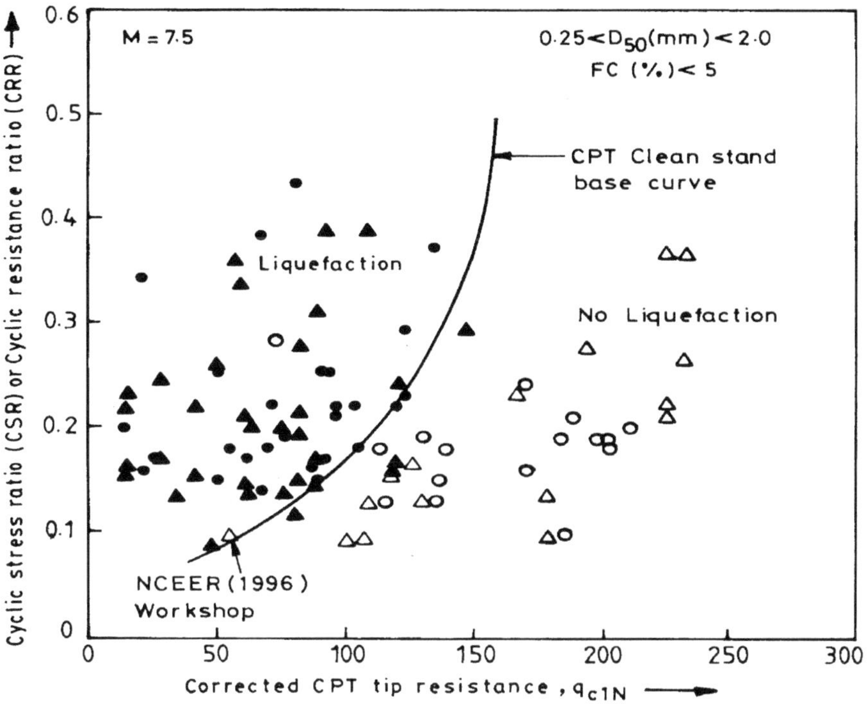

Fig. 2. Calculation of CRR from CPT q_{c1N} (Youd et al., 2001)

3.1.3 Shear wave velocity (V_S) based methods

The shear wave velocity based procedure has advanced significantly in recent years, with improved correlations and more databases as summarized by Andrus and Stokoe (2000). Shear wave velocity can be determined either by subsurface geophysical method or by surface geophysical method as explained earlier Liquefaction potential can be evaluated from the shear wave velocity (V_S) using the following three methods. These procedures can be useful particularly for sites where it is difficult to penetrate or sample soils.

3.1.3.1 Andrus and Stokoe (2000) method

Andrus and Stokoe have carried extensive research into the use of shear wave velocity as an index of liquefaction resistance. Various investigators have developed relationships between shear wave velocity and liquefaction resistance. This section presents detailed guidelines for applying the procedure described in Andrus and Stokoe that was developed using suggestions from two workshops and following the general format of the Seed and Idriss (1971) simplified procedure. Andrus and Stokoe (2000) developed liquefaction resistance criteria from field measurements of shear wave velocity Vs. This method form the basis for the currently accepted shear wave velocity criteria for liquefaction potential assessment. Liquefaction hazard assessment based on this method is done and a liquefaction hazard map of Delhi region hazard map is prepared.

Shear wave velocity Vs is corrected similar to SPT 'N' value using the atmospheric pressure Pa and initial effective vertical stress, σ_0'. The cyclic resistance ratio (CRR) is determined empirically at various depths using the correlation developed between CRR and the shear wave velocity for the liquefaction assessment. The detailed analysis procedure is outlined in the later chapter.

3.1.3.2 Hatanaka, Uchida and Ohara (1997) method

Hatanaka et al. (1997) performed a systematic research relating the undrained cyclic shear strength of high quality undisturbed gravel samples to the shear wave velocity measured insitu V_{S1} is used for correcting the effect of effective confining stress on V_S by using the following equation.

$$V_{S1} = \frac{V_s}{(\frac{\sigma v'}{98})^{3/8}} \quad (22)$$

The value of 3/8 in the above equation is the average value of 0.5 and 0.25 which covered the test result. It is also important to know the K_O value for converting the liquefaction strength obtained in laboratory R_{LAB} to that in the field R_{INSITU} based on the following equation

$$R_{INSITU} = 0.9\{\frac{(1+2K_o)}{3}\} R_{LAB} \quad (23)$$

The variation of shear wave velocity with confining pressure for various soil conditions was given by Hardin (1963).

3.1.3.3 Tokimatsu, Yamazaki, and Yoshimi (1986) method

This method is proposed by Tokimatsu et al. (1986). The working principle in this procedure is that the liquefaction strength has a good correlation with elastic shear modulus for a given soil under a given confining pressure. The procedure is outlined in Fig. 3.The procedure shown on the left corresponds to the shear wave velocity measurement in-situ. Based on the measured shear wave velocity, Vs the elastic shear modulus in the field G_{OF} can readily be determined. The procedure on the right involves laboratory tests on a specimen reconstituted from the sample obtained at the site. Both liquefaction test, the shear modulus at small shear strain G_{OL}, of the specimen is measured and compared with G_{OF}. If they are equal, the liquefaction test has to be done on the same specimen. If they are not equal that usually means that G_{OF} is larger than G_{OL}, cyclic shear stresses are applied to the specimen until the shear modulus reaches the value; then the liquefaction test is conducted. In order to take the effects of void ratio (e) and confining stress (σ_c') on the G_O is normalized with respect to e and σ_c' as described as

$$G_O = \frac{G_o}{\left\{F(e_{\min})(\sigma')^n\right\}} \quad (24)$$

$$F(e_{\min}) = (2.17 - e_{\min})^2 (1 + e_{\min}) \quad (25)$$

where, n is a factor whose value equals to 2/3.

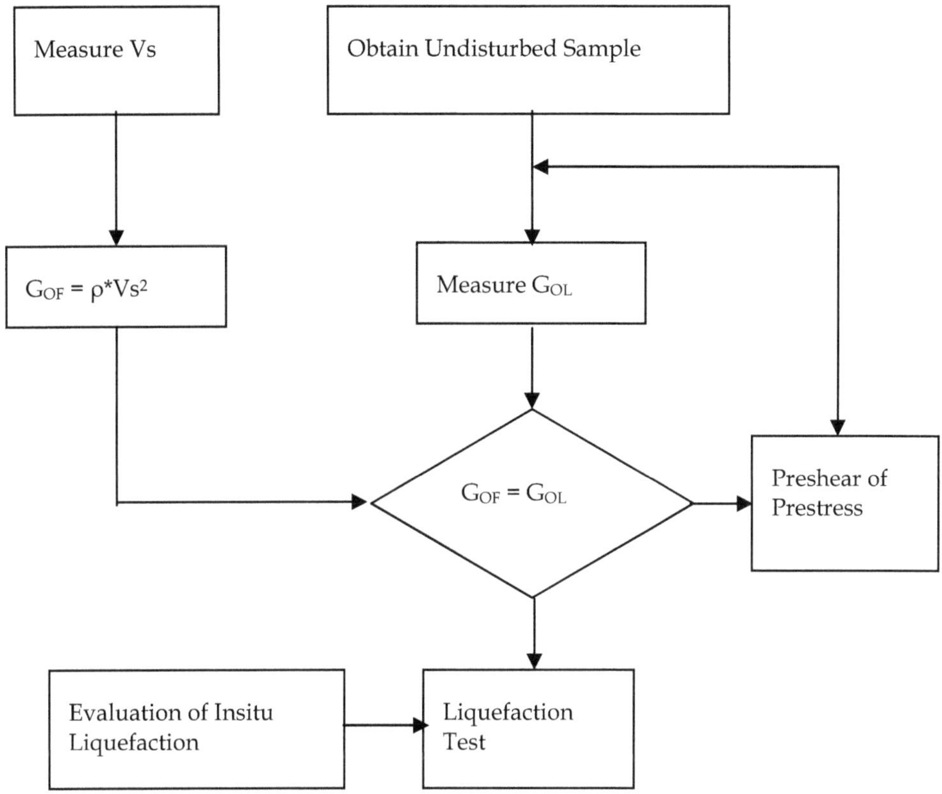

Fig. 3. Outline of Method (Tokimatsu et al., 1986)

In this method, it is important to determine G_{OL} from laboratory test. G_{OL} value highly depends on the confining stress at laboratory. As a result it is an important work to reasonably estimate the k_o value of insitu soils.

Tsurumaki et al. (2003) verified the simplified procedure using new data for different kinds of gravelly soils by undrained cyclic triaxial tests on undisturbed sample of gravelly soils obtained by the in-situ freezing sampling method. The undrained cyclic strength increases with increasing shear wave velocity.

3.2 Laboratory methods

3.2.1 Cyclic triaxial test

The cyclic triaxial test is the most commonly used test for the measurement of dynamic soil properties at high strain levels. This test simulates the liquefaction phenomenon during earthquakes by appling cyclic shear to the saturated sandy soil under undrained condition. Axial loading is applied in steps to the specimen and the shear strain and shear stress at various levels of loading. Five to ten levels of shear strain amplitudes are chosen from the range of 10^{-5} to 10^{-2} for testing. Dynamic deformation characteristics are influenced by the effective confining pressure during the test. When an undisturbed sample of normally consolidated soil is obtained, the effective vertical pressure at the depth of sampling is isotropically applied by cell pressure to avoid the influence of over consolidation. In order to obtain in-situ shear modulus of the soil from the laboratory test results, correction of these results is necessary so that a shear modulus corresponding to the average effective principal stress at the sample depth is obtained.

Seed and Lee (1966) were the first to reproduce liquefaction in a cyclic load triaxial test on loose and dense sands and concluded that liquefaction occurs more easily in sandy soils having higher void ratios and void ratio remaining constant, lower the effective confining pressure higher the liquefaction susceptibility.

3.2.2 Cyclic direct simple shear test

The cyclic direct simple shear test is capable of reproducing earthquake stress conditions much more accurately than the cyclic triaxial test. It is most commonly used for liquefaction testing. In this test, a short cylindrical specimen is restrained against lateral expansion by rigid boundary platens, a wire reinforced membrane or with a series of stacked rings. By applying cyclic horizontal shear stresses to the top or bottom of the specimen, the test sample is deformed in the same way as an element of soil subjected to vertically propagating S waves. In recent years, simple shear devices that allow independent control of vertical and horizontal stresses have been developed. To better simulate actual earthquake conditions, Pyke (1979) used a large-scale simple shear apparatus. It is concluded that cyclic strength is related to the relative density of the soil and cyclic stresses that cause liquefaction in simple shear were less than those causing liquefaction in triaxial shear.

3.2.3 Cyclic torsional shear test

Many of the difficulties with cyclic triaxial and cyclic shear test can be overcomed with cyclic torsional shear test. This is mostly used to determine stiffness and damping characteristics over a wide range of strain levels. It allow isotropic or an isotropic initial stress conditions and can impose cyclic shear stresses on horizontal planes with continuous rotation of principal axes. Dobry et al.(1995) used strain controlled cyclic torsional loading along with stress controlled axial loading of solid specimens and has proven effective for measurement of liquefaction behavior. Torsional testing of solid specimens, however produces shear stresses that range from zero along the axis of the specimen to a maximum value at the outer edge. To increase the radial uniformity of shear strains, a hollow cylindrical cyclic torsional shear apparatus were also developed. While hollow cylinder tests offer perhaps the best uniformity and control over stresses and drainage. Ishihara and Li (1972) developed a torsinal triaxial shear test and conducted strain controlled tests on solid cylinders of saturated sands. These tests helped in establishing relationship between cyclic triaxial tests, cyclic simple shear tests and the torsional triaxial test.

3.2.4 Shake table test

Shake table tests of many sizes are being used for liquefaction studies on saturated soil samples prepared in a container, fixed to a shaking platform and vibrated at the desired frequency for a prescribed time. A surcharge is placed on the sample to provide the confining pressure. The measurements of acceleration, pore water pressure and settlements are made during the test. Shaking tables utilize a single horizontal translation degree of freedom, but shake table with multiple degrees of freedom have also been developed. Kokusho (1987) developed a numerical model based on shake table test.

4. Magnitude scaling factor

Seed and Idriss (1982) introduced a correction factor termed as magnitude scaling factor (MSF). This factor can be used to scale up or scale down the CRR based curves upward or downward, depending upon the earthquake magnitude. Figure 4 gives the curves proposed by various authors for different earthquake magnitudes.

4.1 Seed and Idriss (1982) scaling factor

Seed and Idriss (1982) developed a set of MSF from average number of loading cycles for various earthquake magnitude and laboratory test results. These MSF have been routinely applied in engineering practice.

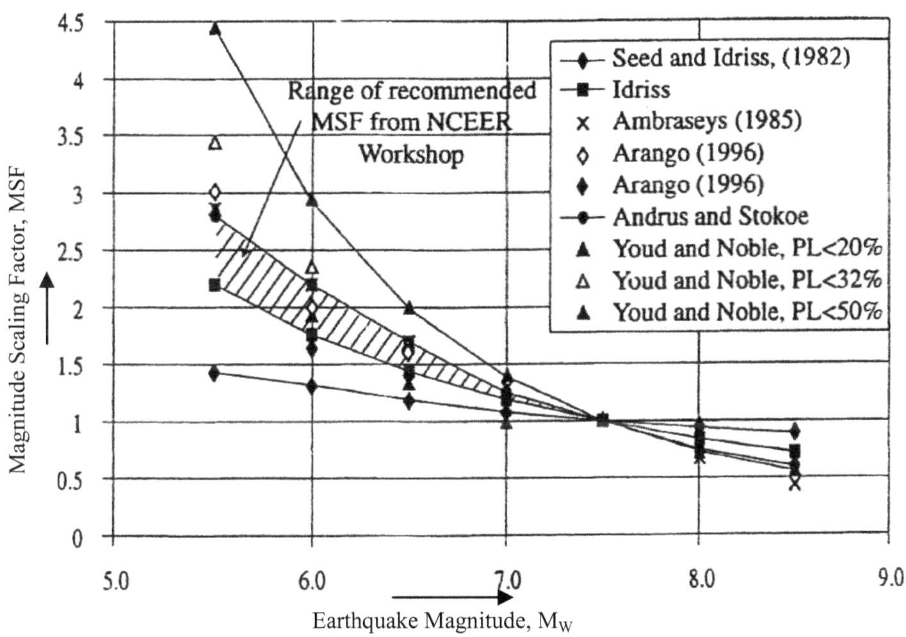

Fig. 4. Magnitude Scaling Factor Derived by Various Investigations (Youd et al., 2001)

Idriss (1995) developed a revised set of magnitude scaling factor and was defined as

$$MSF = \left(\frac{10^{2.24}}{M_w^{\ 2.56}}\right) \tag{26}$$

The revised scaling factors were higher than the original scaling factors for magnitude < 7.5 and somewhat lower than the original factors for magnitude > 7.5 relative to the original scaling factors the revised factors lead to a reduced calculated liquefaction hazard for magnitude < 7.5 but increase calculated hazard for magnitude > 7.5. National Center for Earthquake Engineering Research (NCEER, 2001) workshop participants suggested this revised scaling factors as lower bound for MSF values.

4.2 Ambraseys (1988) scaling factor

Ambraseys (1988) analyzed the liquefaction data and calculated cyclic stress foe sites that did or did not liquefy versus N_{60}. Based on this he developed an empirical equation that define CRR as a function of N_{60} and moment magnitude. For magnitude less than 7.5, MSF s suggested by Ambraseys are significantly larger than both the scaling factors developed by Seed and Idriss. For magnitudes > 7.5, Ambraseys factors are significantly lower and much more conservative. It is not recommended for hazard evaluation because of few data constrain.

4.3 Andrus and Stokoe (1997) scaling factor

Andrus and Stokoe developed scaling factors by drawing bounding curves for sites where surface effects of liquefaction were or were not observed for earthquake magnitudes of 6, 6.5 and 7.0.MSF for magnitude < 6 and > 7.5 were extrapolated from the following equation

$$MSF = \left(\frac{M_w}{7.5}\right)^{-2.56} \tag{27}$$

For magnitude < 7.5 the MSF s proposed by Andrus and Stokoe are rather close in value to the MSFs proposed by Ambraseys. For magnitude > 7.5 Andrus and Stokoe MSFs are slightly smaller than the revised MSFs proposed by Idriss.

4.4 Youd and Noble (1997) scaling factor

Youd and Noble (1997) used a probabilistic and logistic analysis to analyze case history data from sites was or were not reported following past earthquakes. They defined three sets of MSFs for different magnitude ranges and with different probabilities of liquefaction (P_L) occurrence.

$$P_L < 20\,\% \quad MSF = \left[\frac{10^{3.81}}{M^{4.53}}\right] \quad \text{for } M_W < 7 \tag{28}$$

$$P_L < 32\% \quad MSF = \left[\frac{10^{3.74}}{M^{4.33}}\right] \quad \text{for } M_W < 7 \tag{29}$$

$$P_L < 50\% \quad MSF = \left[\frac{10^{4.21}}{M^{4.81}}\right] \quad \text{for } M_W < 7.75 \tag{30}$$

The NCEER (2001) workshop report provides a useful insight in to the choice of MSF. For magnitude < 7.5, the lower bound for the recommended range is the new MSF proposed by

Idriss. The suggested upper bound is the MSF proposed by Andrus and Stokoe (1997). The upper bound values are consistent with MSFs suggested by Ambraseys (1988) and Youd and Noble(1997a) for $P_L < 20\%$. For magnitude > 7.5, the new factors recomnded by Idriss should be used. These new factors are smaller than the original Seed and Idriss (1982) factors. The relations by Ambraseys (1998) and Arango (1996) give significantly larger MSF values for earthquake magnitudes less than 7. Table 1 gives the MSF values given by various researchers.

Magnitude M	Seed and Idriss (1982)	Ambra-seys (1988)	Arango (1966)		Andrus and Stokoe (1997)	Youd and Noble (1997)		
			Distance Based	Energy Based		$P_L < 20\%$	$P_L < 32\%$	$P_L < 50\%$
5.5	1.43	2.86	3.00	2.20	2.80	2.86	3.42	4.44
6.0	1.32	2.20	2.00	1.65	2.10	1.93	2.35	2.92
6.5	1.19	1.69	1.60	1.40	1.60	1.34	1.66	1.99
7.0	1.08	1.30	1.25	1.10	1.25	1.00	1.20	1.39
7.5	1.00	1.00	1.00	1.00	1.00	---	---	1.00
8.0	0.94	0.67	0.75	0.85	0.80	---	---	0.73
8.5	0.89	0.44	---	---	0.65	---	---	0.56

Table 1. Magnitude Scaling Factors Suggested by Various Researchers (Youd and Noble,1997)

During the Niigata earthquake (M=7.5) on June 16, 1964, widespread liquefaction was observed in the low-lying areas. The liquefaction was accompanied by foundation failure and failure of retaining structures. It is important to note, however, that the structures themselves suffered very little damage; they essentially settled and rotated as rigid bodies under the loss of bearing capacity. Several of the apartment buildings were later jacked back to a vertical position and underpinned with new foundations. Nearly 20 years after the earthquake, excavation beneath the building, as part of the construction of upgraded foundations for an increase in the height of the building, revealed that the piles had been extensively damaged during the Niigata earthquake. Large displacements of the buildings were driven by gravity and resulted in performance that would be considered as "failure" by almost any definition, even though the structural elements of the buildings were virtually undamaged. In the 1964 Niigata earthquake, a number of RC buildings settled and tilted due to a loss of bearing capacity.

During the 1990 Luson earthquake, substantial building damage occurred in 2 to 4 story buildings resting on liquefied ground in Daguapan City. During the 1999 Kocaeli earthquake in Turkey, settlement or tilting of buildings due to liquefaction was widespread in Adapazari. Although liquefaction in this case was less severe than in Niigata probably because the soil contained a lot of fines, many 5 to 6 story buildings suffered heavy structural damage due to uneven settlement of at most 2 m .Riverbanks, levees, and dikes suffered liquefaction induced settlement and sliding failures during the 1993 Kushiro-Oki earthquake, the 2003 Tokachi-Oki earthquake and many other earthquakes. During the 2003

Tokachi-Oki earthquake, the right banks of the Tokachi River suffered severe damage due to liquefaction-induced deep sliding and crest settlement . Another failure mode is the uplift of subsurface structures such as pipelines, buried tanks, pools, etc. Concrete manholes and pipes of sewage systems are raised because of liquefaction in surrounding soils during the 1993 Kushiro-Oki earthquake, the 2003 Tokachi-Oki earthquake, and the 2004 Niigata ken-Chuetsu earthquake. Although surrounding ground appeared stable during earthquakes, sands backfilling buried structures liquefied and buoyant pressure in the liquefied fill raised them over 1.5 m. The detailed liquefaction assessment requires several inputs regarding the site specific geological, geophysical, geotechnical, seismotectonic, ground motion parameters and their effects on the structures considering their design aspects, local soil conditions which would enhance the earthquake effects like soil amplification, liquefaction of soils etc.

5. References

Ambraseys N. (1988) Engineering Seismology, *J. Earthq. Eng. and Struct.Dyn.*,17, p.66

Andrus R.D. and Stokoe K.H. (1997) Liquefaction Resistance of Soils from Shear Wave Velocity, *Proc. NCEER Workshop on Evaluation of Liquefaction Resistance of Soils*, 89-128.

Andrus R.D. and Stokoe K.H. (2000) Liquefaction Resistance Based on Shear-Wave Velocity, *J .Geotech. Engg. Div., ASCE.,* (126)11:1015-1025.

Arango I. (1996) Magnitude Scaling Factors for Soil Liquefaction Evaluations, *J.Geoecht. Engg. Div., ASCE.,*.122(11), 929-936.

Borcherdt R.D., Wentworth C.M., Glassmoyer G., Fumal T.,Mork, P. and Gibbs, J (1991) On the Observation, Characterization, and Predictive GIS Mapping of Ground Response in the San Francisco Bay Region,California, *Proc of 4th International Conference on Seismic Zonation*, 3, 545-552.

Boulanger, R.W. 2003. High overburden stress effects in liquefaction analysis. *Journal of Geotechnical and Geoenvironmental Engineering, ASCE*, 129(12): 1071–1082.

Cetin K.O., Seed R.B., Moss R.E.S., Der Kiureghian A., Tokimatsu K. H., L.F. Jr., and Kayen, R.E. (2000). Field Case Histories for SPT-Based In Situ Liquefaction Potential Evaluation, Geotechnical Engg. Research Report UCB/GT-2000/09.

Dobry, R., Tabaoda, V. and Liu.L. (1995) Centrifuge Modeling of Liquefaction Effects During Earthquakes, *Proc. of 1st International Conference on Earthquake Geotechnical Engg.*.3, 1291-1324.

Desai M.D (1968) Subsurface, Exploration by dynamic penetrometers. Nasal Pub. & printers, Surat, 17 pp.

Hatanaka M., Suzuki Y., Kawasaki T. and Masaaki E. (1988) Cyclic Undrained Shear Properties of High Quality Undisturbed Tokyo Gravel, *J. Soils and Foundations,* (28) 4, 57-68.

Hatanaka M., Uchida A. and Ohara J. (1997) Liquefaction characteristics of a gravelly fill liquefied during the 1995 Hyogo-ken Nanbu Earthquake, *Soils and Foundations*, 1, 3,107-115.

IDRISS I.M. and BOULANGER R.W. (2004) Semi-Empirical Procedures for Evaluating Liquefaction Potential During Earthquakes. Proc. of 11th SDEE and 3rd Conference. University of California, Berkeley IS: 2131 (1981) *Method for Standard Penetration Test for Soils*

Iwasaki T., Tokida K., Taksuoko F., Watanabe S., Yasuda S. and Sato H.(1982) Microzonation for Soil Liquefaction Potential using Simplified Methods, *Proc. of 3rd International Conf. on Microzonation,* 3, 1319-1330.

Kayen R. E., Mitchell J. K., Seed R. B., Lodge A., Nishio S., and Coutinho R. (1992) Evaluation of SPT, CPT, and Shear Wave Based Methods for Liquefaction Potential Assessment using Loma Prieta Data, *Proc. of 4th Japan-U.S. Workshop on Earthquake Resistant Design of Lifeline Facilities and Countermeasures for Soil Liquefaction,* Hamada, 177-204.

Kramer S,L. (2000) Geotechnical Earthquake Engineering, *Prentice Hall, NewJersey,* 653 P.

Kokusho T. (1987) In-situ Dynamic Soil Properties and their Evaluation, *Proc.8th Asian Regional Conf. of SMFE,* Kyoto, II: 215-240.

Moss R.E.S., Seed R.B., Kayen R.E., Stewart J.P., Youd T.L., and Tokimatsu, K. (2003) Field Case Histories for CPT-Based Insitu Liquefaction Potential Evaluation. *UC Berkeley Geoengineering Research Report No.UCB/GE-2003/04*

Pyke R. M. (1979) Nonlinear Soil Models for Irregular Cyclic Loadings, *J.Geotech. Engg. Div., ASCE,* 105(6), 715-726

Rao K.S. and Mohanty W.K. (2001a) The Bhuj Earthquake and Lessons for the Damages, IGS News, 33, 3-10.

Robertson P.K. and Wride C.E. (1998) Evaluating Cyclic Liquefaction Potential Using the Cone Penetration Test, *J. Canadian Geotechnical Engg.,*(35)3, 442-459.

Seed H. B. Tokimatsu K. Harder L. F. and Chung, R. M. (1985) Influence of SPT Procedures in Soil Liquefaction Resistance Evaluations, *J.Geotechnical Engg. Division, ASCE,* (111) 12, 1425-1445.

Seed H.B and Idriss I.M. (1971) Simplified Procedure for Evaluating Soil Liquefaction Potential, *J. Soil Mechanics and Foundations, ASCE,* (97) SM9,1249-1273.

Seed H.B. and Lee K.L. (1966) Liquefaction of Saturated Sands During Cyclic Loading, Journal of the Soil Mechanics and Foundation Division, ASCE, 92, 105-134

Seed H.B., and Idriss I.M. (1982) Ground Motions and Soil Liquefaction During Earthquakes: *Earthquake Engineering Institute Monograph,* p134.

Seed H.B., Idriss I.M., and Arango I. (1983) Evaluation of Liquefaction Potential Using Field Performance Data, *J. Geotechnical Engg., ASCE,* 109 (3)458-482.

Seed H.B. and Peacock W.H. (1971) Test Procedures for Measuring Soil Liquefaction Characteristics, J. *Soil Mechanics and Foundations Div., ASCE,* 152-167.

Shibata T. and Teparaksa W. (1988) Evaluation of Liquefaction Potential of Soils Using Cone Penetration Tests, *Soils and Foundations,* (28) 2, 49-60.

Steven L. Kramer.,Ahmed.W. Elgamal (2001) Modelling Soil Liquefaction Hazards for Performance-Based Earthquake Engineering: Pacific Earthquake Engineering Research Centre,p 5-8.

Tokimatsu K., Yamazaki T. and Yoshimi Y. (1986) Soil Liquefaction Evaluations by Elastic Shear Moduli, *Soils and Foundations,* 26, 25-35

Tsurumaki.S K.N., Yuichi K., Uchida A. and Babasaki R. (2003) Study on the Simplified Procedure for Evaluating the Undrained Cyclic Strength for Gravelly Soils, *Proc. of 17th International Conference on Structural Mechanics in Reactor Technology (SMiRT 17)*Prague, Czech Republic, August 17 –22.

Youd T.L., Idriss I.M., Andrus R.D., Arango I., Castro G., Christian J.T., Dobry R., Finn W.D.L., Harder L.F.Jr., Hynes M.E.,Ishihara K., Koester J.P., Liao S.S.C., Marcuson

W.F.III., Martin G.R., Mitchell J.K., Moriwak, Y., Power M.S., Robertson P.K., Seed R.B. and Stokoe K.H.II. (2001) Liquefaction Resistance of Soils: Summary Report from the 1996 NCEER and 1998 NCEER/NSF Workshops on Evaluation of Liquefaction Resistance of Soils, *J. Geotechnical Engg., ASCE,* (127)10,817-833.

Youd, T. L., And Noble, S. K. (1997). Magnitude scaling factors.' *Proc., NCEER Workshop on Evaluation of Liquefaction Resistance of Soils,* Nat. Ctr. for Earthquake Engrg. Res., State Univ. of New York at Buffalo, 149–165.

2

Liquefaction Remediation

Sarfraz Ali
NUST College of Civil Engineering
Pakistan

1. Introduction

Naturally occurring soils are heterogeneous materials due to their varying composition broadly termed as granular and cohesive in view of their grain size distribution, and or contractive and dilative according to physical behaviour. Their engineering behaviour is affected by composition, moisture conditions, stress history, boundary conditions, future loading conditions, seismic conditions, etc. The intensity and magnitude of such variable factors make the engineering behaviour of soils under static and dynamic conditions variable from one point to another and is thus regarded as nonlinear and anisotropic. Behaviour under low and high moisture conditions is significantly different under dynamic loading such as earthquakes, machine vibrations, traffic vibrations, etc.

Engineering behaviour of soils under loads is evaluated in terms of shear strength which comprises of parameters known as cohesion, angle of internal friction, effective stress, and pore pressure. Under static conditions (monotonic loading) when loads reach to limit where pore pressure equals effective stress, saturated loose granular soils loose their strength and behave like a liquid. This phenomenon is called liquefaction. While behaviour of medium to dense saturated sands under static loading is dilative in nature, under cyclic loading of certain magnitude and shaking, undergoes failure similar to liquefaction called "cyclic mobility". Civil engineering structures built on sites of loose saturated and dense cohesionless soils are therefore at risk of liquefaction and cyclic mobility with potential of damage to the structures and need improvement to safeguard against damages by liquefaction or cyclic mobility.

In the wake of failures due to flow slides of loose saturated sand slopes and hydraulic shells of dams, the concept of liquefaction was introduced by Arthur Casagrange in between 1935 and 1938. Large scale damages due to liquefaction came in the lime light during 1964 earthquakes in Anchorage, Alaska, and Nigata in USA and Japan and in Loma Prieta earthquake of 1989 in USA. Geotechnical engineering concepts of shear strength, effective stress, and pore water pressures, etc. were introduced by that time and the geotechnical engineers were able to explain the reasons of such damages; the loss of strength of loose saturated sand soils due to cyclic stresses caused by the earthquake waves called, "liquefaction". Since then, exhaustive research efforts have been made to understand phenomenon of liquefaction due to monotonic and dynamic loadings, procedures to evaluate liquefaction, and remediation potential hazards due to liquefaction.

2. Background

On October, 8, 2005, Kashmir area of Pakistan was struck by a 7.8 magnitude of earthquake. Keeping in view, damages of public life and infrastructure (roads, life lines, etc.), this was one of deadliest earthquakes in the history of Pakistan and the world. The destruction zone remained in the mountainous regions where soil sediments are mostly devoid of loose saturated sands and therefore damages due to liquefaction were not noticed. However, in the backdrop of damages caused by this earthquake, building codes were revised with stringent measures on evaluation of all earthquake related hazards and construction of earthquake resistant structures.

Soon after this earthquake, a firm had planned to construct a heavy workshops complex in an industrial area located some 65 km west of Islamabad. The planned RCC framed workshops had covered area of 25000 m^2 and individual column loads varied from 1200 kN to 2300 kN. Geotechnical site investigation revealed that the soil at construction site is composed of loose pockets of alluvial deposits of Indus River comprising silty and sandy strata with varying degree of fines in different layers. Besides potential hazard of liquefaction, the site had low bearing capacity. Allowable bearing capacity (BC) of 100 kPa was available against 150 kPa required to support the foundation. Deep Dynamic Compaction DDC technique being the cheapest as compared to other techniques was selected to remediate liquefaction hazard and to improve the BC to 160 kPa. Uncertainty both in defining subsurface conditions and soil response to a specific soil improvement technique necessitated detailed testing of the site to evaluate effectiveness of the designed program.

3. Purpose and scope

The DDC project at Attock industrial area being first soil improvement project in the country was viewed as a good opportunity to enhance practical aspects of this technique. While the project was going on, the opportunity was used to carry out research with a view to formulate liquefaction evaluation methodology, devise methods to monitor degree of improvement and effectively measure and manage ground vibrations. This chapter while highlighting literary aspects of this technique provides detailed account of planned objectives and outcomes. The author has proposed a practical procedure for undertaking soil improvement project based on his personal experience of this real time project. The scope of the chapter is limited to practical aspects of case history project and covers following:

- Approach adopted for site characterization.
- Methodology for liquefaction evaluation.
- Liquefaction Remediation.
- Ground Vibrations.
- Quality Controls.
- Conclusions and Recommendations.

4. Site characterization

Evaluation for liquefaction involves identifying nature of soils and loading conditions that will cause the liquefaction to occur. While determining liquefaction susceptibility, there could be two possibilities; firstly the soil deposits found in natural state of stresses and subjected to earthquake shaking and secondly the natural soil deposit subjected to construction of structures (static loading) and expected dynamic loading from machine or traffic vibrations. Evaluation for liquefaction must therefore encompass site characterization and well considered loading conditions (past stress history, stresses due to structures, earthquake, traffic/machine vibration, etc.).

Site characterization comprising geotechnical and geophysical techniques of the potential project site and up to a depth of influence of the planned structures should be performed to evaluate all aspects affecting engineering behaviour of the site during useful life of the planned structures. Various aspects of site characterization are described separately and shown in Figure 1.

4.1 Geology

Geological structures such as faults folds, etc., type of sediment (marine, plains, mountainous, etc.), age of sediments, degree of consolidation, past stress history, stratigraphy, etc.

4.2 Topography

Topographic details such as location of the site, general relief of the site and adjacent areas, flow of storm water towards the site, elevation, etc.

4.3 Hydrology and boundary conditions

Hydrological and boundary conditions such as depth of ground water table and its fluctuation with seasonal variation (if any) due to rains, drainage pattern of the site and adjacent areas, proximity to water bodies affecting moisture conditions of the site such as floods, ground water flow regimes, etc.

4.4 Regional tectonic setting and seismicity

Aspects related to regional tectonic setting and seismicity such as proximity to fault lines, type of fault lines, past earthquake tsunami history, local seismic codes on horizontal and vertical peak ground accelerations, duration of shaking, past history of earthquake damages such as ground ruptures, liquefaction, collapse of building, destruction of life lines, etc. A site specific seismic hazard analysis will be very useful tool in evaluation of more realistic dynamic parameters and is recommended to be performed.

4.5 Field and laboratory investigation

Field and laboratory investigation should be conducted to develop SPT/CPT/DCPT profiles (these would later be used for comparing pre to post compaction degree and depth of improvement), unit weight, moisture content, Atterberg Limits, shear strength parameters, shear wave velocity (this will be used to assessing ground vibrations), etc. Depth is an important aspect for evaluation of liquefaction hazard. Generally it should be equivalent to the depth of influence of the planned structures or 20 m depth whichever is larger. Depth of influence of structures can be calculated using stress distribution theories available in the literature.

4.6 Geotechnical profile

Finally a pre improvement geotechnical profile of the soil deposit at site should be developed consisting of type of layering, percentage fines, moisture content, and SPT/CPT/DCPT variation, shear wave velocity in tabular or graphical pattern as shown in Table 1 and Figure 2 (more columns can be added for pre to post compaction comparison). This format will be a useful ready reckoner during the execution phase of the project and for pre to post compaction comparison.

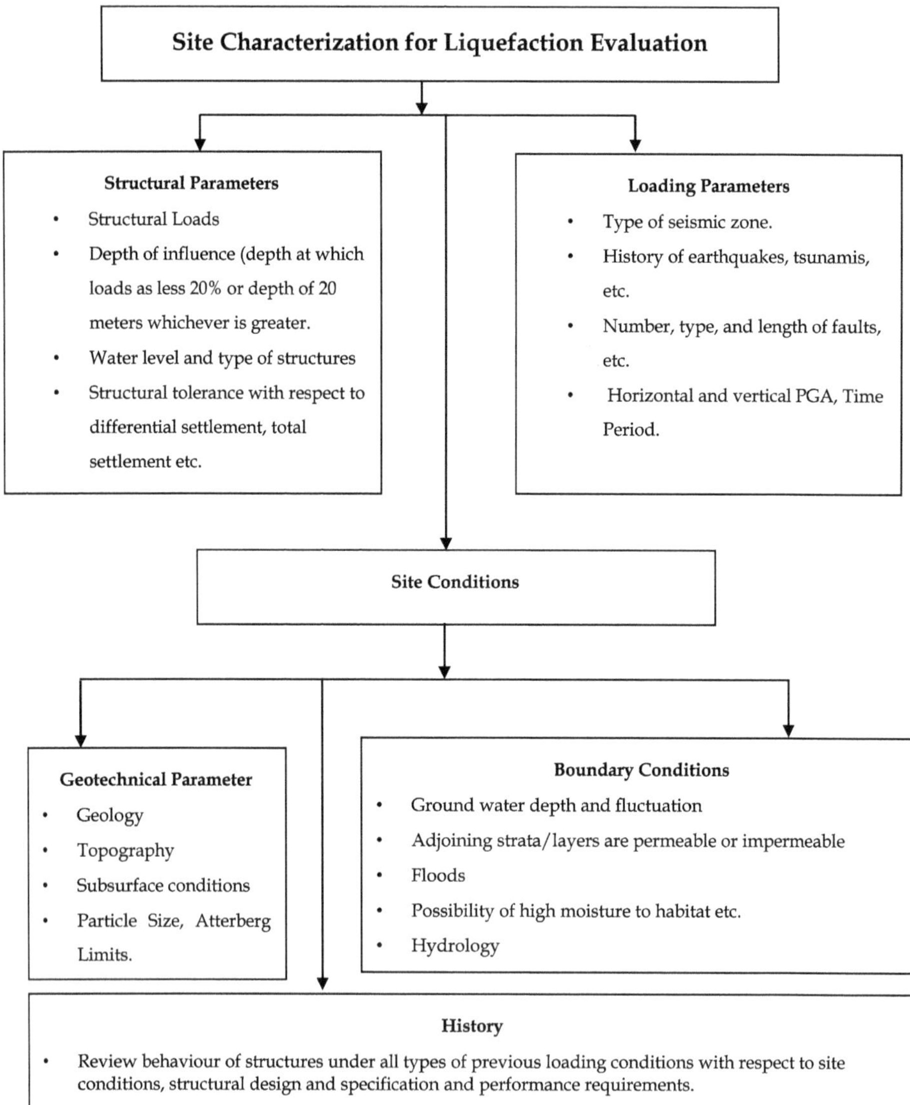

Fig. 1. Parameters fir site characterization

Depth (m)	Type of Soil	Fine Contents (%)	Moisture Content (%)	SPT N-Values		Shear Wave Velocity (m/sec)	
				Pre compaction	Post compaction	Pre compaction	Post compaction

Table 1. Geotechnical Profile

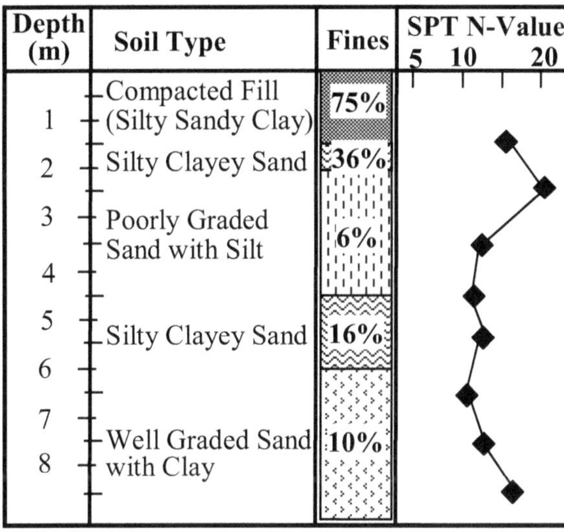

Fig. 2. Geotechnical Soil Profile

5. Methodology adopted for evaluation of liquefaction

Susceptibility of soil deposit at site to liquefaction should be determined through multiple approaches since remediation of liquefaction hazard is a cost intensive event that would raise the overall project cost. Various criterions evolved by researchers. Historical, geological, and compositional criterion suggested by Kramer, (1996) were used to evaluate liquefaction susceptibility.

5.1 Evaluation for liquefaction susceptibility

Evaluation for liquefaction susceptibility can be made according "To" "Historical" "Geologic" "Compositional" criterion proposed by Kramer (1996) and empirical correlations based on SPT N-values. Historically, the construction site of research project is located in Himalayan seismic zone having a long history of earthquakes in the past and prediction of earthquakes in the future therefore the site demanded an evaluation for seismic hazards. Geologically the construction site is composed of poorly sorted an alluvial stratum of recent age with un-corrected SPT N-value of less than 12. Compositionally, the soil layers present at depths from 3.5 to 5.5 m, containing fines as low as 6%, low SPT N-values, and rise of ground water table to a 5 m (from existing 12 m) in monsoon season renders the project site susceptible to liquefaction. Since al sites susceptible may not actually liquefy, therefore evaluation for initiation of liquefaction must be performed to actually establish or rule out the potential hazard of liquefaction.

5.2 Evaluation for initiation of liquefaction

The next step after evaluation of liquefaction susceptibility is the evaluation for initiation of liquefaction because it might happen that a soil at a particular site is susceptible to liquefaction but devoid of monotonic and dynamic loading conditions necessary to initiate it. Its occurrence requires a loading that is strong enough to initiate or trigger liquefaction. Most popular methods for evaluation of initiation of liquefaction are "cyclic stress approach" and "cyclic strain approach" (Kramer, 1996). In this paper, initiation of liquefaction is evaluated using cyclic stress approach. An earthquake of magnitude 7, peak ground acceleration of 0.16g, and ground motion of 0.4 seconds were used for evaluation for initiation of liquefaction for the case history project.

5.3 Characterization of site specific dynamic loading

Site specific analysis should be carried out for the expected monotonic loading and the existing soil conditions. Site specific dynamic analysis for a particular site requires characterization of expected dynamic loading (earthquake, machine, traffic, etc.) in terms of a level of uniform cyclic shear stress that is applied for an equivalent number of cycles (Kramer, 1996). Equation (1), proposed by Seed et al. (1983) can be used for calculation of equivalent cyclic shear stresses. In equation (1), τ_{cyc} = cyclic shear stress at the time of earthquake, $a_{\max}$ = peak ground surface acceleration, σ_v = total vertical stress, g = acceleration due to gravity, r_d = value of stress reduction factor for a given depth (0.98 to 0.96 for depth of 3.5 to 5.5 m depth, Seed and Idriss, 1971).

$$\tau_{cyc} = \frac{0.65 a_{\max}}{g \times \sigma_v \times r_d} \tag{1}$$

5.4 Characterization of liquefaction resistance

Cyclic shear resistance should be calculated at various depths using equation (2), (Kramer, 1996).

$$\tau_{cyc,L} = CSR_L \times \sigma'_{v^o} \tag{2}$$

Where σ'_{v^o} = initial vertical effective stress and CSR_L = cyclic stress ratio. SPT N_{Field} values were corrected to $(N_1)_{60}$ by applying appropriate corrections which include corrections for overburden, energy, borehole diameter, rod length, and sampling.

5.5 Factor of safety

Factor of safety against liquefaction is expressed as $\mathbf{FS}_L = \tau_{cyc,L} / \tau_{cyc}$. Liquefaction occurs when $\mathbf{FS}_L$ is less than 1 (Kramer, 1996). Factor of safety should be checked at various depths to ascertain initiation of liquefaction or otherwise. Factor of safety of the research project was less than 1 prior to DDC, Table 2.

Depth (m)	$(N_1)_{60}$	CSR_L	σ'_{vo} (kN/m³)	$\tau_{cyc,L}$	τ_{cyc}	FS_L
3.5	6	0.06	64.01	3.84	6.52	0.59
4.5	7	0.07	71.43	5.00	8.19	0.61
5.5	8	0.08	80.12	6.41	9.96	0.64

Table 2. Cyclic Shear Resistance of Soil and Factor of Safety (FS_L)

6. Liquefaction remediation

Various techniques are available to improve a potential project site to a desired level of site performance vis-à-vis potential hazards. Selection of site will depend on factors such as geotechnical site conditions, required degree of improvement, availability of technology, cost, time, etc.

6.1 Soil improvement techniques to remediate liquefaction

Some of the techniques suitable for remediation of liquefaction hazards are; soil replacement (if desired depth of improvement is shallow), deep dynamic compaction, vibrocompaction, grouting, deep soil mixing, etc. In order to select a most efficient and economical remediation technique requires a comprehensive review of existing geotechnical site investigation, susceptibility to liquefaction hazard, compactibility of soil at site, cost-benefit analysis of techniques vis-à-vis desired degree and depth of improvement. Since this chapter deals with remediation of liquefaction hazard using DDC, only details pertaining to this technique will be discussed at length hereafter.

6.2 Deep dynamic compaction technique

Deep dynamic compaction technique has effectively been used for remediation of liquefaction hazard in Pakistan and other countries. The technique has also been used successfully for improvement of Municipal Solid Waste sites for construction of buildings, parks, recreational facilities, etc. The technique though simple, requires careful design, construction, and quality controls as described to ensure desired results. It involves repeatedly dropping of heavy weights from a crane on each impact point. The impact energy of the falling weight densifies the soil at depth. DDC program generally consist of two or more passes followed by ironing pass to compact the top ground surface, Figures 3 and 4. The weight of tamper, height of fall, number of drops, and grid spacing is selected to achieve required depth and degree of improvement. The mechanism of densification for unsaturated soil is analogous to large scale proctor compaction. For loose saturated granular soils, the impact from heavy weights liquefies the soils and particles are rearranged to a denser configuration. An impact of heavy falling weights generates ground vibrations which limits its use in built up areas although there are techniques to minimize ground vibrations such as trenches, air bags, etc.

6.2.1 Objective of improvement

The author is of the opinion that a site that requires remediation for liquefaction hazard would also need improvement for soil bearing capacity. Considerations for a soil improvement project should therefore include type of structure, type of foundation, bearing capacity vis-à-vis depth and degree of envisaged improvement. Existing site conditions must be carefully reviewed while selecting and designing a site improvement project. It might happen that the planned site is undulating and requiring fill, or water logged, or may contain vegetation and waste material of old disused structures. While preparing a site for DDC, following must be ensured:

- All fills must be of a material suitable to act as a granular blanket.
- Old pavement and floors of the demolished buildings must be ripped and removed and back filled with granular material to facilitate transfer of impact energy to deeper layers.
- Underlying services and trenches must be removed and suitably back filled.
- Preferably, the site should be fairly levelled.

Multiple options must be considered to improve the site and finalized only after the detailed investigation has been completed for example, for lighter structures or area requiring shallower depth and degree of improvement, different option can be considered than the heavier and sensitive structures. The layout of the planned structures may even be re-adjusted to place critical structures on one segment and remaining ones on the other side. The type of foundation must be considered in post improvement scenario, it might be possible to place the foundation for lighter structures even on the surface thus reducing overall cost of the project.

Fig. 3. Dynamic Compaction Equipment (heavy crane, tamper, release mechanism) and process

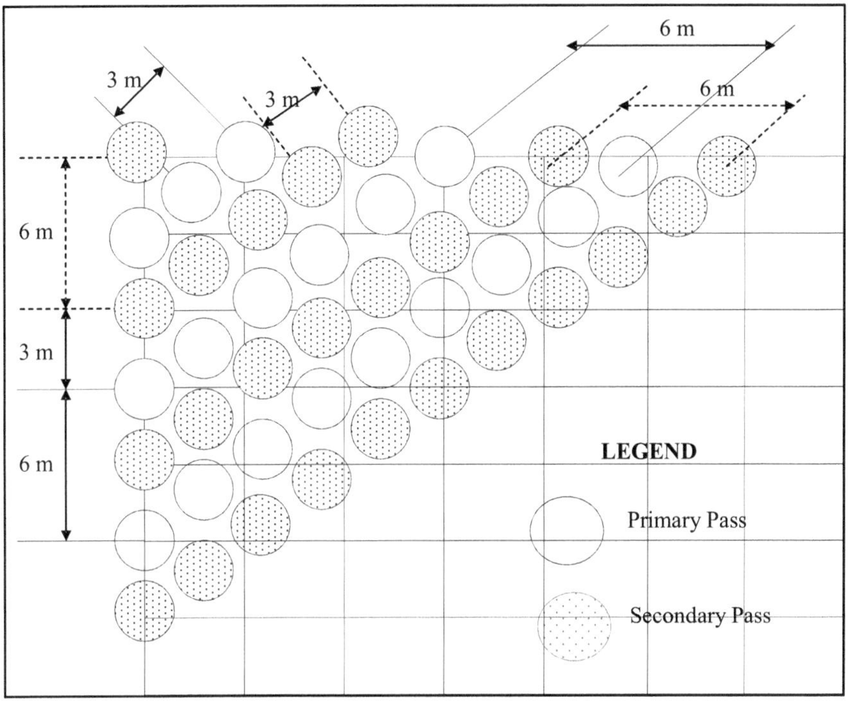

Fig. 4. Diagrammatic layout of primary and secondary pass during DDC

6.2.2 Project performance needs

Candidate civil engineering project requiring evaluation for liquefaction hazard may include low and high rise buildings, life lines (water supply lines, gas pipe lines), bridges, etc. Usually a large project comprises of number of structures with varying specifications, depth of influence and performance needs. It is important to note that all structures within a project may not need soil improvement therefore it would be important to identify structures demanding improvement vis-à-vis their criticality and user needs.

6.2.3 Availability of technology and experience

DDC technique requires heavy duty cranes with lifting and drop arrangements for a tamper from different heights. The tamper can be made of steel, concrete or concrete with steel casing. The tamper should have air vents through its body to enhance impact energy. There could be many unforeseen in such projects therefore an adequate reserve of essential crane spares should be available at site.

6.2.4 Time and cost

Since this technique requires mobilization of very equipment to the site, like all other projects a cost benefit analysis of DDC compared with other options such as avoidance, soil replacement, deep soil mixing, grouting etc. is a pre-requisite for adoption of this technique.

Apart from the cost, DDC is a simple technique but is a time consuming process. Availability of time must therefore be considered right at the outset of the project. Generally this technique is suitable for large projects that do not have very tight timelines.

6.2.5 Preliminary DDC design to evaluate improvement capacity of soil

Determination of compactibility or improvement potential of soil deposit at a site needs no emphasis and depends on factors such as particle size, liquid limit, relative density, type of layers, and depth of ground water table. Although DDC is suitable for most of the soils including municipal solid waste site, its best effectiveness has been reported in granular soils. For cohesionless soils, depth of improvement is proportional to square root of the energy per blow.

A preliminary DDC program can be designed from correlations proposed by Menard and Broise (1975), Lukas (1986) given below as Equation 3 and 4 respectively. Attention must be paid to the applied energy range over the project area for Equation 2 and "n" values for different soils when using Equation 3.

$$D_{max} = \sqrt{WH} \tag{3}$$

$$D_{max} = n\sqrt{WH} \tag{4}$$

As a thumb rule, grid spacing of primary pass of DDC should not be less than the desired depth of improvement with secondary (and tertiary if spacing of secondary pass is more than 3 m) in between the primary pass. An ironing pass with an overlap of 1/3rd of the tamper diameter is good enough to compact the upper soil layer. To finalize the number of passes, number of drops at each impact point, a comprehensive experimental design is required to record both degree and depth of improvement in horizontal and lateral extents. Improvement and vibrations should be verified after pre-selected number of drops, author suggests; firstly after half the number of drops, secondly after next 2/3rd number of drops and finally after all the drops are completed. Keeping in view large number of variables such as site and soil conditions, moisture conditions, depth of ground water table, etc., the author is of the view that best mean to achieve desired improvement is through well planned, executed, and monitored preliminary DDC program. Minimum area for of 24m Xx 24 m is recommended for preliminary DDC.

6.3 Experimental design for evaluation of improvement for field trial of preliminary DDC program

Pre to post compaction improvement can be evaluated by comparing pre to post compaction electrical resistivity profiles, SPT N-values, crater depth, quantum of backfill material, shear wave velocity, etc. As far SPT is concerned, a value should be selected as the criteria to terminate further drops. For alluvial soil deposits with 5 m desired depth of improvement, an SPT value of 15 was selected as limit for discontinuing drops for the case history project. There could be different ways to evaluate improvement during preliminary design and field trial stage of the project. To observe improvement vis-à-vis number of drops or impact energy and intensity of ground vibrations, author used a layout shown in Figure 5 which proved very effective and can be used for future projects.

Evaluation of improvement under impact points, between the craters, and laterally away from impact points is very important, as shown in Figures 6, 7, 8, respectively. With a thoughtful evaluation methodology, actual improvement is possible to assess. For the research project, the layout of test craters and location of boreholes was designed with a view to keep the distance between pre and post compaction boreholes as minimum as practically feasible. In this research, pre to post compactions SPT were performed within a

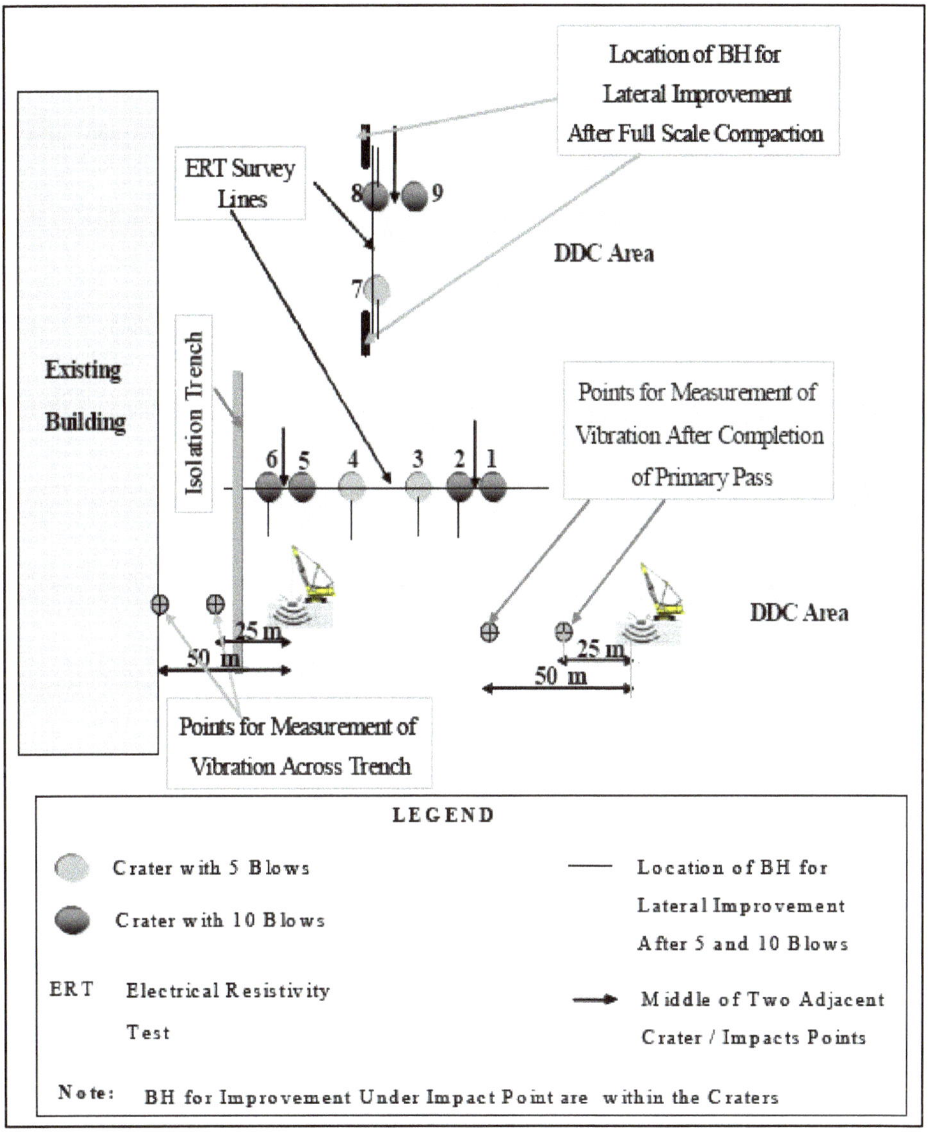

Fig. 5. Layout of Compaction Site, Test Craters, Vibration Measurement Points and Location of ERT Survey

distance of 2 m. Total of 9 test craters; crater no. 1 through crater no. 9, were used in this research. Total of 48 boreholes were drilled and 384 SPT performed to evaluate improvement at different points. Detail of experimental design is given as, refer to Figure 5:

- improvement after 5 blows was evaluated at crater no 3, 4, and 7
- improvement after 10 blows was evaluated at crater no 2, 6, and 8
- lateral improvement after 5 blows was evaluated around crater no. 1, 5 and 7
- lateral improvement after 10 blows was evaluated around crater no 2, 6, and 8
- Improvement at middle of adjacent craters was evaluated between crater no 1 & 2, crater no 5 & 6 and crater no 8 & 9 after full scale compaction i.e. after primary, secondary and ironing pass no compaction was carried out within 9 m of the test craters.
- Locations of all boreholes has been referenced to the center of impact point and are given in terms of tamper diameter "D" (D = 2.4 m)

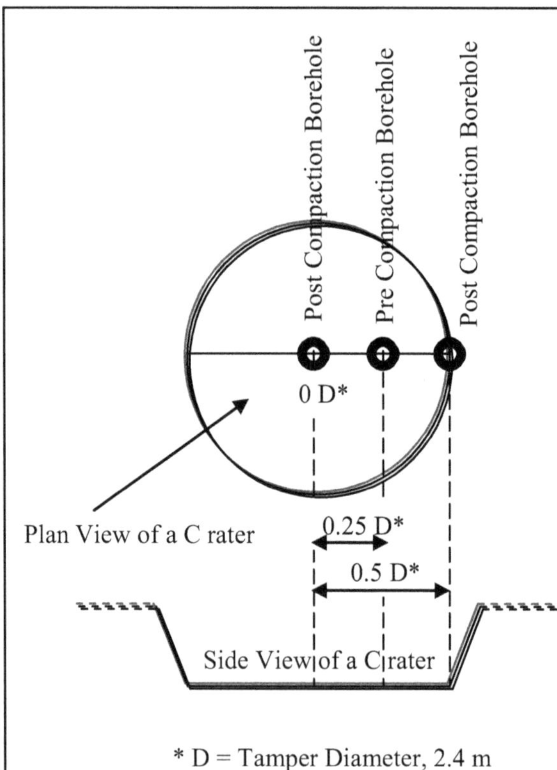

Fig. 6. Location of boreholes for evaluation of improvement under impact point

6.4 Field trial and finalization of dynamic compaction program

Once the field trial is complete, improvement in depth, between impact points and in lateral direction must be evaluated. To assess improvement in realistic manner, details of the field trial of research project is explained in the ensuing paragraphs. SPT were performed two weeks after full scale compaction. Improvement under impact point was evaluated at crater # 2, crater # 6 and crater # 8 (upper halves). Improvement was also evaluated at the middle of crater # 1 and crater # 2, crater # 5 and crater # 6, crater # 8 and crater # 9. Improvement in lateral direction was measured only at two places, i.e., towards lower side of crater # 7 and upper side of crater # 8, refer to Figure 5.

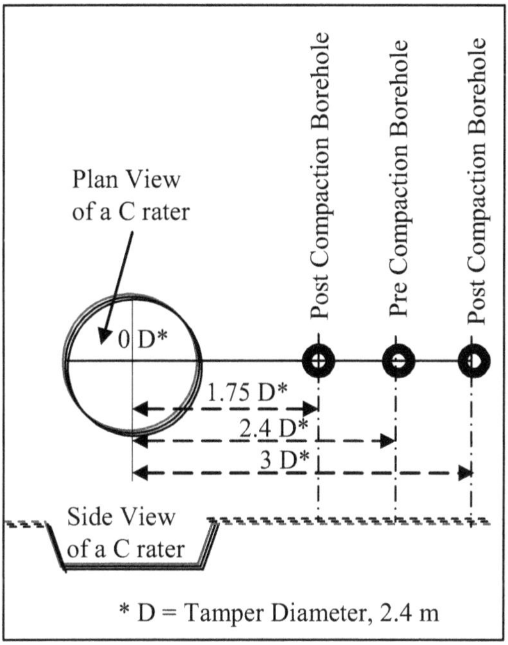

Fig. 7. Location of boreholes for evaluation of improvement in lateral direction

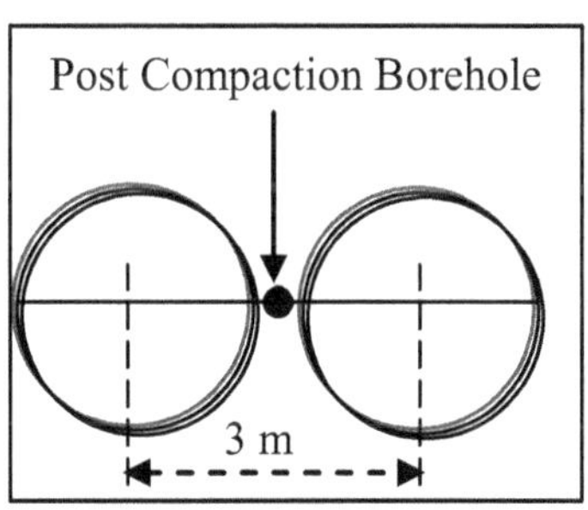

Fig. 8. Location of boreholes for evaluation of improvement at the middle of adjacent craters

6.4.1 Improvement at crater # 1 (upper half) and middle of crater # 1 and crater # 2

Layout of BH for evaluation of improvement under impact point and at the middle of two adjacent craters (crater # 1 and crater # 2) is shown in Figure 9. Pre-to-post compaction SPT profile after full scale compaction for improvement under impact point at crater # 2 and at the middle point of crater # 1 and crater # 2 is shown in Figure 10.

It is evident from Figure 10, that directly under impact point on upper side of crater # 2; significant improvement was noted in top 3.5 m strata. Moderate improvement was observed in strata between depths of 3.5 m to 5 m. Improvement was marginal in strata depth of 5 m to 7 m. Improvement below 7 m depth remained insignificant. At the middle of crater # 1 and crater # 2, improvement was slightly less than the improvement under impact point however still the improvement was significant in upper 3.5 m strata. Improvement below 3 m depth at the middle of these craters was almost same as that of improvement under impact point of crater # 2.

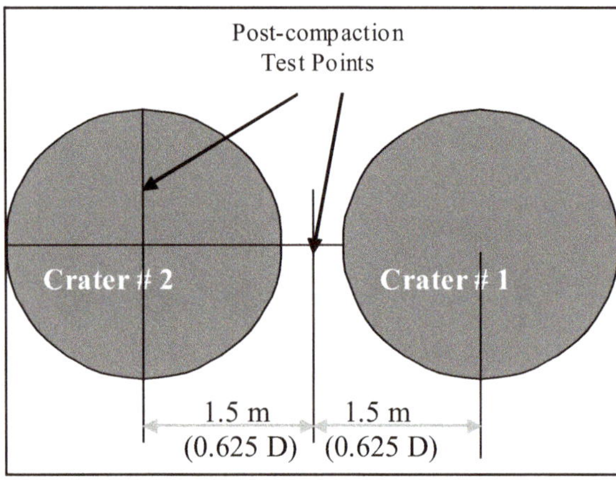

Fig. 9. Layout of BH for Evaluation of Improvement after Full scale Compaction at Crater # 1 (Upper Half) and Middle of Crater # 1 and Crater # 2

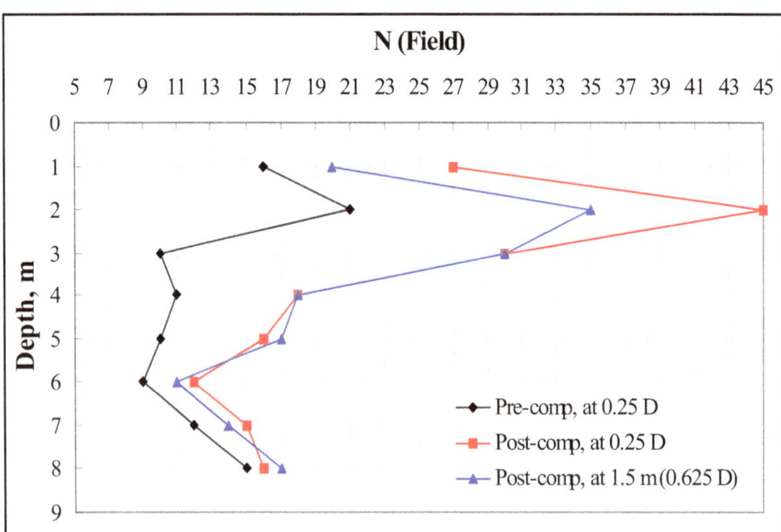

Fig. 10. Pre-to-post Compaction N (Field) Comparison of Improvement at Crater # 1(Upper Half) and Middle of Crater # 1 and Crater # 2

6.4.2 Improvement at crater # 5 (upper half) and middle of crater # 5 and crater # 6

Layout of BH for evaluation of improvement under impact point and at the middle of crater # 5 and crater # 6 is shown in Figure 11. Pre-to-post compaction SPT profile after full scale compaction for improvement under impact point at crater # 6 and at the middle point of crater # 5 and crater # 6 is shown in figure 12.

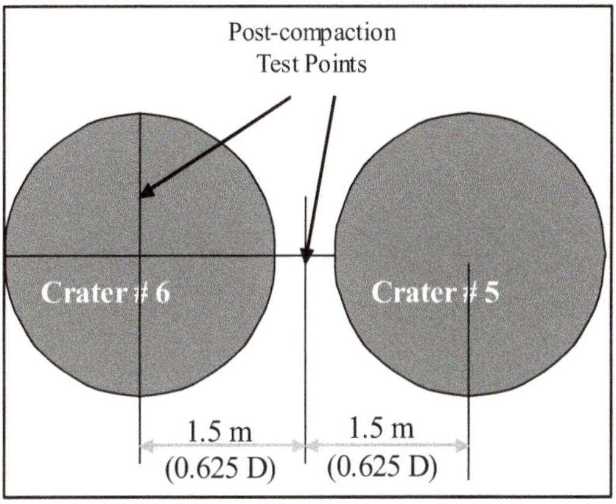

Fig. 11. Layout of BH for Evaluation of Improvement after Full scale Compaction at Crater # 5 (Upper Half) and Middle of Crater # 5 and Crater # 6

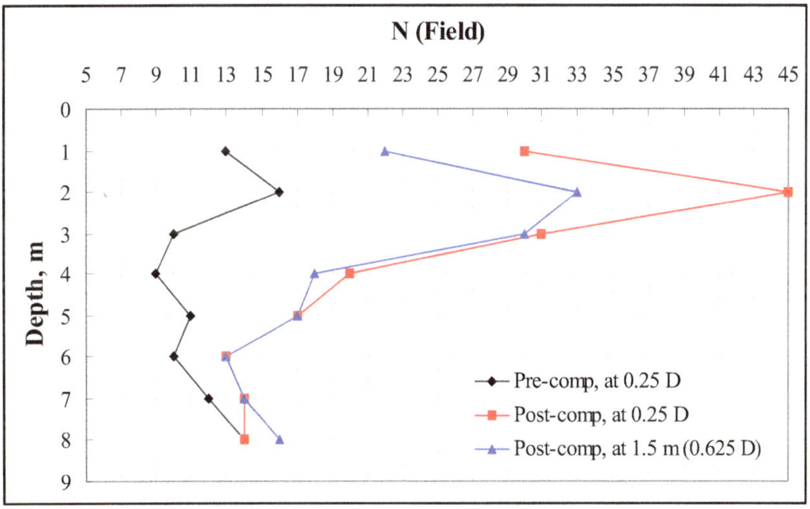

Fig. 12. Pre-to-post Compaction N (Field) Comparison of Improvement at Crater # 5 (Upper Half) and Middle of Crater # 5 and Crater # 6

Under impact point, improvement was significant in upper 4 m strata, moderate in strata depth of 4 m to 5 m and marginal from 5 m to 6 m. At the middle of crater # 5 and crater # 6, improvement was slightly less than the improvement under impact point however still the improvement was significant in upper 3 m strata. Improvement below 3 m depth at the middle of these craters was same as that of improvement under impact point of crater # 6 with slight variation at depths of 4 m and 8 m.

6.4.3 Improvement at crater # 8 (upper half) and middle of crater # 8 and crater # 9

Layout of BH for evaluation of improvement under impact point and at the middle of crater # 8 and crater # 9 is shown in Figure 13. Pre-to-post compaction SPT profile after full scale compaction for improvement under impact point at crater # 8 and at the middle point of crater # 8 and crater # 9 is shown in Figure 14.

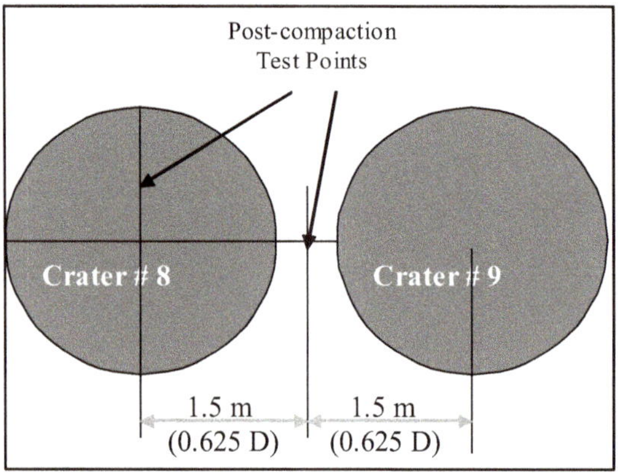

Fig. 13. Layout of BH for Evaluation of Improvement after Full scale Compaction at Crater # 8 (Upper Half) and Middle of Crater # 8 and Crater # 9

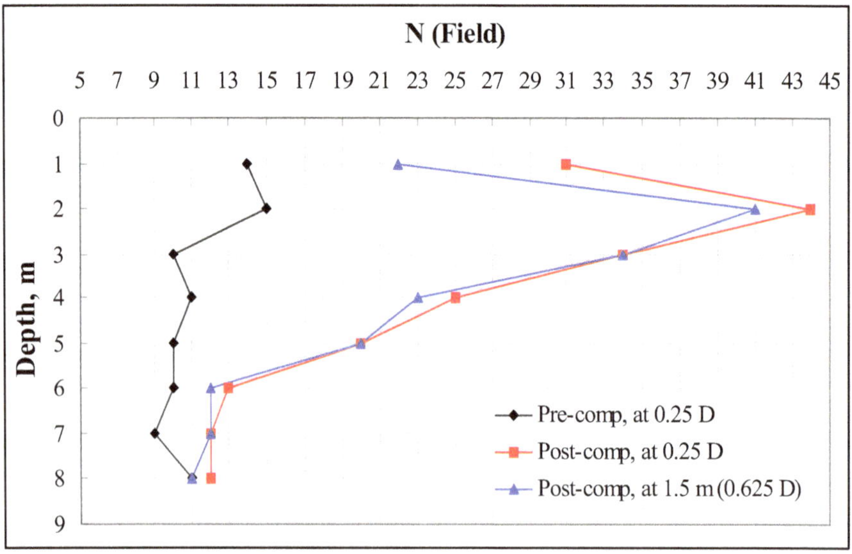

Fig. 14. Pre-to-post Compaction N (Field) Comparison of Improvement at Crater # 8 and Middle of Crater # 8 and Crater # 9 (Figure 5.17)

Directly under impact point on upper side of crater # 8; significant improvement was noted in top 4 m strata. Moderate improvement was observed in strata between depths of 4 m to 5.5 m. Improvement was marginal in strata depth of 5.5 m to 7.5 m. Improvement below 7.5 m depth was insignificant. At the middle of crater # 8 and crater # 9, improvement was slightly less than the improvement under impact point in the upper 2 m strata, still the improvement was significant. Improvement below 2 m depth at the middle of these craters was same as that of improvement under impact point of crater # 8.

6.4.4 Lateral improvement , lower side of crater # 7

Pre-to-post compaction N (Field) comparison after full scale compaction for improvement in lateral direction on lower side of crater # 7 is shown in Figure 15. In lateral direction, improvement was moderate in upper 2.5 m strata and marginal from 2.5 m to 3.5 m strata from centre of impact point to distance of 1.75 D. Improvement was marginal in the upper 1.5 m strata from 1.75 D to 3 D from centre of impact point. Improvement below this depth was insignificant.

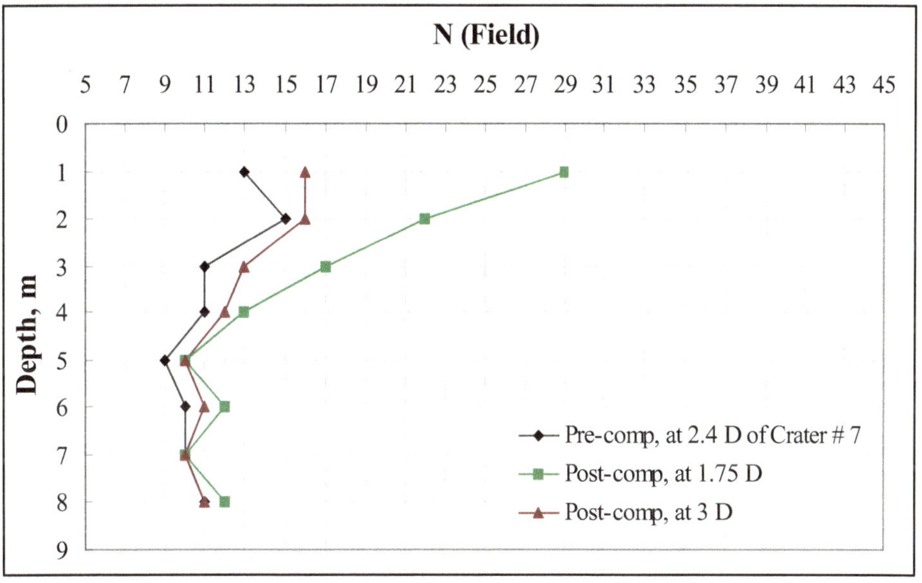

Fig. 15. Pre-to-post Compaction N (Field) Comparison of Lateral Improvement at Lower Side of Crater # 7

6.4.5 Lateral improvement, upper side of crater # 8

Pre-to-post compaction N (Field) comparison after full scale compaction for improvement in lateral direction on upper side of crater # 8 is shown in Figure 16. Improvement was moderate in upper 3 m strata and marginal from 3 m to 5 m strata from centre of impact point to distance of 1.75 D. Improvement was marginal in the upper 1.5 m strata from 1.75 D to 3 D from centre of impact point. Improvement below this depth was insignificant.

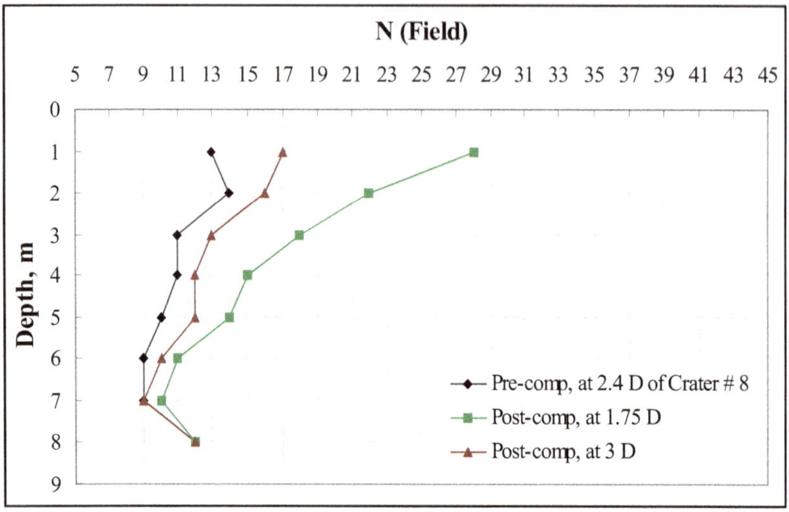

Fig. 16. Pre-to-post Compaction N (Field) Comparison of Lateral Improvement at Upper Side of Crater # 8

7. Quality controls

At the end of field trial, DDC program can be finalized with respect to number of passes, number of drops, grid spacing of all the passes (primary, secondary, tertiary, ironing etc.), and height of fall. During field trial, improvement monitoring mechanism is finalized in terms of crater depth vis-à-vis number of drops, quantum of backfill material, and increase in penetration resistance (SPT, CPT, or DCPT). Since it may not be possible to perform so many penetration tests, more convenient means would still remain crater depth, and back fill material. Author also used pre to post compaction electrical resistivity profiles as a tool to monitor depth and degree of improvement. It is however emphasized that electrical resistivity profile can be tricky in case of rains or increase in moisture conditions at the site. In such circumstances, only an experienced geophysical or geotechnical engineer will be able to interpret profiles. While pre to post compaction resistivity profiles can be compared easily in 2-dimensional perspective, monitoring of improvement through crater depth and back fill will require tabulated format which must be kept at site and filled in carefully.

Consequent to results of field trial of the designed compaction program, 5 cm settlement between any two successive blows was selected as acceptable compaction criteria for achievement of desired improvement. Final crater depths after 5 blows and 10 blows are shown in Figure 17. Observations on crater depths are:

- Almost same trend in crater depths has been observed at all craters
- Crater depths increased with increase in no. of blows
- Crater depths are more for initial 5 blows (around 96 cm) than for next 5 blows (around 37 cm)
- At crater # 7 to crater # 9, crater depths are more for initial 5 blows and less for next 5 blows.

- Slightly more crater depths for 5 blows at crater # 7 to crater # 9 are because strata at this location were relatively loose

Crater depths of research project have been plotted and compared with normalized crater depths of DDC case histories of non-collapsible soils proposed by Mayne et al., - 1984 as shown in Figure 18. Comparison of crater depths indicates that crater depths of research project are towards lower limit of DDC case histories. Since relationship of normalized crater depths and no. of blows is nonlinear therefore best fit curve has been obtained using a second order polynomial equation, given in Figure 18.

Relationship of crater depths and depth of improvement of research project was nearly linear as shown in Figure 19; however the depth of improvement was more for larger crater depths. A best fit line was obtained with linear equation (equation 5) where depth of improvement was dependant variable and crater depth was independent variable. The equation of the best fit line gives somewhat higher values of depth of improvement for crater depths of 0.5 m to 1 m and lower depth of improvement at crater depths of more than 1.5m. In view of the lower depth of improvement of project after full scale compaction, the

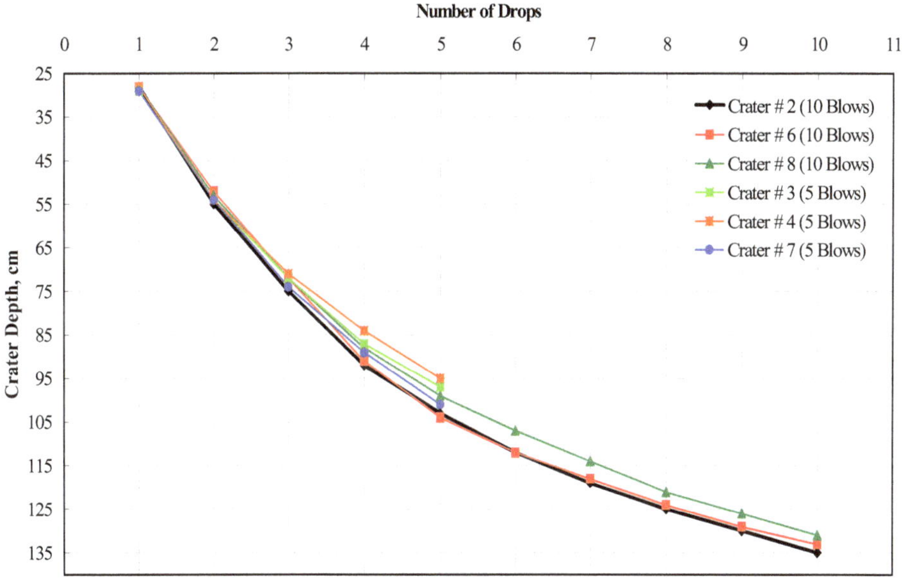

Fig. 17. Crater Depth Measurements after 5 and 10 Blows

results of equation 5 would be towards lower bound of the expected range of improvement of a similar compaction program under similar soil conditions. Since R^2 value of the equation is 0.98, equation 51 shows a good correlation.

$$D = 3.7\,d - 0.11 \tag{5}$$

Where,
D = Depth of improvement, m
d = depth of crater, m

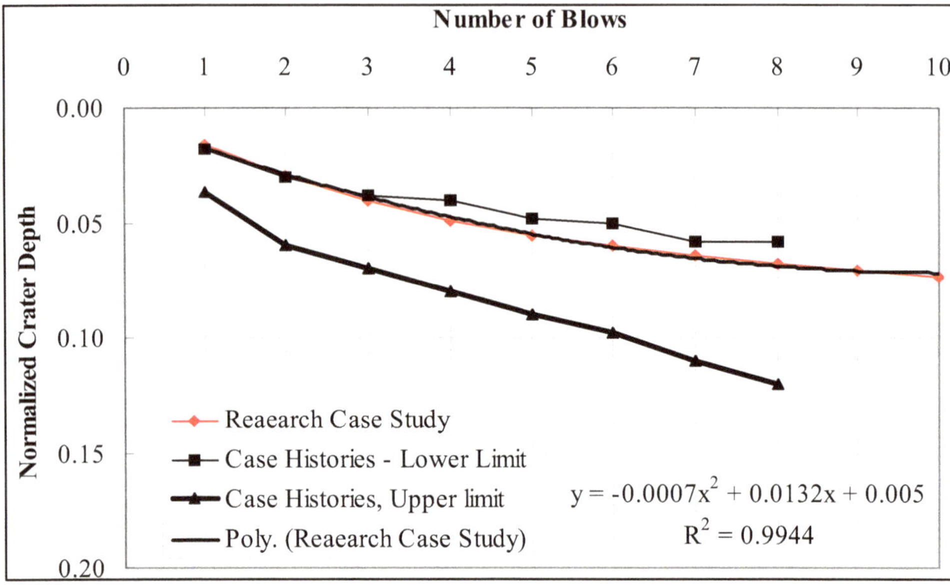

Fig. 18. Comparison of Normalized Crater Depths Vs No. of Drops with the Range proposed by Mayne et al., (1984)

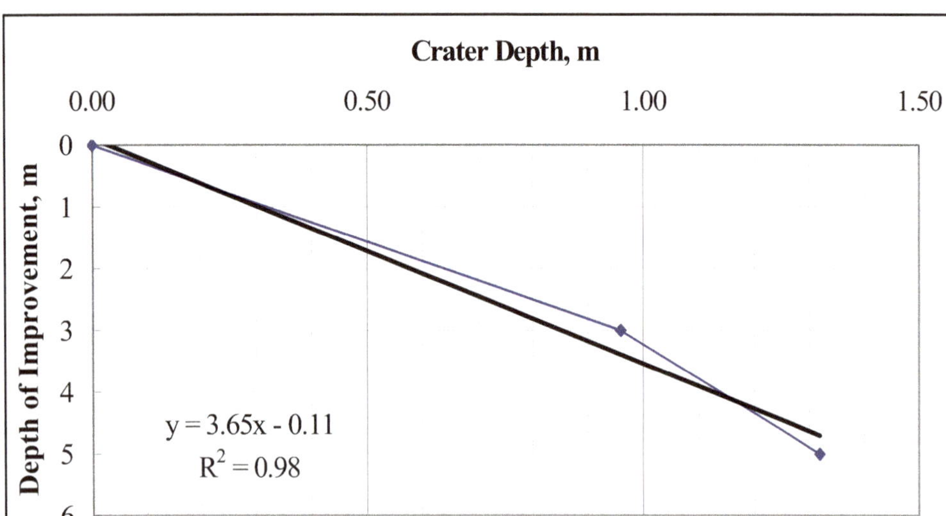

Fig. 19. Plot of Crater Depths vs. Depth of Improvement

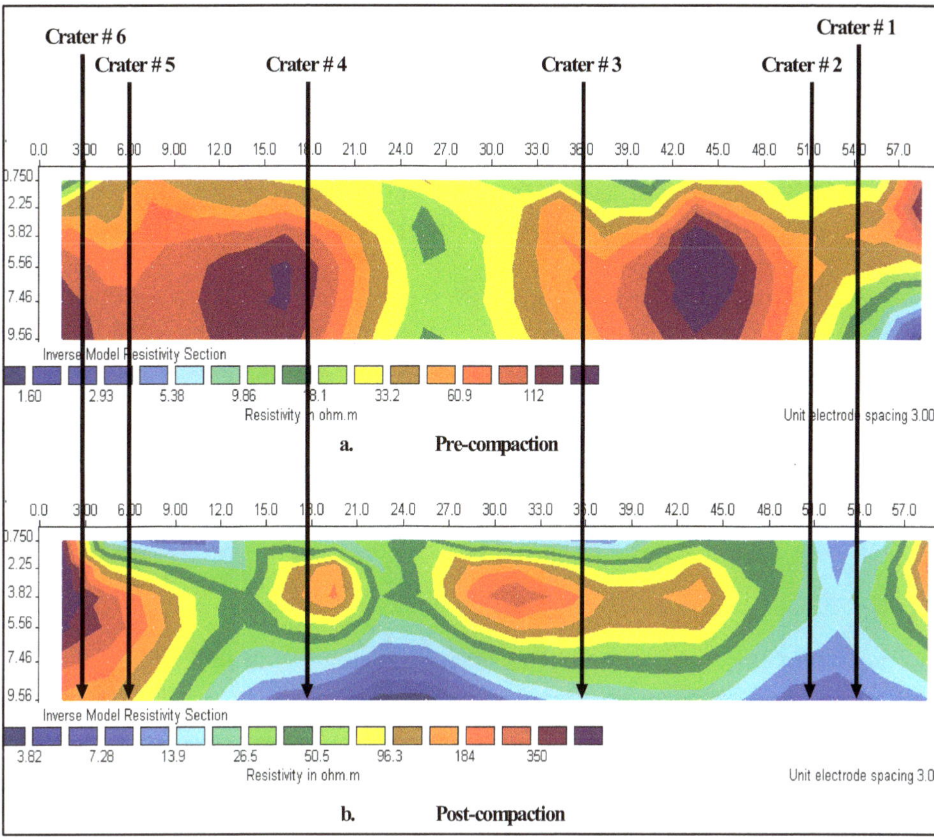

Fig. 20. ERT Profiles, Crater # 1 to Crater # 6

7.1 Evaluation of improvement from ERT profiles

Pre-to-post compaction ERT profiles of strata at crater # 1 to crater # 6 are shown in Figure 20. The profiles are based on relative densities of various layers in the strata therefore the colour legend of pre and post-compaction profiles indicate different resistivity ranges. At crater # 1 to crater # 6, over all pre-compaction resistivity range of 1.60 Ω-m to 112 Ω-m increased to post-compaction range of 3.82 Ω-m to 350 Ω-m.

Pre-to-post compaction ERT profiles of strata at crater # 7 to crater # 9 are shown in Figure 21; the overall increase in resistivity was more in this case. Pre-compaction resistivity range of 0.370 Ω-m to 509 Ω-m increased to post-compaction resistivity range of 6.34 Ω-m to 1054 Ω-m. Salient of pre-to-post compaction ERT profiles are:

- Post-compaction ERT profiles confirm in general, the 5 m depth of improvement evaluated from SPT
- Relatively loose strata are also indicated at crater # 1, crater # 2 and crater # 8
- Keeping in view the same nature of strata and SPT results at all other craters, presence of loose strata could be attributed to some error in recording of resistivity data

- More no. of BH could have better explained the variation in resistivity in between test craters

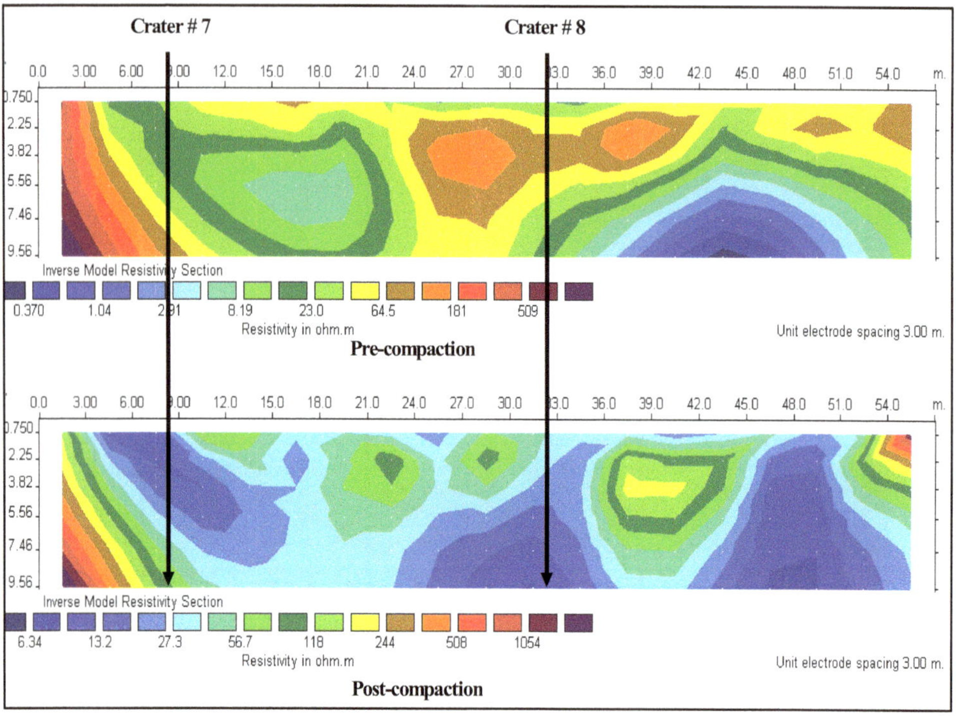

Fig. 21. ERT Profiles, Crater # 7 to Crater # 8

8. Ground vibrations

Ground vibrations from dynamic compaction can be dangerous to the adjacent structures and therefore needs to be determined prior to compaction operation. Effect of vibrations on various structures and human beings is shown in Table 3 and 4. Vibrations were predicted using Scaled Distance Approach proposed by Mayne (1985). To assess attenuation effect of this trench, vibrations were measured across the trench using "Linear Variable Displacement Transducer (LVDT)"-an approach analogous to "Impulse Response Function" approach proposed by Svinkin (1996a, 1997), refer to Figure 5. In a dynamic compaction project in Pakistan, a 2 meter wide and 4 meter deep trench was excavated at the edge of the compaction site to reduce vibrations to avoid any damage to a building some 50 meters away from the site. Vibrations were measured across the trench at a distance of 25 m to assess the attenuation effect of trench and compare with the value obtained from empirical correlations. Vibrations were also measured on compaction site after the primary pass to assess the effect of primary impacts on the magnitude of vibrations.

Linear Variable Displacement Transducers (LVDT) and accelerometer was used for measurement of vibrations. These were installed on top of 1 inch dia, 1.25 m long -

deformed steel bar. The steel bar was drilled into ground with only the top surface visible at the ground surface 2 inch thick foam was placed under the base of the LVDT assembly to minimize inertia effect of the LVDT assembly. Layout of the test is shown in Figure 22 and 23. Data was recorded for 10 consecutive drops at a distance of 25 m from point of impact across the trench and at distance of 25 m at place where primary pass was completed. The data recorded was numerically differentiated to obtain peak particle velocity (PPV) and was further differentiated to obtain acceleration.

Type of Structure and their Natural Frequencies	Safe level of Vibrations (mm/s)
Reinforced or framed structures, industrial and heavy commercial buildings at 4 Hz and above	50
Un-reinforced or light framed structures, residential or light commercial type buildings at 4 Hz -15 Hz	15-20
Un-reinforced or light framed structures, residential or light commercial type buildings at 15 Hz -40 Hz and above	20-50

Table 3. Effect of vibrations on structures (British Standard 7385: Part 2-1993).

Level of Vibrations (mm / sec)	Effect of Vibration
0.1	not noticeable
0.15	nearly not noticeable
0.35	seldom noticeable
1.00	always noticeable
2.00	clearly noticeable
6.00	strongly noticeable
14.00	very strongly noticeable
17.8	severe noticeable

Table 4. Effect of vibrations on human beings (British Standard 7385).

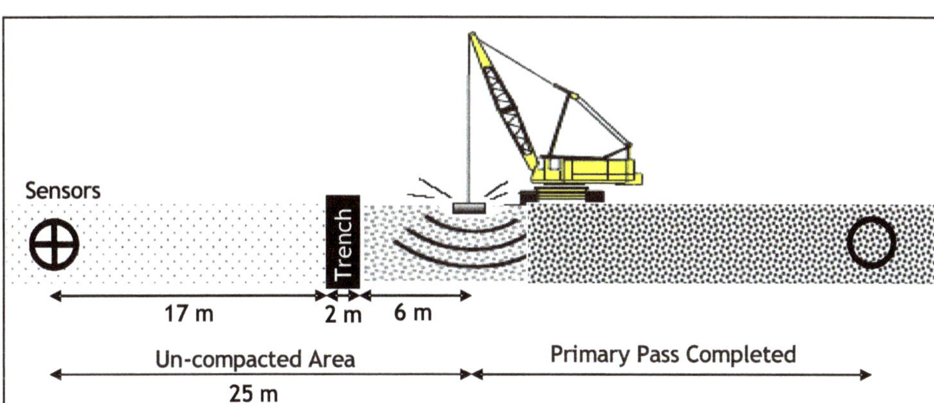

Fig. 22. Layout of Vibration Measurement Points at 25 m Distance

Fig. 23. Installation arrangement of LVDT and accelerometer

The vibrations are estimated through empirical correlations or measured with the help of instruments such as portable seismograph, accelerometers, velocity transducers, linear variable displacement transducers (LVDT), etc. Mayne et al. (1984) proposed an empirical relationship between PPV and inverse scaled distance to estimate minimum and maximum range of PPV against various scaled distances and is shown in Figure 3. Inverse scaled distance is square root of the compaction energy, $\sqrt{WH}$ divided by the distance, d from the impact point. From Figure 24, a minimum PPV of 7 mm/sec and maximum PPV of 70 mm/sec were estimated at a distance of 25 m from point of impact.

Empirical Equation (6), proposed by Rollins and Kim - 1992 [6] for estimation of the PPV, is based on field monitoring data of several dynamic compaction projects. PPV in these projects was measured using portable seismograph. The frequency of these vibrations ranged from 5 - 40 Hz. At a distance of 25 m away from impact point, a PPV of 14.16 mm/sec is estimated using Eq. (6).

$$PPV = 20\left(\sqrt{WH}/d \right)^{1.03} \tag{6}$$

Where,

PPV = Peak Particle Velocity, mm/sec
W = Weight, tons
H = Height of fall, m
d = Distance from point of impact, m

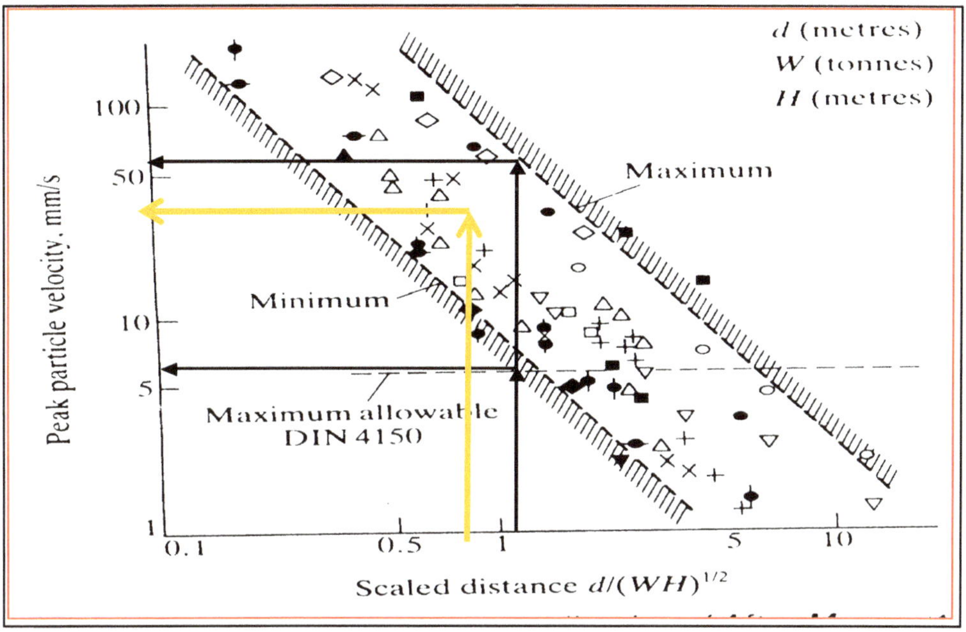

Fig. 24. Peak Particle Velocity (PPV) Vs Scaled Distance, Estimated PPV of Research Project shown in Yellow Lines

8.1 Across isolation trench

Across isolation trench, maximum PPV of 22.5 mm / sec was measured at 25 m distance from the point of impact as shown in Figure 25. This maximum velocity was obtained at 9th consecutive drop. PPV could not be measured at distance of 50 m point either due to very weak signals received by sensors or some malfunctioning of sensors.

8.2 Over partially compacted site

Over partially compacted site (a place where primary pass of compaction was completed), a maximum PPV of 14.1 mm / sec on 10th consecutive drop was calculated at a distance of 25 m away from point of impact as shown in Figure 25 and Table 5 and accelerogram is shown in Figure 26 and 27. Vibrations at a distance of 50 m away from impact point were very low and are not mentioned. The best fit curve is given by exponential equation. Increase in PPV at 25 m point across trench with each successive drop indicates that the PPV increased with increase in soil density due to DDC. An overall decrease in PPV after completion of first pass indicates that the initial pass created several denser / loose soil mediums. The primary impact points at grid spacing of 6 m centre to centre with un-compacted soil in between caused almost 50 % attenuation of PPV.

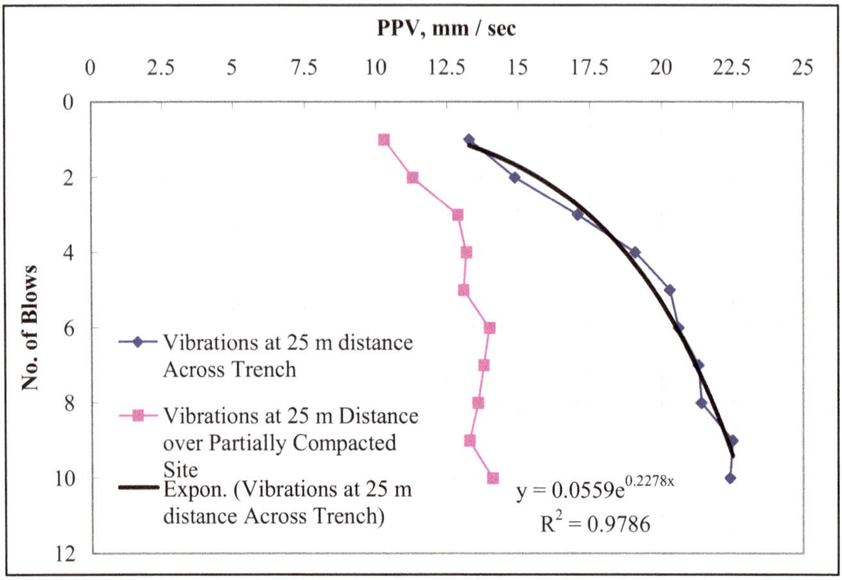

Fig. 25. Measured Ground Vibrations

No of Blow	PPV Across Trench (at 25 m Distance) mm/sec	PPV Over Partially Compacted Site (at 25 m Distance) mm/sec
1	13.3	10.3
2	14.9	11.3
3	17.1	12.9
4	19.1	13.2
5	20.3	13.1
6	20.6	14
7	21.3	13.8
8	21.4	13.6
9	22.5	13.3
10	22.4	14.1

Table 5. Range of PPV for Various Numbers of Blows

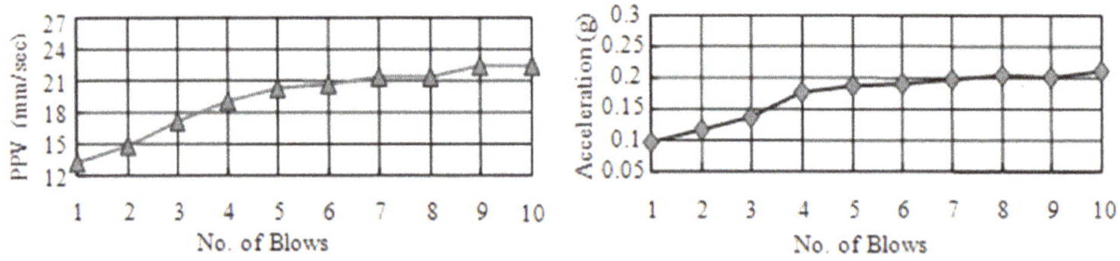

Fig. 26. Accelerograms for successive blows

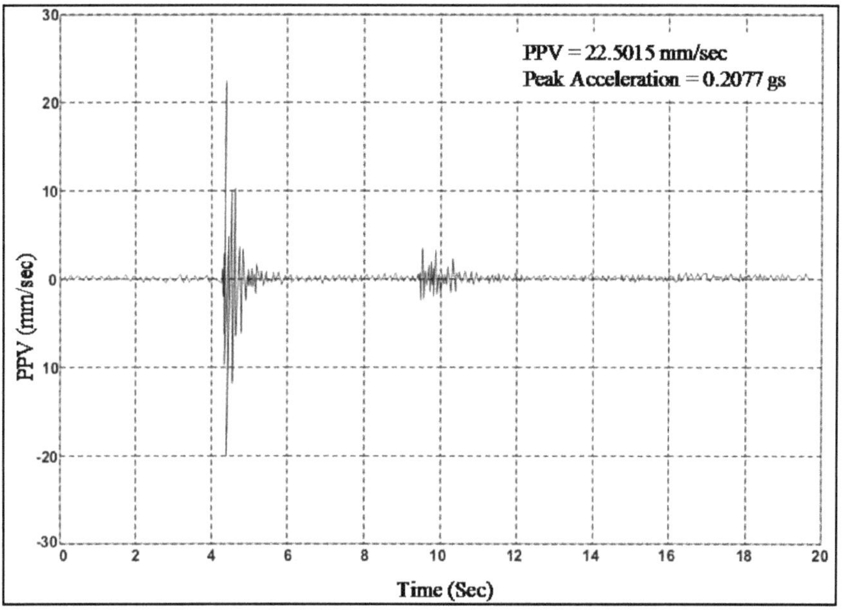

Fig. 27. Accelrogram of PPV at 10th blow

DDC technique can remediate liquefaction potential and enhance bearing capacity up to desired limits at desired depths. Factor of safety evaluated after DDC is given in Table 6 which shows an increased value of factor of safety at three different depths.

Depth (m)	$(N_1)_{60}$	CSR_L	σ'_{vo} (kN/m³)	$\tau_{cyc,L}$ ·	τ_{cyc}	FS_L
3.5	23	0.24	71.43	17.14	7.46	2.3
5.5	14	0.17	82.12	13.96	9.46	1.5
7.5	12	0.15	92.81	13.92	11.41	1.2

Table 6. Cyclic Shear Resistance of Soil after Dynamic Compaction and Factor of Safety (FS_L)

9. Conclusion and recommendations

In the light of research conducted for remediation of liquefaction hazard, improvement under impact points, lateral direction after 5 blows, 10 blows and full scale compaction, following conclusions are presented:

- directly under impact points, the bearing capacity of soil improved to 160 kPa upto a depth of 5 m.
- maximum improvement under impact points occurred at depths from 2 m to 4 m.
- sharp decrease in improvement is observed below 4 m depth.
- improvement was negligible below 6.5 m depth.
- with the increase in no. of drops from 5 blows to 10 blows, the degree of improvement also increased from maximum 35 blows to 45 blows.

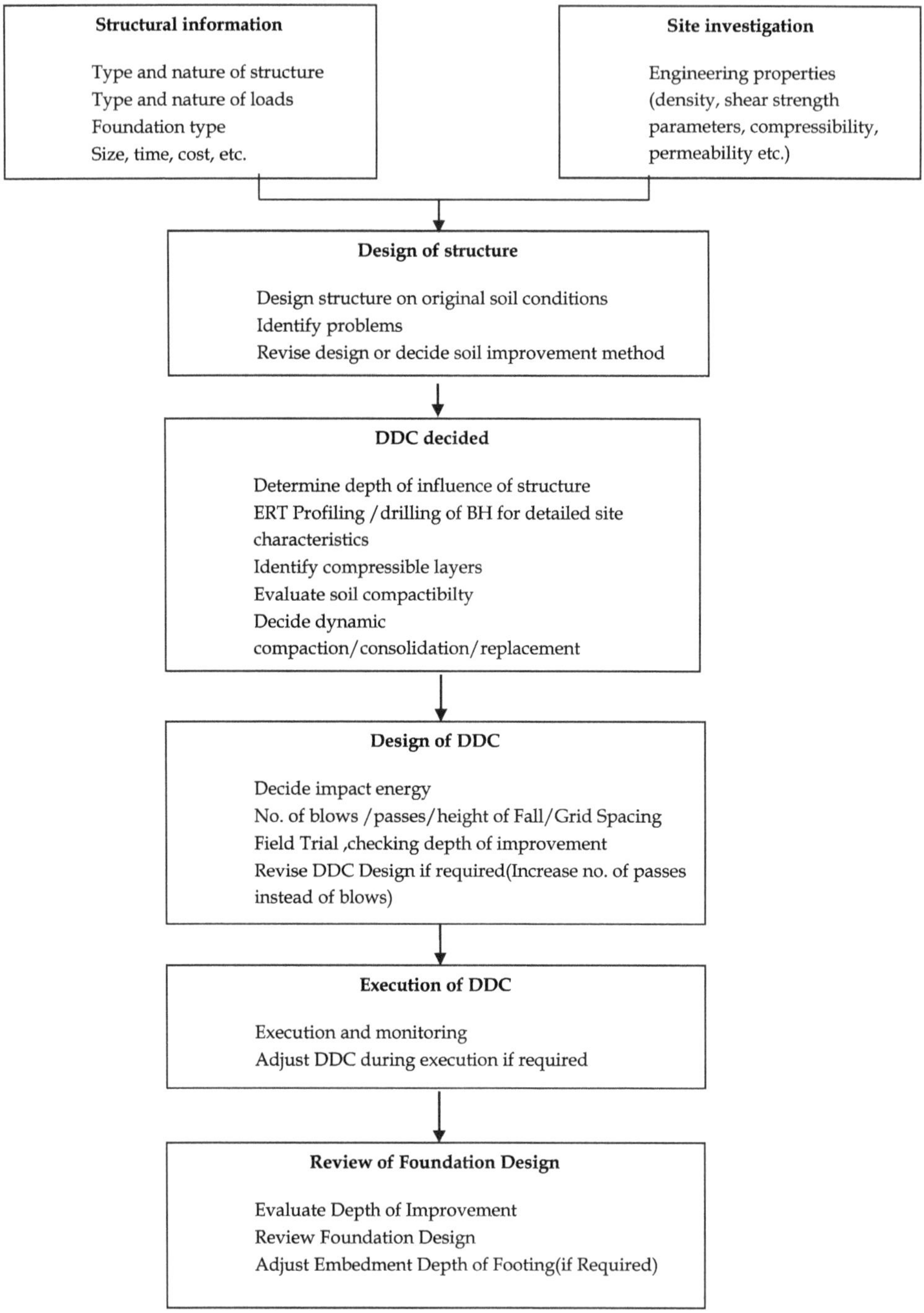

Fig. 28. Suggested Procedures for DDC Project

- increase in no. of blows from 5 to 10, had negligible effect on degree of improvement below 6.5 m depth.
- in upper 2 m of strata, improvement at the middle of any two adjacent impact points was comparatively less than improvement under the impact point.

From results of soil improvement project of alluvial soils, it is concluded that deep dynamic compaction technique can remediate liquefaction potential and enhance bearing capacity upto desired limits at desired depths. The compacted layers at top of strata and reduced grid spacing can significantly reduce the overall depth of improvement. In the light of this research project, a flow chart in Figure 28 is suggested for more efficient and effective DDC program.

10. References

Bowles, E. Joseph [1997]. *"Foundation Analysis and Design"*. McGraw-Hill Book Company, Inc., New York.

Lee, F.H., and Gu, Q., [2004]. "Method for Estimating Dynamic Compaction Effects on Sand". J. of Geotch. and Envmt. Eng., ASCE. No. 130(2), pp. 139-152.

Lukas, R.G. [1986]. *"Dynamic Compaction for Highway Construction"*. FHWA, US DOT, Report No. FHWA/RD-86/133.

Mayne, P.W., Jones J.W., and Dumas, J.C., [1986]. "Ground Response to Dynamic Compaction".J. of Geotech. Eng., ASCE, No. 110(6), pp. 757-774.

Menard, L., and Broise, Y. [1976]. *"Theoretical and Practical Aspects of Dynamic Consolidation"*. Ground Treatment by Deep Compaction, The Institute of Engineers, London. pp. 3-18.

Meyerhof, G. G., [1976]. "Bearing Capacity and Settlement of Pile Foundation". J. of Geotech. Eng. Div. Am. Soc. of Civ. Eng., No. 102(3), pp. 195-228.

Oshima, A., and Takada, N. [1998]. " Evaluation of Compacted Area of Heavy Tamping by Cone Point Resistance". *Proc. Int. Conf. Centrifuge 98.*, Kimura, Kusakabe, and Takemurs, eds., Balkema, Rotterdam, pp. 813-818.

Rollins Kyle, M., and Kim, Ji Hyoung [1994]. "US Experience with Dynamic Compaction of Collapsible Soils", in *In-Situ Deep Soil Improvement, Geotechnical Special Publication,* Am. Soc. of Civ. Eng., New York, pp. 27-43.

Khalid, Muhammad Mumtaz (1995). An investigation into high water table problem at Attock District, an MSc Thesis, National University of Science and Technology (NUST), Pakistan.

Kramer, L., Steven (1996). "Geotechnical earthquake engineering." *Prentice Hall Series.* 351-354, 369-376.

Kyle M. Rollins (1994). In-Situ Deep Soil Improvement, Geotechnical Special Publication No. 45, 37-38.

M/S High Tech, (2004). Report on geotechnical engineering investigation for Attock District, Pakistan.

Seed, H.B., Idriss, I.M. (1971). "Simplified procedures for evaluating soil liquefaction potentials", *J. of Soil Mechanics and Foundation,* ASCE, Vol. 107, 1249–1274.

Seed, H.B., Idriss, I.M., and Arango (1983). "Evaluation of liquefaction potential using field performance data." *Journal of Geotechnical Engineering,* ASCE. 109, 458-482.

Ali, Sarfraz. Evaluation of Deep Dynamic Compaction Technique. Thesis in Master's Degree in Geotechnical Engineering, National University of Science and Technology (NUST), Islamabad, Pakistan, 2007.

Wood R.D. Vibration Screening with Wave Barriers. A look Back for Future Geotechnics, 1999, 325-348.

Svinkin, R. Mark. *Prediction and Calculation of Ground Vibrations.* Proceedings of the 24th Annual Members Conference of the Deep Foundation Institute, Dearborn, Michigan, 1999.

Menard, L. and Broise, Y. *Theoretical and Practical Aspects of Dynamic Consolidation.* Geotechnique, 25, No. 1, 3-18, 1975.

Mayne et al. *Ground Response to Dynamic Compaction.* Journal of Geotechnical Engineering, 1984, 110. No. 6, 345-349.

Rollins, Kyle M. and Kim Ji Hyoung. *US Experience with Dynamic Compaction of Collapsible Soils.* Geotechnical Special Publication No. 1994, 26-43.

3

Lessons Learned from Recent Earthquakes – Geoscience and Geotechnical Perspectives

Robert C. Lo[1] and Yumei Wang[2]
[1]Klohn Crippen Berger Ltd., Vancouver, B.C.
[2]Sustainable Living Solutions LLC, Portland, Oregon
[1]Canada
[2]U.S.A.

1. Introduction

Earthquakes have been occurring long before human development, and will continue to occur with or without human civilization. Nature's forces behind earthquakes are powerful, unstoppable and can be deadly. Recent earthquakes illustrate these destructive forces across the globe. In Japan's March 2011 disaster, nearly 24,000 persons perished or missing in the world's most seismically prepared country with advanced early warning systems for tsunami and earthquake. In January 2010 at Haiti, a developing nation, even worse devastation occurred with about a quarter million fatalities.

Earthquake disasters are often covered in the news media for a short time period. However, after the media blitz fizzles out, the recovery period ensues. Recovery can involve extreme socio-economic hardship - painful emotional losses, physical injuries, public health crisis, widespread environmental contamination and loss of homes and businesses. This readjustment could last for many years. With today's increasing population and economical development in seismic hazard zones (in both developing and developed nations), the global seismic risk is also going up. The field of earthquake science has seen many recent advances, some involving geoscience and geotechnical issues. Synthesized in this chapter are key advances gleaned from literature that can be applied towards risk management decisions to reduce future loss of lives and socio-economic disruptions.

As members of the Earthquake Investigation Committee (EIC) of ASCE Technical Council on Lifelines Earthquake Engineering (TCLEE), the authors have been involved in the investigation for four (Sumatra, Wenchuan, Maule and Tohoku-Oki) of the six recent earthquakes covered in this chapter, focusing on the geoscience and geotechnical aspects. This chapter first highlights the characteristics and damages of these earthquakes: the 2004/2005 Sumatra, Indonesia, 2008 Wenchuan, China, 2010 Haiti, 2010 Maule, Chile, 2010/2011 Christchurch, New Zealand, and 2011 Tohoku-Oki (East Japan) earthquakes (see Table 1). It then discusses some of the geoscience and geotechnical aspects of these earthquakes with references to other relevant seismic events. Finally, it outlines the lessons learned from these events in general as well as with respect to lifelines facilities, and draws some conclusions.

2. Recent earthquakes

2.1 General

Table 1 summarizes the characteristics and damages of the six recent events. It provides a thumb-nail sketch of these events including: date and location, earthquake type and focal mechanism, peak ground acceleration and Modified Mercalli Intensity, special features, casualties, damages and general references. Three of these are tsunami-generating subduction events of magnitude, Mw 8.8 to 9.1-9.3 (see Fig. 1), while the other three are crustal events of magnitude, Mw 6.0 to 7.9, involving blind thrust, strike-slip/thrust or reverse faults. Prominent features of these events are briefly outlined below.

2.2 Prominent features

2.2.1 2004 (Mw 9.1-9.3)/2005 (Mw 8.6) Sumatra, Indonesia earthquakes/tsunamis

The December 26, 2004 Sumatra earthquake was triggered by the rupture of a locked segment of the fault plane at least 500 km long by 150 km wide between the subducting Indo-Australian Plate and the upper Eurasian (Burma) Plate (see Fig. 2). Figure 3 shows the computed vertical and horizontal components of surface displacements of the upper plate based on a finite-fault model (EERI 2005, 2006, ASCE 2007, BSSA 2007).

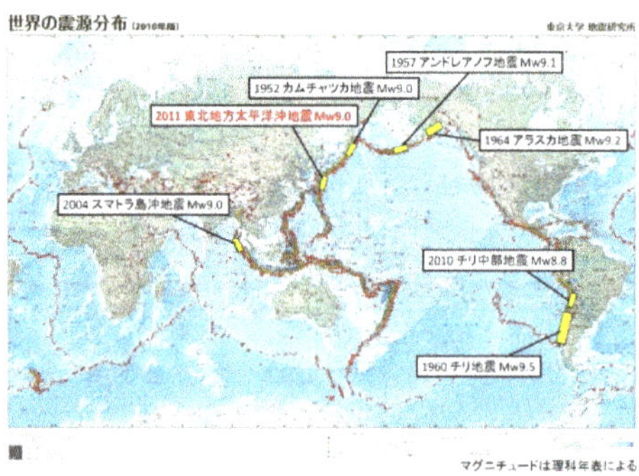

Fig. 1. Major Circum-Pacific Subduction Earthquakes Since 1957. http://outreach.eri.u-tokyo.ac.jp/ eqvolc/201103_tohoku/

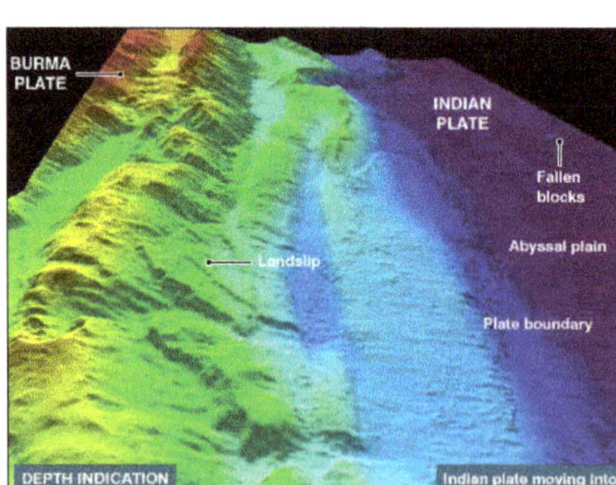

Fig. 2. Three-Dimensional Sonar Imagery of Seabed off the coast of Sumatra Island. (BBC 2005)

Date (Local Time)	Earthquake Location	Type	Focal Mechanism	Peak Acceleration Intensity, MMI	Special Features	Casualties	Damages		General References
							General	Lifelines	
2004 Dec. 26 (7:58 am)	**Sumatra Indonesia (Indian Ocean)**	Interplate	Mw 9.1 to 9.3, depth 30 km 1300 km x 150 km rupture zone defined by aftershocks, min. 500 km rupture length at main shock, max. 20 m fault displacement Sea floor uplifted several metres	Modified Mercalli Intensity ≤ X+	Heavy casualty in countries around Bay of Bengal and Indian Ocean due to no warning of incoming tsunami Tsunami run-up 2 to 5 m, up to a max. of 31 m, max. flow depth >9 m and inundation up to 3 to 4 km inland Earthquake damage in Aceh, North and West Sumatra	Very High death ~180,000+ to 230,000+ injured ~125,000 missing ~46,000 displaced ~1.69 million	Severe Economic loss ~ US $13 billion Impact on tourism	Moderate Tsunami inundation - main cause of lifelines damage in affected areas	ASCE 2007 EERI Newsletter Mar - Jul 2005 EERI Spectra 2006 BSSA 2007 USGS 2004
2005 Mar 28 (11:09 pm)	**Nias Indonesia (Indian Ocean)**	Interplate	Mw 8.6, depth 30 km	Modified Mercalli Intensity ≤ VI	Eq. shaking, Ground tilted with up to 1 m uplift and subsidence Panic due to residual fear about tsunami, which was rather subdued in the event, ranging from 1 to 3 m near epicenter and ≤0.4 m further away	Medium death 1,000+	Moderate Collapse of and damage to buildings	Moderate Damage to roads, bridges, pipelines and power facilities	EERI Spectra 2006 USGS 2005
2008 May 12 (2:28 pm)	**Wenchuan China (Asia Continent)**	Crustal	Mw 7.9, depth 19 km Thrust/strike-slip rupture along main (200 km long, 2-4 m offset) and splay (50 km long, 0.5-2 m offset) fault	≤ 0.98 g Modified Mercalli Intensity up to X to XI	Widespread earthquake shaking, massive landslides, debris flows and formation of landslide dams Maximum offset (main fault 11 m and splay fault 3.5 m) Building code design requirements under-estimate significantly the seismic demand Substantial portion of existing building stocks deficient in seismic resistance	Very High death 69,195 injured 374,643 missing 18,392 displaced 5 million	Severe Economic loss ~US $120 billion Collapse of and damage to buildings	Severe 53,000 km of roads damaged Bridges collapsed, damaged Power lines, communication systems demolished, disrupted 47,643 km of pipelines and 8,426 water treatment plants damaged	EEEV 2008 EERI Newsletter Oct. 2008 BSSA 2010 USGS 2008
2010 Jan. 12 (4:53 pm)	**Haiti (Caribbean Sea)**	Crustal	Mw 7.0, depth 13 km A combination of reverse and left-lateral strike- slip faulting related to the Enriquillo-Plantain Garden fault system, ruptured zone about 30 km by 15 km, with left- lateral slip fault displacement, and max. moment release in the first 10 secs.	Estimated ~0.3 to 0.45 g Modified Mercalli Intensity IV to VIII	Heavy casualties due to collapse of poor-quality (unreinforced masonry or non-ductile reinforced concrete) building stocks Liquefaction and lateral spreading, landslides and rockfalls contributed to damage	Very High death ~233,000 to 250,000 ~ 1,000 from cholera injured ~300,000 homeless ~1 million	Severe Economic loss ~ US $7.8 billion	Main port suffered extensive damages, including collapse of piers and submerged cranes	EERI Newsletter Apr./May 2010 USGS 2010 USGS/EERI 2010

Table 1. Summary of Relevant Earthquake and Damage Data for Six Recent Earthquakes In 2004 to 2011

Date (Local Time)	Earthquake Location	Type	Focal Mechanism	Peak Acceleration Intensity, MMI	Special Features	Casualties	Damages: General	Damages: Lifelines	General References
2010 Feb. 27 (3:34 am)	**Maule (Offshore Bio-Bio) Chile (South Pacific Ocean)**	Interplate	Mw 8.8, depth 35 km Rupture zone 500 km by 100 km Duration over 1 min >0.05 g in general, and over 2 min in Concepcion area Horiz. displacement up to 3 m to west Along coastline from south to north, up to 2 m uplift in the south, 0.5 m subsidence in the middle and no discernable change in the north	Up to 0.65 g Modified Mercalli Intensity up to VI to IX	Tsunami run-up height in the range of 3 to 9 m Coastal uplift and subsidence Numerous aftershocks including 19 in the range of Mw 6.0 to 6.9 within the first month	Low death ~521 more than 200 due to tsunami missing 56	Severe Economic loss ~ US $30 billion	Liquefaction caused significant damage to power transmission towers, water and waste water systems Earthquake caused damage to and extended outage of oil refineries Tsunami damage to ports and bridges	GEER 2010 EERI Newsletter Jun. 2010 USGS 2010
2010 Sep. 4 (4:35 am)	**Darfield (Canterbury) New Zealand (South Pacific Ocean)**	Crustal	Mw 7.0, depth 10 km Reverse fault with strike-slip right-lateral (maximum 4 m) and vertical (maximum 1 m) displacement, about 30 km surficial rupture, and 40 km west of Christchurch	Up to 0.74 g horiz. with even higher vert. in epicentral area M.M. intensity in epicentral area VIII-IX in Christchurch VI to VII	Widespread liquefaction and lateral spreading Poor performance of unreinforced masonry buildings Non-structural damage to building components and contents Repair of many damaged structures in progress when the Mw=6.1 aftershock struck	Nil 1 died of a heart attack 2 injured by fallen objects	Economic loss ~ US $3 billion Structural damages to unreinforced masonry buildings and non-structural damages to other buildings Liquefaction caused significant building damage	Liquefaction and lateral spreading caused significant damage to water and waste water systems Underground voids created by loss of sand due to sand ejection from ground	EERI Newsletter Nov. 2010 USGS 2010
2011 Feb. 22 (12:51 pm)	**Christchurch New Zealand (South Pacific Ocean)**	Crustal	Mw 6.1, depth ~5 km Oblique blind thrust fault with up to 2.5 m slip displacement About 14 km long sub-surface rupture within 10 km of Christchurch city centre	Maximum spike to 1.5 g near epicentre and up to 0.72 g near city centre (for about 8 sec.) Modified Mercalli Intensity up to VIII in Christchurch	Aftershock of Sept. 4, 2010 Darfield Earthquake (about 5 months later) Substantial damage to the central business district (CBD) of Christchurch due to closer distance to earthquake and weakened structures by the main shock Widespread liquefaction in the CBD and eastern suburbs Landslides and rockfalls in Port Hills	Low 184	Economic loss ~ US $ 16 billion Collapse of and damage to buildings Liquefaction caused significant building damage	Extensive lifelines damage due to liquefaction and lateral spreading, including: roads and highways, power distribution systems, water and wastewater lines and drainage facilities	EERI Newsletter May 2011 USGS 2011
2011 Mar. 11 (2:46 pm)	**Tohoku-Oki (East Japan) Japan (North Pacific Ocean)**	Interplate	Mw 9.0 depth 32 km Rupture Zone 300 km by 150 km Preceded by a series of foreshocks ranging from Mw 6 to 7.2 since March 9	Maximum spike to 3.0 g Modified Mercalli Intensity up to VI to VIII	Tsunami run-up height 4 to 8 m, with maximum up to 38 m Tsunami and earthquake damaged Fukushima Dai-Ichi nuclear power plant, power and cooling systems, resulting in reactor core rods meltdown and eventual abandonment of the plant Breach of a 17 m high earthfill dam	High death 15,840 injured 5,951 missing 3,546 homes damaged ~200,000	Economic loss ~ US $300 billion 7,735 schools and >300 hospitals damaged	Fuel rods meltdown at a nuclear power plant Earthquake and tsunami destroyed many communities together with lifelines, such as power, transportation including bridges, communication, water and waste-water system Bridge deck failures due to tsunami uplift Liquefaction and lateral spreading caused significant damages to lifelines	JSCE 2011 PEER 2011 USGS 2011 EERI 2011

Table 1. (continued). Summary of Relevant Earthquake and Damage Data for Six Recent Earthquakes In 2004 to 2011

The stealthy nature of tsunami onslaught of coastal inhabitants and international tourists around the Bay of Bengal and Indian Ocean (see Fig. 4) in the morning after Christmas of 2004 and ensuing heavy casualty focused the world's attention at the time. The tsunami run-up height had considerable variation around the Indian Ocean, but ranged in general from 2 to 5 m, and reaching a maximum of 31 m in Sumatra (see Fig. 5). The event served as an impetus to improve the tsunami-warning system for the countries in the region. Although the subsequent smaller event on March 28, 2005 further south involved nominal tsunami waves reaching 1 to 3 m height locally and did far less damage, it stirred up considerable local fear due to the dreadful earlier event.

No strong-motion acceleration time histories were recorded in the epicentral region. In the near-field northwest and north Sumatra, tsunami compounded earthquake-shaking damage. In the far-field tsunami was the predominant cause of destruction. The severity of tsunami damage was affected by many factors such as bathymetry, shoreline configuration and topography, etc. which influenced the wave focusing, reflection and refraction; tsunami run-up height; extent of inland inundation; flow velocity and scour, wave pressure, uplift and debris impact force. EERI (2006) noted the following tsunami-related phenomena:

- Maldives suffered moderate damage, although the coral-atolls archipelago rises only about 2 m above the mean sea level. Since the islands rise from the seafloor steeply, wave amplification was nominal.
- The Indian mid-ocean ridges served as wave guides, and funnelled the tsunami away from the tip of Africa.
- The tsunami generating capacity of an earthquake is governed by the mass of the water body suddenly displaced by the seafloor movement. The presence of the Nias and Simeulue Islands reduced the affected water body during the 2005 earthquake, thus induced relatively low tsunami.
- Unlike tidal gauges that could be affected by harbour resonance, tsunameters can indicate free-field tsunami height.

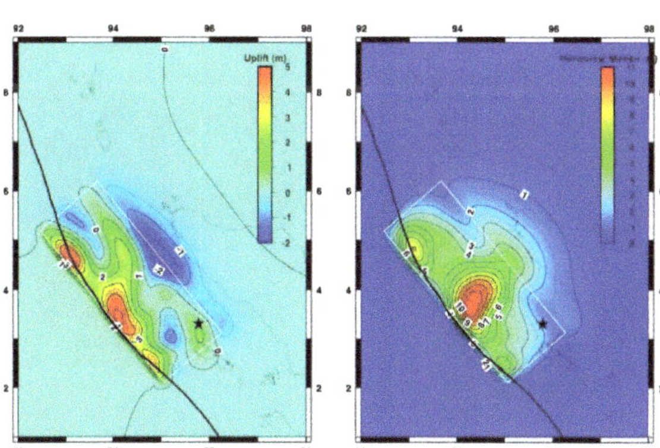

Fig. 3. Modelled Surface Displacements of the Upper Plate in Metres, for Vertical (Left) and Horizontal (Right) Components, 2004. http://neic.usgs.gov/neis/eq_depot/2004/eq_041226/neic_slav_ff.html

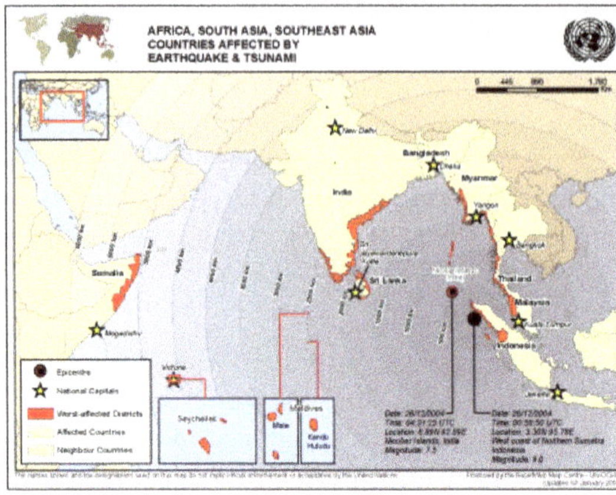

Fig. 4. Tsunami-Damaged Countries Around Indian Ocean, 2004.
(UN OCHA 2005)

- Two types of leading waves of tsunami were modelled back in 1994: a leading depression N-wave (LDN) and a leading elevation N-wave (LEN). Tide-gauge records on Fig. 6 confirmed the validity of the earlier hydrodynamic modelling: LDN wave was shown on Phuket, Thailand record, while LEN wave on Male, Maldives record. Historically, the leading depression N-wave, i.e., that causing the initial receding of the shoreline as the tsunami approaching, has been a death trap for many unwary fishermen and beachcombers.

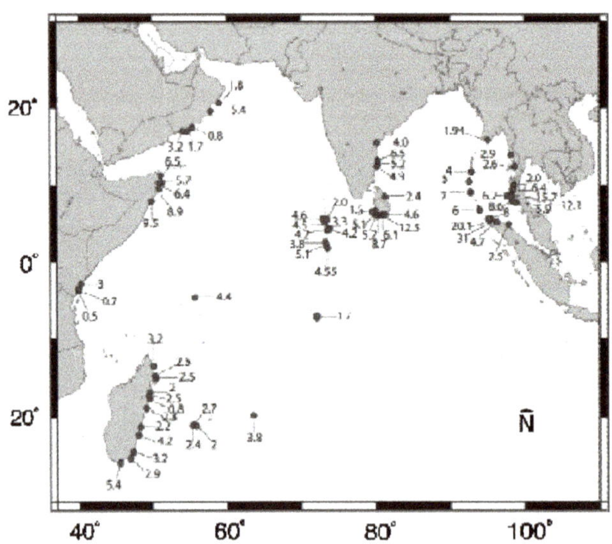

Fig. 5. Representative Tsunami Runup Heights along Shores of Indian Ocean, 2004.
(EERI 2006)

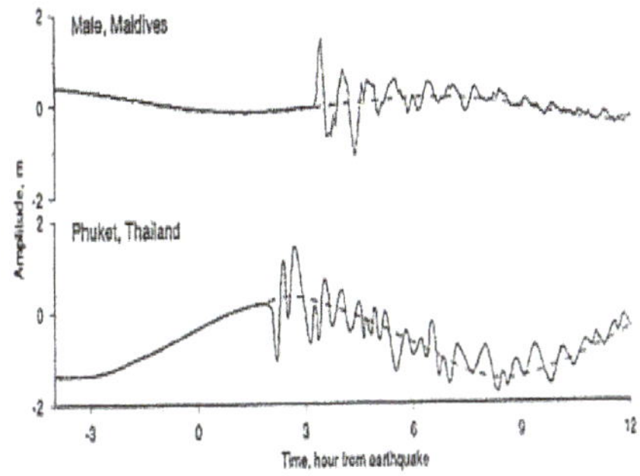

Fig. 6. Tide Gauge Records from Male, Maldives and Phuket, Thailand, 2004. (EERI 2006)

2.2.2 2008 (Mw 7.9) Wenchuan, China earthquake

The latent threat of the causative Longmenshan fault system was formally recognized by the geoscience research community about a year prior to the 2008 event, but this finding did not influence the seismic code at the time. The major seismic event of Mw 7.9 impacted a large region in the southwest China, involving several provinces that were significantly under-designed for the event. Figures 7 and 8 show the conditions of the old Beichuan town before and after the earthquake.

Fig. 7. Old Beichuan Town Before Earthquake. http://www.eeri.org/site/meetings/us-china-us-china-symposium

Fig. 8. Old Beichuan Town After Earthquake. http://www.eeri.org/site/meetings/us-china-us-china-symposium

Due to the steep and rugged topography in the affected mountainous region, wide spread landslides (see Figs. 9 and 10) have been major destructing factors, besides strong earthquake shaking (EEEV 2008, EERI 2008). About 20,000 fatalities, near one-fourth of the total, were caused by 15,000 geohazards in the form of landslides, debris flows and rockfalls, with the largest landslide involving a volume of 1.1 billion m^3. In the high, steep slopes (see Fig. 10) along the 270 km long Longmenshan tectonic belt, the large vertical acceleration and topographic amplification of ground motion have resulted in more than 10,000 potential geohazard sites after the event (Yin et al. 2011).

Fig. 9. Landslides in Proximity of Old Beichuan Town. http://www.agu.org/news

Fig. 10. Landslides along Minjiang River near Wenchuan County. (EEEV 2008)

Up to 34 landslide lakes were formed in Sichuan and one in Gansu provinces, threatening about 700,000 people living downstream. The largest one was located in Tangjiashan, Beichuan County (see Fig. 11), with a 71-m high debris dam blocking the Shitingjian river forming a lake about 800 m long and 600 m wide. The downstream flood-threatened area had to be evacuated, and the debris dam breached by excavation and blasting to remove the secondary flood hazard.

Figure 12 shows the number and severity of landslides per km of National Highway Route #213 over the hanging wall versus foot wall. As expected in a thrust-fault earthquake (Sommerville 2000), there is substantially more damage over the hanging wall as compared to that over the foot wall. Similarly, there is more damage in the area along the earthquake propagation direction than in the opposite direction.

The strong shaking with peak ground acceleration up to 0.98 g, ground failures and fault displacements up to 2 to 4 m caused wide spread destruction of communities and infrastructures. The long duration of strong shaking, over 100 seconds in general, was detrimental to unreinforced masonry buildings, and non-ductile reinforced concrete buildings that form the bulk of the building stock in the affected area. Figure 13 shows the acceleration response spectra at Qingping Station in the epicentral region (Ventura et al. 2008). Superimposed on the figure is the design acceleration spectra for Vancouver, British Columbia with Site Class C local soil condition, according to NBCC (2005) code for comparison purposes. As typical in most earthquakes, vertical peak ground acceleration is of similar value as its horizontal counterparts in the near field.

Fig. 11. Landslide Lake at Tangjiashan Landslide. Engineering Resilient Cities - Mahin - Oct 2008

Initially, accesses to remote areas were handicapped by the disruptions of highways and railways. The prompt and orderly nation-wide rescue and restoration programs were responsible for mitigating the suffering of affected population and the recovery of the region to normalcy. The unique Chinese mechanism for emergency response, recovery and

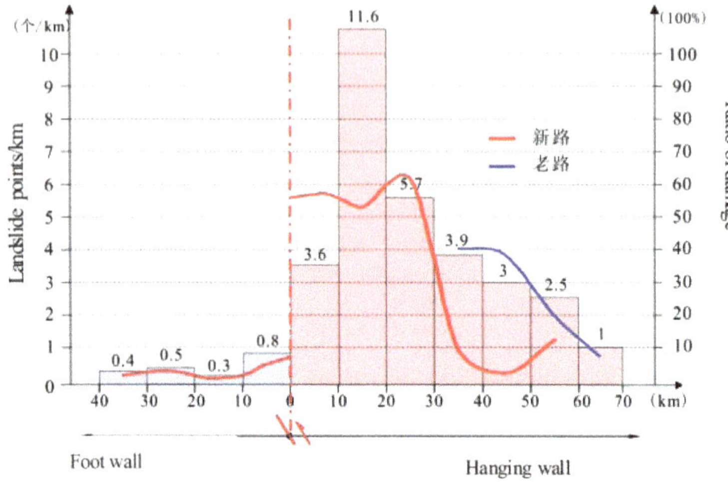

Fig. 12. Landslide Activities above Hanging Wall versus Foot Wall. (EEEV 2008).

reconstruction involved communities and jurisdictions located far away from the damaged areas. This twinning of communities in need and those to help accomplished dual goals: sharing of enormous financial hardship; and cultivating camaraderie among population. Figure 14 shows some of the officials who had spent several months in a donated school to assist relocated residents from outlying communities including the neighbouring province. Temporary dwelling units were set up across the earthquake damaged region. Figure 15 shows that the communities thus set up have become new settlements with all amenities to conduct normal life. Residents were finding work both within the settlement and outside. Grain drying activity was seen in the foreground of the figure (Lo 2009).

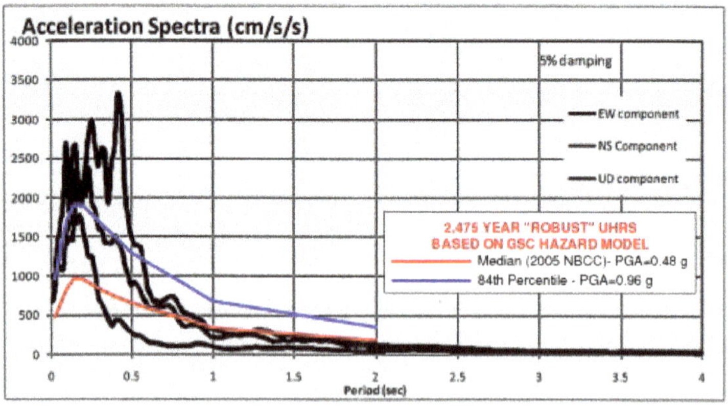

Fig. 13. Acceleration Response Spectra for Qingping Station in Epicentral Region. (Ventura et al. 2008)

Fig. 14. School Donated for Temporary Accommodation of Survivors in Chongzhou, 2008.

A comprehensive three-year reconstruction program covered the management organization and socio-economical structure for regional revitalization, and was carried out by the twinned communities. A new Beichuan town was constructed in Yongchang City about 25 km downstream of the destructed town with many traditional architectural elements of the local Qiang minority (see Fig. 16).

Fig. 15. Relocated Community in Temporary Accommodation of Chongzhou, 2008.

Fig. 16. City Scene Along Yongchang Boulevard, Yongchang, Sichuan - New Beichuan Town. http://www.skyscrapercity.com/ showthread.php?t=1222823

Figure 17 shows the Yongchang River Bank, and Figure 18, an apartment building constructed in the new town with the support of a twinned city located near the northeast coast of China, about 1,400 km away.

Fig. 17. Yongchang River Bank, Yongchang http://www.skyscrapercity.com/ showthread.php?t=1222823

Fig. 18. New Apartment Building Constructed with Support from Linyi City, Shandong Province, 2010. http://www.eeri.org/site/meetings/ us-china-symposium

2.2.3 2010 (Mw 7) Haiti earthquake

Haiti has suffered devastating earthquakes similar to the 2010 event in the past, despite recent seismic quiescence. The 2010 earthquake is caused by a combination of reverse and left-lateral strike-slip faulting related to the Enriquillo-Plantain Garden fault system. The threat of this specific fault to the population was not recognized prior to the event (EERI 2010, USGS/EERI 2010). There is no strong motion record for the main shock in Haiti. Peak ground acceleration was estimated in the range of 0.3 to 0.45 g in the affected area. The building stocks, consisting mainly of unreinforced masonry and non-ductile reinforced concrete structures (see Fig. 19) including government buildings (see Fig. 20) and buildings

Fig. 19. Damaged Buildings Located on Hill Slope. http://www.worldcatastrophe.com

used by the United Nation Stabilization Mission Headquarters, are generally inadequate to withstand this level of shaking (Fierro and Perry 2010). The earthquake fatality was estimated at about 233,000 to 250,000, with 300,000 injured and 1 million homeless. Subsequent to the earthquake outbreaks of cholera caused about 1,000 deaths as of October 2010 due to the poor hygiene condition.

Fig. 20. Damaged Presidential Palace in Port-au-Prince. http://alanrdennis.wordpress.com

After the event, USGS (2010) developed initial seismic hazard maps for Haiti to assist the management of earthquake response and reconstruction effort. The increased seismic hazard along the adjacent portion of the Enriquillo fault directly to the south of Port-au-Prince must be considered in the post-earthquake reconstruction.

2.2.4 2010 (Mw 8.8) Maule, Chile earthquake/tsunami

In the 2010 Maule, Chile earthquake, the tsunami contributed close to half of the total casualty of about 521 due to the timing of the event occurring in the early morning. Tsunami damage to the over 500 km long coastline varied, as it came in during low tide and involved 3 to 4 major surges. Tsunami run-up height generally ranged from 3 to 9 m in the epicentral region with the maximum reaching about 20 m along cliff faces.

The relative low casualty speaks for the seismic-resistant capacity of the general building stocks. The fact that high-occupancy buildings were vacant at the time of earthquake also helped to reduce casualty. Damaged or collapsed buildings suffer design flaws such as lack of vertical continuity of load-bearing structural elements, relatively thin shear walls and lack of detailing requirements for reinforcing steel in special boundary elements, etc. (GEER 2010). After the earthquake, looting has been reported in the affected region until the army was deployed to maintain law and order.

Figure 21 shows the modelled slip distribution over the fault plane in the land area. The maximum fault slip along the coast line is about 300 cm. The earthquake occurred in the summer with prevailing low water table. It reduced the occurrences of landslides and liquefaction. The relatively long duration of earthquake shaking did trigger slope slumps in marginally stable or wave-undermined natural slopes, and liquefaction and lateral spreading in low-lying areas with high water table. Figure 22 shows a concrete pier used for unloading fish catches from small vessels at the Coronel fishing port. Its failure appears to be caused by the combination of earthquake shaking, foundation liquefaction and possibly tsunami wave force.

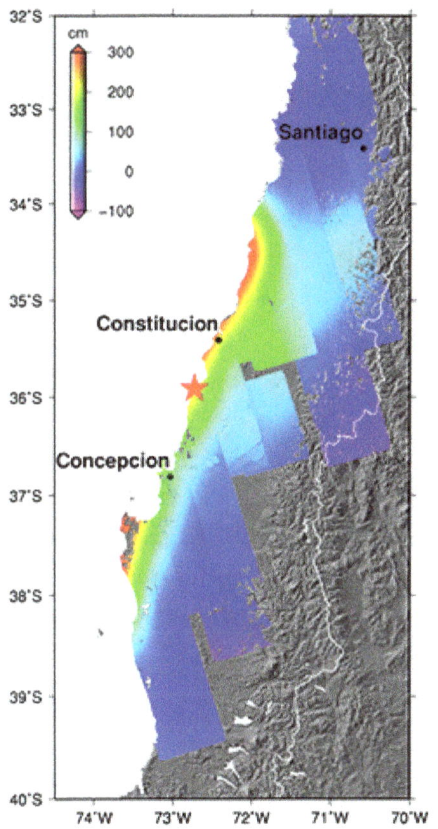

Fig. 21. Fault Plan Slip Distribution over land area - Maule, Chile.
http://tectonics.caltech.edu/sliphistory/2010_chile/index.html

The flow failure of a tailings dam triggered by the earthquake is shown in Figs. 23 and 24 (GEER 2010). Compared with a large number of tailings dam failures in the 1965 La Ligua earthquake and two failures in the 1985 Santiago earthquake, tailings dam performance in this event seems to show some improvement. Compacted engineering fills and structures supported by improved ground using stone columns and micro-piles were reported to perform well.

Fig. 22. Distorted Fish-Unloading Concrete Pier at Coronel Port, Chile. (TCLEE 2010)

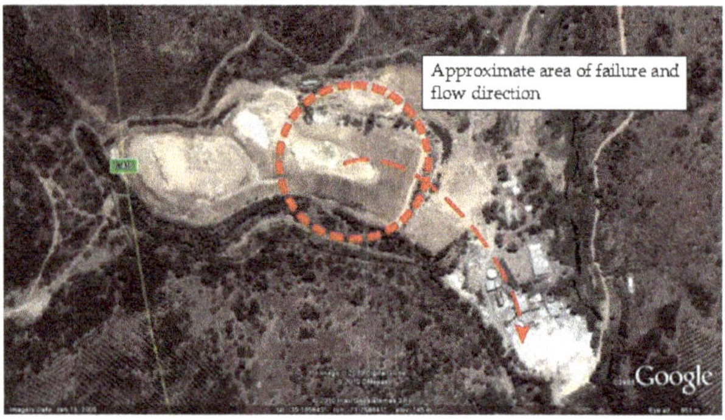

Fig. 23. Bird's-eye View of Las Palmas Tailings Impoundment Prior to Failure. (GEER 2010)

Fig. 24. Upper Scarp of Failed Tailings Impoundment. (GEER 2010)

2.2.5 2010/2011 (Mw 7.0) Darfield (Canterbury) / (Mw 6.1) Christchurch New Zealand earthquakes

Both the September 2010 Mw 7.0 Darfield and February 2011 Mw 6.1 events were shallow crustal earthquakes, which triggered wide-spread liquefaction in the fluvial and diluvial deposits in the epicentral region (see Figs. 25 and 26). Peak ground accelerations spiking above 1 g, often with higher vertical than horizontal acceleration, were reported in the epicentral areas of both events. In the central business district of Christchurch, the largest city on the South Island, peak acceleration reached about 0.7 g on February 22, 2011 as compared to about 0.3 g on September 4, 2010. Figure 27 shows the horizontal acceleration response spectra for Christchurch Hospital with 5% damping. It illustrates that the aftershock is reflected by the current New Zealand 2,500-year design spectrum, while the main shock is corresponding to the 500-year design spectrum. Proposal to increase the design spectrum for periods less than 1.5 sec is being considered (EERI 2010/2011).

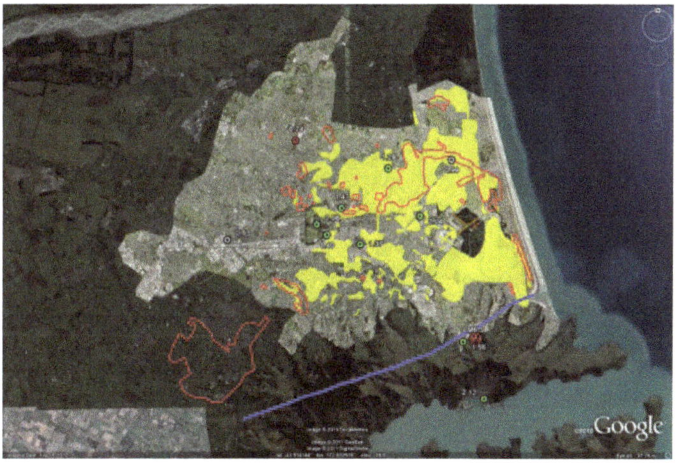

Fig. 25. Liquefaction Maps: Sept. 2010 (Red) and Feb. 2011 (Yellow). (GEER 2011)

Fig. 26. Liquefied Sand Deposited on Street, 2011. http://otilya.com/view/ News/Asia-Pacific/New-Zealand

The smaller aftershock caused significant casualties and heavier structure damages in Christchurch (see Figs. 28, 29 and 30) than the main event, because it occurred much closer to the central business district, where many buildings had already suffered varying degrees of damage due to the prior event.

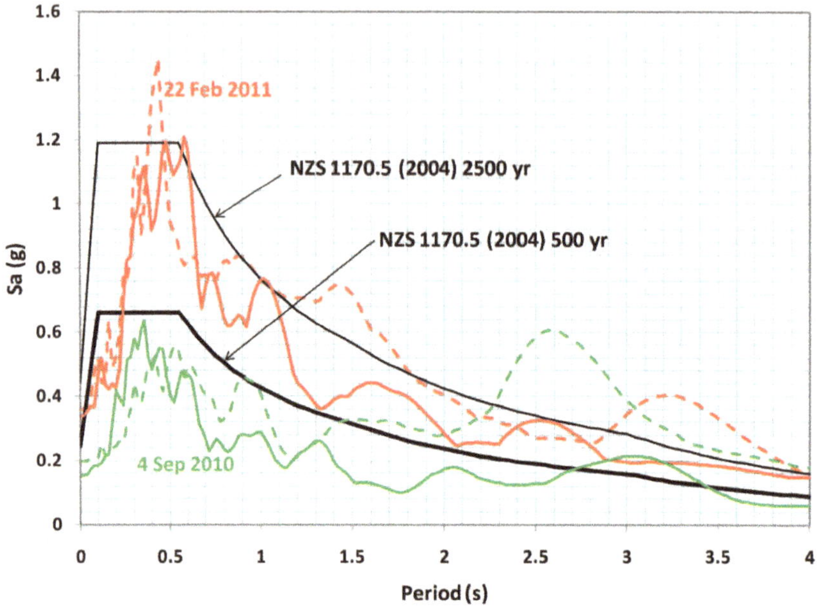

Fig. 27. Horizontal Acceleration Response Spectra for Christchurch Hospital. (EERI 2011)

Fig. 28. Scene of Downtown Christchurch Captured by a Tourist During Earthquake, 2011. http://otilya.com/view/christchurch_ as_the_quake_hits.html

Unreinforced masonry and non-ductile reinforced concrete buildings suffered most structural damages, while liquefaction and lateral spreading contributed to foundation failures or differential settlement of buildings and pipeline breaks. Non-structural damages to building components and content caused food-supply shortage and slowed down business recovery. Other damages were caused by rock falls in the Port Hills area.

Fig. 29. Two Survivors Comforting Each Other During Rescue Operation at Pyne Gould Building, 2011. http://www.kiwiblog.co.nz/2011/03/News/Asia-Pacific/New-Zealand

Fig. 30. Topographic Amplification May Play a Role in Damaging an Unreinforced Masonry Building, 2011. http://otilya.com/view/NewsAsia-Pacific/New-Zealand

2.2.6 2011 (Mw 9.0) Tohoku-Oki (East Japan) earthquake/tsunami

Figure 31 shows the modelled slip distribution over the fault plane which triggered the Tohoku-Oki earthquake. The slip on the offshore rupture plane was about 30 m, but the slip along the east coastline reduced to below 5 m. Figure 32 shows acceleration time histories recorded in the earthquake region (NIED 2011). These records have been used to analyze structures such as the study of the response of high-rise buildings to long-period ground motions (Takewaki et al. 2011). It is worth to note that some of the records in Fig. 32 show multi-sequences of earthquake excitations. Moustafa and Takewaki (2010) indicated that such earthquake sequences would result in more severe structural damages due to the accumulation of inelastic deformations. Substantial damages were suffered by the buildings and infrastructures in such a major subduction event of long duration due to the combined effect of earthquake shaking and tsunami onslaught. Earthquake subsidence caused some coastal areas submerged in tidal water (see Fig. 33). JSCE (2011) provided an overview of the characteristics of the earthquake and tsunami; and damage reconnaissance including: geotechnical and structural damage, liquefaction, damage to dykes and levees, and bridges.

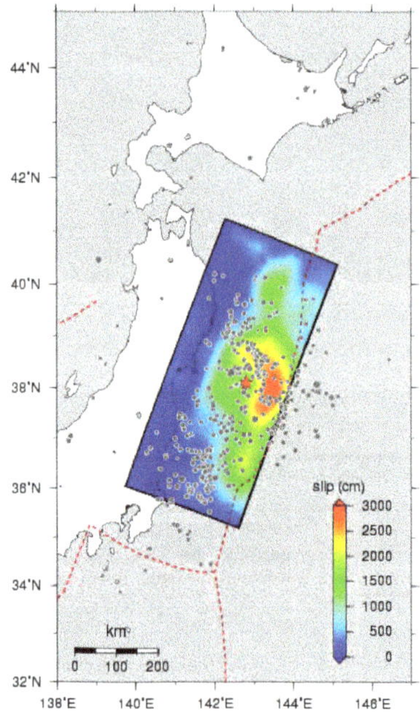

Fig. 31. Fault Plane Slip Distribution Tohoku-Oki, Japan. http://tectonics.caltech.edu/slip_history/2011_taiheiyo-oki/index.html

Tsunami run-up height was in a range of 4 to 8 m, with a maximum of up to 38 m. The existing tsunami-protection seawalls and gates in the coastal region were overrun in places causing significant destruction. As a result, the numbers of casualty (15,840) and missing people (3,546) tend to be high for Japan, a seasoned country having invested heavily in the preparation for and defence against the frequent seismic and tsunami events. Emergency response efforts were hampered by fuel shortage, telecommunication and transportation disruption, damage to fire and police stations and hospitals and nuclear radiation.

The destructive power of earthquake and tsunami has long been experienced and recognized by human race since its existence. However, 2011 Tohoku-Oki earthquake has revealed another relatively modern threat, i.e., the trigger of a long-lasting and potentially deadly nuclear incident. Salt water intrusion into pumping systems resulting in equipment failures has been reported earlier such as in the 1993 Hokkaido and 2004 Sumatra events. This relatively minor problem usually only led to delay in the restoration of the water and sewer systems. Nevertheless, at the Fukushima Dai-Ichi nuclear generating station (see Fig. 34), the disruption of the cooling systems of nuclear reactors and power failure resulted in the meltdown of reactor fuel rods.

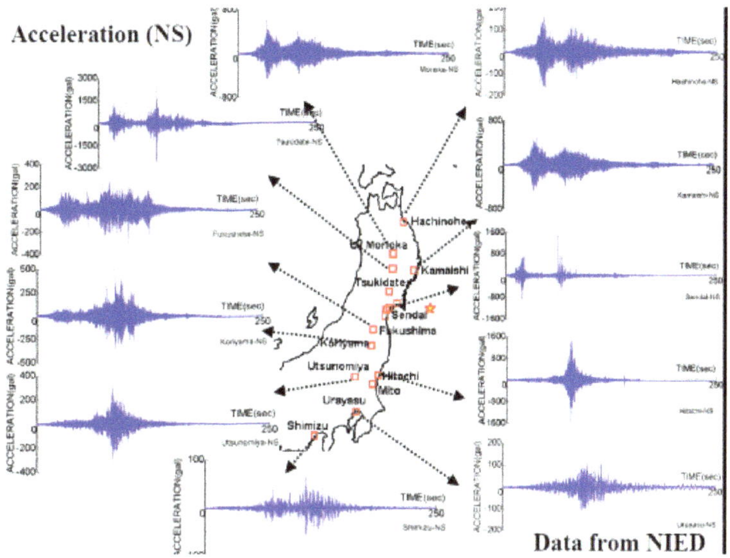

Fig. 32. Ground Acceleration Time histories Recorded in Earthquake Region. (NIED 2011)

It is reported that during risk assessment of the nuclear power station prior to the earthquake, a tsunami expert was overruled by an executive, when the survivability of the reactors in a potential tsunami event was raised. Thus, the design flaw at this plant was not corrected, while another plant with a newer design survived the current event without problem. Construction records also revealed that in 1967, the owner, Tokyo Electric Power, excavated 25 metres off the 35-metre high natural ground where the reactors were to be located. The lowering of the site grade appeared to facilitate equipment transportation as well as pumping of seawater for cooling the reactors.

The current solution is to abandon the power station, and to eventually entomb the damaged reactors in concrete after their cool-down, even though ongoing radiation leak continues to contaminate the plant site, its surrounding land and sea five months after the earthquake (August 2011).

A mandatory 15 per cent cut of peak-time power consumption is in force this summer (2011) for large users in and around Tokyo. Only nineteen of the 54 Japanese reactors that were in service prior to the March 11 earthquake are working in July 2011, because of local opposition. It is uncertain how long will it take to bring the radiation problem at the Fukushima Dai-Ichi plant site under control. Germany and Switzerland have indicated their plan to phase out nuclear power plants, while other countries have also starting the reassessment of their future nuclear-power development plans. Thus, the radiation concern for all nuclear power plants has to be carefully evaluated, and robust defence measures against natural and man-made hazards have to be implemented around the world.

Fig. 33. Waterfront Buildings in Onagawa, Japan Inundated by Tidal Water Due to Earthquake Subsidence. http://tectonics.caltech.edu/slip_history/2011_taiheiyo-oki/index.html

Fig. 34. Blast-Damaged Fukushima Dai-Ichi Nuclear Power Plant Unit 3 (left) and Unit 4 (right).http://pinktentacle.com/

3. Geoscience and geotechnical aspects

Geoscience and geotechnical aspects of those earthquakes covered in Section 2 and Table 1 are discussed here together with other relevant events (Lo and Wang 2008).

3.1 Geoscience aspect

- Earthquake events inflicted heavy loss of life and property as well as disrupting businesses. They illustrate the high risk of structure inadequacy in populated and developed areas vulnerable to seismic hazard.
- Tsunami hazard to coastal population can be mitigated by advanced warning and physical barriers such as sea walls and gates. However, when warning system is either not established (Sumatra 2004) or insufficient evacuation time is available due to proximity to the fault rupture, and the physical tsunami barriers are overrun (Tohoku-Oki 2011), the tragic loss of lives and properties is severe.
- Earthquakes usually occur according to a typical pattern, starting with a single main event, maybe preceded by foreshocks, followed by a series of smaller aftershocks that would diminish in magnitude and number with time. Exceptions to this pattern are not uncommon. In fact, the concept of "earthquake conversations" (Stein, 2003 and 2005) suggests that the hypothesis of stress-triggering of earthquake may be the mechanism controlling earthquake occurrence. As population centres around the world grow from isolated metropolitans to more or less continuous industrialized belts, the following earthquake patterns observed in recent years deserve more attention in order to mitigate damages.
    - Numerous large aftershocks of the Sumatra 2004, Maule 2010 and Tohoku-Oki 2011 subduction events, and the rupture of an adjacent segment of the locked plate boundary offshore such as the Nias 2005 event or the increased threat to the Tokyo region after the Tohoku-Oki 2011 event, and a Mw 6.9 crustal (or intraslab) normal-fault event on March 11, 2010 in the region of Libertado O Higgins in Chile triggered by the February 27, 2010 subduction event;
    - Several large regional seismic events occurred within several months after the Wenchuan 2008 event due to stress readjustment in the region;
    - The Christchurch 2011 aftershock of the Darfield 2010 event occurred much closer to the population centre;
    - Turkey Kocaeli (August) and Düzce (November) earthquakes in 1999 - triggered by the rupture of two adjacent segments of the North Anatolian transform fault; and
    - El Salvador January and February earthquakes in 2001 - an offshore intra-slab event followed by an inland crustal event.

  Thus, major seismic events of various source mechanisms could occur in adjacent areas within a short time interval. This phenomenon has significant implications in earthquake design and restoration. First, there is a potential threat to the safety of personnel engaging in the emergency response. Secondly, any retrofitted structure may have to undergo severe seismic test during or shortly after its repair. It may no longer be acceptable to lower the design criteria for retrofitted structures, even though it has been currently practiced in many countries as the experience of Christchurch in 2010 and 2011 shows.

- Earthquake investigation tools and methods have continually been evolved with significant recent advances, such as:
    - Global Positioning System (GPS) – GPS has been increasingly deployed along major faults to provide continuous deformation data over the tracked GPS stations;
    - Interferometric Synthetic Aperture Radar (InSar) – InSar is a satellite remote-sensing technique used to track areal ground deformation adjacent to faults. Both GPS and InSar are used to monitor the ground deformation along major faults and/or in highly seismic areas before and after earthquake occurrence; and
    - Satellite imagery, such as Google Earth and high-resolution imagery, and Light Detection and Ranging (LiDAR) have become standard tools used for the study of post-earthquake damages. A workshop was conducted by the Global Earth Observation Catastrophe Assessment Network (GEO-GAN 2010) to review the use of high-resolution satellite imagery with the coordinated efforts by volunteering professionals to conduct rapid structure damage assessments, geotechnical studies and landslide surveys. The development started in 2008 for studying the Wenchuan earthquake, and expanded in 2010 for the Haiti earthquake, involving verification of the assessment results by field reconnaissance on the ground (see www.virtualdisasterviewer.com).
- Digital seismographs – Newly installed digital seismographs capture a wide spectrum of seismic ground motions accurately and efficiently and can be easily transmitted to regional centers for further data processing than their earlier analog counterparts. The number of strong-motion accelerograms available for seismic design analyses have increased drastically in recent years including from four of the six events discussed here with the exception of the 2004/2005 Sumatra and 2010 Haiti events.
- Coral as natural record of relative sea-land movement (see Fig. 35) – Coral has been successfully used by researchers to study co-seismic and inter-seismic ground movements caused by inter-plate subduction earthquakes (Caltech 2011).
- Internet – The internet has been used by the United States Geological Survey for conducting earthquake intensity survey, and by the earthquake investigators to coordinate international investigation efforts through dedicated websites, links, individual blogs and social media. Such widespread use of electronic communication could potentially be harnessed as an effective, informal means to spread tsunami warning, especially to distant coastal areas to reduce the likelihood of repeating the 2004 Sumatra tsunami tragedy.
- Information technology - Modern equipment such as digital camera, GPS, laptop and satellite communication, etc. all contributes to the efficiency of an individual investigator.

3.2 Tsunami and geotechnical aspects

Earthquake secondary effects refer to non-tectonic surface processes that are related to earthquake shaking. They are often spectacular in expression and are main causes for loss of life and property. These effects include: tsunami, landslide, rockfall, turbidite (dense, sediment-laden flow offshore), liquefaction and lateral spreading, and site amplification, etc. (Lo et al. 1996). Their preserved geologic signatures sometimes serve as paleoseismic evidence for strong ground motion of prehistoric earthquakes (Yeats et al. 1997).

3.2.1 Tsunami effects

- Forces associated with tsunami – In addition to the inundation effect of rapidly moving waves, tsunami also exerts additional physical forces to structures such as hydrodynamic pressure including uplift (see Figs. 36, 37 and 38), debris impact and wave scour (see Figs. 39 and 40). The overturning of seawalls could be caused by the combined effect of tsunami hydrodynamic pressure, uplift and scour. Both passive and active measures are used to mitigate tsunami damage. Passive measures are use of buffer zone and/or natural barriers such as mangroves and sand dunes to avoid building structures in near shore areas. Where structures are required in tsunami-vulnerable zone, both inundation effect and tsunami-related forces have to be considered in the overall design and operation of these structures. Well designed concrete buildings, which have crawl space on the first floor to allow tsunami flowing through, performed well (see Figs. 41 and 42). However, the "soft-story" effect under seismic loading condition should be carefully evaluated.
- The important impact of tsunami hydrodynamic pressure on the performance of port facilities and bridges have been recognized for some time, vertical restrainers to counteract the tsunami uplift force against bridge decks such as experienced in the Maule and Tohoku earthquakes are proposed by Kawashima (2011). Policy to require vessels leaving the ports and sailing to the sea upon receiving tsunami warning would prevent damage to the vessels as well as port infrastructures from vessel impact.

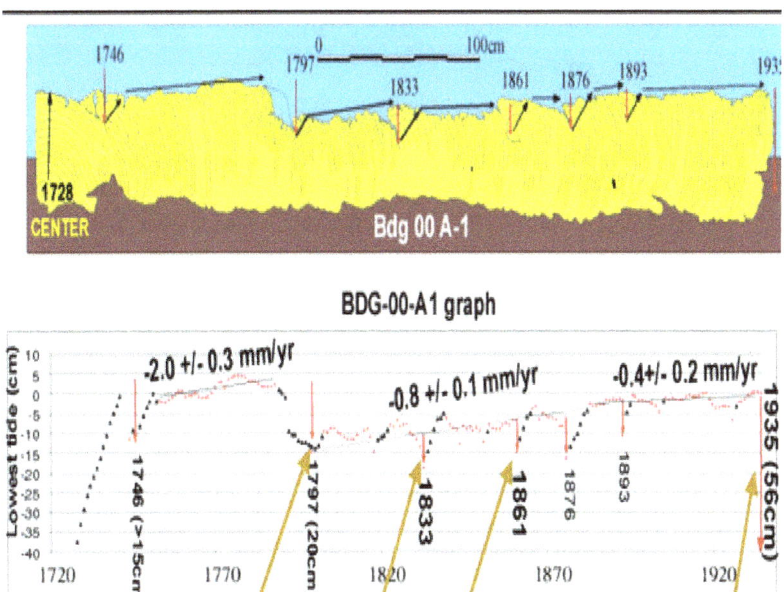

Fig. 35. Sea Level Changes Inferred from Truncated Coral Growth Rings. (Caltech 2011) http://www.tectonics.caltech.edu/images/sumatra/coral_record_web.jpg

Fig. 36. Wharf Damaged by Wave Impact and Uplift, Bam Nam Kem, Phang Nga Province, Thailand, 2004.

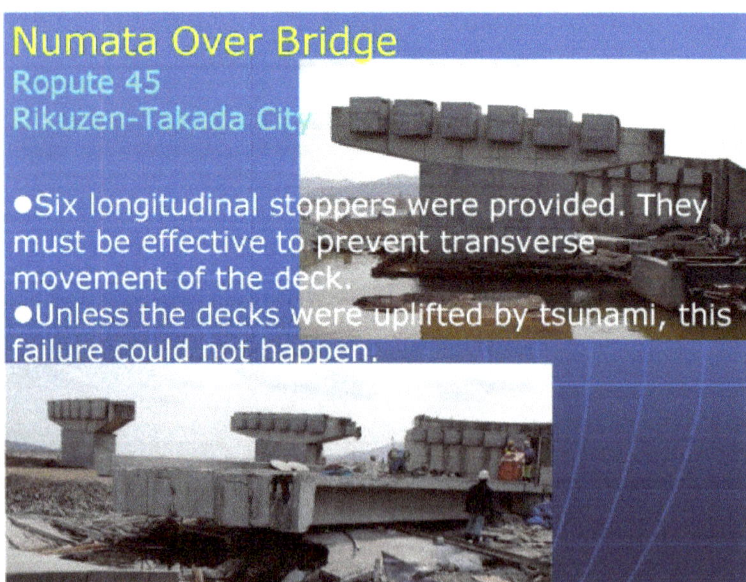

Fig. 37. Uplift of Bridge Deck In Spite of Longitudinal Stoppers - Tohoku Earthquake. (Kawashima 2011)

Fig. 38. Overturned Police Station, Onagawa, Japan, 2011.

Fig. 39. Pump Station Damaged by Wave Scour, Patong Beach, Phuket Province, Thailand, 2004.

Fig. 40. Scouring of Foundation Soils and Severing of Building's Utility Lines, Phi Phi Island Hospital, Thailand, 2004.

Fig. 41. Building with Crawl Space on the First Floor, Kamala Beach, Phuket Province, Thailand, 2004.

3.2.2 Ground failure effects

Ground failure includes landslide, slope failure, liquefaction, lateral spreading, bearing capacity failure, underground failure and ground displacement:

- Landslide - Correlations between the number of landslides and earthquake magnitude have been established around the world. In regions with pyroclastic deposits, the number of landslides for a given magnitude far exceeds those in other regions due to the relative large void ratio and weak strength of these deposits. The areas threatened by a landslide include the sliding land mass as well as the area traversed by the fast-moving sliding debris (see Fig. 43). The presence of water in the soil mass, as reflected in its degree of saturation, increases the probability of triggering a slide during an earthquake. Thus, whether the occurrence of earthquake is in dry or wet season could have significant difference in its impact. Heavy precipitation in earthquake stricken area could trigger post-earthquake landslides, debris flows and/or lahars (mudflows of volcanic debris). Moreover, the amount of water within the sliding mass and/or being entrained along its moving path increases its run-out distance and, hence, downstream impact. The local topographical detail along the debris travel path dictates the area being buried by the slide debris. The formation of a landslide lake, caused by temporary blockage of river by unstable slide debris, poses a special flood hazard. Timely construction of a spillway to drain the lake is required to ameliorate the downstream flood hazard such as the case with the Tangjiashan Landslide-Lake after Wenchuan 2008 earthquake (see Fig. 11).

Fig. 42. Close-up of Fig. 41, Showing Inter-connected Grade Beams and Columns.

- Slope failure - Failure in natural or man-made slope occurs where the available effective shear strength is exceeded by prevalent shear stress due to the earthquake-induced pore pressure increase and/or presence of inertia force during earthquake shaking. Slope failure of highway and/or railway embankment could disrupt vital transportation link and wrought havoc in the community life and regional economy as evidenced in all major earthquakes. Cracks and upstream slope slump of the 26-m high Fena earth and rockfill dam after Guam earthquake in 1993 required emergency repair by U.S. Navy Seebees.
- Liquefaction and lateral spreading - Liquefaction and lateral spreading could occur in much gentle slopes or level ground, where the soil deposits involve relatively loose cohesionless materials such as sand and silt and occasionally gravelly soils. Loss of shear strength and associated ground movements would affect structures located in

areas with high groundwater table, such as low-lying areas, river banks and ports, and lead to slope failure, lateral spreading, foundation horizontal displacement and settlement or uplift, bearing-capacity failure, and rupture of lifelines. Emergency repair of flood-control dykes or levees damaged by liquefaction of subsoils was often needed such as after Tohoku-Oki earthquake (see Fig. 44). The Tohoku-Oki earthquake caused levee damage at over 2,000 locations.

Fig. 43. Debris Flow at Huosigou, Chongzhou, China, 2008.

- Bearing-capacity failure – Foundation bearing-capacity failure also occurs in cohesive soil due to the softening effect of earthquake shaking. The punch shear failure of footings is associated with soil heave in the adjacent area. For tall and narrow buildings with an aspect ratio (height/width) greater than 2, it could lead to severe tilting or overturning of the building (see Fig. 45). This type of foundation behavior was observed in Mexico City due to the presence of sensitive Mexico City clay during the 1985 Michoacan earthquake, and again in Adapazari City due to the presence of Holocene alluvium during the 1999 Kocaeli earthquake.

Fig. 44. Levee Slope Repair and Sand Boils at Toe, Japan. (JSCE 2011)

Fig. 45. Overturned Building of High Aspect Ratio in Adapazari City, Turkey, 1999.

- Surface expression of underground failure - Sinkhole may occur during an earthquake. The phenomenon appears to be puzzling until the cause is identified. Usually, sinkhole occurs where pre-existing underground voids already formed prior to the earthquake. These voids may be formed by solution of limestone, abandonment of mine workings or "piping" erosion of fine soils by groundwater flow. In other situation, earthquake-triggered failure of underground structure could also result in the formation of sinkholes. The former examples are sinkholes occurred along highway culverts in El Salvador and Nisqually earthquakes where "piping" erosion of backfill around the culverts was present (see Fig. 46), and sinkholes occurred at the abutment of Matahina Dam in New Zealand (Gillon and Newton 1991). The latter example is the sinkholes formed due to the collapse of Bolu Tunnels after the 1999 Düzce earthquake. At Christchurch, concern of voids formed underground due to massive loss of sand to the surface because of liquefaction was raised (see Fig. 26).

Fig. 46. Sinkhole Formed by Collapse of Roadway During Earthquake, Washington, 2001.

- Ground displacement - Ground displacement includes tectonic movement or non-tectonic surface processes. When it occurs within the footprint of structures, it affects the structural stability and integrity. Where lifelines intersect fault traces, the impact of potential ground displacement related to fault rupture is a major design challenge as the experiences in Wenchuan, Kocaeli and Dücze earthquakes show. On the other hand, differential settlement of approach fill and pile supported bridge deck at bridge abutments is a frequently recurring scene, e.g., at Christchurch (see Fig. 47). Subsoil volume decrease would lead to foundation settlement with no adjacent soil heave.

3.2.3 Ground shaking effect

- Site amplification - Site amplification of ground motion through soft soil deposits has been well recognized and incorporated in the modern seismic design codes. However, severe building damages suffered at soft sites at Adaparazi and Avcilar during 1999 Kocaeli earthquake reiterate the importance of this phenomenon. Avcilar is near Istanbul and about 80 km west of the epicenter. The distance attenuation of ground motion was more than compensated by local site amplification. Similarly, topographic amplification of ground motion was also evident during Christchurch (see Fig. 30), and all other recent earthquakes.

4. Lessons learned from earthquakes

4.1 General

Earthquake damage in a given area tends to reflect its social-economic condition. Thus, wide spread damage to sub-standard structures and heavy loss of life seem to recur unfortunately too frequently. Comartin et al. (2004) considered that reducing high loss of life from earthquakes is the most important challenge facing the global earthquake engineering community. Efforts underway to meet this challenge included: improving construction practices in individual countries such as India and Columbia and promoting international collaboration and information sharing. The EERI/IAEE World Housing Encyclopedia is a valuable platform to spearhead this ongoing effort.

Conversely, in areas where earthquake-resistant design codes and engineering practices are kept up-to-date, damage tends to be limited to older and inadequately designed structures, or areas with special site conditions. However, this general observation may have exceptions such as in Christchurch, New Zealand and Tohoku, Japan, when potential seismic hazards are under-estimated.

Outlined in the following are some of the lessons learned from the recent earthquakes:

- Tsunami warning - Timely warning of tsunami and evacuation drill are prerequisites for saving lives in such an event. Sumatra earthquake-tsunami is an unprecedented tragedy to illustrate this point, and provided an impetus for the set up of a tsunami-warning system for countries around the Indian Ocean. Adequate warning requires the implementation of a system involving instrumentation on land and on ocean floor, and international organization and communication protocol. The system has to be maintained vigilantly to be effective. While warning against tsunamis triggered by distant earthquakes is adequate in North America, further work is required for warning and protection against tsunamis caused by nearby earthquakes as demonstrated in the experience of 2010 Maule, Chile and 2011 Tohoku-Oki, Japan events.

- For inhabitants living along vulnerable shorelines, running to high grounds or elevated structures, when feeling earthquake shaking or seeing unusual signs of waves including receding sea water or advancing wave, remains to be practical means to escape drowning by tsunami. Tragedy in the recent Chilean and Japanese events, however, remind us that the success of escape is not guaranteed, when the time window for survival is measured in minutes rather than hours, especially for elder or handicapped members of the population. The need for reliable near field (local) tsunami warning was further demonstrated in the case of 2005 Nias earthquake. About three months after the Sumatra earthquake, the tsunami-scared populace was all too eager to run for their life after given the warning. Nevertheless, no significant tsunami was realized this time. The ensuing evacuation commotion, and other false tsunami alerts in the wake of the 2004 Sumatra earthquake served to emphasize that the conduct of timely and reliable tsunami warning and orderly evacuation remains both a technical and societal challenge. Similar confusions to a lesser extent were also reported in some of the tsunami alerts in the North America.
- Tsunami vertical evacuation refuges - are an engineering option for enhancing public safety against near-field earthquakes (FEMA 2008/2009). These structures should be clearly identified as tsunami-proof refuges. They need to be designed for both earthquake and tsunami loads, as well as to a safe height above the tsunami run-up, allowing for rapid public ingress and upward movement (see Fig. 48). As a cost-effective measure, any elevated structure in a tsunami-vulnerable zone such as water tower, lighthouse or fire-lookout tower could be designed with additional capacity and features to accommodate tsunami-escapees looking for an escape path.

Fig. 47. Settlement of Approach Fill at Bridge Abutment Formed a Bike-Jump for a Youngster. http://otilya.com/view/News /Asia-Pacific/New-Zealand

Fig. 48. Vertical Evacuation Structure at Shirahama Beach Resort in Japan. (FEMA 2009)

- Instrumentation for strong ground motion - Installation and maintenance of modern instrumentations for recording strong ground motion are important to characterize earthquake motion. Further instrumentation to measure the response of critical structures would improve the understanding and analysis of seismic behavior of the instrumented structures. However, lack of strong motion data in epicentral areas at Port-au-Prince, Haiti; Sumatra and Nias, Indonesia serve to indicate that it is not always feasible to achieve this goal due to economic, political and/or technical constraints. Moreover, potential tsunami inundation should also be considered when selecting sites for strong motion instrumentation.
- ASCE (2011) published a special issue on the selection and modification of earthquake ground-motion for nonlinear dynamic analysis of structures. The suite of 17 papers provides useful information about the ground-motion input in the nonlinear response-history analysis of structures. In particular, Moustafa (2011) outlined a new framework for modelling design earthquake loads for inelastic structures, using the method of critical excitations and adopting damage indexes to reflect the structure damage level.
- Land use - Microzonation is to be prepared for identifying areas susceptible to tsunami run-up, strong earthquake shaking, ground failures such as landslides, fault displacements, slope failures and liquefaction, etc. Determination of appropriate land use should then be guided by the identified hazards to mitigate potential loss.
- Seismic design - Update of seismicity hazard and upgrade of seismic design are to be carried out on a regular basis. Although the replacement of vulnerable structures before they are decimated by earthquake could prevent avoidable loss of life, funding required for such an undertaking seems always difficult to come by, especially during the current weak financial condition around the world. Up-to-date seismic design measures should be implemented in new structures. Whenever possible, retrofit of existing structures should be facilitated by legal requirements and/or financial incentives.

- Emergency response and public education - Drills of emergency response need to be practiced regularly to enhance the capability of response agencies and personnel, and public awareness. Preservation of certain damaged or failed structures in the earthquake-stricken areas as memorial sites would go a long way to remind and educate the public about the importance of adequate seismic design and earthquake preparedness. Figure 49 shows the memorial park established at the preserved old Beichuan townsite in honour of the victims, and Figure 50 shows the public taking a sober earthquake lesson at the park.

Fig. 49. Memorial Park at Old Beichuan Town. http://www.eeri.org/site/meetings/ us-china-symposium

Fig. 50. Educational Tour of Memorial Park. http://www.eeri.org/site/meetings/us-china-symposium

- Potential impacts on secondary damages by earthquakes such as tsunamis, fires, landslides, debris flows, dam breaches, disruptions of lifelines and release of hazardous material including radiation are to be considered in order to devise realistic response strategy, and to plan for orderly implementation of long-term upgrade of existing facilities.

4.2 Lifelines facilities

Lifelines are those systems required for the survival and function of societies, and include: water and wastewater, electric power, communications, transportations, natural gas and liquid fuels, etc. Human societies often evolved on the nodal points of the rivers and ports transportation networks located on flood plains. Thus, various geotechnical issues reviewed in Section 3.2 affect the seismic performance of lifelines facilities, such as rupture of water and waste water pipelines, disruption of power and communication lines, and damage to bridges, power transmission and communication towers, etc. due to foundation failures, liquefaction and lateral spreading.

Lessons applicable to lifelines facilities include the following:

- Fault Displacement - Fault displacement caused significant structure damages in the Wenchuan earthquake. Earlier examples are the collapse of Arifiye overpass across the Trans-European Motorway (TEM) in the Kocaeli earthquake (with about 3 to 3.5 m fault displacement) and significant displacement and damage to the Bolu Viaducts in the Düzce earthquake (with 1.5 m displacement of adjacent piers and 13° rotation of pile caps across the fault). The impact of potential fault displacement on the performance of the sub- and super-structures crossing active faults remains to be a serious challenge to the designers.
- Review and Upgrade - The reliable service of a lifeline depends on proper functioning of all its important components at all times. It requires regular review and upgrade of these facilities in normal time as well as prompt repair of failed components after an earthquake.
- Redundancy - Robustness of the lifelines relies on the built-in redundancy. As a certain line or node of a given lifeline is down, the availability of alternate lines or nodes is critical for maintaining the lifeline function during the critical period of emergency response and restoration.
- Two 380-kV transmission lines bypassed the Adapazari substation, which suffered circuit-breaker damage during the Kocaeli earthquake. These two lines were critical to power restoration after the earthquake. This type of redundancy requires careful planning and consideration of a variety of potential damage scenarios based on knowledge of potential earthquakes and physical conditions of the lifelines facilities.
- Lack of redundancy in the cooling system of nuclear reactors at the Fukushima Dai-Ichi nuclear power plants is the fatal flaw. It eventually led to the meltdown of the fuel rods. It not only stopped power generation at the plants, but also disrupted and/or curtailed power generation at other nuclear plants due to the shadow that its incidents casted on the safety of nuclear power in general.
- Tsunami Inundation - One side effect of tsunami inundation during Sumatra earthquake was to cause pump failures due to salt water intrusion. This delayed the

restoration of those water and sewer lines whose function relies on the pumps. Thus, sea water intrusion needs to be considered and guarded against or avoided in the design of water and wastewater systems. Moreover, contamination of agricultural irrigation and groundwater supply systems by sea water intrusion deserves careful attention.

5. Conclusions

Earthquake, with its potent destructive power associated with the sudden release of enormous energy stored over an indefinite period of seismic acquiescence, strikes terror in people's mind around the world. This chapter reviews six recent earthquakes, which occurred in 2004 to 2011, including three tsunami-generating subduction earthquakes (along the coastlines of the Indian Ocean, South and North Pacific Ocean) and three crustal events (in the interior of the Asia Continent and two islands in the Caribbean Sea and South Pacific Ocean). These earthquakes affected both developed countries (New Zealand, Chile and Japan) and developing countries (Indonesia, China and Haiti). Due to the economic and resources constraint, the ability of a country to direct its resources to cope with natural hazards inherent to its location and setting varies. Thus, both self development and international cooperation and assistance are important to enhance the overall capacity of each country to protect its population and economy against the ravage of earthquake and its associated secondary hazards. Summarized in the following are some of the observations gleaned and lessons learned and/or re-learned from recent and other relevant earthquakes:

5.1 General observations

- Post-earthquake investigation has evolved with the technology advancement over the years. The institutional support of this activity has been augmented many times by the individual voluntary efforts of professionals of various disciplines. The collaboration of research and practicing experts in these critical occasions promotes the development of appropriate technology that improves the survival chance of people and sustaining performance of structures for different regions of the world.
- The occurrence of a series of earthquakes in a given region, be it a main shock followed by aftershocks, or a foreshock followed by the main shock, or a sequential rupture of adjacent faults, deserves special attention in seismic design and post-earthquake response and reconstruction.
- The concurrent and/or secondary effect of earthquake, such as tsunami, ground failure, breach of man-made or landslide dam, release of hazardous material and fire hazard should always be considered. An even more serious and long-lasting hazard, i.e., the meltdown of fuel rods and radiation leak at nuclear power plant is now added to this list.
- The level of casualty and structure destruction, in general, reflect the engineering practice such as seismic building code and its enforcement, design and construction practice and workmanship in an area. Numerous and large-scale landslides further contributed to the destruction of the 2008 Wenchuan earthquake. Unreinforced

masonry and non-ductile concrete buildings were the main structure type that suffered heavy damage in the 2008 Wenchuan, 2010 Haiti, 2010 Chile and 2011 Christchurch earthquakes.
- The heavy tsunami casualty in the 2004 Sumatra event was due to the absence of a warning system, while the higher than normal casualty in the 2011 Tohoku-Oki event was due to the unusually high tsunami run-up in places and relatively short time window available for evacuation of people from the low-lying areas.

5.2 Lessons learned

- The building code is a living document, which requires periodical review and update to reflect current understanding of the regional and local site seismicity, structural earthquake performance and latest design and technology. Any important new findings, such as the recognition of new or higher seismic hazard than that envisaged by the current code, should be carefully reviewed and implemented in the revised code as soon as practical.
- The high cost of re-building collapsed, demolishing unsafe, and retrofitting damaged buildings has caused some re-examination of the main objective of building code, i.e., to protect life safety of occupants and minimizing their injury. Beyond the life-safety objective, each jurisdiction may find its own optimal economic balance between the cost of initial design and construction, versus the cost and delay of re-construction after an earthquake. The current trend of structural design based on performance requirement will provide some flexibility to allow the adoption of a design higher than the minimum code standard based on life safety.
- Schools and emergency-response facilities, including fire and police stations, and hospitals, especially those located on weak subsoils prone to liquefaction, tsunami-vulnerable zone, or site subject to geologic hazard such as landslides, rock falls, etc., should be relocated or retrofitted to meet the requirement of maintaining their function after an earthquake.
- Critical facilities, with high occupancies, or containing hazardous materials or susceptible to radiation leak, or serving important lifeline functions to society (e.g., power generation and distribution and telecommunication facilities), or have other sensitive parameters, should be designed and maintained at higher seismic standard.
- Major lifelines that are either co-located and/or are interdependent with respect to maintaining their proper function should have special performance requirements in order to avoid multiple and/or cascading lifelines failures.
- Emergency response and resilience plans for the region, and local jurisdictions should be developed and exercised. These plans should focus on robustness, redundancy, resourcefulness, and rapidity.
- Mutual-aid plans among adjacent jurisdictions, both locally and internationally, should be pre-arranged prior to the occurrence of earthquake so that the precious time required for effective rescue operations in the initial golden period of 72-hours will not be wasted.

6. Acknowledgements

The authors wish to acknowledge the assistance they received from the librarians and administration staff of Klohn Crippen Berger for their effort in literature search and manuscript preparation, and from colleagues of ASCE TCLEE. The critical review and helpful suggestions by Professor A. Moustafa is greatly appreciated.

7. References

ASCE (2007). Sumatra-Andaman Islands Earthquake Tsunami of December 26, 2004, Technical Council on Lifeline Earthquake Engineering, Monograph No. 30.

ASCE (2011). Special Issue on Earthquake Ground-Motion Selection and Modification for Nonlinear Dynamic Analysis of Structures, J. Struct. Eng. 137 (3).

BBC (2005). http://news.bbc.co.uk/2/hi/science/nature/4247409.stm

BSSA (2007). The 2004 Sumatra-Andaman Earthquake and the Indian Ocean Tsunami, Bull. Seism. Soc. Am. 97, No. 1A, Jan.

BSSA (2010). The 2008 Wenchuan, China Earthquake, Bull. Seism. Soc. Am. 100, No. 5B, Nov.

Caltech (2011). http://www.tectonics.caltech.edu/

Comartin, C. et al. (2004). A Challenge to Earthquake Engineering Professionals, Earthquake Spectra, Vol. 20, No. 4, pp. 1049-1056.

EEEV (2008). General Introduction of Engineering Damage of Wenchuan Ms 8.0 Earthquake, Jour. of Earthq. Eng. and Eng. Vib., Vol. 28. Supplement, October.

EERI Special Earthquake Reports. (2005 Mar. to Jul. Sumatra 2004/2005 Eqs.; 2008 Oct. Wenchuan 2008 Eq.; 2010 April and May Haiti 2010 Eq.; 2010 Jun. and Oct. Maule 2010 Eq.; 2010 Nov. Darfield 2010 Eq.; and 2011 May Christchurch 2011 Eq.). EERI Newsletters.

EERI (2006). Summary Report on the Great Sumatra Earthquakes and Indian Ocean Tsunamis of 26 December 2004 and 28 March 2005, W.D. Iwan Tech. Ed., Aug. and Earthquake Spectra, Vol. 22, Special Issue 3, June.

EERI (2011). The March 11, 2011 Great East Japan (Tohoku) Earthquake and Tsunami: Societal Dimensions, August.

FEMA (2008/2009). Guidelines for Design of Structures for Vertical Evacuation from Tsunamis, P646, May 2008, and Vertical Evacuation from Tsunamis: A Guide for Community Officials, P646A, June 2009.

Fierro, E. and Perry, C. (2010). Preliminary Reconnaissance Report - 12 January 2010 Haiti Earthquake, PEER.

GEER (2010/2011). Geo-engineering Reconnaissance of the 2010 Maule, Chile Earthquake, Version 2, May 25, 2010. and Geotechnical Reconnaissance of the 2011 Christchurch, New Zealand Earthquake, http://www.geerassociation.org/, 2011.

GEO-CAN (2010). Remote Sensing and the GEO-CAN Community: Lessons from Haiti and Recommendations for the Future, Global Earth Observation Catastrophe Assessment, Network Workshop Report, June.

Gillon, M. and Newton, C. (1991). Abutment Repairs at Matahina Dam, Proc. of the 17th Congress on Large Dams, ICOLD, Vienna.

JSCE (2011). Preliminary East-Japan Earthquake Damage Investigation Report, Japanese Society of Civil Engineers, April 11. http://committees.jsce.or.jp/report/node/40 (in Japanese).

Kawashima, K. (2011). Damage of Bridges Resulted from East-Japan Earthquake of March 11, 2011, http://peer.berkeley.edu/news/wp-content/uploads/2011/04/East-Japan-Earthquake-21.pdf (in English).

Lo, R.C., Henderson, P.W. and Morrison, K.I. and Finn, W.D. Liam (1996). Seismic Geotechnical Considerations - An Overview, 11th WCEE Paper 1853.

Lo, R.C. and Wang, Y. (2008). Geoscience and Geotechnical Aspects of Recent Earthquakes, 4th Conf. on Geotechnical Earthquake Engineering and Soil Dynamics, Sacramento, Calif. May.

Lo, R.C. (2009). May 12, 2008 Wenchuan Earthquake - Lessons Learned from Its Impact, Emergency Response and Recovery, ASCE TCLEE Conf. on Lifeline Earthquake Engineering in a Multihazard Environment, Jun 28-Jul 1, Oakland, Calif.

Moustafa, A. and Takewaki, I. (2010). Modeling Critical Ground-Motion Sequences for Inelastic Structures, Advances in Structural Engineering, Vol. 13, No. 4, pp. 665-679.

Moustafa, A. (2011). Damage-Based Design Earthquake Loads for Single-Degree-of-Freedom Inelastic Structures, ASCE J. Struct. Eng. 137 (3) pp. 456-466.

NBCC (2005). National Building Code of Canada, National Research Council.

NIED (2011). National Research Institute for Earth Science and Disaster Prevention, http://www.bosai.go.jp/e/index.html.

PEER (2011). Preliminary Briefing on Japan's Tohoku Earthquake, Aftershocks, and Tsunami of March 11, 2011.

Somerville, P. (2000). Seismic Hazard Evaluation, 12th WCEE, Paper 2833.

Stein, R.S. (2003, 2005). Earthquake Conversations, Scientific American, pp. 72-79, January, 2003 and pp. 82-89, Sp Vol. 15, 2005.

Takewaki, I. Murakami, S., Fujita, K., Yoshitomi, S., and Tsuji, M. (2011). The 2011 off the Pacific coast of Tohoku earthquake and response of high-rise buildings under long-period ground motions, Journal of Soil Dynamics and Earthquake Engineering.

UN OCHA (2005). Relief Web Map Center, United Nation Office for the Coordination of Humanitarian Affairs, Jan. 3.

USGS (2004-2011). Significant Earthquake and News http://earthquake.usgs.gov/earthquakes/eqinthenews/

USGS/EERI (2010). The Mw 7.0 Haiti Earthquake of January 12, 2010, USGS/EERI Advance Reconnaissance Team Report, USGS Open-File Report 2010-1048.

USGS (2010). Documentation for Initial Seismic Hazard Maps for Haiti, Open File Report 1067.

Ventura, C., Elwood, K., and Huffman, S. (2008). Preliminary Report - Wenchuan Earthquake of May 12, 2008 - Reconnaissance Trips.

Yeats, R.S., Sieh, K. and Allen, C. R. (1997). The Geology of Earthquakes, Oxford Univ. Press.

Yin, Y., Zheng, W., Li, X., Sun, P. and Li, B. (2011). Catastrophic Landslides Associated with the M8.0 Wenchuan Earthquake, Bull. Eng. Geol. Environ. 70:15-32.

4

Lateral In-Situ Stress Measurements to Diagnose Liquefaction

Richard L. Handy
Iowa State University
U.S.A.

1. Introduction

Death by earthquake usually involves toppling and crumbling of masonry walls, falling roofs, and pancaking floors in multistory buildings. Residents of underdeveloped countries are particularly vulnerable because structures only rarely are designed to resist earthquakes. The safety of masonry construction could be improved with buttressing or cellular design, but those procedures are little known and seldom used, suggesting that effective charity should include education and innovation.

Even earthquake-resistant structures remain vulnerable to a hidden danger, that posed by liquefaction. Most susceptible are saturated loose sands and silts where shaking movements can cause individual soil grains to become dislodged and temporarily suspended in water, turning the soil into a dense liquid analogous to quicksand. The process occurs at depth due to increased horizontal shearing stresses, so gentle slopes can turn into landslides, buried tanks can float up out of the ground, and buildings can sink or tip over. Fortunately the destruction is relatively slow compared with walls crumbling and roofs falling, and can allow enough time for people to escape with their lives. A famous example in Niigata, Japan, where as shown in Fig. 1 multistory apartment buildings remained intact and in one

Fig. 1. Apartment buildings at Kawagishi-Cho, Niigata, Japan, tipped after the foundation soil liquefied during the 1964 earthquake.

instance tipped far enough that residents could crawl out of windows and walk to safety down the exterior wall.

Perhaps the most serious situation is where a potentially liquefiable soil supports a dam, where a failure would be sudden and could take many human lives.

Liquefaction is soil specific, so building sites in earthquake areas can be tested to determine their susceptibility. However, the tests are largely empirical and based on case histories. This chapter examines the role of a more fundamental soil property, lateral in-situ stress, on soil collapse and liquefaction, and emphasizes the use of a relatively new tool to measure that stress.

2. The nature of liquefaction

Since both lateral and vertical stresses in soils tend to increase with depth, they are conveniently expressed in terms of a ratio of effective stresses, K_o. During liquefaction, soil grains are temporarily suspended in water: Effective or intergranular stresses are zero, and there is no K_o. Then the grains settle and displace water so it seeps upward, often taking sand particles with it. The reconstituted soil density therefore is higher than it was prior to liquefaction, so liquefaction can be a one-time event, mainly limited to geologically youthful sandy deposits such as deltas, alluvial fans and floodplains, and to loose, man-made sandy fill soil that is dumped without being compacted.

Liquefaction probably occurs with all major earthquakes, but in severely damaged areas the evidence lays buried underneath the devastation. Prehistoric earthquakes can be dated from sedimentary sequences that contain sand carried upward by the displaced water and deposited as shallow cones at an earlier ground surface.

Stress ratios existing in soils prior to and after liquefaction have not yet been fully explored, the major restraint being the difficulty of obtaining reliable in-situ stress data.

3. Current methods for evaluating the liquefaction potential of a soil

The pioneering work of Seed and Idriss (1971) developed criteria for identifying potentially liquefiable soils from Standard Penetration Test (SPT) data. The method has since been improved through the use of empirical corrections to the data, and is supplemented by cone penetration tests (CPT) and measurements of shear wave velocity (Youd et al. 2001).

Schmertmann (1971) suggested that the SPT is essentially a side friction test, implying a dependence of blow counts on horizontal stress existing and developed in the soil. Kelly and Lutenegger (2004) show close relationships between blow counts and side friction measured by torquing the SPT sampler, with the exact relationship depending on the type of soil. The lateral stress ratio K_o appears in several formulas for cone resistance aced on cavity expansion theory (Yu and Mitchell 1998). Thus even though the lateral in-situ stress changes in the neighborhood of a test instrument, it still may play an important role in a liquefaction process.

4. Measuring K_o

Vertical effective stress in soil can be obtained from depth, soil density, and elevation of the groundwater table, but lateral stress must be measured. Empirical estimates of the lateral stress ratio K_o based on depth, density, friction angle, and overconsolidation ratio, do not address effects or likelihood of liquefaction.

One approach for measuring lateral in-situ stress in soil is to bore a hole and insert a measuring device without disturbing the soil. This is the concept of the self-boring pressuremeters. However, tests are time-consuming, and some plucking or shoving aside of individual soil particles during drilling is unavoidable.

Other devices such as the Cone Penetration Test (CPT) and Dilatometer laterally displace the soil to the extent of creating a lateral bearing capacity failure. Empirical relationships can be developed based on pressure chamber testing, but the changes in lateral stress caused by disturbance involve an unknown combination of soil responses including elastic compression, consolidation and shearing. These occur simultaneously and create changes in lateral stress that are larger and more variable than the stresses that are intended to be measured. A third approach is to leave a flat pressure cell in place long enough for stresses to equilibrate, but that ignores the permanence of soil arching action.

A different approach is used with the K_o Stepped Blade.

4.1 The K_o stepped blade

The methodology used with the K_o Stepped Blade was adapted from analytical chemistry, to intentionally introduce different levels of disturbance and extrapolate the results to a condition of zero disturbance. This is accomplished with four blade steps ranging in thickness from 3.0 to 7.5 mm, Fig. 2.

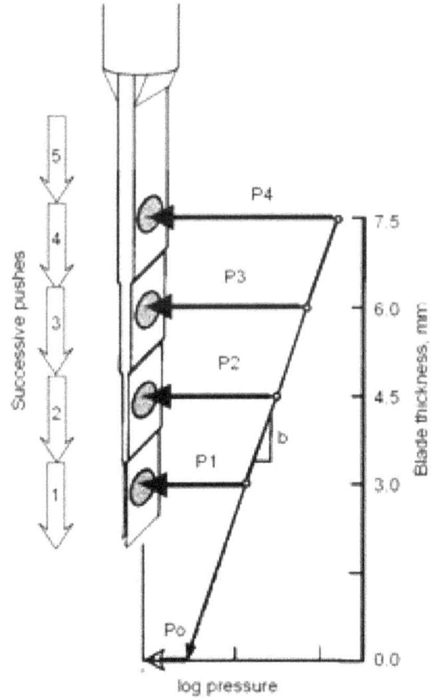

Fig. 2. Schematic showing the method used to interpret Stepped Blade data. A linear *e-log p* relationship indicates consolidating behavior. Reprinted with permission from Handy and Spangler (2007). The McGraw-Hill Companies, New York.

The blade is advanced incrementally into the bottom of a boring so each measurement is made on a progressively increasing step thickness at the same depth. While the test is rapid, enabling as many as 10 determinations in an hour, the blade cannot be pushed continuously, but must be withdrawn and the boring extended deeper for each new test series. A blank space at the top of the Blade allows four multi-stress measurements to be made at 127 mm (5 in.) depth increments in each series. Measurements are directional owing to the flat shape of the Blade.

4.2 Interpretation of K_o stepped blade data

Extensive measurements in the field and in laboratory simulations show a consistent relationship between step thickness and the logarithm of pressure, illustrated by the graph in Fig. 2 (Handy et al. 1982; White and Lutenegger 2008). Analogy to the familiar *e-log p* linear relationship suggests that soil penetrated by the Blade must be laterally consolidating. Data that do not meet this consolidation criterion are not included in the interpretations.

Stepped Blade steps are thin in order to minimize the tendency to create a lateral bearing capacity failure. The two most common causes for data rejection are (1) elastic response that increases the pressure on the thinnest step prior to a breakdown of the soil structure, and (2) a lateral bearing capacity failure that relieves stress on the thickest step(s). The three behaviors, elastic response, consolidation, and lateral bearing capacity failure, are illustrated in Fig. 3. Elastic and bearing capacity responses must not be used in the extrapolations, and are readily recognized from pressure decreases instead of increases on successive blade steps.

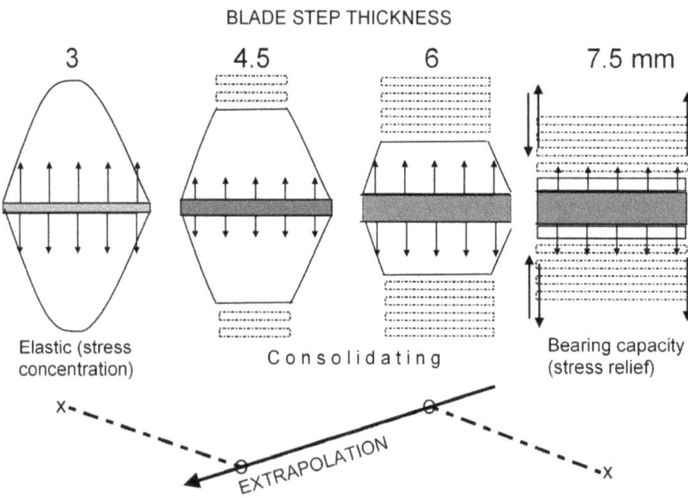

Fig. 3. Stress diagrams inferred from pressure measurements on different thickness steps, showing the critical role of consolidation and reasons for rejecting specific data points.

An elastic concentration of stress most commonly occurs with the thin step in stiff soils, and a relief of pressure from a bearing capacity failure most commonly occurs with the thick step in soft soils. Usable data sequences therefore may have two, three or four data points. *A discriminating evaluation of each data point is required before using a regression analysis to extrapolate the lateral in-situ stress, or the analysis will yield incorrect and erratic interpretations including a low value for* R^2.

A linear regression analysis of the logarithm of measured pressures versus blade step thickness yields a slope and intercept. The intercept is the lateral stress that existed prior to insertion of the blade, and the intercept can be examined for consistency. A test performed close to a hole or hard inclusion in the soil will change the slope and can be another reason to question some data. However, since such a condition cannot be confirmed, questionable data points are shown on graphs but are labeled with question marks.

The distribution of soil pressure across the face of the Blade was investigated by pulling narrow friction strips (Handy et al. 1982) and emphasized the importance of focusing measurements in the central area of each step.

4.3 The K_o stepped blade in relation to other in-situ tests

In Fig. 4, soil response mechanisms interpreted from Stepped Blade data are applied to other in-situ displacement soil tests. For example, it appears likely that the width/thickness ratio of the Dilatometer causes a lateral bearing capacity failure, so empirical correlations are used to estimate both stress and modulus from expansion of a metal diaphragm. A more direct measure of modulus might be obtained from elastic response of soil to penetration by the thinnest step of the Blade, a possibility that has been examined but not yet fully investigated.

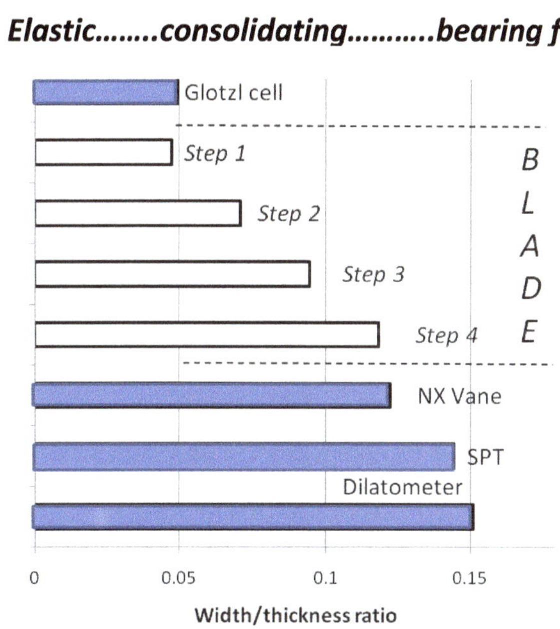

Fig. 4. Width: thickness ratios for various in-situ displacement test devices and (at top) relationships to soil behavior determined from Stepped Blade tests. Consolidating behavior is required to interprete lateral in-situ stress from Blade tests. The *w/t* ratio for the cone is 1.0. Modified from Handy et al. (1990)

4.4 Pneumatic pressure cells

The several requirements for pressure cells used in the Blade are (1) that they be as thin as the thinnest Blade step, (2) that they measure pressure on a discrete centralized area, (3) that they be durable and reliable, and (4) that when damaged they can be easily replaced or repaired in the field. A back-pressured pneumatic cell was developed to meet these criteria and is shown schematically in Fig. 5. An important advantage of the pneumatic cells is that measurements involve gas pressures and do not require calibrations.

The flat surface of each cell is covered with Teflon over an elastomeric membrane to minimize friction with the soil, and to create an internal seal between the two cell chambers. The chambers are slowly pressurized with gas pressure while maintaining a small differential between the two chambers of the cell. At the instant when the gas pressure equals the soil pressure anywhere around the threshold, the membrane is lifted sufficiently to cause an internal leak so pressures in the two chambers equalize. At that time pressure is read and recorded, and pressures to both chambers are simultaneously vented to prevent bulging of the membranes and ready the cell for the next pressure reading. Selector valves are turned to direct the gas pressures to the next Blade step, and the test procedure is repeated by simply closing the two relief valves. If desired, a reading can be repeated for additional confirmation.

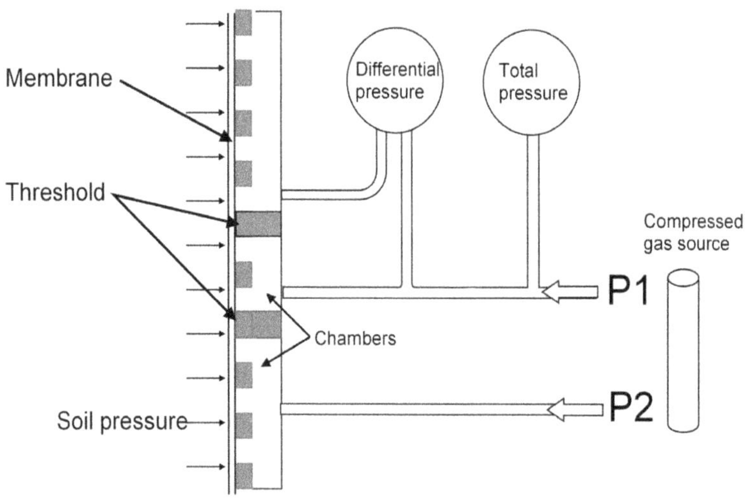

Fig. 5. Part of the secret of the K_0 Stepped Blade is the back-pressured pneumatic cells. At the instant when gas pressure equals soil pressure, the differential pressure goes to zero so the total pressure is read from the appropriate pressure gauge. After Handy and Eichner, U.S. Pat. No. 4,662,213, 1987.

Experiments have shown that the Teflon cover is much more durable than thin sheet metal but nevertheless can be scraped or punctured. If a leak occurs, positive gas pressure prevents water and soil from entering the cell, and since cells plug in they can be quickly replaced in the field. Experiments with electrical strain-gauged metal membranes showed that they were not reliable under the severe test conditions imposed in the field.

A hollow rib attached to the back of the Blade carries the multiple gas lines and reinforces against bending.

5. Test sites

Several test sites were selected in potentially liquefiable soils:

1. The Jackson Lake Dam was built to increase the water level in a natural glacial lake located on the Snake River in northwestern Wyoming, U.S.A. The log-crib dam was constructed in 1907 and failed three years later. It was replaced between 1911 and 1916 by a concrete and earth dam.

In 1964 the destructive nature of liquefaction was emphasized both by the Niigata Earthquake in Japan and the Alaska Earthquake, and in 1976 the Jackson Lake Dam foundation soil was determined to be a potentially liquefiable on the basis of Standard Because a failure could be catastrophic, in 1986-89 the dam was removed and the gravelly sand foundation soil was densified by Deep Dynamic Compaction, shown in Fig. 6. Stepped Blade Tests were conducted to measure lateral pressures in the sand after the dam was removed, both before and after the site was compacted.

Fig. 6. Deep Dynamic Compaction to prepare a gravelly sand foundation soil for the reconstructed Jackson Lake Dam, Wyoming. Reprinted with permission from Handy and Spangler (2007), The McGraw-Hill Companies, New York.

2. Considerable areas of poorly compacted sand fill exist in many bay areas including San Francisco, California, where there is a history of liquefaction dating from the 1906 earthquake. A potentially liquefiable sand was identified at the San Francisco Naval Shipyard, so experiments were conducted by the U.S. DOT-FHWA to determine the effectiveness of driving displacement piles to densify the sand. K_o Stepped Blade tests

were conducted to measure lateral pressures in the sand both before and after pile driving.

3. A relatively recent development called Rammed Aggregate ® Piers, or the Geopier ® method for soil reinforcement, is becoming increasingly popular for improving foundation soils. As shown in Fig. 7, the rammer head is shaped to push aggregate outward into the soil and thereby increase lateral stress. K_o Stepped Blade tests were conducted to measure those stresses at various depths and distances from the piers.
4. K_o Stepped Blade tests conducted in collapsible loess soil to determine relationships between collapsibility and lateral in-situ stress, and suggested a history of liquefaction in a basal, saturated zone. This should affect slope stability, and massive landslides in the thick loess in China have taken many human lives. The trigger usually is excessive rains, but also can involve earthquakes.

Fig. 7. The Geopier hydraulically actuated rammer is shaped to direct part of the ramming energy outward and increase lateral stress in the surrounding soil.

5.1 Test results from the Jackson Lake Dam site, Wyoming

The Deep Dynamic Compaction shown in Fig. 6 involved dropping an approximate 100 tonne weight about eight times at each location on a grid pattern. This left a pattern of surface depressions that were 2 to 3 m deep. The results from K_o Stepped Blade tests conducted before dynamic compaction are shown in the left diagram in Fig. 8, and afterwards in the right diagram in the same figure. The site was difficult because of angular gravel-size rock fragments in the soil.

At the left in Fig. 8, below about 5.5 m (18 ft) depth Stepped Blade tests gave K_o in the neighborhood of 0.5 without giving consideration to the weight imposed by the dam. The Jáky relationship indicates that this corresponds to an angle of internal friction of 30°, which is reasonable for normally consolidated sand.

In the same figure, the weight of the dam compressed the soil and increased lateral stress down to a depth of about 4m (13 ft), indicating that normal consolidation penetrated to that depth. Below that depth there is a transition to a K_o of about 0.3, indicating underconsolidation and a potential for collapse that can be attributed to the weight of the dam.

At the right in Fig. 8, Deep Dynamic Compaction is shown to cause a substantial increase in lateral stress such that the average K_o is 1.0, perhaps from seepage forces created when the weight of the overlying soil was supported by pore water pressure (Handy 2012). The uneven distribution of lateral stress after compaction may be symptomatic of localized shearing that concentrated stress in one area while reducing it in another.

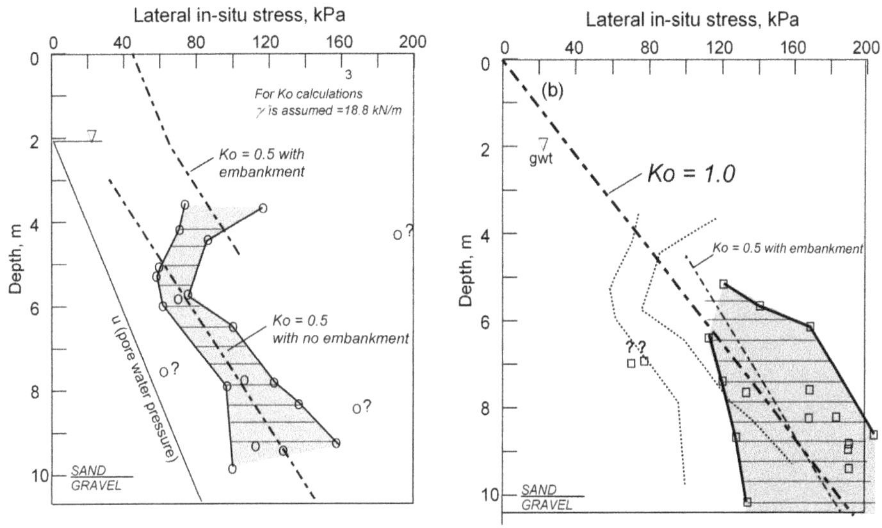

Fig. 8. Lateral stresses measured in foundation sand at the Jackson Lake dam site, (left) before and (right) after compaction. At the left, the soil was normally consolidated until after the embankment was built, when the weight of the dam compacted the upper part but left the lower soil underconsolidated and potentially liquefiable. At the right, compaction increased the average K_o to 1.0, indicating a history of induced liquefaction, and K_o = 0.5 after the dam was replaced. From Handy (2012), modified from Handy et al. (1990).

In this connection the narrow range of lateral stresses measured before compaction, in a range as low as ±8 kPa or ±10%, compared with ±50% after compaction, indicates that the variability is real and should not be assigned to random experimental error. Large variations in lateral stress also have been measured within a compacted earth embankment (C. Anderson, personal communication).

After compaction and restoration of the weight of the dam, K_o is about 0.5, indicating that compaction was successful in creating a quasi-normally consolidated condition, and the dam should be safe from future collapse and liquefaction of the foundation soil.

5.2 Test results from the San Francisco Naval Shipyard, California

The San Francisco Naval Shipyard was first established in 1870 at Hunters Point in the southeast part of the city. Starting in the early 1900's the area was expanded by dumping fill sand, making this an appropriate site to measure the effectiveness of driving piles to compact the sand. The discovery of contamination with nuclear and other wastes led to the shipyard being closed in 1994.

Lateral stresses measured prior to pile driving were quite consistent, shown by data points and the shaded area to the left in Fig. 9. A trend line drawn through the data and extended upward to the y axis indicates a prior surcharge equal to the weight of about 5 m (16 ft) of the same soil.

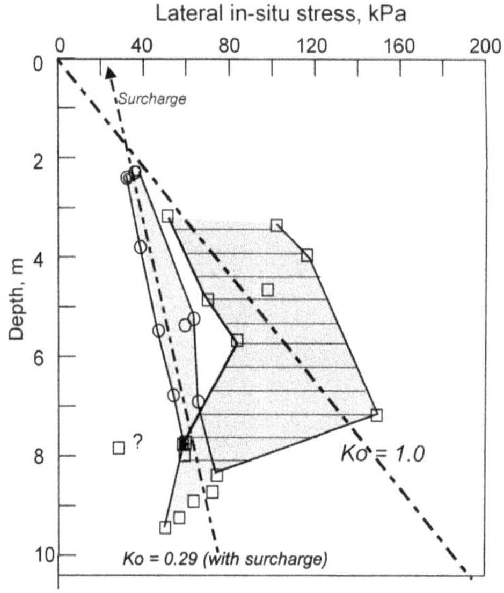

Fig. 9. Tests in loose sand at the San Francisco Naval Shipyard the slope of the lateral stress line gives $K_o = 0.29$ prior to pile driving, indicating a potential for liquefaction. As in the case of DDC, pile driving increased the average K_o to 1.0, indicating temporary liquefaction down to the bottoms of the piles. Data are from Handy et al. (1990)

The K_o value obtained from the slope of the trend line is 0.29, compared with a value of 0.5 that can be expected if the soil is normally consolidated. (A K_o calculated without taking into account the surcharge is meaningless as it becomes infinite at the ground surface.) The measurements of lateral stress confirm that the soil is underconsolidated and potentially liquefiable, and a suitable candidate for remediation.

Lateral stresses measured after pile driving were more erratic, but nevertheless cluster around K_o = 1.0 down to the bottom level of the piles. Below that depth, lateral stress remained unaffected.

The K_o value of 1.0 may be from seepage forces during temporary liquefaction that occurred during pile driving. Liquefaction would be beneficial for enabling a wider distribution of lateral stress, but also would tend to relieve passive pressures that otherwise would act to increase skin friction. A liquid condition cannot sustain a passive ratio any more than it can sustain an active ratio. The lack of passive pressure is consistent with a group effect of 1.0.

5.3 Test results after Geopier® soil reinforcement

The Geopier method for soil reinforcement involves boring holes that typically are 760 mm (30 in.) in diameter, and incrementally filling the holes with layers of rammed coarse aggregate. As previously noted, the hydraulic rammer head, Fig. 7, is shaped to direct a substantial portion of the ramming effort laterally outward into the soil.

5.3.1 K_o no longer constant

Fig. 10 shows lateral stress measurements at different depths and distances from rammed piers at the site of a stadium in Memphis, Tennessee. The soil was saturated loessial silt. Because of the variability of the pier diameter caused by incremental ramming, data are plotted versus distance from pier centers. K_o is not constant with depth, which is quite a different distribution of lateral stress from those in the preceding examples. Instead of K_o it is the lateral stress at particular radial distances that is constant with depth. This probably relates to the uniform distribution of ramming energy throughout the length of a pier, and may be influenced by compaction that proceeds from the bottom up instead of from the top down.

5.3.2 Another evidence for temporary liquefaction

In Fig. 10 the pier contact stress is obtained by extrapolation of data using plastic theory. The open arrow pointing to the right in the figure indicates an undiminished transfer of radial compaction stress outward to a radial distance of almost 2 m ((6 ft). This radial distribution of stress is possible only if the soil temporarily behaved as a liquid.

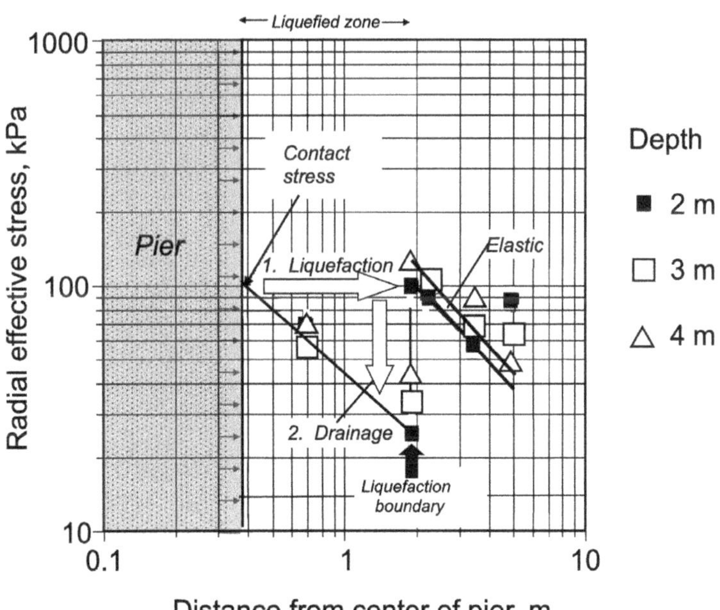

Fig. 10. Radial stresses measured in saturated loess soil near Rammed Aggregate Piers in Memphis, Tennessee. The data suggest a succession of behavioral modes that includes temporary liquefaction. Modified from Handy and White (2006).

5.3.3 Where does the water go during Geopier ramming?

Rammed Aggregate Piers are constructed incrementally by compacting successive layers of aggregate such that each layer has a final compacted thickness of about 0.3 m (1 ft). During ramming, the data in Fig. 10 indicates that within the liquefied zone the pore pressure exceeds 100 kPa (14.5 psi, 1 Tsf). Beyond a radial distance of 1.9 m, which is outside of the liquefied zone, the decrease in radial stress is linear with distance. Elastic theory indicates that this can occur only if there a lack of restraint from circumferential tension, which can be explained by radial tension cracking. Radial cracks have been observed emanating outward from piers, and can extend down to a depth where the fluid pressure exceeds the existing lateral stress. In a normally consolidated soil below the groundwater table that can be as deep as 20 m (60 ft).

Open cracks are an open invitation for injection by a soil-water mixture, and soil particles carried in with the water can prop the cracks open and create a drainage gallery. That is the concept used in the "Hydrofrak" process to enhance oil and gas production from wells. After the fluid pressure dissipates, the injected cracks should create a ring of compression that preserves stress in the elastic zone.

Cavity expansion theory allows an estimate of the post-drainage friction angle from the slope of the line in the drained zone (Baguelin et al. (1978). This gives a friction angle of about 40°, which represents a substantial increase over friction angles commonly measured in loess (Handy and White 2006).

It will be noted in Fig. 10 that identical behaviors were observed at all three test depths.

5.3.4 How lateral stress affects preconsolidation pressure

Fig. 11 (a) represents a normally consolidated where intergranular friction acts to support the grains and reduce the lateral pressure. If a high lateral stress is applied, it can reverse the directions of the friction arrows, as shown in Fig. 11(b). Then when a vertical load is applied, before the grains can slip and the soil can consolidate, the load must be sufficient to again reverse the friction arrows, Fig. 11(c). This concept is further illustrated by the succession of Mohr circles in the lower part of the figure. It means that imposing a high lateral stress can cause a substantial increase in the preconsolidation pressure so within specific limits the soil will tend to compress elastically instead of by consolidating. This can reduce total settlement of a structure by an order of magnitude. This has been confirmed by load tests and is an important feature in Geopier design.

It should be noted that in a conventional laboratory consolidation test the lateral stress is not controlled, and in the conventional triaxial it is difficult to apply a lateral stress that exceeds the vertical stress, so in both cases the stress ratio remains in a normally consolidating mode.

What if lateral stress later is released, for example by trenching? During compaction the soil cohesion is greatly reduced by remolding, and then is recovered, a source of "setup factor" for driven piles. Schmertmann (1991) suggests that even the strength of sand can be doubled by aging, and a time-related recovery of strength is characteristic of liquefaction (Youd et al. 1996).

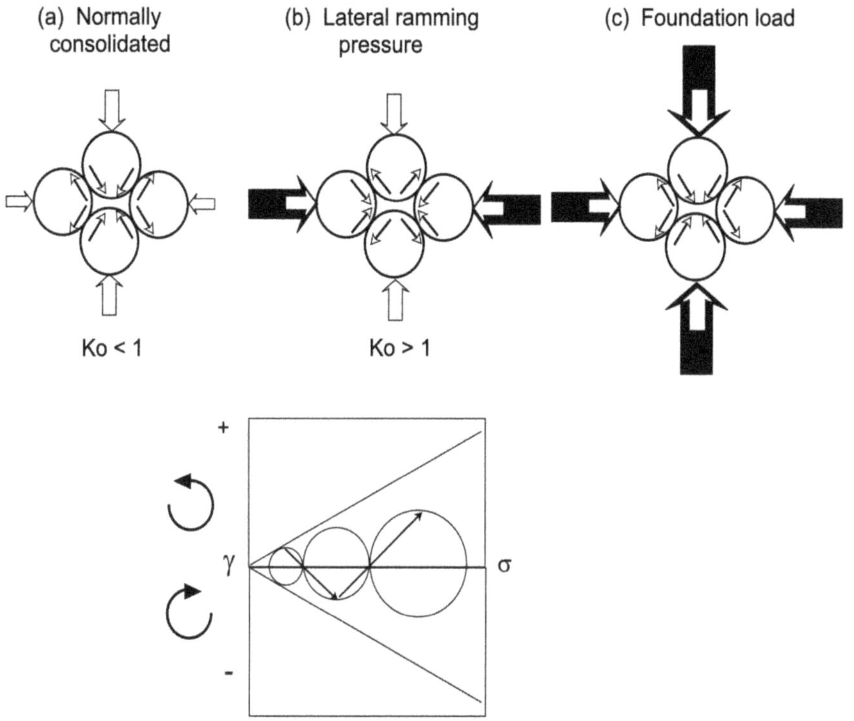

Fig. 11. Model illustrating how a high imposed lateral stress can reverse friction directions at grain contacts and increase the preconsolidation pressure without actually consolidating the soil. Modified from Handy (2001).

5.4 Test results in collapsible loess

Loess is wind-deposited silt, and comes from "löss," which is German for loose. Thick loess deposits that are close to a river floodplain source are so porous that it does not retain a groundwater table. Only lower parts of deposits are saturated, where the groundwater table is perched on an underlying less permeable stratum.

The strength and ability to stand in high steep banks such as shown in Fig. 12 is attributed to negative pore water pressure that is enhanced by a small content of active clay minerals, The strength therefore is sensitive to the moisture content, and sudden collapse can occur upon wetting. A concentration of surface water therefore can contribute to rapid and intense gullying and can crate vertical "chimneys." Severe conditions can result in an almost karst-like terrain (Derbyshire et al. 1994).

5.4.1 Loess K_o before collapse

Some of the early tests of the Blade were in collapsible loess, and gave K_o values that were so low, 0.1 to 0.2, that they were met with considerable skepticism. However, the data were repeated and found to be consistent.

Low lateral stress probably results from dust settling into a nest of grasses and then being washed down by rains until grains come into contact, where they are held in place by negative pore water pressure or suction. As the loess accumulated, vegetation migrated upward and left root channels that although often mineralized still may enhance drainage. Densities as low as 100 kg/m^3 (64 lb/ft^3) have been measured in loess actively being deposited in Alaska.

The loess deposit investigated in this study is thinner and farther from the source than that depicted in Fig. 12, in a relatively flat area in the western part of Omaha, Nebraska. It has a weathered soil profile and probably has been subjected to modification of the lateral stress through compaction. K_o values in Fig. 13 indicate that except for near-surface modifications the loess is collapsible down to a depth of about 5 m (16 ft).

Fig. 12. Western Iowa loess deposits are thick close to the Missouri River floodplain, where glacial outwash was the major source of wind-blown silt. The author in his younger days is sampling about half-way down the cut slope.

5.4.2 Loess K_o after collapse

In Fig. 13 in the lowest 2 m (6 ft) of the deposit, four independent data points give $K_o = 1.0$. The reproducibility confirms the precision of the measurements, and the value may be an indication of accuracy since it characterizes a liquid condition.

The highest K_o in Fig. 13, at a depth of 1 m (3 ft) translates into a vertical pressure of about 100 kPa (14 psi), which can be explained by ordinary tire pressures.

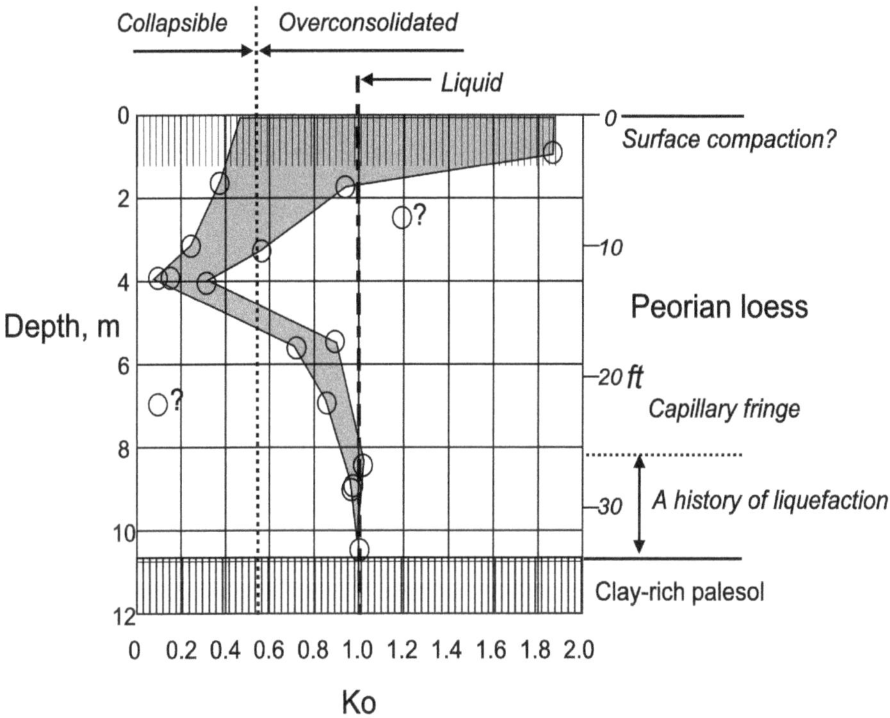

Fig. 13. K_o Stepped Blade data from Nebraska loess about 17 km (11 miles) from the Missouri River floodplain source. From Handy (1994).

5.4.3 K_o and partial collapse

Between depths of about 4 and 8 m (13 and 26 ft) in Fig. 13 there is a transition zone from the lowest value of K_o to $K_o = 1.0$. Ordinarily values that are higher than those resulting from normal consolidation are attributed to overconsolidation, but that appears to be unlikely at this site since it would require that loess deposition would be interrupted by a major erosion cycle.

The explanation may relate to a capillary fringe that includes saturation in the lower part, where pore water pressures are negative and diminish to zero at the groundwater table. The implication is that the soil may be only *partly* collapsed, increasing lateral stress to a value that is higher than normally would occur but not high enough to create a $K_o = 1.0$ liquid condition. To the author's knowledge this may be the first indication of a partial collapse.

5.4.4 Geography of collapsible loess

Loess deposits become thinner and finer grained with increasing distance from a source. Larger near-source thicknesses such as shown in Fig. 12 can be attributed to randomly variable wind directions that pile up more soil when winds are blowing nearly parallel to a source (Handy 1976).

A commonly used predictor for loess collapsibility is the saturation moisture content, which can be calculated from the soil unit weight and mineral specific gravity, in relation to the liquid limit. Loess containing less than 16 percent 2 μm clay is very likely to be collapsible, and one containing more than about 32 percent clay is not (Handy 1973).

5.4.5 Loess collapse and slope stability

A relationship between liquefaction of loess and slope stability is instantly appreciated if one walks along the toe of an active landslide involving saturated basal loess. The soil squeezing out of the basal zone only appears to be firm enough to walk on. The most likely condition for sliding is after extensive rains that raise the groundwater elevation to a new level in underconsolidated loess.

6. Conflicting observations

A K_o that is less than about 0.5 for sand and 0.55 for silt indicates an underconsolidated condition that is susceptible to collapse and liquefaction. This can be triggered by an earthquake or other ground vibrations such as those induced by pile driving or compaction, or in the case of loess soils, by saturation.

However, a low K_o is not required for lateral compaction used in the Geopier process.

Whereas compaction from the top down can result in a constant K_o with depth, Geopier compaction results in a variable K_o and constant lateral stress with depth, with the developed lateral stress depending on radial distance from the piers.

What may explain this difference? One possibility is that the pier ramming process involves compaction energy that is uniformly distributed with depth, resulting in a different stress distribution than when compaction is from the ground surface down.

A Rammed Aggregate Pier is surrounded by a zone of highly compacted soil that should help to contain the ramming energy for the next lift. That may not be the case for pile driving, although this possibility is yet to be investigated. Pile driving from the top down may generates a peripheral zone of sheared and dilated soil that is less able to confine the pressure created by driving.

With normal consolidation, K_o normally is in a range 0.5-0.6, whereas the K_o acquired after drainage of a liquefied soil is consistently 1.0. Why the difference? In both cases soil grains are settling out of water that is under pressure.

One difference is that normal consolidation that occurs as vertical stress gradually is increased activated intergranular friction as shown in Fig. 1(a), where it acts to partly support the individual soil rains and reduce lateral stress. During liquefaction that friction does not exist, and during drainage there are large seepage forces and a transfer from fluid pressure to effective stress. The process would be difficult to simulate but might be examined with laboratory tests.

7. Conclusions with regard to K_o, collapse and liquefaction

A low K_o is an indicator of a potential for collapse and liquefaction, provided that there is a sudden change in conditions that is sufficient to break down the soil structure.

The defining value for collapse and liquefaction is a K_o less than about 0.5-0.55. The angle of sliding friction of quartz is about 25°, which according to the Jáky relationship should result in $K_o = 0.58$ without a contribution from positive or negative dilatancy.

During liquefaction as water is ejected from soil its density increases, tending to make liquefaction a one-time event that is mainly limited to geologically young or poorly compacted man-made sand or silt deposits.

Lateral in-situ stress is a manuscript for stress history, as certain values are indicative of normal consolidation, overconsolidation, and underconsolidation. For example, the amount of removal of overburden often can be determined from a linear relationship between lateral stress and depth. In a preciously collapsed soil K_o is consistently found to equal 1.0.

Artificial compaction results in a much higher variability of lateral stress than does consolidation involving natural processes.

8. Conclusions with regard to the instrumentation

Lateral stress has been called the "Holy Grail" of soil mechanics (M. Jamiolkowski, personal communication). It plays a major role in shallow and deep foundation bearing capacity, pile skin friction, slope stability, and pressures on retaining walls.

The stated goal of the research sponsor, the U.S. DOT-FHWA, was to support development of an instrument that would rapidly determine lateral *in situ* soil stress within an arbitrary margin of ±1 psi (7 kPa). In Fig. 11 the several data points for $K_o = 1.0$ occupy a much narrower range, within about ± 0.4%. The narrow range of data in Fig. 8 fall in a range ± 6 kPa, which still is within the target range even though it includes a soil variable.

The reproducibility and speed of the test make it useful not only to determine average values, but also to define standard deviations that may be equally meaningful, for example by diagnosing man-made compaction as opposed to natural consolidation.

The possible use of lateral in-situ stress to indicate a potential for collapse or liquefaction was not anticipated during development of the test. However, it is not unusual for a new approach to open new vistas and paths that can unexpectedly lead in diverse directions.

The easy path to lateral stress is through computer modeling, and before that there was mathematical modeling. However, every model should be tested against the real world because otherwise it may be accepted but nevertheless flawed. Measurements inspired Terzaghi to develop consolidation theory; it was not the other way around. Measurements made in the 20th Century of pressures on retaining walls revealed inadequacies of 19th and 18th Century Rankine and Coulomb theories (Terzaghi 1943; Handy 1985). Measurements of soil loads on buried pipes made over 100 years ago led to a theoretical model that requires empirical corrections and only recently was shown to incorporate a simple and obvious error (Handy 2004).

Lateral stress existing in soil at the toe of an embankment or earth dams should be an index to their stability against sliding, by allowing a determination of an existing factor of safety. Lateral stress existing in soil next to a shallow foundation can test and evaluate the safety against a foundation failure, and can be helpful to determine if an additional load may be allowable. Measurements of lateral pressures in soils behind a retaining wall target the existing factor of safety and how it may change under varying conditions. With so many potential uses the K_o Stepped Blade may be underutilized.

9. Acknowledgements

Administrators who see a need and are guided more by their dreams than by their worries tend to remain anonymous and seldom get their due. The following organizations are listed chronologically according to their participation in the research and development of the K_o Stepped Blade: The U.S. DOT-FHWA; Soil Systems, Inc., Marietta, GA; Spangler Geotechnical Laboratory of Iowa State University, Ames, IA; Eichner Engineering, Ames, IA; Handy Geotechnical Instruments, Inc., Madrid, IA; GeoSystems, Inc., Lenexa, KS., Geopier Foundation Co., Inc., Moorsville, NC, and the SBIR program of the U.S. Army Corps of Engineers WES, Vicksburg, MS. Many graduate students worked with, tested and modified the Blade and made it work. It was a group effort.

10. References

Baguelin, F., Jézéquel, J. F., and Shields, D. H. (1978). *The Pressuremeter and Foundation Engineering*. Trans Tech Publications, Clasuthal, Germany.

Derbyshire, E., Meng, X., Wang, J., Zhou, Z., and Li, B. Collapsible Loess on the Loess Plateau of China. In Proc. Of the NATO Advanced Research Workshop on Genesis and Properties of Collapsible Soils. Kluwer Academic Publishers, Netherlands, pp. 267-293.

Handy, R. L. (1973). Collapsible Loess in Iowa. *Proceedings of the Soil Science Society of America Vol. 37, No. 2, pp. 281-284.*

Handy, R. L. (1976). Loess Distribution by Variable Winds. *Geological Society of America Bulletin*. Vol. 87, pp. 915-927.

Handy, R. L., Remmes, B., Moldt, S, and Lutenegger, A. J., and Trott, G. (1982). In-situ Stress Determination with the K_o Stepped Blade. *Journal of the Geotechnical Engineering Division of the American Society of Civil Engineers*. Vol. 108, No. GT11, pp. 1405-1422.

Handy, R. L. (1985). The Arch in Soil Arching. *Journal of the Geotechnical Engineering Division of the American Society of Civil Engineers*. Vol. 111, GT 3, pp. 302-318.

Handy, R. L., Mings, C., Retz, D., and Eichner, D. (1990). Field Experience with the K_o Stepped Blade. *Transportation Research Record*. Issue 1278, pp. 125-134.

Handy, R. L. (1994). A Stress Path Model for Collapsible Loess. In *Proc. of the NATO Advanced Research Workshop on Genesis and Properties of Collapsible Soils*. Kluwer Academic Publishers, Netherlands, pp. 33-49.

Handy, R. L. (2001). Does Lateral Stress really Control Settlement? *American Society of Civil Engineers Jour. of Geotechnical and Geoenvironmental Engineering*. Vol. 1127, No.GT7, pp. 623-626.32,

Handy, R. L. (2004). Anatomy of an Error. *American Society of Civil Engineers Jour. of Geotechnical and Geoenvironmental Engineering*. Vol. 130, No. 7, pp. 768-771.

Handy, R. L., and White, D. J. (2006). Stress Zones near Displacement Piers: I. Plastic and Liquefied Behavior. *American Society of Civil Engineers Jour. of Geotechnical and Geoenvironmental Engineering*. Vol. 132, No. GT1, pp. 63-71.

Handy, R. L., and Spangler, M. G. (2007). *Geotechnical Engineering*. The McGraw-Hill Companies. ISBN-13:978-0-07-48120-5 and 10-0-07-148120-6. New York.

Handy, R. L. (2012, in review). Ko and Liquefaction at Jackson Lake Dam, Wyoming. American Society of Civil Engineers Jour. of Geotechnical and Geonenvironmental Engineering.

Kelley, S. P., and Lutenegger, A. J. (2004). Unit Skin Friction from the Standard Penetration Test Supplemented with the Measurement of Torque. *American Society of Civil Engineers Jour. of Geotechnical and Geoenvironmental Engineering*. Vol. 130, No. 5, pp. 540-543.

Schmertmann, J. (1979). Statics of the SPT. *American Society of Civil Engineers Jour. of the Geotechnical Engineering Divisioin*. Vol. 105, No. GT5, pp. 655-670.

Schmertmann, J. (1991). The Mechanical Aging of Soils. *American Society of Civil Engineers Jour. of Geotechnical and Geoenvironmental Engineering*. Vol. 117, No. 9, pp. 1288-1330.

Terzaghi, K. (1943). Theoretical Soil Mechanics. John Wiley & Sons, New York.

White, D. J., and Lutenegger, A. J., eds. (2008). *The Papers of R. L. Handy*. CTRE, Iowa State University. ISBN 978-0-9820144-0-0, Ames, Iowa, U.S.A.

Youd, T. L., Idriss, I. M., Andrus, R. D., Arango, I., Castro, G., Christian, J. T., Dobry, R., Finn, W. D. L., Harder, L. F., Jr., Hynes, M. E., Ishihara, K., Koester, J. P., Liao, S. S. C., Marcuson, W. F., III, Martin, G. R., Mithcell, J. K., Moriwaki, Y., Power, M. S., Robertson, P. K., Seed, R. B., and Stokoe, K. H., II (2001). "Liquefaction Resistance of Soils: Summary report from the 1996 NCEER and 1998 NCEER/NSF Workshops on Evaluation of Liquefaction Resistance of Soils." *Journal of the Geotechnical Engineering Division of the American Society of Civil Engineers.* Vol. 127, No. GT 10, pp. 817-833.

Yu, H. S., and Mitchell, J. K. (1998). Analysis of Cone Resistance: Review of Methods. *American Society of Civil Engineers Jour. of Geotechnical and Geoenvironmental Engineering.* Vol. 124, No. 2, pp. 140-149.

5

Simplified Analyses of Dynamic Pile Response Subjected to Soil Liquefaction and Lateral Spread Effects

Lin Bor-Shiun
Geotechnical Enigineering Research Center, Sinotech Engineering Consultants, Taiwan

1. Introduction

Pore water pressures greater than effective soil stress and subsequent liquefaction are known to occur in saturated sand deposits subjected to earthquake excitations. Liquefaction of soils can result in a reduction of soil strength and yields large settlement via lateral spreading. For superstructures supported on pile foundations embedded in such soils, these effects can be devastating. For example, the 1964 Niigata earthquake in Japan damaged the foundation piles under one of the piers of the 12 spans, 207 meter long Showa Bridge. After the earthquake, an excavation survey of damaged piles indicated that bending failure occurred due to the lateral spreading of river bed soils (Hamada, 1992). Similarly, in the 1994, when Northridge earthquake occurs, river bank areas between Santa Clarita and Fillmore, Highway 23 crosses over the Santa Clara River, where sand boils were observed near a bridge pier. Cracks induced by lateral spreading were found approximately 4.5 m away from the pier (Stephen et al., 2002). Afterward, in the 1995 Kobe Earthquake, quay walls along the coastline of Kobe moved up to several meters toward the sea as a result of lateral spreading (Tokimatsu and Asaka, 1998). Some papers also discuss liquefaction-induced lateral spread under the foundations of long-span bridges subjected to spatially-varying ground motions. (Abbas and Manohar 2002; Zerva and Zervas, 2002; Wang et al., 2004; Zerva 2009). More recent devastating earthquakes such as the March 2011 Tohoku earthquake in Japan and the January 2010 Haiti earthquake can be found and reported by EERI (Earthquake Engineering Research Institute) and USGS (United States Geological Survey).

Recent research has focused on understanding the transfer of forces between a pile and the surrounding layered soil during liquefaction (Hamada, 1992; Meyersohn, 1994; Tokimatsu, 2003, Bhatachaya et al., 2002、2004; Jefferies and Been, 2006). Excavation surveys by Hamada (1992) clearly showed that foundation piles are especially susceptible to damage at the interface between liquefied and non-liquefied layers. This observation was also verified by Meyersohn (1994) and Lin et al. (2005) with static numerical techniques.

In terms of static design for pile foundations, the current mechanism of failure assumes that the soil pushes the pile. The Japanese Road Association Code (JRA, 1996) has incorporated this concept. The code advises civil engineers that the non-liquefied layer acts passive pressure on the piles and liquefied layer offers thirty percents of overburden pressure when designing piles against bending failure due to lateral spread. Other codes such as USA code (NEHRP, 2000) and Eurocode 8, part 5 (1998) also have specifications about the problem (Bhattacharya et al., 2005). In the other hand, Tokimatsu (2003) investigated that the equivalent earth pressure acting on the pile during liquefaction in shaking table tests can be defined as the seismic passive pressures subtracting the seismic active pressures. This concept was also verified with the centrifuge tests by Haigh and Madabhushi (2005) and Madabhushi et al (2010). In a design process, engineers need the limit states to define the serviceability of members according to the safety of performances to structures (Priestley et al., 1996; Kramer and Algamal, 2001).

When piles are subjected to lateral spreading, lateral forces are exerted directly on the embedded depth of piles within liquefied layer. There are generally two methods to analyze this phenomenon. The first method is called the "Force-based method". Using an explicit numerical procedure, earth pressure is applied onto the piles based on a viscous flow model (Chaudhuri et al. 1995; Hamada and Wakamatsu, 1998; Lin et al., 2010). In order to effectively use the force-based method, several soil parameters must be known. Also, the force-based method can account for the effect of soil topography. In the second method, known as the "Displacement-based method", observed or computed lateral ground displacements are transmitted by theoretical soil springs on the whole pile system (Tokimastu and Asaka, 1998; Ishihara, 2003; Chang and Lin, 2003; Cubrinovski and Ishihara, 2004; Preitely et al., 2006). The second method has several advantages such as being able to choose a soil spring model that matches the complexity of the soil stratum. Also, nonlinear material effects can be considered.

This chapter investigates pile response to loading caused by liquefaction using the EQWEAP (Earthquake Wave Equation Analysis for Pile) numerical analysis procedure (Chang and Lin, 2003; Chang and Lin, 2006; Lin et al., 2010). Both a displacement and forced based form of EQWEAP are used. Methodology and case study comparisons with results of these two procedures are presented separately. The chapter ends with a final synthesis of observations and conclusions drawn from the two methods.

2. Methodology

2.1 Brief overview

The Winkler foundation model is often used in analyzing the deformation behaviors of the pile foundations. For solutions of the dynamic Winkler foundation model, or the so-called beam on dynamic Winkler foundation (BDWF) model, the wave equation analysis, initially proposed by Smith (1960), has been suggested for the driven piles. To make the wave equation analysis more accessible at the time-domain, the author (Chang and Yeh, 1999; Chang et al., 2000; Chang and Lin, 2003) has suggested a finite difference solution for the deformations of single piles under superficial loads. Such formulations can be extended for the case where the piles are subjected to seismic ground shaking. Prior to analysis of the pile system shown in Figure 1, the seismic induced free-field excitation behavior of the soil stratum needs to be obtained. A description of the soil stratum behavior during excitation provides a one-dimensional soil amplification solution for the site. For the site of interest, time dependent earthquake records are used with the modified M-O method to calculate the dynamic earth pressure coefficients (Zhang et al., 1998). Liquefaction potential at various depths of the site is evaluated for the limited pore water pressure ratio (Tokimatsu and Yoshimi, 1983). and numerical methods such as the finite element method or the mechanical model which models the discrete model of the pile system (Bathe, 1982) .

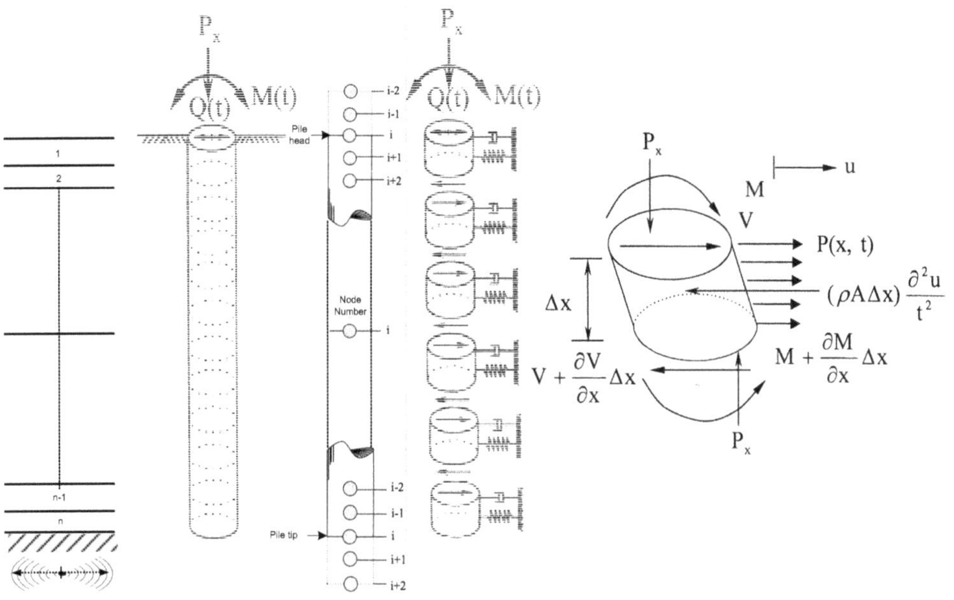

Fig. 1. Discrete system of the single pile

The earthquake motions can be decomposed into vertical and horizontal components. Pore water pressure effects are accounted for using an excess pore water pressure model. Soil deformation, seismic loading, resistance, damping and the inertia forces of the soil relative to time are applied to the pile segments and used to solve for the corresponding pile displacements. Figure 2 shows the layout of the described superposition procedure.

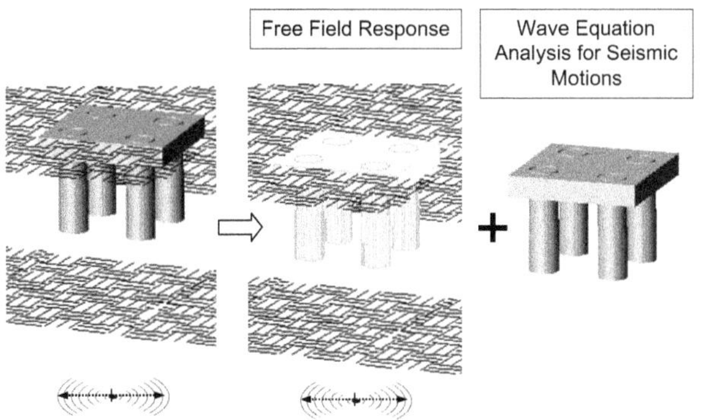

Fig. 2. Superposition of the free-field analysis and WEA

Formulations can be derived from the wave equations of the piles. Analysis of the foundations can be performed assuming unloaded or time dependant sustained loading conditions. With proper boundary conditions at the pile head, interactions of the structural system can be modeled. The above procedure is known as EQWEAP, which mainly concerns the nonlinear behaviour of liquefied soil induced permanent ground displacement rather than piles. Figure 3 illustrates the flow chart for the EQWEAP

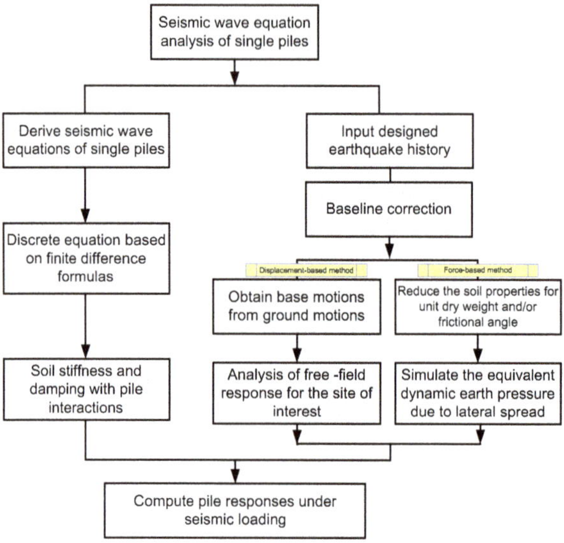

Fig. 3. Flowchart summarizing the numerical procedures of the analysis

2.2 EQWEAP: Displacement-based method

2.2.1 Wave equation of pile foundations concerning soil liquefaction

To make the wave equation analysis of the deformations of single piles under superficial loads more accessible in the time-domain, several authors (Chang and Yeh, 1999; Chang et al., 2000; Chang and Lin, 2003) have suggested a finite difference solution. Such formulations can be extended to the case where the piles are subjected to seismic ground shaking. Assuming force equilibrium, the governing differential equations of the pile segment exciting laterally can be written as:

$$E_p I_p \frac{\partial^4 u_p(x,t)}{\partial x^4} + \rho_p A_p \frac{\partial^2 u_p(x,t)}{\partial t^2} + P_x \frac{\partial^2 u_p(x,t)}{\partial x^2} + C_s \frac{\partial u(x,t)}{\partial t} + K_s u(x,t) = 0 \tag{1}$$

where $u(=u_p - u_s)$ = relative pile displacements, u_p = absolute pile displacements, u_s =the absolute soil displacements, E_p =Young's modulus of the pile; I_p = moment of inertia of the pile, ρ_p = uniform density of the pile, A_p = cross-section area of the pile; P_x = superstructure loads, C_s and K_s = damping coefficient and stiffness of the soils along the pile, and x is ordinate variable, and t represents for time.. For earthquake loading transmitting from the soils, Eq. (1) can be expanded using the central difference formula as shown below:

$$u_p(i,j+1)=\frac{1}{A+C}\begin{bmatrix}-u_p(i+2,j)+(4-B)u_p(i+1,j)\\ -(6-2A-2B+D)u_p(i,j)\\ +(4-B)u_p(i-1,j)-u_p(i-2,j)\\ -(A-C)u_p(i,j-1)+C[u_s(i,j+1)-u_s(i,j-1)]+Du_s(i,j)\end{bmatrix} \tag{2}$$

where $A=\dfrac{A_p\rho_p\Delta x^4}{E_pI_p\Delta t^2}$; $B=\dfrac{P_x\Delta x^2}{E_pI_p}$; $C=\dfrac{C_s\Delta x^4}{2\Delta tE_pI_p}$; $D=\dfrac{K_s\Delta x^4}{E_pI_p}$

Eq. (2) indicates that the absolute pile displacements under the earthquake excitations can be solved directly from the absolute displacements of the adjacent soil. A major advantage of this method is that the matrix analysis is not required in solving for the pile deformations. One can simply use a free-field analysis to obtain the liquefied soil displacements using the excess pore water pressure model(as described in Section 2.22) and then substitute the displacements into Eq. (2) to obtain the desired solutions. This is similar to those suggested in the multiple-step analysis of the soil-structure interaction problems. In addition, equations describing the lateral excitations of the highest and lowest elements of the pile should be modified using proper boundary conditions listed as follows.

Top of the pile:

a. Free head:

$$\frac{\partial^3 y_p(x,t)}{\partial x^3}=\frac{P_t}{E_pI_p};\frac{\partial^2 y_p(x,t)}{\partial x^2}=\frac{M_t}{E_pI_p} \tag{3}$$

b. Fixed head:

$$\frac{\partial^3 y_p(x,t)}{\partial x^3}=\frac{P_t}{E_pI_p};\frac{\partial y_p(x,t)}{\partial x}=0 \tag{4}$$

At the tip of the pile:

$$\frac{\partial^2 y_p(x,t)}{\partial x^2}=0\,;E_pI_p\frac{\partial^3 y_p(x,t)}{\partial x^3}+P_X\frac{\partial y_p(x,t)}{\partial x}=0 \tag{5}$$

where M_t and P_t are the external moment and load applied at the pile head. The discrete forms of these equations can be derived with the central difference schemes. Detailed derivations can be found in Lin (2006).

2.2.2 Soil stiffness and damping

For discrete models of the various soil types (sand, clay, etc.), both of the stress-displacement curves (t-z, q-z and p-y equations) and the Novak's dynamic impedance functions are used popularly in practice. The former, which is established empirically from the in situ pile load tests, can be used for substantial load applied slowly. The later, initially suggested by Novak (1972, 1974 and 1977) for the soils around the piles subjected to small steady-state vibrations, is able to capture the dynamic characteristics of the soil resistances and energy dissipations. Soil displacements close to a pile subjected to dynamic loading are nonlinear (Prakash and Puri, 1988; Nogami et al, 1992; Boulanger et al., 1999; El Naggar and Bently, 2000). El Naggar and Bently (2000) used a nonlinear soil model that incorporated a p-y curve approach to predict dynamic lateral response of piles to soil movement. The computed responses were found compatible with the results of the statnamic pile test. The nonlinear stiffness of the p-y equations is adapted in this investigation. The corresponding soil stiffness is described as below:

$$k_{NL} = \frac{8\pi G_m(1-v)(3-4v)\left[\left(\frac{r_0}{r_1}\right)^2 + 1\right]}{\left(\frac{r_0}{r_1}\right)^2 + (3-4v)^2\left[\left(\frac{r_0}{r_1}\right)^2 + 1\right]\ln\left(\frac{r_1}{r_0}\right) - 1} \tag{6}$$

where r_0 is the pile radius, r_1 is the outer radius of the inner zone, v is Poisson's ratio of the soil stratum, and G_m is the modified shear modulus of the soils. A parametric study shows that a ratio r_0 / r_1 of 1.1-2.0 yields the best agreement.

For the damper, a transformed damping model is used. Equivalent damping ratios, D, of the soils at steady-state excitations are first computed from the Novak's dynamic impedance functions, K* where

$$K^* = K_{real} + i\, K_{imag} = K(\omega) + i\, \omega C(\omega) \cong K(\omega)\,(1+2iD) \tag{7}$$

$$D \cong \frac{\omega C(\omega)}{(2K(\omega))} \tag{8}$$

In the above equations, $K(\omega)$ and $C(\omega)$are the frequency-dependent stiffness and damping coefficient of the impedance. For simplicity, the computed damping ratios are incorporated with the static stiffness Kst to model the kinematics of the soil. The revised damping coefficient $c(\omega)$can be written as:

$$c(\omega) \cong 2DK_{st} / \omega \tag{9}$$

Decomposing the actual load-time history into a series of small impulses, the damping coefficient $c(t)$ can be obtained by integrating a damping function c(t) to a set of unit impulses of the actual load-time history. Knowing that $D=C(\omega)\omega/2K(\omega)$, the associated geometric damping ratios can be computed. Modeling the values of $D(\omega)/\omega$ and assuming that they are symmetric with respect to the ordinate, a mathematical expression of the damping can be written as:

$$c(t) = AK_{st}t^{-B} \tag{10}$$

where A and B are the model parameters (Chang and Yeh, 1999; Chang and Lin, 2003).

2.2.3 Modeling soil liquefaction

Soils affected by induced pore-water pressure reduce the lateral resistance of the piles. This study utilized an excess pore water empirical model to complete effective stress analysis (Martin et al., 1975; Finn et al, 1977; Finn and Thavaraj, 2001), and obtain free-field motions under liquefaction. Kim (2003) successfully predicted the excess pore-water pressure resulting in soils subjected to earthquake shaking by verifying results with laboratory tests. This model can be divided into undrained conditions and drained conditions as follows:

a. Undrained condition:

$$\Delta u_w = \frac{\Delta \varepsilon_{vd}}{\left(\frac{1}{\overline{E}_r} + \frac{n_p}{K_w}\right)} \tag{11}$$

where Δu_w = an increase in pore water pressure; $\Delta \varepsilon_{vd}$ = an increment in volumetric strain; $\overline{E}_r$ = one dimensional rebound modulus at an effective stress (σ'_v); n_p = porosity, and K_w = bulk modulus of water.

For saturated sand $K_w >> \overline{E}_r$ and therefore

$$\Delta u_w = \overline{E}_r \Delta \varepsilon_{vd} \tag{12}$$

According to simple shear test, the volumetric strain increment ($\Delta \varepsilon_{vd}$) is a function of the total accumulated volumetric strain (ε_{vd}) and the shear strain (γ). The relationship is given by

$$\Delta \varepsilon_{vd[i]} = C_1(\gamma - C_2 \varepsilon_{vd[i-1]}) + \frac{C_3 \varepsilon_{vd[i-1]}^2}{\gamma + C_4 \varepsilon_{vd[i-1]}} \tag{13}$$

$$\varepsilon_{vd[n]} = \sum_{i=1}^{n} \Delta \varepsilon_{vd[i]} \tag{14}$$

where [i] = ith time step or cycle; and C_1, C_2, C_3, and C_4 are constants depending on the soil type and relative density. An analytical expression for rebound modulus ($\overline{E}_r$) at any effective stress level (σ'_v) is given by

$$\overline{E}_r = \frac{(\sigma'_v)^{1-m}}{mk_2}(\sigma'_{v0})^{m-n} \tag{15}$$

where σ'_{v0} is initial value of the effective stress; and k_2, m and n are experimental constants for the given sand.

b. Drained condition:

If the saturated sand layer can drain during liquefaction, there will be simultaneous generation and dissipation of pore water pressure (Sneddon, 1957; Finn et al. 1977). Thus, the distribution of pore-water pressure at time (t) is given by

$$\frac{\partial u_w}{\partial t} = \overline{E}_r \frac{\partial}{\partial z}(\frac{k}{r_w}\frac{\partial u}{\partial z}) + \overline{E}_r \frac{\partial \varepsilon_{vd}}{\partial t} \tag{16}$$

where u = the pore-water pressure; z = the corresponding depth; and k = the permeability ; and r_w is the unit weight of water. Before conducting the free-field analysis, the adequate shear modulus (Seed and Idriss, 1970) may be determined from the following equation

$$G = 1000K_2(\sigma'_m)^{0.5} \tag{17}$$

where K_2 is a parameter that varies with shear strain and σ'_m is the mean effective stress. Pore water pressure will increase during shaking and leads to a decrease of effective stress. In some situations, pore-water pressure equals overburden stress in sand deposits and may liquefy. The initial shear modulus can be calculated from the initial effective stress. Then, G is modified due to the shear strain and pore water pressure under liquefaction. The modified value is substituted in place of the former one and convergence of solutions is obtained using an iterative manner.

In addition, to avoid over-predicting the excess pore water pressure and ensure compatibility with practical observations, it is suggested to use the pore water pressure ratio (r_u) to accurately control soil liquefaction levels (Lee and Albaisa, 1974; DeAlba et al., 1976; Tokimatsu and Yoshimi, 1983). The equation is given by

$$r_u = \frac{1}{2} + \frac{1}{\pi}\sin^{-1}(2F_L^{\frac{1}{\alpha\beta}} - 1) \tag{18}$$

where α, β are the experienced constants, and F_L is the safety factor of liquefaction. In order to use the above formulas, the liquefaction potential analysis of the site needs to be conducted prior to the analysis.

2.2.4 Free field analysis

The one-dimensional seismic excitations of soils onto the piles are computed from a free-field response analysis for the site of interest. Such an analysis can be conducted using the finite element technique, or be simply solved for using the 1-D wave propagation model and the lumped mass analysis. For simplicity, the lumped mass model is selected. To analyze the equations of motion of the soil layer under the earthquake excitations, the relative deformations of the structural system are obtained with the base accelerations induced by the earthquake. Base motions of the site are obtained by modifying the seismic accelerogram recorded at the ground surface of that site. This is done simply by obtaining the frequency-spectrum of the accelerogram, and then multiplying it with the analytical 'transfer function' represented for the ratios of the accelerations occurring at the base (bedrock) and those at the ground surface of that site (Roesset, 1977). This computation would complete a frequency-domain convolution and prepare a base-acceleration spectrum to solve for the corresponding accelerogram. To have consistent results for a specific site, one must be very cautionous about the wave velocities and the thickness of the soil layers used in the analyses. Crosschecks are required for vertical and horizontal excitations to ensure that the analytic parameters are rational. Notice that the discrete solutions of the wave equations are in terms of the displacements only. To obtain the time–displacement history of the soils, a baseline correction procedure (Kramer, 1996) is suggested to eliminate the integral offsets of the velocities and displacements appearing after the quake excitations. The responses of the free-field using the above procedure have been checked with the solutions of FEM as shown in Figure 4. Using this simplified model just be only computed one-way ground response depending on the inputted seismic motions. And, despite the simplicity of the geometry, an exact solution of the full model, and a detailed analysis of the phenomenon, have not perfectly been achieve (Schanz and Cheng, 2000).

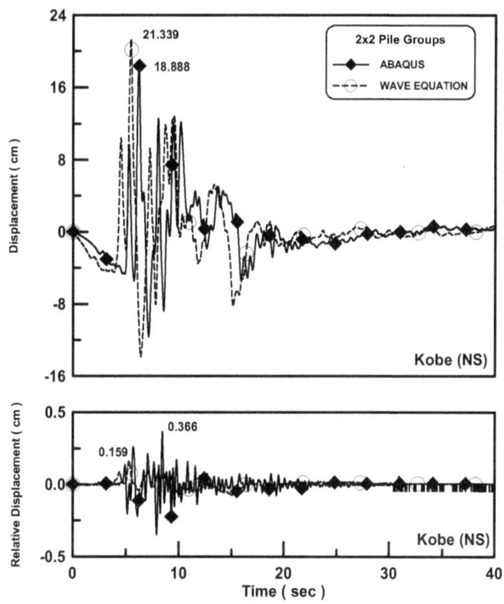

Fig. 4. Comparison of numerical results from WEA and FEM

2.3 EQWEAP: Force-based method

2.3.1 Wave equation of pile foundations concerning lateral spread

The wave equation describing a single pile under lateral loads can be derived based on a force equilibrium of the pile segments shown in Figure 1 as follows,

$$EI\frac{\partial^4 u(x,t)}{\partial x^4}+\rho A\frac{\partial^2 u(x,t)}{\partial t^2}+P_x\cdot\frac{\partial^2 u(x,t)}{\partial x^2}=P(x,t) \tag{19}$$

where u is the lateral pile displacement relative to the soil, E is the Young's modulus of the pile, I is the moment inertia of the pile, ρ is mass density of the pile, A is the cross-sectional area of the pile, P_x are the superstructure loads, $P(x,t)$ is the time-dependent loading due to laterally spreading at various depths, x is ordinate variable, and t represents for time. Using explicit finite difference schemes, the discrete form of Eq. (19) can be written as

$$u(i,j+1)=\frac{1}{A_1}\begin{bmatrix}-u(i+2,j)+(4-B_1)\cdot u(i+1,j)\\+(2A_1+2B_1-6)\cdot u(i,j)\\+(4-B_1)\cdot u(i-1,j)-u(i-2,j)\\-A_1\cdot u(i,j-1)+C_1\end{bmatrix} \tag{20}$$

where $A_1=\frac{\rho A\Delta x^4}{EI\Delta t^2}$; $B_1=\frac{P_x\Delta x^2}{EI}$; $\frac{P(x,t)\Delta x^4}{EI}$.

For the initial condition, $u(i,j)$ and $u(i,j-1)$ are set to zero. Equation 20 can only calculate the responses of piles under lateral loads along the lenfth of the pile. The head and tip can not be solved for. With proper boundary conditions (see Eq. 3~4), the other equations can then be derived. While the liquefaction-induced dynamic earthquake pressures are computed, the pile responses at various depths can be solved through the above formulations.

2.3.2 Dynamic earth pressure

Since Okabe (1926) and Mononobe and Matsuo (1929) introduced the concept of dynamic lateral pressure, many reports and practical works have been conducted in this manner (Ishibashi and Fang, 1987; Richard et al., 1990; Ishibahi et al., 1994; Budhu and Al-karni, 1993; Richard et al., 1993; Soubra and Regenass, 2000). Tokimatsu (1999, 2003) and Uchida and Tokimatsu (2005) determened several factors that affect the response of a pile in saturated sand by using a shaking table tests. They suggested that the total earth pressure acting on the foundation, when neglecting the friction between foundation and soil (see Fig. 5), is define as:

$$P_E=P_{EP}-P_{EA}=Q-F \tag{21}$$

where P_E is total earth pressure, P_{EP} and P_{EA} are earth pressures on the active and passive sides, Q is shear force at the pile head, and F is total inertial force from the superstructure

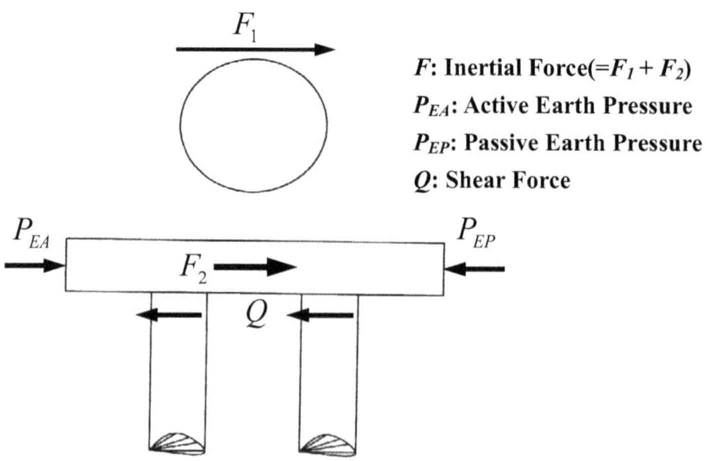

Fig. 5. Schematic layout of forces acting on Foundation (from Uchida and Tokimatsu, 2005)

and foundation. In addition, Haigh and Madabhushi (2005) have verified that the adjacent stresses of single piles subjected to lateral spreading forces would range between the active state and the passive state through centrifuge modeling.

Based on the Mononobe-Okabe method, Zhang et al. (1998) successfully derivates the time-dependent coefficients of earth pressure under active and passive states that involves the motions of soils and foundations. One can also modify the plane strain model of soil wedge to extend it to be three dimensional analysis. The descriptions and formulations of the coefficients of active and passive earth pressures are referred to Zhang et al. (1998).

2.3.3 Modeling lateral spread

For lateral spread induced by liquefaction, the soil properties such as the unit weights and the friction angles of the soils could be corrected based on the calculated pore water pressure ratios. There are two ways depicting the weakness of soils during liquefaction (Matsuzawa et al. 1985; Ebeling and Morrison, 1993). Those equations are given by

$$\gamma'_s = \gamma_s(1 - r_u) \tag{22}$$

$$\phi'_{eff} = \tan^{-1}[(1 - r_u)\tan\phi'] \tag{23}$$

where γ'_s is the unit weight of the soil, γ_s is the effective unit weight of the soil, ϕ' is the friction angle of the soil, and ϕ'_{eff} is the effective friction angle of the soil.

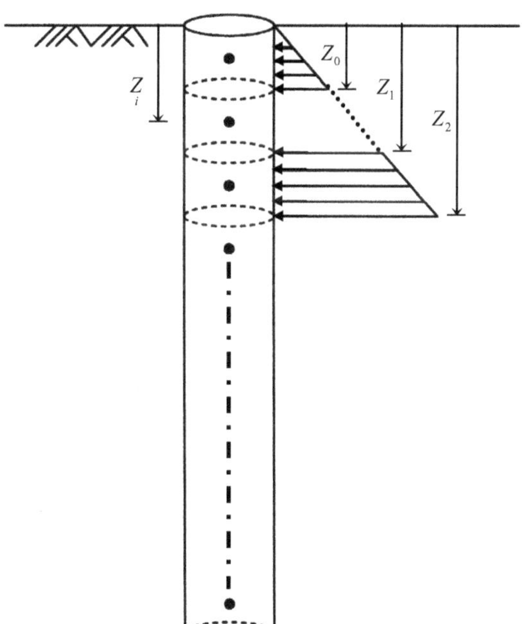

Fig. 6. Distribution of earth pressure along a pile

Figure 6 illustrates the distributions of earth pressures along the pile with the discrete blocks and nodes. According to the geometry of pile (see figure 7) and Eq. (24), the lateral forces at various depths are determined by

$$P_E = (\gamma'_s \underset{i}{Z} K_E) \cdot (B) \tag{24}$$

where $\underset{i}{Z}$ is the corresponding depth of node, K_E is the equivalent dynamic coefficients of earth pressure (i.e. $K_E = K_{EP} - K_{EA}$), and B is the loaded width of the pile body ($= \pi\, d / 2$, where d is the pile diameter).

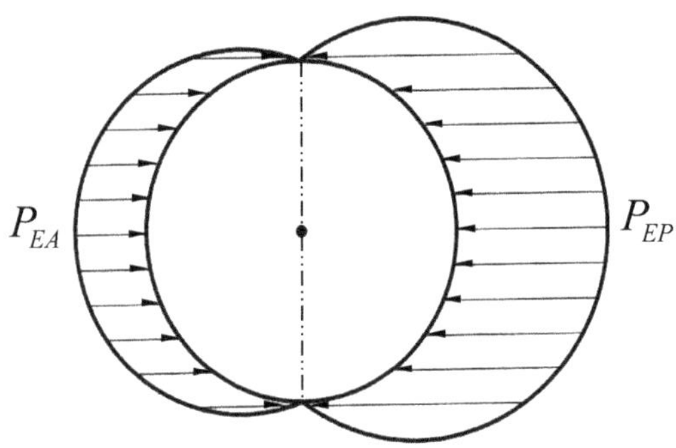

Fig. 7. The loaded width of the pile body due to lateral spreading

3. Practical simulation

In the following section, two case studies are presented, one of which focuses on pile foundation damagess caused by the Niigata earthquake in Japan (Hamada, 1992) and the other which focuses on foundation pile cases damaged during the 1995 Kobe earthquake. The Niigata earthquake case study utilize the displacement-based EQWEAP method, in which the free-field and the wave equation analysis are both performed to calculate the dynamic responses of piles under liquefaction. In The Kobe earthquake case studies, the force-based EQWEAP method is utilized to assume lateral flow induced forces on the piles. Dynamic earth pressures caused by lateral spreading of the liquefied layers are first generated and used to model forces exerted on the piles where the deformations of piles occur. These results show the pile failure pattern validate the applied methodology.

3.1 Case study: Pile damages due to soil liquefaction

The Niigata Family Court House was a four-story building located on the left bank of the Shinano River. The building was supported on a concrete pile foundation (Figure 8) each pile of the foundation having a a diameter of 35 cm and length of 6 to 9m. During the earthquake, the pile foundations were damaged by liquefaction-induced ground displacement. Excavation surveys showed that two piles (No.1 pile and No.2 pile) had severe cracks (Figure 9). They were conjecturally crushed by excessive bending moments at the interface between liquefied and non-liquefied layers as shown in Figure 9. According to aerial photographs of the area, the permanent ground displacement in the vicinity of building moved approximately 1.1m and the maximum displacement of No.1 pile and No.2 pile were respectively 50 cm and 70cm. For simplification, the entire soil system could be assumed as an upper layer and a lower layer. The upper layer from the ground surface to the depth of 8m is classified as medium-dense sand. The lower layer from the depth of 8 to 11m is classified as dense sand. The time history of earthquake record adopted the NS-component of the 1964 Niigata Earthquake as illustrated in Figure 10.

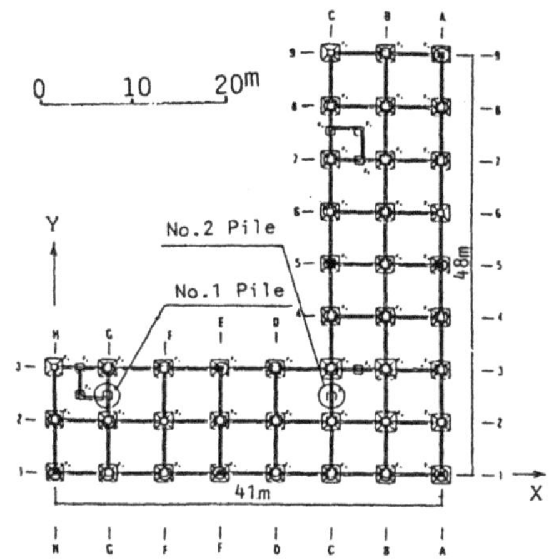

Fig. 8. Footing and foundation beams of Niigata Family Court House (from Hamada, 1992)

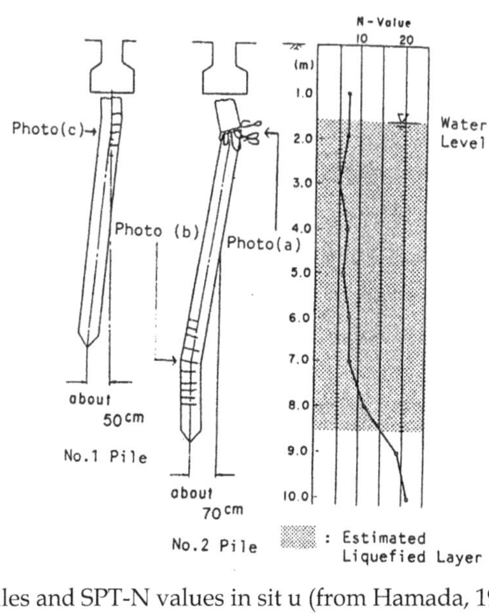

Fig. 9. Damage to piles and SPT-N values in sit u (from Hamada, 1992)

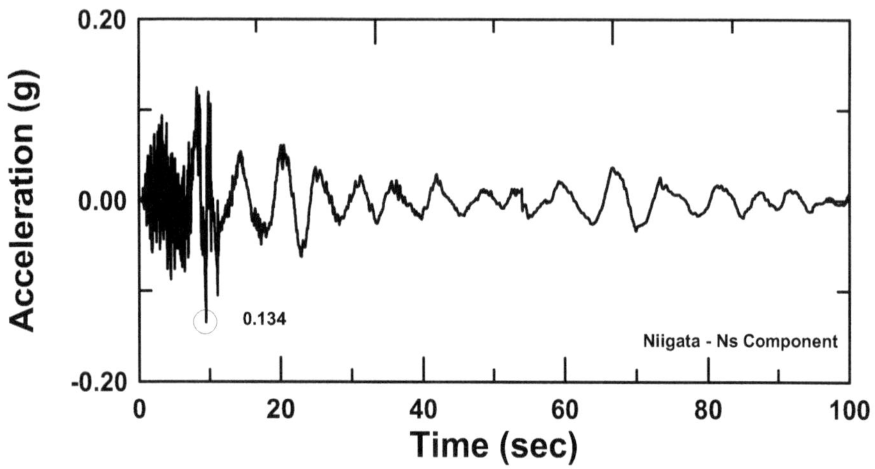

Fig. 10. Time history of Niigata Earthquake (NS Component)

The initial shear modulus of the soils at the any depth can be calculated by Eq. (17). The distribution of shear modulus is similar to the hyperbolic form observed in gibson soils and increases with the depth. The determination of pore water ratio pressure (r_u) and reduction factors (D_E) versus depth can be estimated by the liquefaction potential method suggested by Tokimatsu and Yoshimi (1983) with Eq. (18) for various levels of liquefaction. Moreover, one can conduct EQWEAP analysis to obtain the liquefied free-field response considering the effect of pore water. The excess pore pressure ratios at different depths are shown in Figure 11. It was found that the soil layer reached a liquefied state gradually after 2.8 seconds.

Figure 12 shows the time histories of ground motions. The maximum displacement of the ground which takes place at the surface is 47.3 cm at about 10 seconds. The liquefied layer ($r_u = 100\%$) ranging between the depths of 2 m to 8m displaces by 30 cm to 45 cm (see figure 12). The displacements reduce to about 3 cm below the liquefied layer for $r_u = 14 \sim 45\%$ as shown in Figure 12. Figure 13 indicate the maximum displacements of piles at various depths from wave equation analysis. Based on the results form Figure 13, the peak value occured at the pile head and the relative displacements between the pile head and pile tip are 50 cm and 69 cm. The maximum bending moments of piles are shown in Figure 14 and those peak values would also occur approximately at the interface between liquefied and non-liquefied layers. Comparing the numerical results by Meryersohn (1994), the computed values are nearly consistent with the ones reported. In the meantime, the peak shear forces of piles also occur at this zone. Therefore, the excessive bending moment and shear zone of the pile is again revealed in this study using the suggested procedures.

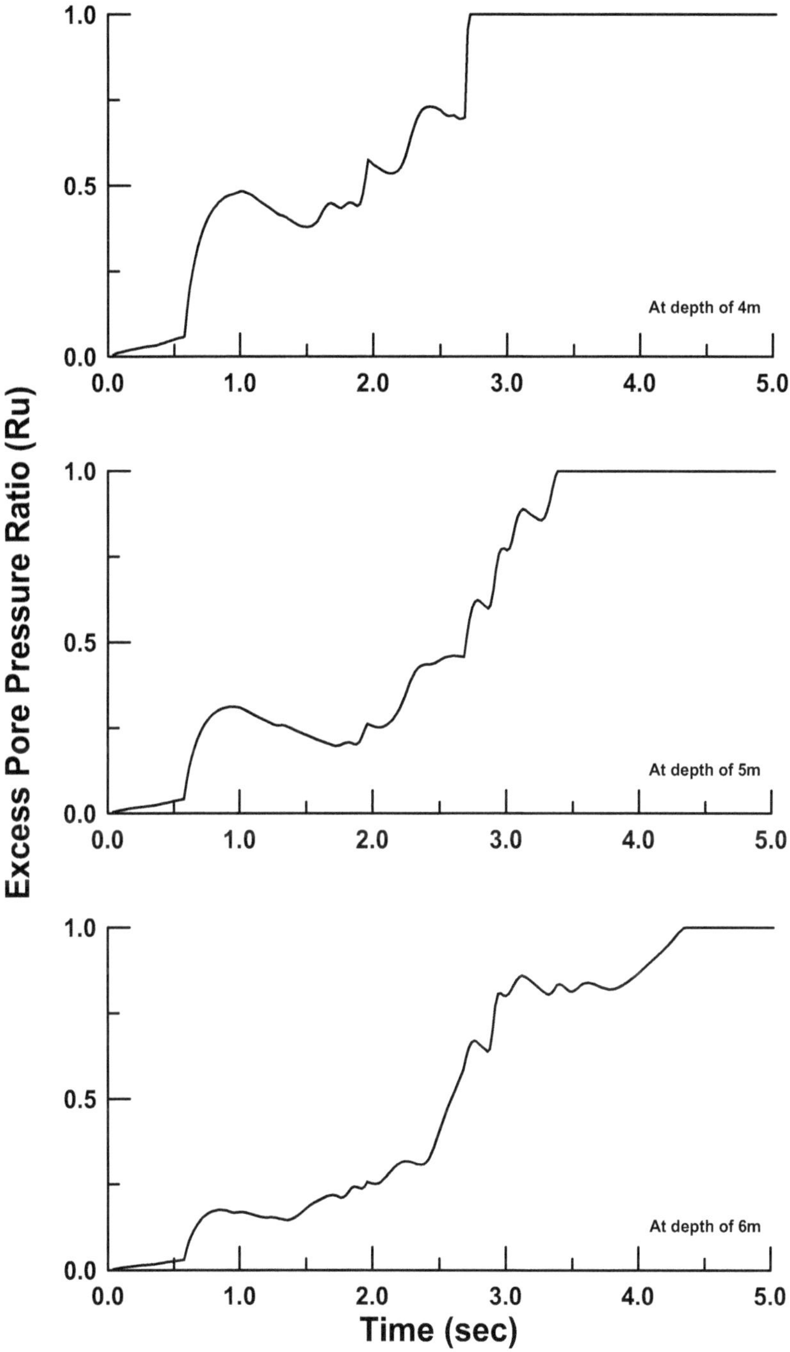

Fig. 11. The time history of excess pore pressure ratios at different depths

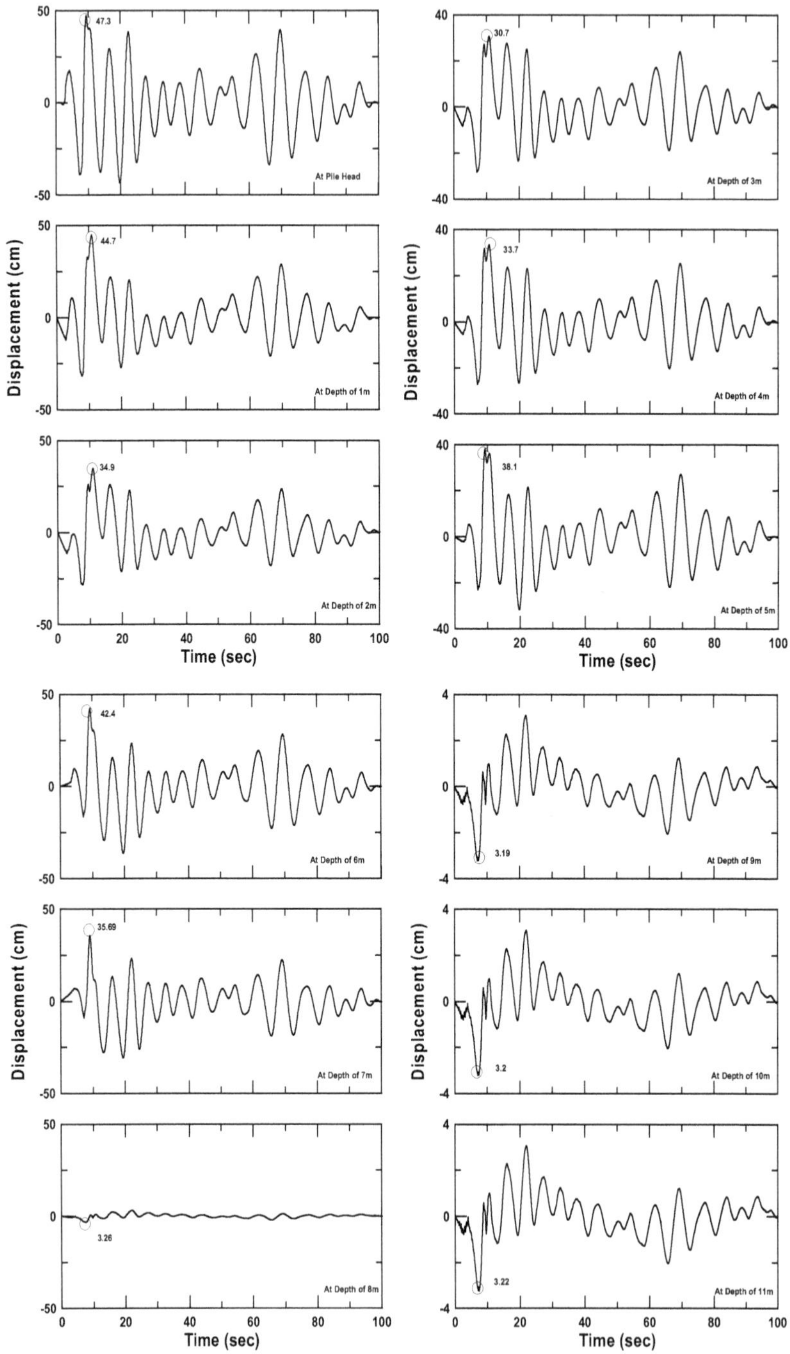

Fig. 12. Time histories of ground motions at different depths

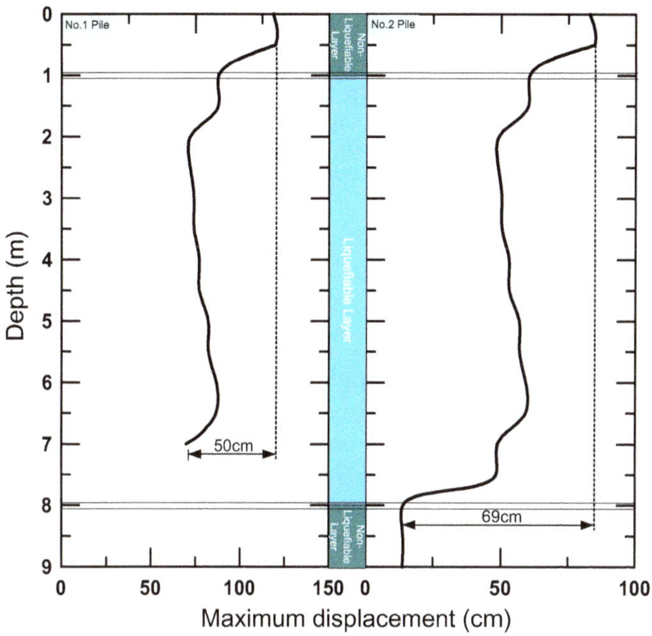

Fig. 13. Maximum pile displacements for No.1 Pile and No.2 Pile

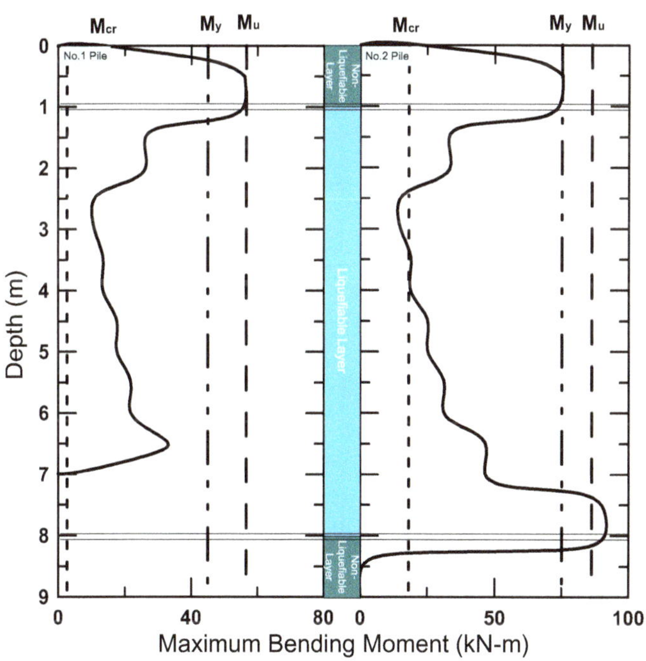

Fig. 14. Maximum pile bending moments for No.1 Pile and No.2 Pile

3.2 Case study: Pile damages due to lateral spread

Mikagehama is a man-made island in the port area of Kobe home to a number of liquefied propane gas (LPG) and oil tanks. During the 1995 Kobe Earthquake, the soils underlying the foundatinos of tanks liquefied. A quay wall moved seaward and lateral spreading of the backfill soils damage the piles supporting the tanks. Oil-storage tank TA72 is chosen to be a target, which is located in the west part of the island about 20m from the waterfront. Figure 15 illustrates the cross sectional view of tank and underlying pile foundations. The tank has a diameter of 14.95 m and its storage capacity is about 2450 kl. It is supported on 69 precast concrete piles each with the length of 23 to 24 m and diameter of 45 cm. The water table is estimated at the depth of 2 to 3 m. Sand compaction piles were conducted to increase the SPT-N values of the Masado layer around the outside of Tank TA72.

According the relation between the bending moment (M) and curvature (φ) where D_0 is the diameter of pile and N is axial load on pile head, one can know that the cracking bending moment (M_{cr}), the yield bending moment (M_y) and the ultimate bending moment (M_u) are 105 kN-m, 200 kN-m, and 234 kN-m respectively. The ultimate shear strength is 232 kN with regards to ACI (1998). Ishihara and Cubrinovski (2004) have utilized bore-hole cameras and inclinometers to inspect the damages of the piles. Their results for pile No. 2 are shown in Figure 15. The main failure field was located at a depth of 8 to 14 meters where the piles were found to have developed many cracks. Moreover, pile No. 2 had wounds due to large deformations where lateral spreading of liquefied soils developed along the weak interface. Quantifying damage of structures caused by earthquakes in terms of Park and Ang damage indices, an index that provides a measure of structure damage level, gave a value of 0.8, signifying the piles were in a near state of collapse. (Park and Ang, 1985; Moustafa, 2011).

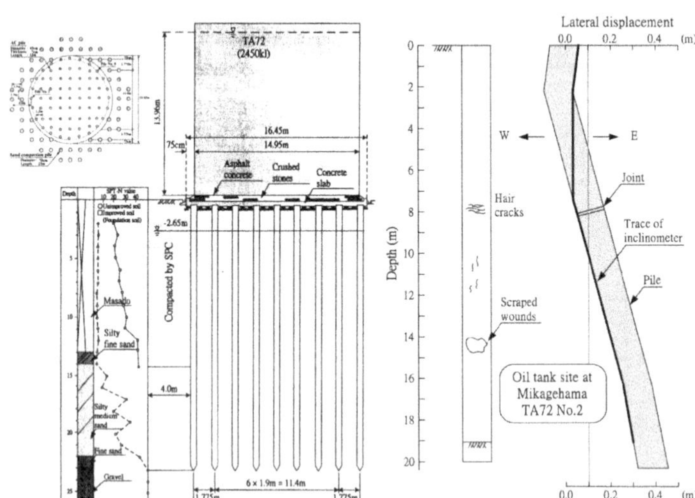

Fig. 15. Cross sectional view of Tank TA 72 and its foundation (from Ishihara and Cubrinovski, 2004)

In this study, the length of pile is assumed to be 24 m with a diameter of 45 cm. Seismic record of the NS-component of 1995 Kobe Earthquake is adopted. According to the field data, distributions of pore water ratio pressure (r_u) versus the depths can be estimated by evaluating the liquefaction potential of that site. With all the required data and incorporating with the modified M-O model (Zhang et al, 1998), the dynamic coefficients of earth pressure are computed as shown in Figure 16. Also, the unit weight of the soil is reduced by r_u (refer to Eq. 22). When obtaining those dynamic earth forces to insert and execute the wave equation analysis, the time histories of displacements along the pile can be illustrated as shown in Figure 17. The displacement of the pile head oscillates significantly with time, but the peak value is smallest. As the depth increases, the peak displacement of pile becomes larger. Those peak displacements along the pile are shown in Figure 18(a). The maximum value among them occurs at the pile tip about 52.7 cm and the maximum relative displacement between the pile top and the pile bottom is estimated about 44.7 cm. The deformed shape of the pile is similar to pile No. 2. It can be found that the maximum bending moments which exceed the ultimate bending moment at depths of 2 to 23 m and that this zone is the mose dangerous zone.. With regards to the shear failure, the weak interface exists at a depth of 11 m, in which the maximum shear force is close to the ultimate (Figure 18b~18c). The above observations are agreeable to field investigations reported by Ishihara and Cubrinovski (2004).

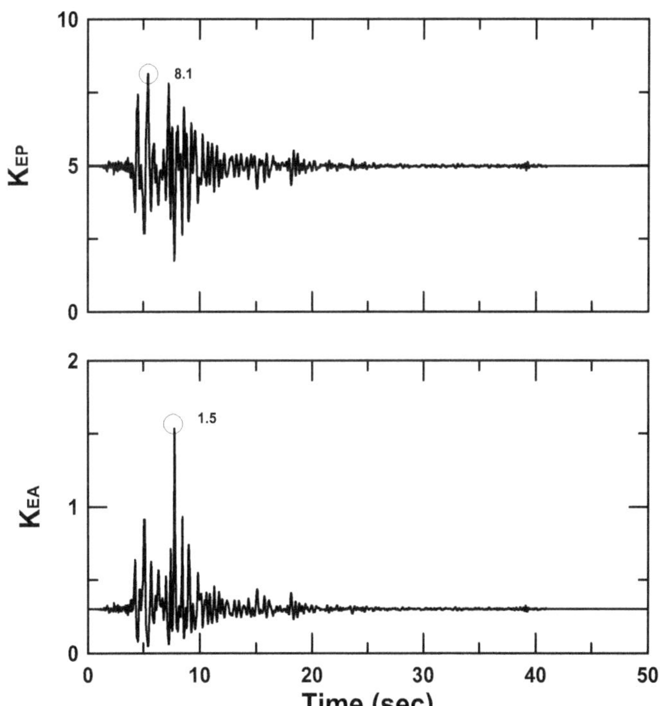

Fig. 16. Dynamic coefficients of earth pressure

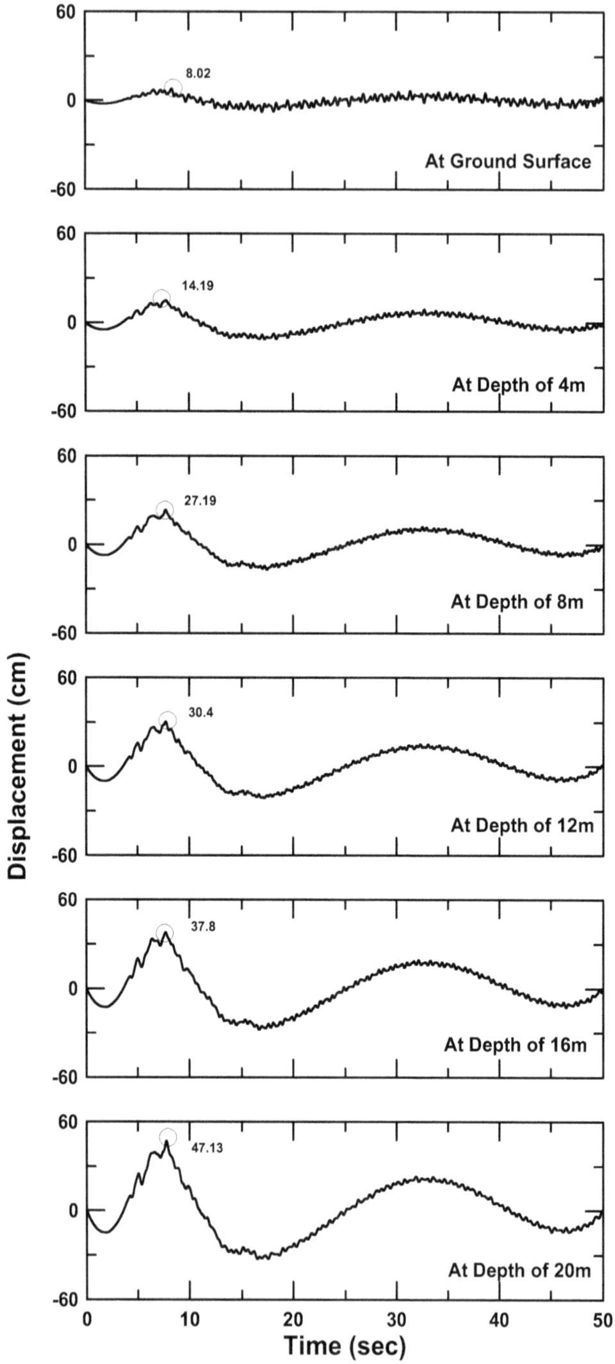

Fig. 17. Time histories of lateral displacement along the Tank TA72 No.2 pile

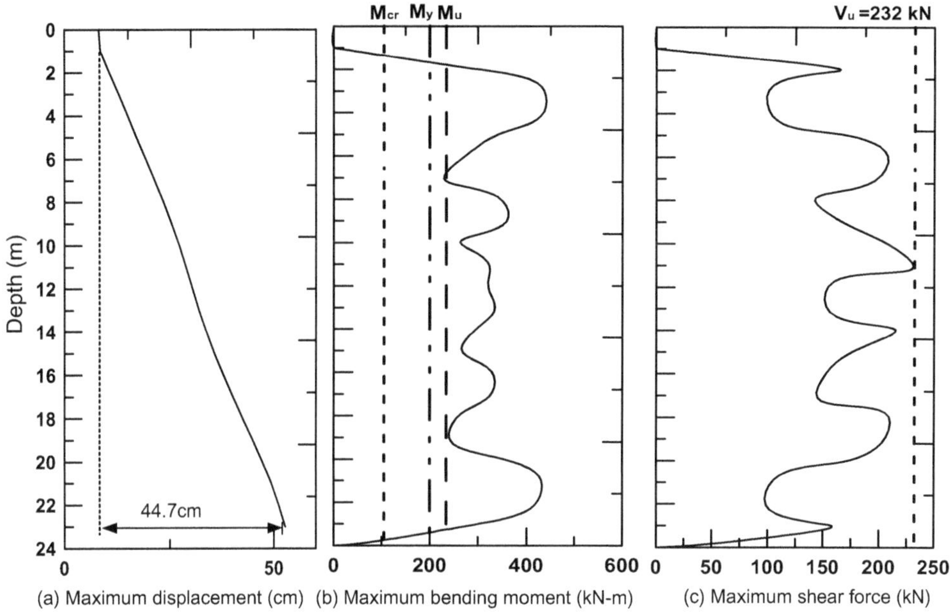

Fig. 18. Maximum structural response along the Tank TA72 No.2 pile
(a) Maximum displacement; (b) Maximum bending moment; (c) Maximum shear force

4. Conclusions

EQWEAP is a simplified but effective procedur to analyze the dynamic pile-soil interaction under the earthquake. In the analysis, the pile deformations are obtained solving the discrete wave equations of the pile, where the seismic ground motions are pre-calculated from one-dimensional lumped mass model assuming a free-field condition or dynamic earth pressure are directly exerted onto the pile. This chapter presented both displacement- and forced-based form of the EQWEAP analysis method along with two comparative case studies: Using wave equation analysis and the EQWEAP method, pile response to liquefaction has been computed and compared to the case histories of the Niigata earthquake records. Case histories of the Kobe earthquake show that the lateral spreading can be a major cause to damage the piles. Specifically conclusions for the displacement and forced based EQWEAP methods can be summarized as follows:

1. Based on the suggested numerical procedure using EQWEAP (Chang and Lin, 2003; Lin et al., 2011), one can evaluate the motions of the soil stratum and the pile foundations at various depths to estimate the occurrence of pile damages and patterns of failure. This procedure provides a simplified but rational dynamic analysis to the pile foundation design work.
2. The use of the empirical excess pore pressure model for liquefaction can be applicable to soils underneath the liquefiable layers using a minimum pore pressure ratio. The pore pressure ratio should be calculated using the empirical formula suggested by Tokimatsu and Yoshimi (1983) providing that the factors of safety against liquefaction are known.

3. Not only the interfaces between the liquefied and non-liquefied layers can exert excessive bending moments and shear stress, but also the layer contrast of the soils can yield similar effects. Engineers need to be more careful in designing pile shafts that are susceptible to fail due the liquefaction resulting from earthquakes and the layer contrast.
4. The wave equation analysis can be used to model the pile responses under lateral spread due to earthquake. The modified M-O model (Zhang et al., 1998) incorporating reduction methods for soil parameters were successfully used to represent the dynamic earth pressures of the lateral spread. The numerically determined pile deformations were similar to deformations discovered at piles actually affected by lateral spread. In advance, if nonlinear behaviour of pile such as the moment-curvature relationship and complexity of pile geometries can also be considered simultaneously in this method, the results would be enhanced to capture detailed mechanism and definite performance of piles foundations.

5. Acknowledgement

The special thank goes to my colleague, Mr. Jeff Keck. The typeset and revision that he gave truly help to complete this task. The assistance is much indeed appreciated.

6. References

Abbas, A.M. and Manohar, C.S. (2002). Critical spatially varying earthquake load models for extended structures. *Journal of Structural Engineering*, Vol. 9, No.1, pp. 39-52.

American Concrete Institute (1995). *Building Code Requirements for Structural Concrete*, ASCE, USA

Bathe, K.J. (1982). *Finite Element Procedure in enigeering analysis*, Prentice Hall, Englewood Cliffs, New Jerwey, 735 p.

Bhattacharya, S., Bolton, M.D. and Madabhushi, S.P.G. (2005). *A Reconsideration of the Safety of Piled Bridge Foundation in Liquefiable Soils*, Soils and Foundations, JGS, Vol. 45, No. 4, pp. 13-25.

Bhattacharya, S., Madabhshi, S.P.G. and Bolton, M.D. (2002). Pile Instability during Earthquake Liquefaction, *Engineering Mechanics Conference*, ASCE , Seattle, July.

Bhattacharya, S., Madabhushi, S.P.G. and Bolton, M.D. (2004). An alternative Mechanism of Pile Failure in Liquefiable Deposits during Earthquakes, *Geotechnique*, Vol.3, No.54, , pp. 203-213.

Boulanger, R.W., Curras, C.J., Kutter, B.L., Wilson, D.W. and Abghari, A., (1999). Seismic Soil-Pile-Structure Interaction Experiments and Analyses, *Journal of Geotechnical and Geoenvironmental Engineering*, ASCE, Vol.125, No.9, pp. 750-759.

Budhu, M. and Ai-karni, A. (1993). Seismic Bearing Capacity of Soils, *Geotechnique*, Vol.43, No.1, pp.187-187.

Chang, D.W. and Lin, B.S. (2003). Wave Equation Analyses on Seismic Responses of Grouped Piles", Procds., *12th Asian Regional Conf. on Soil Mechanics and Geotechnical Engineering*, Singapore, August, pp. 581-586.

Chang, D.W. and Yeh, S.H., (1999). Time-domain Wave Equation Analyses of Single Piles Utilizing Transformed Radiation Damping, *Soils and Foundations*, JGS, Vol. 39, No. 2, pp. 31-44.

Chang, D.W., Roesset, J.M. and Wen, C.H., (2000). A Time-domain Viscous Damping Model Based on Frequency-dependent Damping Ratios, *Soil Dynamics and Earthquake Engineering*, Vol. 19, pp. 551-558.

Chaudhuri, D., Toprak, S., and O'Rourke, T.D. (1995). Pile Response to Lateral Spread: a Benchmark Case, Lifeline Earthquake Engineering, *Procds., 4th U.S. Conference, San Francisco*, California, pp. 1-8.

Curbrinovski, M and Ishihara, K. (2004). Simplified Method for Analysis of Pile Undergoing Lateral Spreading in Liquefied Soils, *Soils and Foundations*, JGS, Vol. 44, No. 5, pp. 119-133.

DeAlba, P., Seed, H.B. and Chan C.K. (1976). Sand Liquefaction in Large-Scale Simple Shear Tests, *Journal of Soil Mechanics and Foundations Division*, ASCE, Vo1.102, GT9, pp. 909-927.

Ebeling, R.M., and Morrison, E.E. (1993). The Seismic Design of Waterfront Retaining Structures, Office of Navy Technology and Department of the Army, *NCEL Technical Report R-939*, Naval Civil Engineering Laboratory, Port Hueneme, CA, 256 p.

El Naggar, M.H. and Bentley, K.J. (2000). Dynamic Analysis for Laterally Loaded Piles and Dynamic p-y Curves, *Cananian Geotechnical Journal*, Vol.37, pp. 1166-1183.

Eruocode 8 (1998). Design Provisions for Earthquake Resistance of Structure-Foundations, Retaining Structures and Geotechnical Aspects, *European Committee for Standardization*, Brussels.

Finn, W.D.L. and Thavaraj, T. (2001). Practical Problems in the Seismic Analysis of Bridge Pile Foundations, *Procds., 10th Int. Conf. Computer Methods and Advances in Geomechanics*, Tucson, Arizona, USA, pp. 1011-1018.

Finn, W.D.L., Lee, K.W., and Martin, G.R. (1977). An Effective Stress Model for Liquefaction, *Journal of the Geotechnical Engineering Division*, ASCE, Vol. 103, No. SM7, pp. 657-692.

Haigh, S.K. and Madabhushi S.P.G. (2005). The Effects of Pile Flexibility on Pile-Loading in Laterally Spreading Slops, *Proc. Int. Workshop Simulation and Seismic Performance of Pile Foundations in Liquefied and Laterally Spreading Ground*, ASCE, 14 p.

Hamada, M. (1992). Large Ground Deformations and Their Effects on Lifelines : 1964 Niigata Earthquake, in Case Studies of Liquefaction and Lifeline Performance During Past Earthquakes, Vol. 1, Japanese Case Studies, *Technical Report NCEER-92-0001*, NCEER, Buffalo, NY, USA.

Hamada, M. and Wakamatsu, K. (1998). A Study on Ground Displacement Caused by Soil Liquefaction, JSCE, *Journal of Geotechnical Engineering*, No. 596/Ⅲ-43, pp. 189-208.

Ishibashi, I. and Fang, Y.S., (1987). Dynamic Earth Pressures with Different Wall Movement Modes, *Soils and Foundations*, JGS, Vol. 27, No. 4, pp. 11-22.

Ishibashi, I., Osada, M. and Uwabe, T. (1994). Dynamic Lateral Pressures due to Saturated Backfills on Rigid Walls, *Journal of Geotehnical Engineering*, Vol.120, No.10, pp. 1747-1767.

Ishihara, K. (2003). Liquefaction-Induced Lateral Flow and Its effects on Foundation Piles, *5th National Conference on Earthquake Engineering*, Istanbul, Turkey, May, 28 p.

Ishihara, K. and Cubrinovski, M. (2004). Case Studies on Pile Foundations undergoing Lateral Spreading in Liquefied Deposits, *Proceeding of 5th International Conference on Case Histories in Geotechnical Engineering*, New York, Paper SOAP 5.

Japanese Road Association (JRA) (1996). *Specification for Highway Bridges*, Part V, Seismic Design.

Jefferies M. and Been K. (2006). *Soil Liquefaction*, Taylor and Francis, New York.

Kim, S.I. (2003). Liquefaction Potential in Moderate Earthquake Regions, Procds., *12th Asian Regional Conf. on Soil Mechanics and Geotechnical Engineering*, Singapore, August, pp.1109-1138.

Kramer, S.L. (1996). *Geotechnical Earthquake Engineering*, McGraw-Hill Book Co., Inc., New York.

Kramer S. and Algamal A. (2001). *Modeling Soil Liquefaction Hazards for Performance-Based Earthquake Engineering*, PEER Report 2001/13.

Lee, K. L. and Albeisa, A. (1974). Earthquake Induced Settlements in Saturated Sands, *Journal of Soil Mechanics and Foundations Division*, ASCE, Vo1.100, GT4, pp. 387-406.

Lin, B. S., (2006). *Structural Analyses for Pile Foundations Subjected to Soil Liquefaction and Lateral Spreading*, Doctor's Dissertation, Tamkang University, Taiwan, 487 p.

Lin, B. S., Chang, D.W, and Ho, H. C. (2010). Pile Responses under seismic earth pressures of Lateral Spread Using Wave Equation Analysis, ASCE, *Geotechnical Special Publication, N0. 201, Dynamic Soil-Structure Interactions: Part I - Proceedings of Sessions of GeoShanghai 2010*, pp. 109-117.

Lin, S.S., Tseng Y. J., Chiang C.C. and Hung C.L. (2005). Damage of Piles Caused by Laterally Spreading – Back Study of Three Cases", *Workshop on Simulation and Seismic Performance of Pile Foundation in Liquefied and Laterally Spreading Ground*, University of California, Davis, March.

Madabhushi, G., Knappett, J., Haigh, S. (2010). *Design of Pile Foundation in Liquefiable Soils*, World Scientific, Inc.

Martin, G.R., Finn, W.D.L., and Seed, H.B. (1975). Fundamentals of Liquefaction under Cyclic Loading, *Journal of the Geotechnical Engineering Division*, ASCE, Vol. 101, No. GT5, pp. 423-438.

Matsuzawa, H., Ishibashi, I, and Kawamura, M. (1985). Dynamic Soil and Water Pressure of Submerged Soils, *Journal of the Geotechnical Engineering*, ASCE, Vol. 111, No. 10, pp. 1161-1176.

Meyersohn, W.D. (1994). *Pile Response to Liquefaction-induced Lateral Spread*, Doctor's Dissertation, University of Cornell, USA.

Mononobe, N. and Mastuso, H. (1929) On the Deternation of Earth Pressure During Earthquake, *Proceedings, World Engineering Congrss*, 9 p.

Moustafa, A. (2011). Damage-Based Design Earthquake Loads for Single-Degree-Of-Freedom Inelastic Structures, *Journal of Structural Engineering*, ASCE, Vol. 137, No. 3, pp. 456-467.

National Earthquake Hazard Reduction Program (NEHRP) (2000). *Commentary (Federal Emergency Management Agency, USA, 369) for Seismic Regulations of New Buildings and Other Structures.*

Nogami, T., Otani, J., Konagai, K. and Chen, H.L. (1992). Nonlinear Soil-pile Interaction Model for Dynamic Laterally Motion, *Journal of Geotechnical Engineering*, ASCE, Vol.118, No.1, pp. 89-106.

Novak, M. (1974). Dynamic Stiffness and Damping of Piles, *Canadian Geotechnical Journal*, Vol.11, pp. 574-598.

Novak, M. (1977). Vertical Vibration of Floating Piles, *Journal of the Engineering Mechanics Division*, ASCE, Vol. 03, No. M1, pp. 153-168.

Novak, M. and Beredugo, Y.O. (1972). Vertical Vibration of Embedded Footings, *Journal of Soil Mechanics and Foundations Division*, ASCE, Vol. 98, No. SM12, pp. 1291-1310.

Okabe, S. (1926). General Theroy of Earth Pressures, *Journal of the Japan Society of Civil* Engineering, Vol.12, No.1.

Park Y.J., and Ang, A.H.S. (1985). Mechanistic Seismic Damage Model for Reinforced Concrete, *Journal of Structural Engineering*, ASCE, Vol.111, No.ST4, pp. 722-739.

Prakash, S. and Puri, V.K. (1988), *Foundations for Machines: Analysis and Design*, John Wiley and Sons, Canada, pp. 438-458.

Priestley, M.J.N., Seible F. and Calvi, G.M. (1996). *Seismic Design and Retrofit of Bridges*, John Wiley & Sons, Inc.

Richard, R., Elms, D.G. and Budhu, M. (1990). Soil Fluidization and Foundation behavior, *Proc. 2nd International Conference Recent Advance in Geotechnical Engineering and Soil Dynamics*, pp. 719-223.

Richard, R., Elms, D.G. and Budhu, M. (1993). Seismic Bearing Capacity and Settlement of Foundations, *Journal of Geotechnical Engineering Division*, ASCE, Vol. 119, No. 4, pp. 662-674.

Roesset, J.M. (1977), Soil Amplification of Earthquake, *Numerical Methods in Geotechnical Engineering*, McGraw-Hill Book Co., Inc., New York, pp. 639-682.

Schanz, M. and Cheng, A.H.-D (2000). Transient Wave Propagation in a One-Dimensional Poroelastic Column, *Acta Mechanica*, Vol.145, No.1-4, pp. 1-18.

Seed, H.B., and Idriss, I.M. (1970). *Soil Moduli and Damping Factors for Dynamic Response Analysis*, Report NO.EERC 75-29, Earthquake Engineering Research Center, University of California, Berkeley, California.

Smith, E.A.L. (1960). Pile Driving Analysis by the Wave Equation, *Journal of Soil Mechanics and Foundation Divisions*, ASCE, Vol.86, No.SM4, pp. 35-61.

Sneddon, I.N. (1957), *Elements of Partial Differential Equations*, McGraw-Hill Book Co., Inc., New York, pp. 274-275.

Soubra, A. H. and Regenass, P. (2000). Three-dimensional Passive Earth Pressure by Kinematical Approach, *Journal of Geotechnical and Geoenvironmental Engineering*, Vol. 126, No. 11, pp. 969-978.

Stephen, E. D., Nason, J. M., Mark, G. B., and Bryan, J. W. (2002). *Assessment and Mitigation of Liquefaction Hazards to Bridge Approach Embankments in Oregon*, Oregon State University, FHWA-OR-RD-03-04, Nov. 2002.

Tokimatsu K. and Yoshimi Y. (1983). Empirical Correlation of Soil Liquefaction Based on SPT N-Value and Fines Content, *Soil and Foundations*, JSSMFE, Vol. 23, No.4, pp. 56-74.

Tokimatsu, K. (1999). Performance of Pile Foundations in Laterally Spreading Soils, *Procds, 2nd Int. Conf. Earthquake Geotechnical Engineering*, Vol.3, pp. 957-964.

Tokimatsu, K. (2003). Behavior and Design of Pile Foundations Subjected to Earthquakes, *Procds., 12th Asian Regional Conf. on Soil Mechanics and Geotechnical Engineering*, Singapore, August.

Tokimatsu, K. and Asaka, Y. (1998). Effects of Liquefaction-induced ground displacement on pile performance in the 1995 Hyogoken-Nambu earthquake, *Soils and Foundations*, Special Issue, No. 2, pp. 163-178.

Uchida, A. and Tokimatsu, K. (2005). Comparison of Current Japanese Design for Pile Foundations in Liquefiable and Laterally Spreading ground, *Proc. Int. Workshop Simulation and Seismic Performance of Pile Foundations in Liquefied and Laterally Spreading Ground*, ASCE, 10p.

Wang, J.-N., Mesa , L., and Sizemore J. (2004). Seismic Design of the Cooper River Bridge, *Proceedings of Geo-Trans 2004*, ASCE, California, July.

Zerva, A. (2009). *Spatial Variation of Seismic Ground Motions,* CRC Press.

Zerva, A., and Zervas V. (2002). Spatial Variation of seismic ground motions: an overview, *Applied Mechanics Reviews*, Vol.55, No.3, pp. 271–297.

Zhang, J. M., Shamoto, Y., and Tokimatsu, K. (1998). Seismic Earth Pressure for Retaining Walls under Any Lateral Displacement, *Soils and Foundations,* Vol. 38, No. 2, pp. 143-163.

6

Earthquake Response Analysis and Evaluation for Earth-Rock Dams

Zhenzhong Shen, Lei Gan, Juan Cui and Liqun Xu
State Key Laboratory of Hydrology-Water Resources and Hydraulic Engineering, Hohai University
China

1. Introduction

An earthquake is any shaking of the ground, usually intense shaking of the ground - caused by either natural sources or by humans. There are many different types of earthquakes. The most common is tectonic earthquake, which occur when rocks in the earth's crust break due to geological forces created by movement of tectonic plates. Another type of earthquake is called volcanic earthquake, occurring in conjunction with volcanic activity. And also there are various types of earthquakes caused by man-made activities, which caused directly by human involvement, has been recorded as a result of water filling large dams, development of mineral, geothermal and hydrocarbon resources, waste injection, underground nuclear explosions and large-scale construction projects. This type of earthquake includes collapse earthquake, explosion earthquake and induced earthquake due to water filling large dams (Talwani, 1997; Chen & Talwani, 1998) and so on. Earthquake is the rapid vibration of the earth surface whose acceleration motion can easily cause the destruction of hydraulic structures. The relative deformation of the hydraulic structures increases when the inertial force caused by the earthquake increases suddenly. For the hydrodynamic pressure acting on the reservoir, dam foundation may appear relative deformation or different stage deformation. For example, those parts of concrete dams, including dam crest, fold slope, orifice inside the dam and corridor, are easily cracked, which may damage and then cause the leakage of the dam body, even the collapse of the structures after the earthquake. As it to the earth-rock dam, it is easy to appear cracks in dam in the process of the earthquake, meanwhile, dam foundation or dam body may collapse caused by earthquake liquefaction if the sands inside them are not fully roller compaction. Thus earthquake may cause the unsafe of the hydraulic structures, even dam breach which may threatened the life and property safety of the people downstream the river.

On May 12, 2008, a magnitude destructive earthquake (M_s=8.0), with the epicenter in Wenchuan County, struck Sichuan province, in southwestern China, killing more than 100,000 people. The earthquake was the largest and most destructive to strike China since the 1976 Tangshan earthquake, which killed more than 240,000 people (Pei et al., 2010; Wang et al., 2010; Cao et al., 2011). On January 12, 2010, a massive 7.3-magnitude earthquake has struck the Caribbean nation of Haiti. The earthquake struck about 15km (10 miles) south-west of Port-au-Prince which is the capital city of Haiti. Approximately 230,000 people died immediately or during ensuing weeks, mostly due to acute trauma (Kenneth, 2011). The most devastating earthquake (M_s=9.0) in Japan after the 1923 Great Kanto earthquake hit the eastern Japan in the afternoon of March 11, 2011. Nearly 30,000 people were killed or missed in the Tohoku earthquake and the ensuing monster tsunami (Takewaki et al., 2011). In this paper, the non-linear FEM is adopted to calculate the Wenchuan earthquake response. Then, the earthquake-resistance safety of the Bikou dam is evaluated.

1.1 Bikou earth core dam

The Bikou hydropower station is located on the Bailong River in Gansu province of China, where is about 240km away from the epicenter of Wenchuan earthquake, China, May 12, 2008(Chen et al., 2008; Yu et al., 2009; Lin et al., 2009). The Bikou earth core rockfill dam is a 100m graduate earth-rock dam (Gu, 1989). According to the earthquake loss survey and preliminary report of the Bikou hydropower project suffered "5.12" Wenchuan earthquake, some damage appeared inside Bikou dam, but no dam slope slide and instability damaged phenomenon happened after the earthquake (Zhou et al., 2009; Pan, 2009).The normal water level of the reservoir is 707.0m, and its relative reservoir capacity is 5.21×10^8 m^3. The dam is earth core rockfill dam with height of 101.8m, where the typical cross section of the dam is shown as in Fig.1. The ratio of the upstream slope of the dam are 1:1.8 and 1:2.3 respectively, and the downstream slope are 1:1.7, 1:1.85, 1:2.2 and 1:1.5 respectively. Its width of dam crest is 7.6 m at elevation 710.00m, and the crest length of dam is about 297.36 m. The dam is located in a roughly symmetric valley of V-type, the slope of river bank is about 50-60 degree, and the one of right bank is about 35-45 degree. The depth of overburden layer on the riverbed is about 25-34 m. The core wall in the riverbed area is built on the natural sandy gravel layer, the width of core wall crest is 4.0m at elevation 709.00-710.00m, and the ratio of the upstream slope of the core wall are 1:0.2 and 1:0.25 respectively, the section of the dam along the dam axis is shown as in Fig.2. Two concrete cut-off walls are adopted to connect the soil core wall and the bedrock. The first one built along the dam axis inserts the bedrock with the thickness 1.30m and crest the elevation 630.00m. And the second concrete one is located at the dam axis downstream 12m, inserting the bedrock with the thickness 0.80m and the crest 653.00m. The spillway buildings are made up of one spillway on the right bank, one desilting tunnel and two spillway tunnels in the left bank.

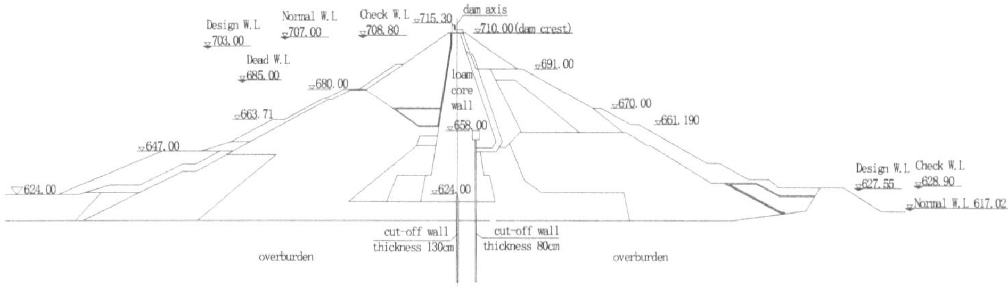

Fig. 1. Typical cross section of the dam

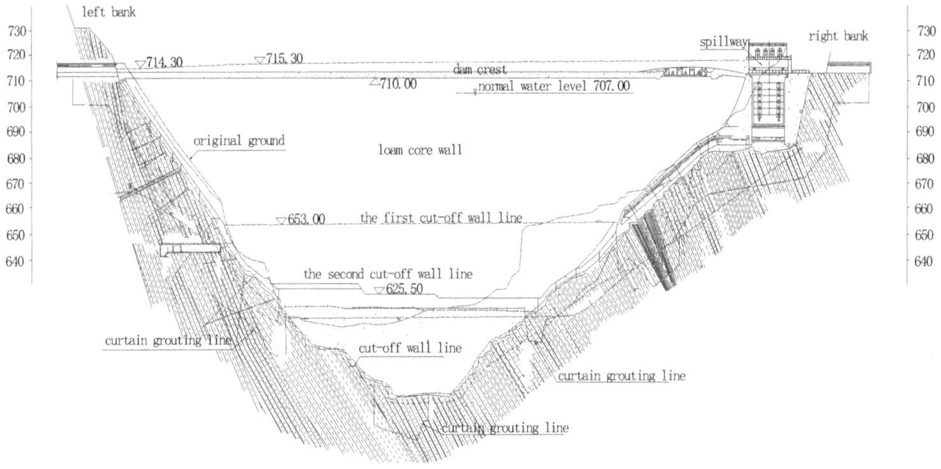

Fig. 2. Section of the dam along the dam axis

1.2 Geological conditions

In the dam site area, the fractures are developed, mainly are the bedding extrusion fracture zones and the slow obliquity faults. Most of the fractures are the tension and twist fractures of steep obliquity, and the slow obliquity fractures in the SN, NNE and EW directions. The faulted structures are mainly as follow: the fault F9, F1, F6 and F14, broken bandwidth 0.3 ~ 3.0m, mainly for fault mud and rock. The fracture zones are mainly made up of fault mud and rock, with width of 0.3-3.0m.

2. Calculation model

2.1 Static analysis model

2.1.1 Control equation of FEM

According to the displacement solution, the basic balance equation of the nonlinear finite element method is as follows

$$\left[K(u)\right]\{u\}=\{R\} \tag{1}$$

Where $\left[K(u)\right]$ is overall strength degree matrix, $\{u\}$ node displacement array and $\{R\}$ node load array.

This equation can be solved by the incremental early strain gauge method, where the basic balance equation holds:

$$[K]\{\Delta u\}=\{\Delta R\}+\{\Delta R_0\} \tag{2}$$

Here, $\{\Delta u\}$ is nodal displacement incremental array, $\{\Delta R\}$ node load incremental array and $\{\Delta R_0\}$ the equivalent node load array of initial strain.

2.1.2 Constitutive model

In the static analysis, the soil materials and overburden layer materials are regarded as elastic-plastic object, where the Duncan-Zhang's nonlinear-elastic (E-B) model (Duncan & Chang, 1970) is carried out. The concrete and bedrock are regarded as linear elastic objects, which obey the generalized Hook's law.

2.2 Dynamic analysis model

2.2.1 Control equation of FEM

The dynamic equilibrium equation can be described as follows after the calculation domain is discretized by the finite element method

$$[M]\{\ddot{\delta}\}+[C]\{\dot{\delta}\}+[K]\{\delta\}=\{F(t)\} \tag{3}$$

Here, $\delta,\dot{\delta},\ddot{\delta}$ are nodal displacement, nodal velocity and nodal acceleration respectively, $\{F(t)\}$ nodal dynamic loading, $[M]$ mass matrix, $[K]$ strength degree matrix, $[C]$ damping matrix obtained by $[C]=\lambda\omega[M]+\lambda[K]/\omega$, where ω is the first mode for vibration frequency and λ the damping ratio.

Eq.(3) can be gradually integral solved by the Wilson linear acceleration method ($\text{Wilson}-\theta$ method), which can instead by the following equations.

$$[\bar{K}]\{\delta\}_t=\{\bar{F}\}_t \tag{4}$$

$$[\bar{K}]=[K]+\frac{6[M]}{dt^2}+\frac{3[C]}{dt} \tag{5}$$

$$\{\bar{F}\}_t=\{F\}_t+[M]\{A\}_t+[C]\{B\}_t \tag{6}$$

$$\{A\}_t=\frac{6}{dt^2}\{\delta\}_{t-dt}+\frac{6}{dt}\{\dot{\delta}\}_{t-dt}+2\{\ddot{\delta}\}_{t-dt} \tag{7}$$

$$\{B\}_t=\frac{3}{dt}\{\delta\}_{t-dt}+2\{\dot{\delta}\}_{t-dt}+\frac{dt}{2}\{\ddot{\delta}\}_{t-dt} \tag{8}$$

$$\{\dot{\delta}\}_t=\frac{3}{dt}\{\delta\}_t-\{B\}_t \tag{9}$$

$$\{\ddot{\delta}\}_t=\frac{6}{dt^2}\{\delta\}_t-\{A\}_t \tag{10}$$

The iterative method is adopted with the consideration of the variations of dynamic shear modulus and damping ratio in calculation process after the average dynamic shear strain change. In the iteration process the convergence criterion is

$$\left|\frac{G^{i-1} - G^{i}}{G^{i}}\right| < 0.1 \tag{11}$$

Here, G^i is the new shear modulus and G^{i-1} the last shear modulus. The maximum iteration times is taken for 6 to lest the iterative calculation into dead circulation.

2.2.2 Constitutive model

In the dynamic analysis (Mejia et al., 1981a, 1982b; Shen et al., 2006a, 2010b), the equivalent nonlinear viscoelastic model is applied in the dynamic calculation and analysis, the soil materials and the overburden layers are assumed to be viscoelastic bodies. They reflect nonlinearity and hysteretic nature of the dynamic stress-strain relation by using of the equivalent dynamic shear modulus G and equivalent damping ratio λ, which can be expressed as the function between the equivalent shear modulus and damping ratio and the dynamic shear strain amplitude. The key point of this model is to confirm the relationship between the maximum dynamic shear modulus $G_{\max}$ and the average effective stress σ_0'. In this paper, the Hardin-Drnevich model (Hardin & Denevich, 1972a, 1972b) is used, where the dynamic shear modulus and damping ratio can be calculated as follows

$$G = \frac{G_{\max}}{1 + \gamma/\gamma_{\gamma}} \tag{12}$$

$$\lambda = \lambda_{\max} \frac{\gamma/\gamma_{\gamma}}{1 + \gamma/\gamma_{\gamma}} \tag{13}$$

$$G_{\max} = K' p_a \left(\frac{\sigma_0'}{p_a} \right)^n \tag{14}$$

Where σ_0' is average effective stress, P_a atmospheric pressure, K' modulus coefficient and n is modulus exponent. Here, $G_{\max}$, σ_0' and p_a have the same dimension. The relation curves of dynamic shear strain γ to dynamic shear modulus and damping ratio can be obtained by dynamic tri-axial test (Xenaki & Athanasopoulos, 2008; Zegha & Abdel-Ghaffar, 2009). When having dynamic calculation, the related relation curve can be inputted directly, and then interpolate and extent the values according to the strain values for the calculation.

The dynamic model of contact face elements referred to the test results of Hohai University, China. The relationship between shear rigidity K and dynamic shear strain γ is described as follow

$$K = \frac{K_{\max}}{1 + \dfrac{MK_{\max}}{\tau_f}\gamma} \tag{15}$$

The shear rigidity K and damping ratio λ have the following relationship:

$$\lambda = \left(1 + \frac{K}{K_{max}}\right)\lambda_{max} \tag{16}$$

$$K_{max} = C\sigma_n^{0.7}\ ,\ \tau_f = \sigma_n \tan\delta \tag{17}$$

Where σ_n is the normal stress on contact face, δ the angle of internal friction on contact face, λ_{max} the maximum damping ratio and M , C are the test parameters.

2.3 Stability evaluation

The stress of element is used to calculate the factor of safety and evaluate its stability, thus the position of slide surface can be obtained. According to the Mohr-Coulomb criterion, the regions whose local safety factors are less than 1.0 are combined together to obtain the most dangerous multi-slip surface. The factor of safety on the surface is defined as the ratio of anti-sliding force to sliding force, and then the relationship between factors of safety and time can be obtained during the earthquake period. In this way, the anti-sliding stability is evaluated by stress when considering the unstable duration of earthquake.

During an earthquake, the dynamic strength of rockfill material is not always less than its static strength, at least we can assume it is equal to static strength. Taking compressive stress as positive and tensile stress as negative, after obtaining the static stress and dynamic stress under an earthquake by finite element method, the local factor of safety of an element can be calculated by the following formula

$$LF_s = \frac{2c\cos\varphi - (\sigma_1 + \sigma_3 - 2u_d)\sin\varphi}{\sigma_1 - \sigma_3} \tag{18}$$

Where c is cohesion, φ the angle of internal friction and σ_1 , σ_3 are the maximum and minimum principal stresses.

The u_d can be calculated by the formula as follow

$$u_d = \frac{(1+\mu)(\sigma_{1d} + \sigma_{3d})}{3} \tag{19}$$

Here, μ is Poisson's ratio and σ_{1d} , σ_{3d} are dynamic stresses respectively.

3. Working behavior of Bikou earth-rock dam

According to the actual engineering conditions of the Bikou earth core rockfill dam, the 3-D non-linear FEM static and dynamic models of the dam are set up to calculate and analyze the stress and deformation characteristics of dam, including the deformation and stress of the dam shell, core wall, two cut-off walls, overburden layers, bedrock and so on. In order to consider the construction process, the process of filling the dam is divided into 15 levels to simulate, and the reservoir impoundment process is divided into 3 levels.

3.1 Initial stress field of dam

In order to calculate the earthquake responses of dam, the initial stress field of dam must be determined firstly. Thus, the 3-D finite element model is created to simulate the construction process of dam and reservoir impounding, and the initial stress field of dam before earthquake can be obtained.

3.1.1 Finite element model

According to the actual situation of the Bikou dam, the 3-D finite element model of the dam and its foundation is set up, which has simulated the geometry and material partition of the dam and its foundation. The control section super-element finite element mesh automatic division technique is adopted to generate the information of finite element model, and furthermore the super-element can be encrypted to form finite element. Based on requirements of structural characteristics, stage loading and forming the super-element grid, a set of control sections with 19 sections horizontal are selected. Then the dam and its foundation are discretized, and the supper-element grid is built up. After the super-element gird is discretized further more, the super-element gird is created whose total nodes are 31523 and total element numbers are 30087. The finite element mesh of the dam and its foundation and its core wall are shown as in Fig.3.

Select domain of the calculation model. 1) The distance from the upstream boundary to the dam axis of the river bed section is about 363.40m (approximately 1.0 times of dam height). 2) The distance from the downstream boundary to the dam toe of the river bed section is around 401.00m. 3) The distance from the left boundary to the right boundary is about 698.18m (approximately 2.0 times of dam height). 4) The vertical distance from the bottom boundary to the dam foundation surface is about 241.80 m.

The coordinate system is set up as following. The X-axis is along stream from up to down with zero at dam axis, the Y-axis is along dam axis from right bank to left bank, and the Z-axis is vertical corresponding to the elevation.

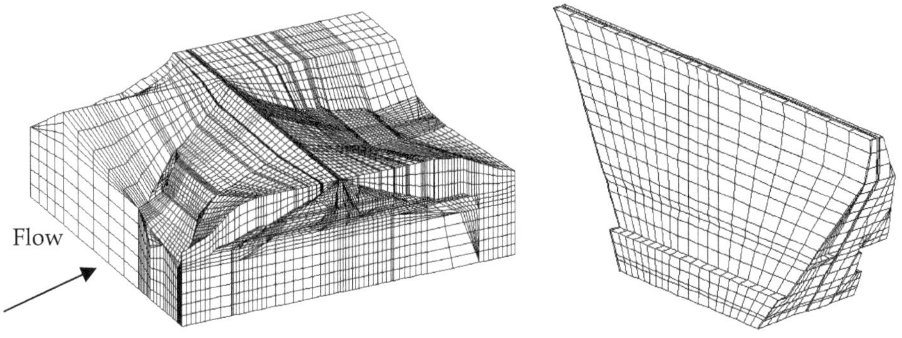

(a) dam body and foundation (b) core wall

Fig. 3. Finite element grid

3.1.2 Material parameters

The Duncan-Chang's nonlinear-elastic (E-B) model is carried out for the core wall and each district materials of the dam shell. According to indoor tests and engineering experience, the static material parameters of the dam are shown as in Table 1 and Table 2. And the parameters of the contact surface are shown as in Table 3.

Material types	$\gamma/kN \cdot m^{-3}$	E/kPa	ν
Concrete	24.5	1.8×10^7	0.20
Bedrock	26.0	1.0×10^7	0.25
Core wall settlings	24.0	8.0×10^3	0.35

Table 1. Parameters of linear elastic model

Material types		γ /kN·m^{-3}	c /kPa	K	n	R_f	K_b	m	$\Delta\varphi$ /°	K_{ur}	n_{ur}
core wall	W	20.7	100	190	0.42	0.81	120	0.86	2.1	450	0.30
	S	21.1	100	150	0.42	0.81	115	0.50	1.6	150	0.33
filter	W	22.1	0	750	0.40	0.75	460	0.80	4.8	600	0.33
	S	23.3	0	650	0.40	0.75	460	0.77	4.4	600	0.33
rockfill	W	21.2	0	1000	0.45	0.80	400	0.75	6.0	1500	0.33
	S	23.6	0	800	0.45	0.80	380	0.70	5.6	1500	0.35
ballast	W	21.4	0	600	0.45	0.80	350	0.69	4.1	860	0.30
	S	23.3	0	400	0.45	0.80	300	0.67	3.6	860	0.30
gravel (dam)	W	22.1	0	900	0.50	0.75	460	0.80	7.5	1700	0.25
	S	23.3	0	850	0.50	0.73	460	0.77	7.5	1700	0.23
gravel (mud)	W	22.5	0	500	0.45	0.74	300	0.79	3.3	900	0.30
	S	23.3	0	300	0.45	0.72	200	0.77	2.8	900	0.30
pebble	W	17.9	0	850	0.40	0.81	400	0.70	5.2	1300	0.33
	S	21.3	0	850	0.40	0.81	400	0.70	5.2	1300	0.33
gravel	S	23.9	0	700	0.50	0.70	320	0.78	7.6	1400	0.45

Table 2. Parameters of Duncan-Chang's model

Material	δ /°	K_s /×10[4]	n_s	R_{fs}	C_s /kPa m[-3]
Loam /concrete	14.0	2.30	0.69	0.75	0
Sandy gravel / concrete	32.2	4.00	0.65	0.75	0
Loam /Loam	17.3	1.80	0.70	0.88	68
Bedrock /concrete	35.0	6.00	0.20	0.80	0
Bedrock /the settlings	30.0	3.00	0.70	0.80	0

Table 3. Parameters of contact surface

3.1.3 Displacement and stress field of dam

By use of the stage loading method to simulate the construction process of dam and reservoir impounding, the displacement field and stress field of dam under normal water level condition are obtained. Here, as examples, the displacement and stress distribution on the maximum transverse cross section of dam (Y=210m) and dam axis maximum longitudinal section of dam (X=-10m) are shown as in Fig.4 ~Fig.6.

Here, the displacement along the coordinate axis direction is positive, that is, the horizontal displacement along the flow with the direction of upstream point to the downstream is positive, the positive horizontal displacement along the dam axis direction is from the right bank to the left, and the positive vertical displacement is from bottom to up. The unit of the displacement is "mm" in some figures. The compressive stress is positive, and the tensile stress is negative. The unit of the stress is "kPa" in some figures.

1. Dam shell

Under normal water level condition, the maximum horizontal displacement pointing to the upstream and the downstream respectively is -289mm and 1132mm.The maximum value appears in the lower crust, close to the slope. The dam horizontal displacement distribution along the dam axis direction shows that the horizontal displacement value along dam axis direction is small. And the maximum settlement occurs in the upstream dam shell which is near the dam axis in the middle of the river. The maximum vertical displacement of the dam body is 1273mm, accounts for about 1.2% of the maximum height.

The maximum first principal stress of the dam body is 2534kPa, the maximum second principal stress is 2445kPa and the maximum third principal stress is 1842kPa. The maximum principal stress of the upstream and downstream dam shell occurs at the bottom of dam near the dam axis. The shear stress level of most rock-fill units are less than 0.85, no the shear failure zone appears in the dam body. It indicates that the dam is stable.

2. Core wall

The maximum horizontal displacement pointing to the downstream of the loam core wall is 812mm.The maximum horizontal displacements of core wall along dam axis direction is 181mm, pointing to the left bank. And the maximum vertical displacement of core wall is 1413mm, appearing in the upstream of the core wall near the dam axis where the elevation is 657.00m at the deepest valley section. The maximum first principal stress of the core wall is 1839kPa, the maximum second principal stress is 1373kPa, and the maximum third principal stress is 1256kPa.

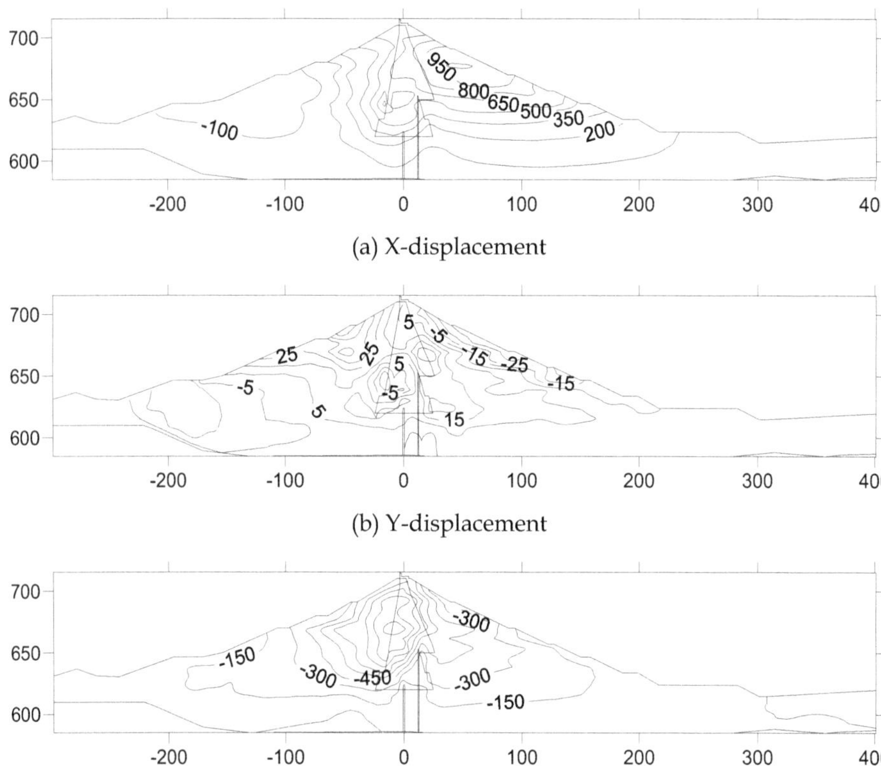

Fig. 4. Displacement distribution on section Y=210m (unit: mm)

Under normal water level condition, although core wall stress is smaller than depletion layer stress, but no tensile stress appeared, so the loam core wall won't produce pull crack. Meanwhile, for the strength and modulus of core wall is lower and the core wall is easy to adapt to deformation, core wall stress levels are lower. It indicates that the core wall is stable.

3. Cut-off wall

The maximum horizontal displacement of the first cut-off wall is 469mm, pointing to the downstream, and the vertical displacement is -112mm, and that of the second cut-off wall are 583mm and -129mm respectively. The maximum first principal stress of the first concrete cut-off wall is 9806kPa, the maximum third principal stress is 714kPa, and that of the second one are 16844kPa and 1094kPa respectively.

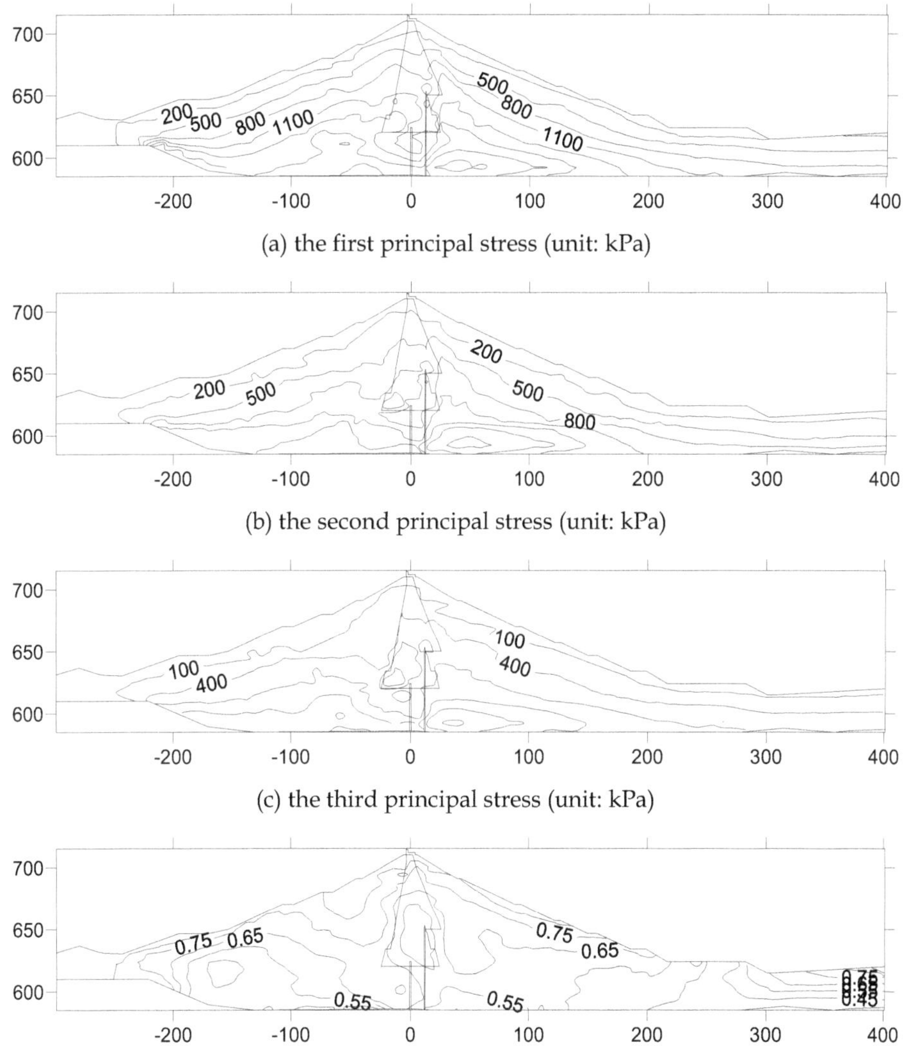

(d) stress level

Fig. 5. Stress distribution on section Y=210m

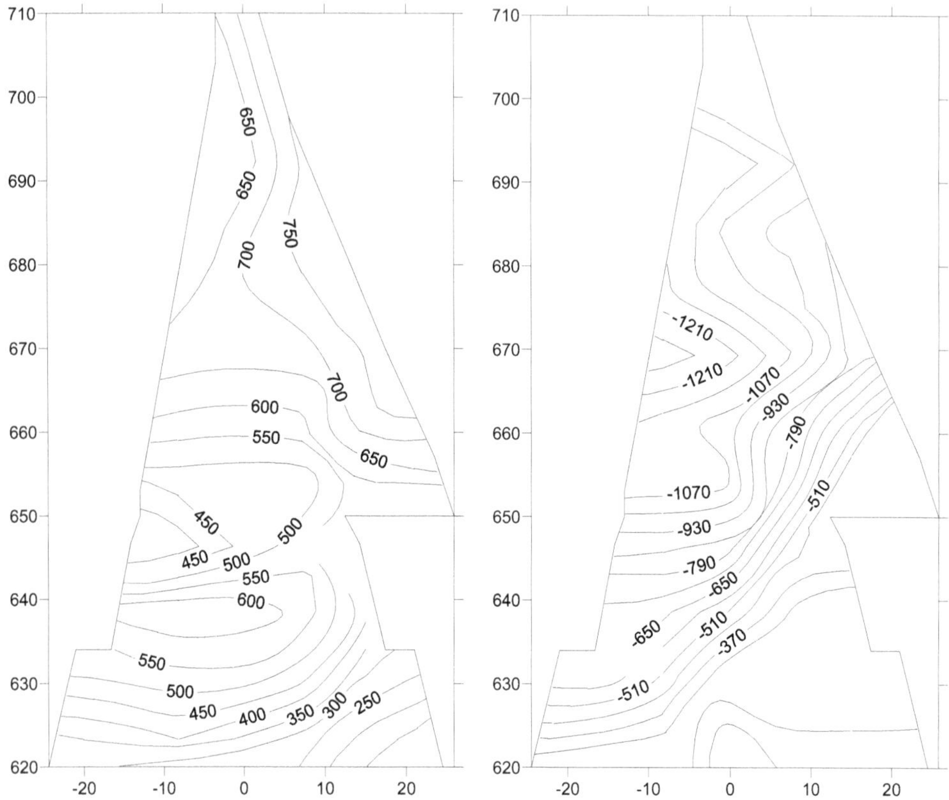

(a) X-displacement (b) Z-displacement

Fig. 6. Displacement distribution of core wall on section Y=210m (unit: mm)

3.2 Earthquake responses of dam

3.2.1 Finite element model

The finite element model what can be used to analyze earthquake responses of the dam is build, which is the same as the model for the static analysis.

3.2.2 Dynamic parameters and calculation condition

The time history analysis method is used and the equivalent nonlinear viscoelastic model is chosen as the constitutive model of soil, which assumes that the soil of dam body and foundation overburden are considered as viscoelasticity, and equivalent shear modulus G and equivalent damping ratio λ are applied to reflect the nonlinearity and hysteretic nature this two properties of soil dynamic stress-strain relation. The dynamic parameters of dam are shown as in Table 4 and Table 5.

Material parameters	Soil Material							
	loam	filter	rockfill	ballast	Gravel (dam)	Gravel (mud)	pebble	Gravel (foundation)
K	375	1875	2282	1147	2052	696	1701	1603
n	0.63	0.61	0.55	0.58	0.54	0.57	0.55	0.54

Table 4. Maximum dynamic shear modulus parameters of dam materials

Material parameters	Soil Material							
	loam	filter	rockfill	ballast	Gravel (dam)	Gravel (mud)	pebble	Gravel (foundation)
K_a	0.4653	1.5232	2.2814	1.6118	1.6653	1.4226	2.1450	1.6603
n_a	1.1883	1.2100	2.0871	1.5464	1.5553	1.2002	2.0322	1.4876
K_v	1.8252	2.0632	2.3221	1.7877	1.7623	1.5308	2.2895	1.7222
n_v	1.7119	1.9121	2.2007	1.6206	1.5989	1.3359	2.1659	1.5098

Table 5. Residual strain parameters of dam materials

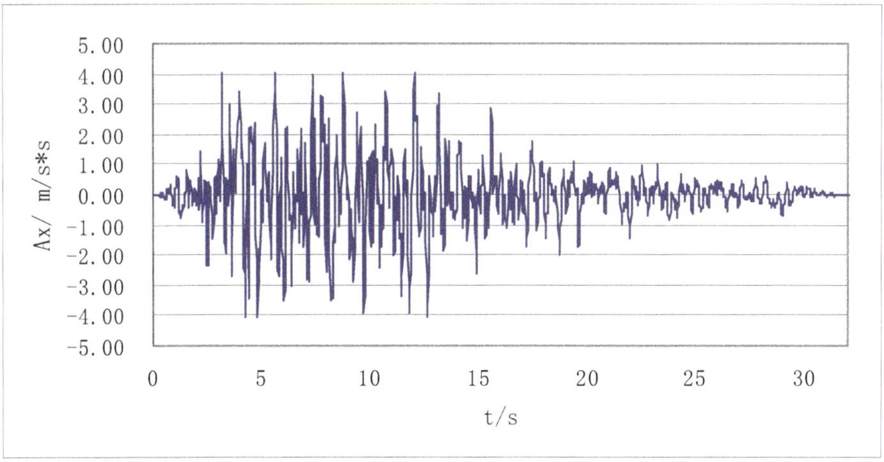

Fig. 7. The input curve of earthquake acceleration of bedrock in horizontal direction

Under the Wenchuan earthquake condition, in the dam site of Bikou hydropower station, the seismic intensity is comprehensively evaluated as degree 9. Correspondingly, the bedrock in the dam site has a peak level acceleration of 404cm/s^2 and its seismic acceleration curve of bedrock in horizontal direction is shown as Fig.7, which is in up-down stream direction and in dam axis direction. When calculating, the seismic acceleration value of vertical direction is regarded as two thirds of the horizontal one. Meanwhile, the dynamic water pressure is evaluated by the additional mass method called Westergaard method, that using equivalent additional mass instead of the dynamic water pressure to stack with the quality of the dam body during the earthquake.

3.2.3 Earthquake responses

Under the conditions of Wenchuan earthquake intensity, the earthquake response of the dam are analyzed, including the maximum acceleration response, displacement response and stress response of dam body, loam core and cut-off walls, and the earthquake induced permanent deformation (Serff et al., 1976; Taniguchi et al., 1983; Kuwano & Ishihara, 1988; Cascone & Rampello, 2003; Elia et al, 2011) of the dam is also obtained. The maximum values of earthquake responses analysis results are shown as in Table 8.The distributions of the dynamic analysis main features (Idriss et al., 19773; Ahmed-Waeil et al., 1990a, 1993b; Zhu et al., 2003) are shown as in Fig.8~Fig.14.

1. Acceleration response

Calculation results show that the natural vibration period of Bikou dam is about 0.74 seconds. The maximum acceleration response of the rockfill respectively is 4.08m/s^2 along the river direction, 4.43m/s^2 along the dam axis direction and 3.20m/s^2 along the elevation direction. For the core, the maximum acceleration response along the above three directions are 3.89m/s^2,4.14m/s^2 and 3.14m/s^2 respectively. Along the dam axis direction, the acceleration response of the dam is the greatest, followed by the acceleration response along the river direction and the acceleration response along the elevation direction is the smallest.

When earthquake comes, the upstream and downstream dam shell materials, which have a relatively long distance to the cut-off wall, are slightly affected. The corresponding horizontal acceleration response is basically proportional to the dam height and the maximum one appears near the upstream and the downstream slope surfaces which are close to the dam crest. The seismic acceleration response of a point is related to the vertical distance between the point and the bottom of the river valley; the height is greater, the corresponding seismic response is greater. As the slide slope of the right bank is slower than the left bank, the maximum horizontal acceleration response appears much closer to the right bank. From the middle of river bed to the both banks, the horizontal acceleration response decrease gradually which is related to the dam body type and boundary conditions. From the distribution of acceleration response on cross section, it can be seen that near the right bank, as the base of the downstream dam body is higher than the upstream and overburden layer exist at the bottom of the upstream dam body, the acceleration response of the upstream dam body is greater than the downstream dam body. The situation is opposite near the left bank, as deep overburden layer exists at the bottom of the downstream dam body and the downstream dam slope is much steeper than the upstream, the acceleration response of the downstream dam body is greater than the upstream dam body.

2. Displacement response

The maximum displacement response of the rock-fill is 25.92mm along the river direction, 16.99mm along the dam axis direction and 12.39mm along the elevation direction. For the loam core, the maximum acceleration response along the above three directions are 25.61mm, 16.77mm and 12.33mm respectively.

The displacement responses are not so great. Along the river direction, the displacement response of the dam is the greatest, followed by the displacement response along the dam axis direction. The smallest displacement response is along the elevation direction. The distribution of displacement response is almost the same with acceleration response. Under the seismic action, the displacement response is basically proportional to the dam height and the maximum one appears near the upstream and the downstream slope surfaces which are close to the dam crest. The seismic acceleration response of a point is related to the vertical distance between the point and the bottom of the river valley, the height is greater, corresponding seismic response is greater. The maximum horizontal displacement response appears near the dam crest above the deepest site of river valley. As overburden exists at the left bank, the maximum vertical displacement appears much closer to the left bank. From the center of river bed to the both banks, the displacement response decreases gradually which is related to the dam body type and boundary conditions. From the distribution of displacement response on cross section, the displacement response of the dam is not only related to the grade of upstream and downstream slope, but also the depth and distribution of overburden. Close to the right bank, as the base of the downstream dam body is higher than the upstream and overburden layer exists at the bottom of the upstream dam body, the displacement response of the upstream dam body is greater than the downstream dam body. The situation is opposite near the left bank. As deep overburden layer exists at the bottom of the downstream dam body and the downstream dam slope is much steeper than the upstream, the displacement response of the downstream dam body is greater than the upstream dam body.

3. Stress response

For the rockfill of dam, the maximum first principal stress is 425kPa, the maximum second principal stress is 352kPa and the maximum third principal stress is 203kPa. For the core, the maximum first principal stress is 265kPa, the dynamic tensile stress is 221kPa; the maximum second principal stress is 215kPa, the dynamic tensile stress is 178kPa; the maximum third principal stress is 192kPa, the dynamic tensile stress is 157kPa. For the first concrete cut-off wall, the maximum first principal stress is 3031kPa, the maximum second principal stress is compressive stress with a value of 591kPa and the maximum third principal stress is compressive stress with a value of 363kPa. For the second concrete cut-off wall, the maximum first principal stress is 4482kPa, the maximum second principal stress is compressive stress with a value of 683kPa and the maximum third principal stress is compressive stress with a value of 422kPa.

The stress response of rockfill and core along the river direction is the greatest, followed by the displacement response along the dam axis direction and the displacement response along the elevation direction is the smallest. The strong stress response of the core happens on the major river bed where the loam core connects to the both banks. The big and small principal stress response of the rock-fill become stronger with the increase of cover depth, thus the contour lines are parallel to the dam slope, and the maximum dynamic compressive stress and maximum dynamic tensile stress appear at the bottom of the dam near the deepest river valley. At the same elevation, the stress response of the dam body is much stronger than the core. As the great difference of deformation modulus between concrete and loam, stress concentration appears near the cut-off wall. Besides, at the downstream side of cut-off wall, dynamic stress response is very strong. From the distribution of stress response on longitudinal profile, it can be seen that the stress response at the bottom of the dam on the right bank is a bit stronger than on the left bank, the dynamic stress response is great at the connecting part between dam body and both bank slopes, especially near the place where the section of bank slope varies, stress concentration of strong dynamic stress response appears; and with the increase of peak acceleration, the stress response becomes stronger. So, more attentions should be paid to these weak areas.

Great dynamic stress appears at the top of the concrete cut-off wall, so does in the connection part between bedrock and the cut-off wall close to the bank slope. The stress response of concrete is proportional to the peak seismic acceleration, thus with the increase of peak acceleration, the dynamic tensile stress and compressive stress become greater. The maximum dynamic tensile stress of the cut-off wall is shown as in Table 6. Overall, the dynamic compressive stress of concrete is small and the dynamic tensile stress is relatively much greater, so the monitoring to the strength of the cut-off wall should be strengthened to ensure the safety operation of the dam.

During the earthquake, the maximum shear stresses of the dam are respectively 236kPa. On the cross section of the dam, the maximum dynamic shear stress and the dynamic shear stress become greater gradually from the upstream side and downstream side to the dam axis; but in the middle of the core, the dynamic shear stress decreases greatly, and at the bottom of dam, the dynamic shear stress response of rock-fill and transition material are very great. So, as the significant difference of the filling materials, the maximum shear stress responses are very significant in the connection parts of materials. The shear stress response of rock-fill is very great on the bank slope where the sections vary, and shear stress concentration appears in the local areas of where the sections vary at both banks; in the main river bed where the core connects to both bank, the shear stress is great.

4. Earthquake induced permanent deformation

The maximum permanent horizontal displacement is 74mm along the river direction, 47mm along the dam axis direction and the maximum permanent vertical displacement or settlement is 239mm. Taking no account of the thickness of the overburden on the dam foundation, the maximum dam height is 101.8m, then the permanent settlement induced by earthquake is about 0.23 percents of the dam height.

Due to the upstream water pressure, the earthquake induced permanent deformation of the core along the river orientation points to the downstream, so does the upstream dam shell materials near the dam crest. Along the dam axis direction, the earthquake induced permanent deformation is not so great and the maximum one appears near the upstream dam slope. It can be seen that in the dam abutments of both banks, the displacements of the dam body point to the center of the river valley; near the dam axis, the displacement and its variation gradient at the right bank are greater than the left bank. Affected by the concrete

Item			Value
maximum absolute acceleration response / m·s^{-2}	Rockfill	up-down stream direction	4.08
		dam axis direction	4.43
		vertical direction	3.20
	Core wall	up-down stream direction	3.89
		dam axis direction	4.14
		vertical direction	3.14
maximum displacement response /mm	Rockfill	up-down stream direction	25.92
		dam axis direction	16.99
		vertical direction	12.39
	Core wall	up-down stream direction	25.61
		dam axis direction	16.77
		vertical direction	12.23
maximum stress response of rockfill /kPa	Rockfill	1st principal stress	425/-422
		2nd principal stress	352/-347
		3rd principal stress	203/-188
	Core wall	1st principal stress	265/-221
		2nd principal stress	215/-178
		3rd principal stress	192/-157
	1st cut-off wall	1st principal stress	3031/-2908
		2nd principal stress	591/-589
		3rd principal stress	363/-316
	2nd cut-off wall	1st principal stress	4482/-3759
		2nd principal stress	683/-665
		3rd principal stress	422/-387
Earthquake induced permanent deformation/mm		stream direction (up/down stream)	74/-49
		dam axis direction (left /right bank)	47/-26
		vertical direction(settlement)	-239
maximum shear stress response / kPa			236

Table 6. Earthquake responses of Bikou dam by 3-D dynamic FEM

cut-off wall, the ultimate settlement of dam body below the elevation of 650m is small. And the settlement of the dam body above the river valley is basically uniform from the right side to the left, and the permanent deformation increases with the elevation increases. The maximum settlement appears near the downstream dam crest which is close to the dam axis. In general, the earthquake induced permanent deformations in both conditions are not great.

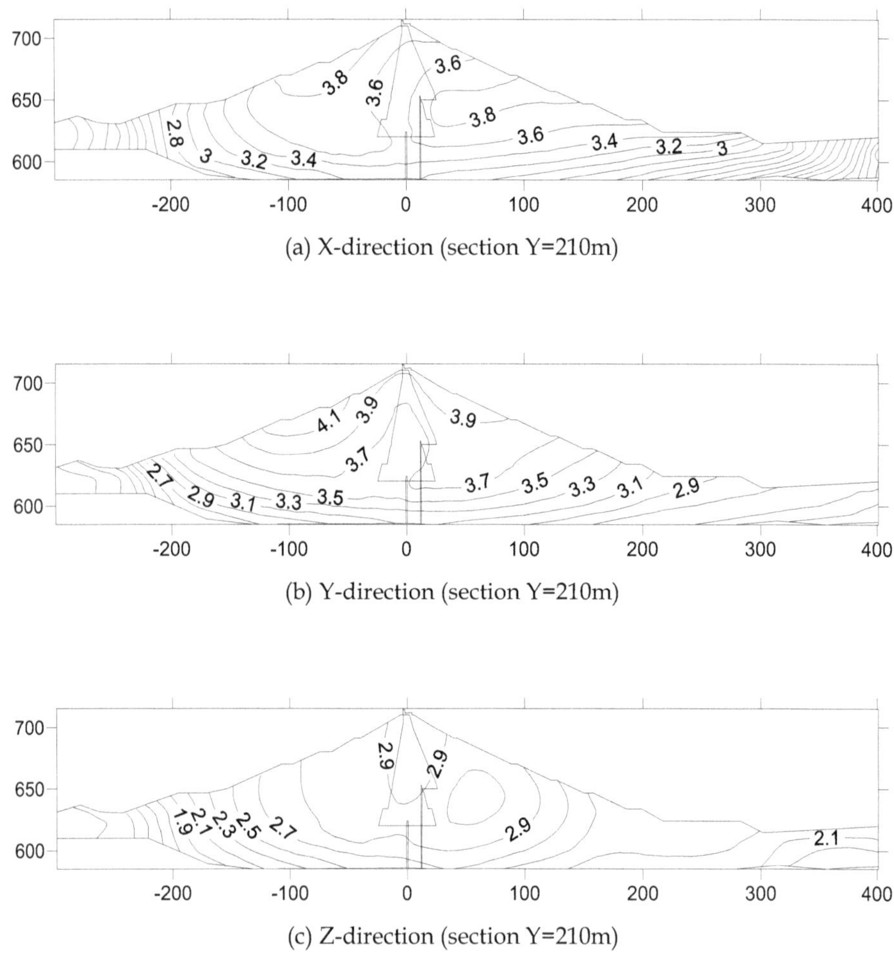

Fig. 8. Distribution of maximum absolute acceleration response (m/s^{-2})

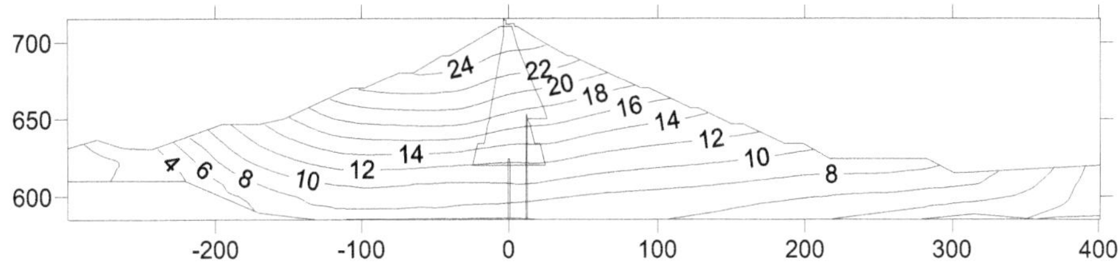

(a) X-displacement (section Y=210m)

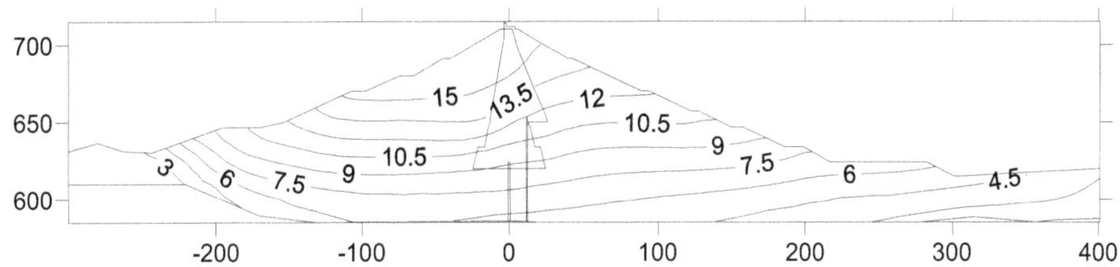

(b) Y-displacement (section Y=210m)

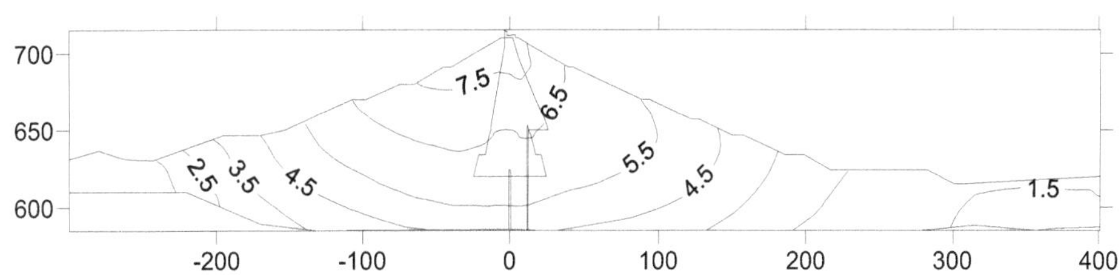

(c) Z-displacement (section Y=210m)

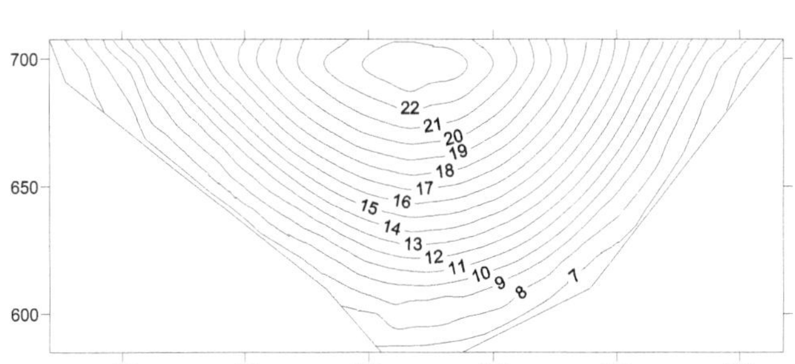

(d) X-displacement (section X=-10m)

Fig. 9. Distribution of maximum displacement response (mm)

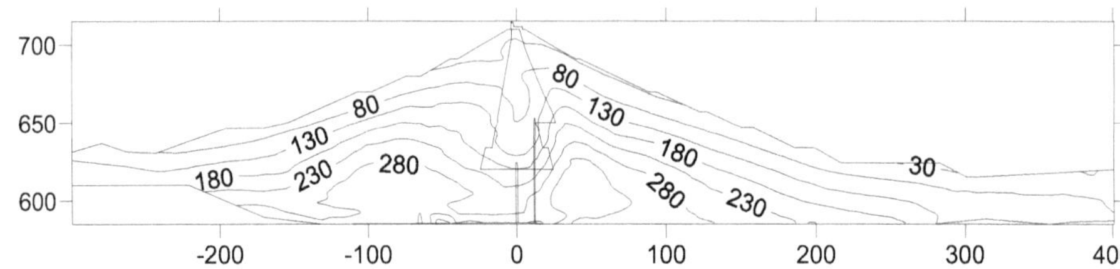

(a) the first principal stress

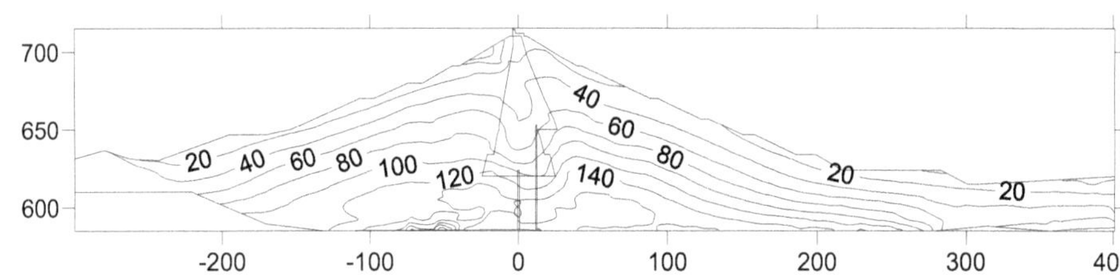

(b) the second principal stress

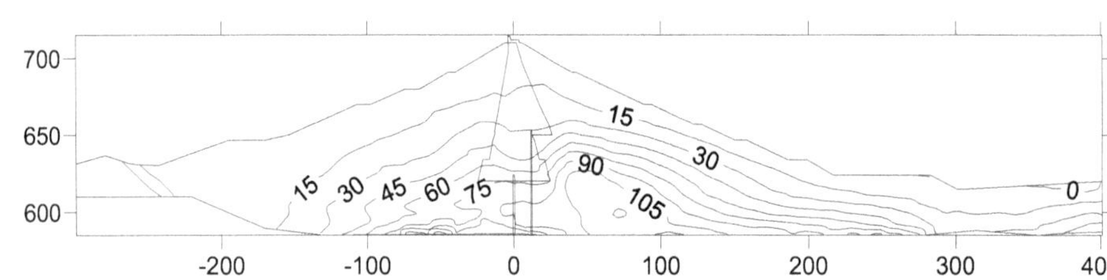

(c) the third principal stress

Fig. 10. Distribution of maximum stress response on section (Y=210m) (unit: kPa)

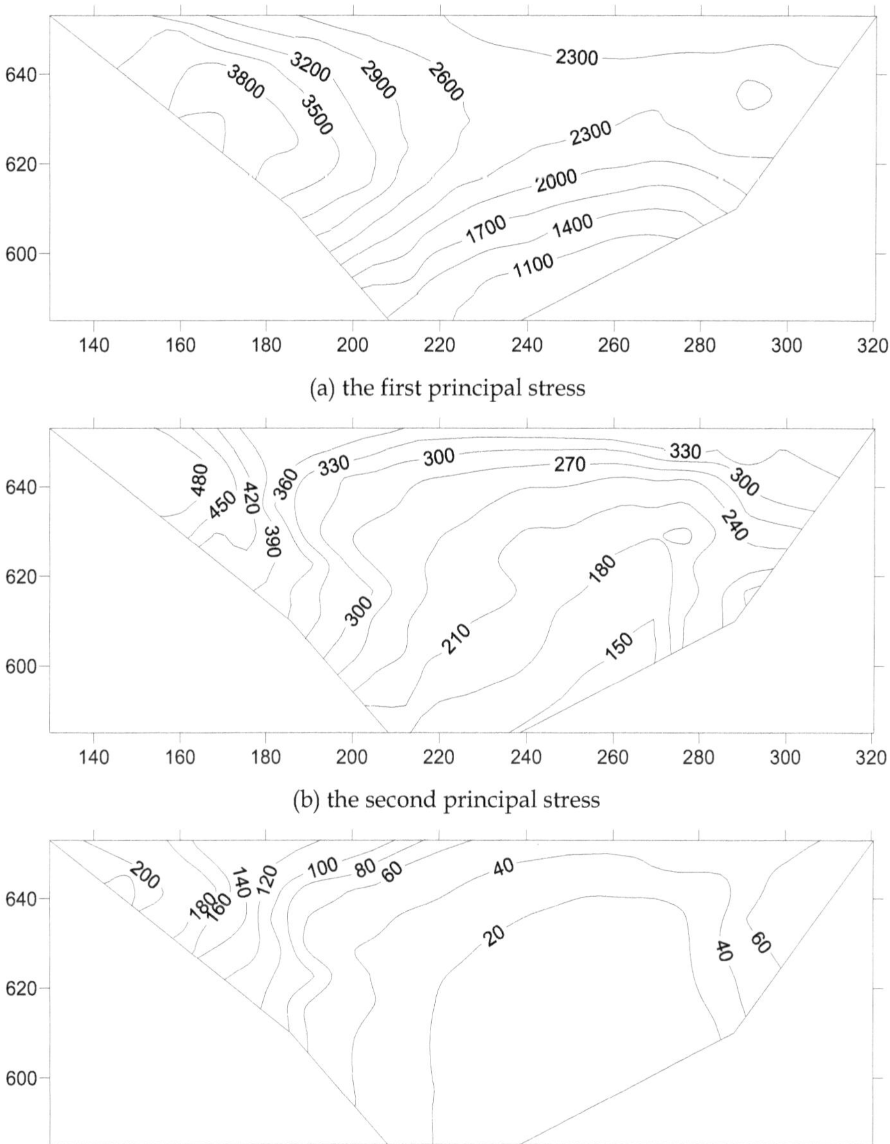

(a) the first principal stress

(b) the second principal stress

(c) the third principal stress

Fig. 11. Distribution of maximum stress response of the second cut-off wall (unit: kPa)

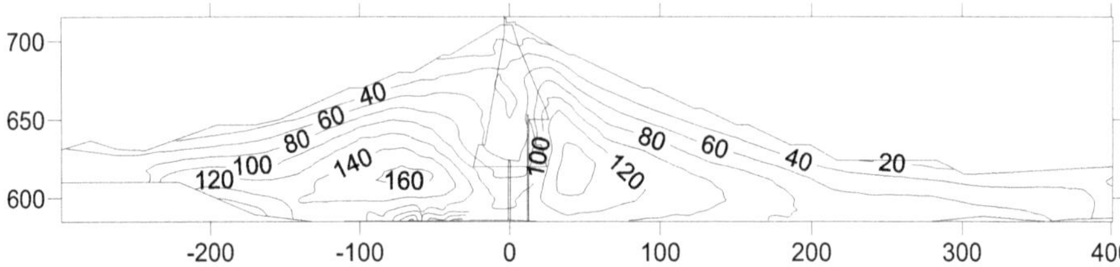

Fig. 12. Distribution of maximum dynamic shear on section (Y=210m) (unit: kPa)

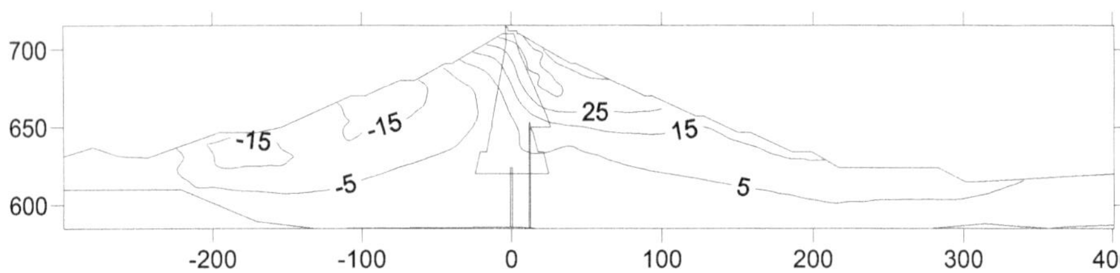

(a) X-direction

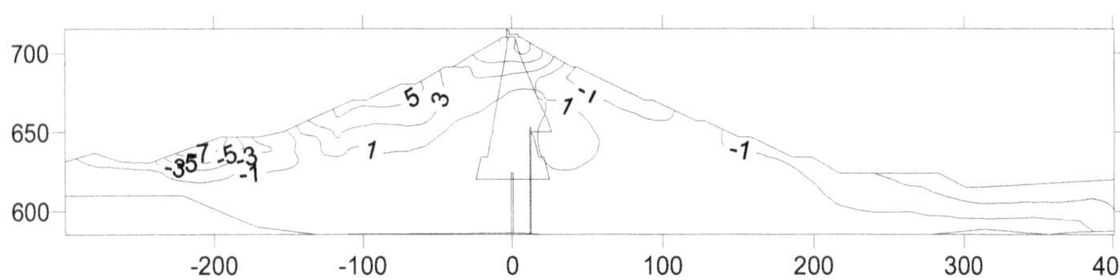

(b) Y-direction

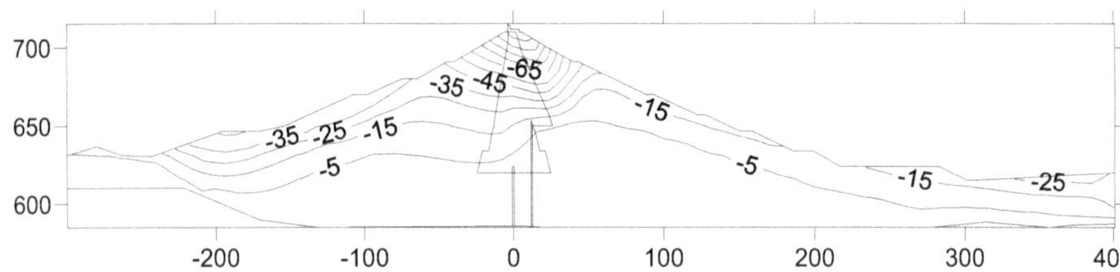

(c) Z-direction

Fig. 13. Distribution of earthquake induced permanent deformation on section(Y=210m) (mm)

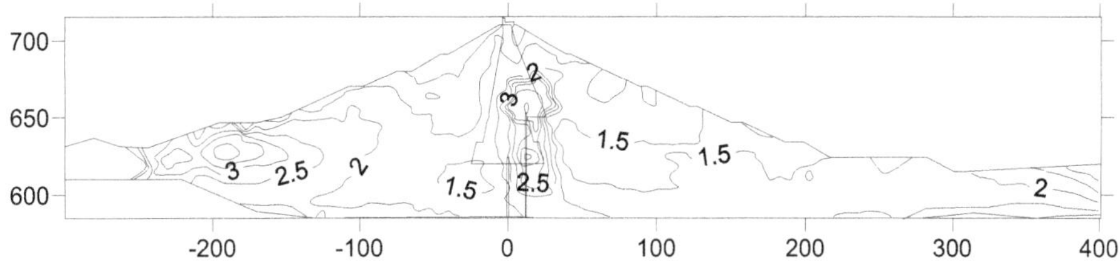

(a) t=5s

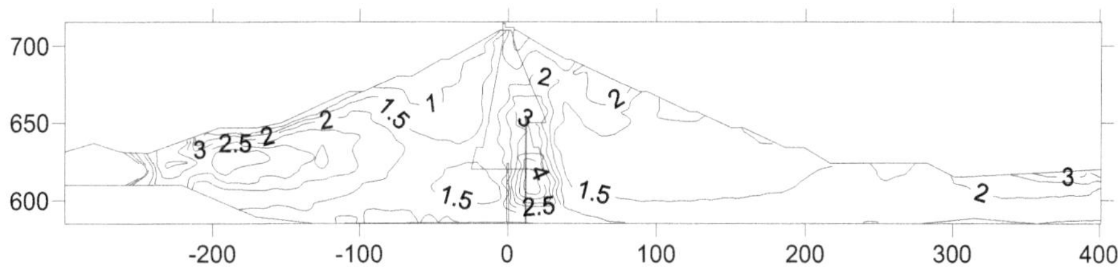

(b) t=15s

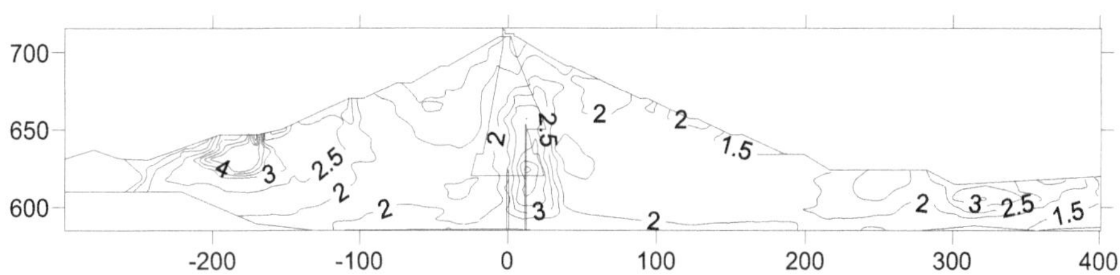

(c) t=25s

Fig. 14. Distribution of safety factors at different times during earthquake on section (Y=210m)

4. Earthquake-resistance safety evaluation

4.1 Comparing the static calculation results and monitoring data

In order to monitor the consolidation settlement in construction and operation, 5 collimating lines are set for the Bikou Hydropower Station soil core dam(Fig.18), which are respectively up dam 0-010.1m(708.00m elevation), under dam 0+007.8m(709m elevation), under dam 0+041.8m(691m elevation), under dam 0+093.0(670m elevation) and under dam 0+142.7(650m elevation). Only up dam section 0-010.0m got observed on December 17, 1975, and other sections started observation two years later after storage for the construction of crest parapet wall and downstream slope and drainage. Monitoring date of dam surface horizontal displacement and settlement before earthquake are shown as in Fig.16 and Fig.17.

Among all the monitoring sites, only section 0-010.0m started observation once storage, while other sections started observation relatively late. Therefore, displacement of these 5 sections cannot directly be compared with that of finite element calculation. As section 0-

010.0m got monitored earlier, its monitoring date should be compared with modified results of finite element calculation, which can be shown as in Fig.16 and Fig.17.

According to the settlement distribution (Fig.18) along dam axis direction of section 0-010.0m, settlements of monitoring points on crest are approximately proportional to their corresponding rock-fill thickness at the same place. Crest position where locates on the deepest valley floor has the biggest settlement, and settlement on right bank is larger than that on left bank. Settlement distribution discipline of finite element calculation is roughly the same with that of monitoring results. Additionally, calculation values are larger than monitoring data, since dam settlement started monitoring later than corresponding calculation situation, then displacement before monitoring was included in the calculation results.

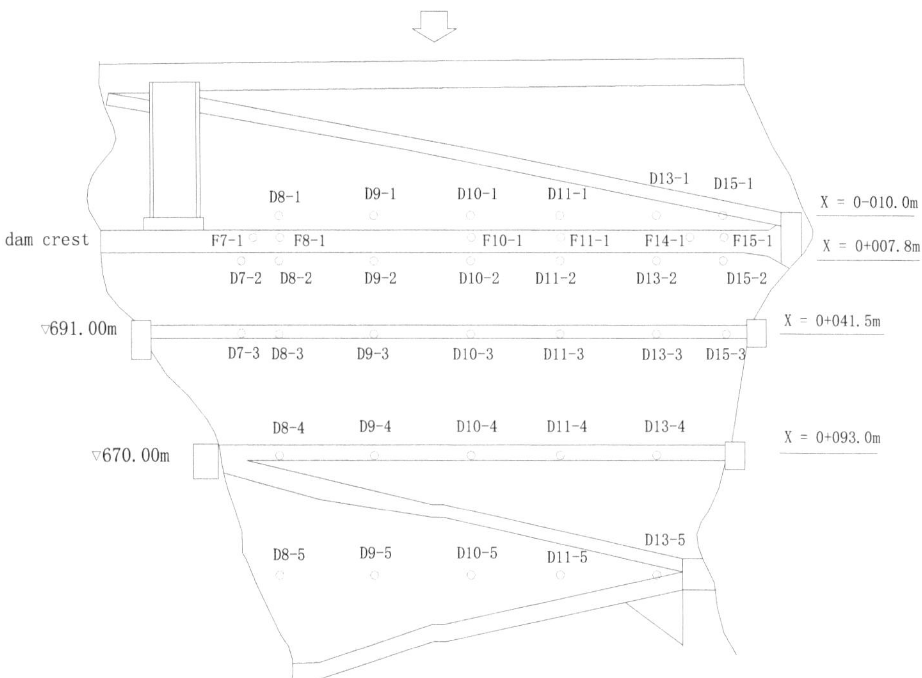

Fig. 15. Layout of monitoring point

Considering the starting observation time of section 0-010.0m, settlements of monitoring points are values after reservoir started storage. Thus, calculation results of finite element analysis should include the settlement values when dam body filling had been finished, that is, settlements incremental in the operation and completion periods. The calculation values are smaller than values of monitoring. It's because that calculation model only considers principal deformation and ignores the rheological of dam body. From the comparison of crest settlement above, calculation model and parameters are reasonable and fundamentally match the reality.

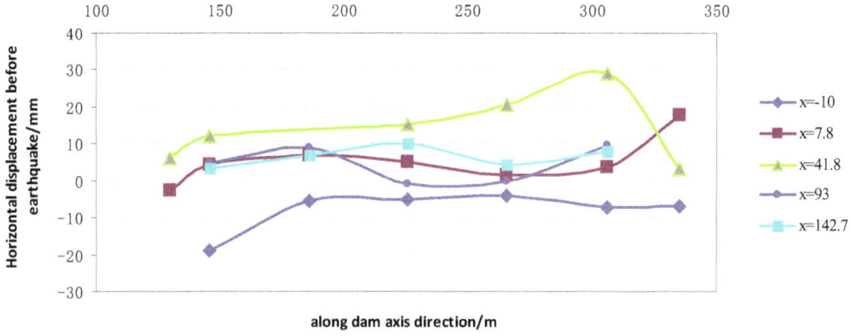

Fig. 16. Distribution of horizontal displacement measured value before the earthquake

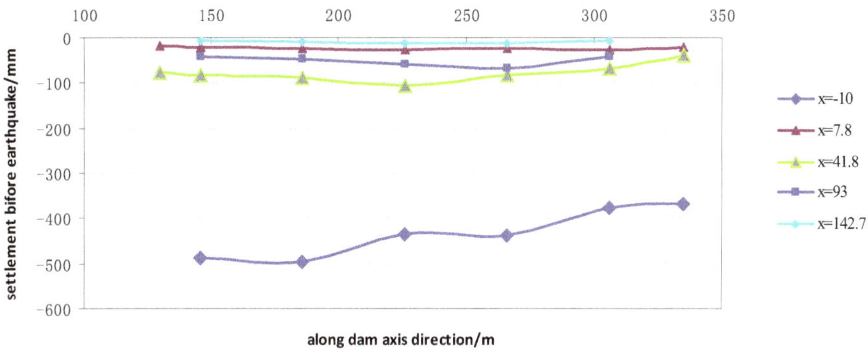

Fig. 17. Distribution of settlement measured value before the earthquake

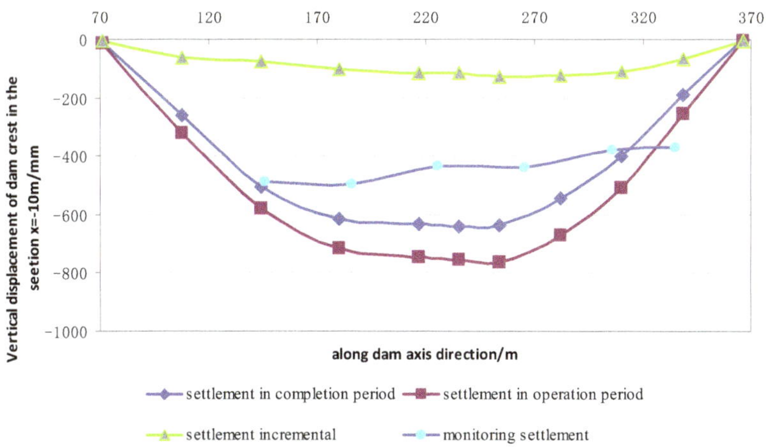

Fig. 18. Comparison of the settlement measured value and calculated value before the earthquake on the section (X=-10.0m)

4.2 Results of earthquake-resistance analysis

In the process of dynamic calculation, safety factor development of every element, excluding bed rock, concrete cut-off walls, wave wall and joints, has been completely recorded throughout earthquake. The safety factor is defined as the ratio of shear strength to shear stress (including both static and dynamic shear stress) of element potential failure surface. As it to analyze the seismic safety of upstream and downstream, the anti-slide stability coefficient is calculated on pseudo-static method. Sweden Slice Method is adopted for slope stability analysis, where both horizontal and vertical seismic actions are considered.

Calculation condition for slope anti-slide stability can be seen in Table 7. The smallest safety factors of upstream and downstream slopes are shown as in Table 8, and the potential sliding surfaces are shown as in Fig.19. As the compacted earth dam design specification (SL274-2001) says that the smallest anti-slide stability factor under seismic action should be no less than 1.15. Under the condition BK-S1, slope stability of Bikou dam cannot meet the present criterion. However, on the conclusion of Bikou station injuring survey under "5.12" Wenchuan earthquake and its initial analysis report, only some joints connecting crest and body to embankment were damaged and needed amending, yet the whole dam was safe in general. Under the seismic action, safety factor of Bikou dam does not satisfy demand, however, the dam did not suffer from sliding or slope instability during the "5.12" Wenchuan earthquake. So, it is suggested that smaller safety factors should be adopted. For example, a level 2 dam has a safety factor ranging from 1.05 to 1.10. In a word, the safety factor distribution suggests that the Bikou dam has some partial scopes where the factors are less than 1 near the crest and upstream and downstream slopes. But the scales are sporadic and will not lead the failure of dam slopes.

Condition	Dam slope	Peak acceleration /g	Upstream water level/m	Downstream water level/m
BK-S1	Upstream and downstream	0.404	707.00	617.02

Table 7. Condition for dam stability analysis

Condition	Peak acceleration/g	Dam slope	Fs
BK-S1	0.404	upstream	1.11
		downstream	1.19

Table 8. Results for dam stability analysis

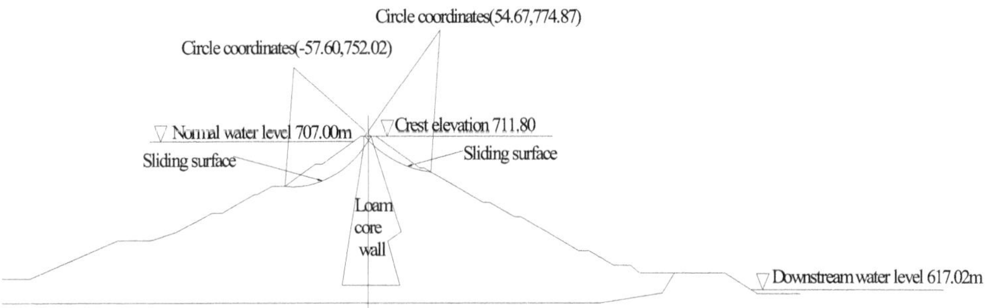

Fig. 19. Dangerous sliding surface position schemes of dam slopes (condition BK-S1)

5. Conclusions

In this paper, according to the actual engineering conditions of Bikou earth core rockfill dam, the 3-D non-linear FEM model of the dam is set up for calculating earthquake response by the dynamic time-history analysis method. By simulating the process of the filling of dam body and reservoir impounding, firstly the original static stress field of dam body is obtained under the normal water level. Then by the dynamic time-history analysis method, the earthquake responses of the dam, including acceleration response, displacement response, stress response and the earthquake induced permanent deformation of the dam are obtained for inputting the earthquake with peak acceleration of bedrock 404cm/s^2. Under the action of the seismic with peak acceleration 404cm/s^2, the acceleration response along the up-downstream direction is maximal with the value of 4.08m/s^2, the one along the dam axis is secondary with the value of 4.43m/s^2, and the one along vertical direction is the minimum with the value of 3.20m/s^2. The maximum displacement response of dam in the three directions are 25.92mm, 16.99mm, and 12.39mm respectively, and the maximum principle stress responses of rockfill body are 425kPa, 352kPa, and 203kPa respectively. The maximum earthquake induced permanent settlement is 239mm, about 0.23 percents of the dam height. It is shown that the earthquake responses of the dam are close to the recorded data. The theories and methods for analyzing the earthquake responses of the earth dam here are feasible, and the results are consistence with the real situation of post-earthquake.

6. Acknowledgement

This paper was supported by the Key Project of National Natural Science Foundation of China (Grant NO. 40930635) and Non-profit Projects Research of Ministry of Water Resources of China (Grant NO. 200901070). Grant NO 2009586012, entitled "State Key

Laboratory of Hydrology-Water Resources and Hydraulic Engineering" provided partial support for this work.

7. References

Ahmed-Waeil, M. E., & Ramana, V. G. (1993). Dynamic behaviour and seismic response of El Infiernillo dam. *Earthquake Engineering & Structural Dynamics*, 1993, Volume 22 Issue 8, pp. 665-684

Ahmed-Waeil, M. E., Ronald, F. S., Mohamed, F. S., et al. (1990). La Villita Dam Response During Five Earthquakes Including Permanent Deformation. *Journal of Geotechnical Engineering (ASCE)*, Vol.116, No.10, October 1990, pp. 1443-1462

Cao, Z. Z., Youd, T. L., and Yuan X. M. (2011). Gravelly soils that liquefied during 2008 Wenchuan, China earthquake, Ms=8.0, *Soil Dynamics and Earthquake Engineering*, 2011, Vol. 31, pp.1132-1143.

Cascone, E., & Rampello, S. (2003). Decoupled seismic analysis of an earth dam, *SOIL DYNAMICS AND EARTHQUAKE ENGINEERING*, 2003, Vol. 23, pp.349-365

Chen, H. Q., Xu, Z. P., and Lee, M. (2008). Wenchuan Earthquake and seismic safety of large dams, *Journal of Hydraulic Engineering*, 2008, Vol.39, No.10, pp.1158-1167

Chen, L. Y., & Talwani, P. (1998). Reservoir-induced seismicity in China, *PURE AND APPLIED GEOPHYSICS*, 1998, Vol. 153, No. 1, pp.133-149

Duncan, J. M., and Chang, C. Y. (1970). Nonlinear analysis of stress and strain in soil, *Journal of the Soil Mechanics and Foundation Division*, *ASCE*, 96(SM5), pp.1629-1653

Elia, G., Amorosi, A., Chan A, H. C., et al. (2011). Fully coupled dynamic analysis of an earth dam, *GEOTECHNIQUE*, 2011, Vol.61, pp.549-563

Gu, G. C. (1989). *Earthquake Engineering for Earth-rock Dams*, Hohai University Press, Nanjing , 1989

Hardin, B. O., & Denevich, V. P. (1972). Shear moduls and damping in soil: design equations and curves. *ASCE*, 98 (SM7), pp.667-92

Hardin, B. O., & Denevich, V. P. (1972). Shear modulus and damping in soil: measurements and parameter effects. *ASCE*, 98 (SM6), pp.603-624

Idriss, I. M., Lysmer, J., Huang, R., et al. (1973), A computer program for evaluating the seismic response of soil structures by variable damping finite element procedures, Berkeley, College of Engineering University of California.

Kenneth, G. (2011). Dermatology Aboard the USNS COMFORT: Disaster Relief Operations in Haiti After the 2010 Earthquake, *Dermatologic Clinics*, 2011, Vol. 29, No.1, pp.15-19

Kuwano, J., & Ishihara, K. (1988). Analysis of permanent deformation of earth dams due to earthquake, *Soils and Foundations*, Vol, 28(01), pp.41-55

Lin, P., Wang, R. K., and Li, Q. B. (2009). Effect Analysis of Structural Safety of Typical Large Dams in Wenchuan 8.0 Earthquake, *Chinese Journal of Rock Mechanics and Engineering*, 2009,Vol.28, No.6, pp. 1261~1268

Mejia, L. H. (1981). Three dimensional dynamic response analysis of earth dam, *Report No. UCB/EERC-81/15*. Berkeley, University of California

Mejia, L. H., Seed, H. B., & Lysmer, J. (1982). Dynamic analysis of earth dams in three dimensions, *Journal Geotechnical Engineering Division (ASCE)*, 1982, 108(GT12), pp. 1354-1376

Pan, R. Y. (2009). *Analysis of the recheck of the earthquake damage of Bikou Core Rockfill Dam*, Hohai University, Nanjing,2009

Pei, S. P., Su, J. R., Zhang, H. J., et al. (2010). Three-dimensional seismic velocity structure across the 2008 Wenchuan M_s 8.0 earthquake, Sichuan, China, *Tectonophysics*, 2010, Vol.491, pp.211-217

Serff, N., Seed, H. B., Makdisi, F. I., & Chang, C. K. (1976). Earthquake induced deformations of earth dams, *Report No. EERC/76-4*, Earthquake Engineering Research Center, University of California, Berkely, 1976

Shen, Z. Z., Wen, X.Y., and Lv, S.X. (2006). Analysis of earthquake responses for Jiudianxia concrete face rockfill dam, L. Berga et al eds, *Dams and Reservoirs, Societies and Environment in the 21st Century*, Published by Taylor & Francis/Balkema, London, 2006, Vol. 1, pp.925–930

Shen, Z. Z., Xu, L.Q., and Wang, W. (2010). Earthquake response of Xika concrete face rockfill dam by EFM, *Engineering, Science, Construction, and Operations in Challenging Environments, 2010 ASCE*, pp.463-472

Takewaki, I., Murakami, S., Fujita, K., et al. (2011). The 2011 off the Pacific coast of Tohoku earthquake and response of high-rise buildings under long-period ground motions, *Soil Dynamics and Earthquake Engineering*, 2011, (In press)

Talwani, P. (1997). On the nature of reservoir-induced seismicity, *PURE AND APPLIED* GEOPHYSICS, 1997, Vol. 150, pp.473-492

Taniguchi,E., Whiteman, R.V., & Warr, W. A. (1983). Prediction of earthquake induced deformation of earth dams, *Soils and Foundations*, 1983, Vol. 23(4)

Wang, W. X., Sun, W. K., and Jiang, Z. S. (2010). Comparison of fault models of the 2008 Wenchuan earthquake (M_s8.0) and spatial distributions of co-seismic deformations, *Tectonophysics*, 2010, Vol.491, pp.85-95

Xenaki, V. C., & Athanasopoulos, G. A. (2008). Dynamic properties and liquefaction resistance of two soil materials in an earthfill dam - Laboratory test results, *SOIL DYNAMICS AND EARTHQUAKE ENGINEERING*, 2008, Vol. 28, pp. 605-620

Yu, H. Y., Wang, D., Yang, Y.Q., et al. (2009). The preliminary analysis of strong ground motion records from the M_s 8.0 Wenchuan Earthquake. *Journal of Earthquake Engineering and Engineering Vibration*, 2009, Vol.29, No.1, pp.1~13

Zeghal, M., & Abdel-Ghaffar, A. M. (2009). Evaluation of the Nonlinear Seismic Response of an Earth Dam: Nonparametric System Identification, *JOURNAL OF EARTHQUAKE ENGINEERING*, 2009, Vol.2009. pp. 384-405

Zhou, J. P., Yang, Z. Y., Fan, J. X., et al. (2009). Seismic Damage Investigation on Large and Medium Sized Hydropower Projects in Wenchuan Earthquake Area. *Chinese Journal of Water Power*, 2009,Vol.35,No.5, pp.1-6

Zhu, S., Wen, S.Q., and Huang, Y. M. (2003). Deformation and Stress Calculation for A 200m High CFRD, *Journal of Hohai University (Natural Sciences)* ,2003, 31(6):631-634

7

Non-Linear Numerical Analysis of Earthquake-Induced Deformation of Earth-Fill Dams

Babak Ebrahimian
University of Tehran
Iran

1. Introduction

From January 2010 to March 2011, eight major earthquakes of magnitude 6.1 or greater, one of which was 9.0, have shaken the globe. These destructive events, such as January 2010 Haiti earthquake and March 2011 Tohoku Japan earthquake, have resulted in numerous impacts on different lifeline facilities such as dams (Takewaki, 2010; Eberhard et al., 2010; Rathje, 2010; Ashford et al., 2011; Harder et al., 2011; Matsumoto et al., 2011). During these recent earthquakes as well as the past strong ones, numerous earth-fill dams have failed and enormous loss of life and significant property damage have been caused (Sherard et al., 1963; Pinto, 1993; Ozkan, 1998; Krinitzsky & Hynes, 2002; Basudhar et al., 2010; Matsumoto et al., 2011). Therefore, evaluation of stability conditions of earth-fill dams is a major issue of concern in seismic areas. Earthquake-induced deformations in an earth-fill dam may lead to overtopping and consequently to severe losses in terms of property and human lives. Hence, an accurate evaluation of seismic stability of earth-fill dam should be employed to guarantee safer conditions for dams during earthquakes.

The performance of earth-fill dams, subjected to seismic action, can be evaluated through different approaches including force-based pseudo-static methods, displacement-based sliding block methods and dynamic analysis (Sherard, 1967; Seed et al., 1975). Since the 1971 San Fernando earthquake in California, understanding the effects of earthquake actions on dams has been significantly progressed (USCOLD, 1992). Gazetas (1987) discussed the historical developments of theoretical methods for estimating the seismic response of earth-fill dams to earthquake. He outlined important features, advantages and limitations of the methods. Pseudo-static approach, the most common method, is widely used in engineering practice to assess the seismic stability of earth-fill dams. This approach is quite simplistic by which the complex aspects of seismic behaviour are represented in terms of static forces and the dam stability is expressed in terms of an overall factor of safety. The response of dam to earthquake may be related to several factors such as dam geometry, mechanical properties of construction soil materials, distributions of pre-seismic stresses and pore water pressures inside the dam body, and input motion characteristics. Most of these factors are partially or totally neglected by the approaches traditionally adopted for assessing the seismic safety of earth-fill dams. For instance, some earthquake parameters such as frequency content and duration which significantly affect the soil response are neglected in the pseudo-static approach (Ambraseys, 1960). On the other hand, studying the seismic response of earth-fill dam is complex and, in general, requires dynamic analysis methods with different levels of sophistication in terms of proper problem formulation, characterization of material properties and modelling of soil stress–strain behaviour. In dynamic analysis methods using numerical simulation techniques, comprehensive analysis of earth-fill dam responses to dynamic loading is allowed. The development of geotechnical computation and numerical modelling offers interesting facilities for dam response analysis, considering complex issues such as soil non-linearity, evolution of pore water pressures and real earthquake records. In this regard, Prevost et al. (1985) presented 2D and 3D non-linear dynamic finite element (FE) analyses of an earth-fill dam, based on non-linear hysteretic analysis using multi-surface plasticity theory. Lacy & Prevost (1987) proposed a general and efficient numerical procedure for analyzing the seismic response of earth-fill dams. In their procedure, the dams were considered as non-linear two-phase systems. They outlined appropriate coupled dynamic field equations for the response of two-phase soil system. Abouseeda & Dakoulas (1998) studied the non-linear seismic behaviour in earth-fill dam-foundation interaction using boundary element (BE) and finite element (FE) methods. Chen & Harichandran (2001) studied the stochastic response of Santa Felicia earth-fill dam, in southern California, to spatially varying earthquake ground motion (SVEGM). They used SVEGM model in which the propagation of seismic waves is taken into account. Cascone & Rampello (2003) investigated the seismic stability of an earth-fill dam using decoupled displacement analysis. Ming & Li (2003) conducted a full coupled analysis of failure and considered the remediation of Lower San Fernando Dam. They used a critical state model, incorporating the concept of state dependent dilatancy for describing the soil behaviour over full loading ranges. Adalier & Sharp (2004) studied the seismic behaviour and remediation of an embankment on a liquefiable foundation. Papalou & Bielak (2004) studied the non-linear seismic response of earth-fill dams with canyon interaction. In their developed FE-based method, the dam was idealized as a shear beam and the surrounding medium as a half-space. The dam's non-linearity was considered using multi-yield surface plasticity theory. Ebrahimian & Vafaeian (2005) considered the seismic response of earth-fill dams during earthquake using 2D full non-linear dynamic analysis. They used finite difference (FD) method and adopted the Mohr-Coulomb elastic-perfectly plastic constitutive model to describe the stress-strain relation of the soil. They focused on the seismic behaviour of very high earth-fill dams. Wang et al. (2006) presented the dynamic analyses in which a non-linear, effective-stress-based soil model is employed. They used bounding surface hypoplasticity model for sand and implemented the model into a 2D finite difference (FD) program. The advantages of the proposed non-linear approach, conducted for a rock-fill dam, were illustrated by comparing the obtained results with those of equivalent linear approach. The model's capability was demonstrated by evaluating the seismic performance of an earth-fill dam. Siyahi & Arslan (2008) carried out the transient dynamic time history FE simulations to investigate the performance of earth-fill dams under seismic excitation. Then, they studied different failure modes of earth-fill dams as the earthquake aftermath. Sica et al. (2008) studied the effect of loading history on the seismic response of earth-fill dams. They considered the static and seismic behaviours of a real case-history using coupled dynamic approach. The approach was solved numerically by FE method. Rampello et al. (2009) studied the response of an earth-fill dam to seismic loading using the displacement-based analysis and the FE effective-stress dynamic analysis. The FE analysis was carried out using a constitutive model which was capable to reproduce the soil non-linearity and calibrated against laboratory measurements. They also investigated the effects of assumed input motion and bedrock depth on the seismic response of earth-fill dam. Ebrahimian (2009) presented a numerical modelling of seismic behaviour of an earth-fill dam rested on liquefiable foundation. The numerical simulation was carried out using effective-stress-based, full coupled non-linear dynamic analysis. In this regard, Finn-Byrne model with extended Masing rules was used to model the pore water pressure generation in the liquefied soils. Ebrahimian (2011) investigated the dynamic behavior of earth dams by using a full non-linear dynamic finite difference analysis. The effects of input motion characteristics and dam reservoir condition on the dynamic response of earth dams were identified in this study. For this purpose, three real earthquake records with different levels and PGAs were used as the input motions.

In many parts of the world, the repetition of medium–strong intensity earthquake ground motions at brief intervals of time has been observed. The design philosophies for dams in seismic regions are based on multi-level design approaches, which take into account more than a single damageability limit state. According to these approaches, a sequence of seismic actions may produce important consequences on the dam safety. In fact, dams have been among the first structures that have been designed systematically against different earthquake levels. Since 1989, the ICOLD guidelines have introduced several levels of seismic loading, namely the Operating Basis Earthquake (OBE), Maximum Credible Earthquake (MCE), Maximum Design Earthquake (MDE) and Safety Evaluation Earthquake (SEE). However, the terms MDE or SEE are used as substitutes for the MCE. In order to analyze the behaviour of dams for specified levels of seismic hazard, several requirements should be considered. The seismic input and performance levels associated with serviceability, damage control, and collapse prevention are also defined. A thorough review about the different earthquake levels for dam design has been given in Wieland (2008). Amadio et al. (2003) analyzed the effects of repeated earthquake ground motions on the response of single-degree-of-freedom systems (SDOF) with non-linear behaviour. Accordingly, a comparison study was performed to investigate the effect of a single seismic event on the originally non-damaged system for different hysteretic models in terms of pseudo-acceleration response spectra and damage parameters. They showed that the elastic–perfect plastic system is the most vulnerable one under repeated earthquake ground motions. Moustafa & Takewaki (2010) modelled ground motions of multiple sequences that produce the maximum damage in the structure. It was shown that critical repeated acceleration sequences produce larger structural damage compared to single critical earthquakes. Afterwards, Moustafa (2011) developed a new framework to model the design earthquake loads for inelastic structures. New measures of the structure performance that were based on energy concepts and damage indices were introduced in his paper.

Concerning the seismic-resistant design of dams, several international standards have been developed by scientific communities in the past 25 years. However, only few countries have their own guidelines and regulations for seismic design of dams. Therefore, the ICOLD Bulletins and the local seismic building codes (e.g., Eurocode 8) are used as references. In brief, other famous international codes are USCOLD (United States Committee on Large Dams), US Army Corps of Engineers (USACE), ANCOLD, (Australian National Committee on Large Dams), IITK-GSDMA Guidelines for Seismic Design of Earth Dams and Embankments and Canadian Dam Association Guidelines for Dam Safety.

In this chapter, a numerical study of seismic behaviour of earth-fill dams overlaying bedrock subjected to real earthquake records is presented. For this purpose, full non-linear dynamic finite difference (FD) analysis is employed incorporating a simple elastic perfectly plastic constitutive model and Rayleigh damping. The former is used to describe the stress-strain response of the soil and the latter to increase the hysteretic damping level. The effect of nonlinear soil behaviour is then considered in the analysis from the very beginning of earthquake loading. In order to precisely explain the soil response under general cyclic loading, Masing rules (Masing, 1926) are implemented into the constitutive model. Soil stiffness and hysteretic damping change with loading history. Firstly, the procedures of calibrating the constructed numerical models with centrifuge test data as well as real case history are presented and explained. Moreover, some important aspects of model calibration are discussed. Long Valley earth-fill dam, subjected to the 1980 Mammoth Lake earthquake, is analyzed for the real case history and the obtained numerical results are compared with the real ones, measured at the site in both time and frequency domains. The computed values show relative good agreements with the measured ones. It is shown that the Masing rules, combined with the simple elastic-plastic model, offer reasonable numerical predictions. A comprehensive parametric study is also carried out to identify the effects of dam height, input motion characteristics, soil behaviour and strength of shell materials on the seismic response of earth-fill dams. It is demonstrated that the fundamental aspects of seismic behaviour of earth-fill dams can accurately be captured by the current numerical procedure. It should be mentioned that this study does not consider the fluid-skeleton interaction, which may have significant effects on the seismic response of earth-fill dams.

2. Methodology

2.1 Numerical modeling procedure

Here, numerical analysis is conducted using FLAC program, based on a continuum finite difference discretization applying Lagrangian approach (Itasca, 2004). Every derivative in the set of governing equations is directly replaced by algebraic expression written in terms of field variables (e.g., stress or displacement) at discrete point in space. Regarding dynamic analysis, explicit finite difference scheme is applied to solve the full equation of motion using the lumped grid point masses derived from the real density of surrounding zone. The calculation sequence first invokes the equations of motion for deriving new velocities and displacements from stresses and forces; then, strain rates are derived from velocities, and new stresses from strain rates. Every cycle around the loop corresponds to one time step. Each box updates all grid variables from known values which are fixed over the time step being executed (Fig. 1).

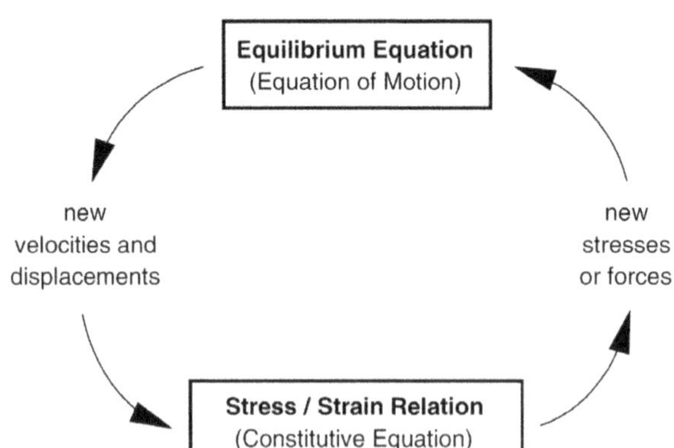

Fig. 1. Basic explicit calculation cycle (Itasca, 2004)

The equation of motion, in the simplest form, relates the acceleration ($d\dot{u}/dt$) of a mass (m) to the applied force (F) which may vary with time. Newton's law of motion for the mass-spring system is:

$$m\frac{d\dot{u}}{dt} = F \tag{1}$$

In a continuous solid body, Eq. (1) is generalized as follows:

$$\rho\frac{\partial \dot{u}_i}{\partial t} = \frac{\partial \sigma_{ij}}{\partial x_j} + \rho g_i \tag{2}$$

where, ρ = mass density; t = time; x_j = components of coordinate vector; g_i = components of gravitational acceleration (body forces); σ_{ij} = components of stress tensor; i = components in a Cartesian coordinate frame.

For problem analysis, the strain rate tensor and rotation rate tensor, having the velocity gradients, are calculated by the following equations:

$$e_{ij} = \frac{1}{2}\left[\frac{\partial \dot{u}_i}{\partial x_j} + \frac{\partial \dot{u}_j}{\partial x_i}\right] \tag{3}$$

$$\omega_{ij} = \frac{1}{2}\left[\frac{\partial \dot{u}_i}{\partial x_j} - \frac{\partial \dot{u}_j}{\partial x_i}\right] \tag{4}$$

where, e_{ij} = components of strain rate; ω_{ij} = components of rotation rate; $\dot{u}_i$ = components of velocity.

The specific mechanical relationship is used in order to obtain the stress tensor as below:

$$\sigma_{ij} = M\left(\sigma_{ij}, \dot{e}_{ij}, \kappa\right) \tag{5}$$

where, M = specific rule of behaviour; κ = history parameters (based on the specific rules which may or may not exist).

The problem selected here is the simplified representation of typical earth-fill dam geometry. The dam section is a symmetric zone section with central clay core rested on bedrock,

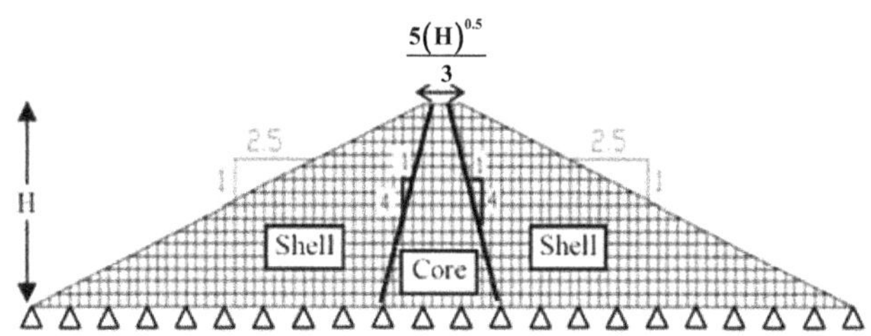

Fig. 2. Geometry of dam

as shown in Fig. 2. Five earth-fill dam cross sections of different heights (H = 40, 80, 120, 200 and 280 *m*) are analyzed.

2.1.1 Constitutive model

Mohr-Coulomb constitutive relation is used to model the behaviour of soil. The failure envelope for this model corresponds to a Mohr-Coulomb criterion (shear yield function) with tension cutoff (tensile yield function). Stress-strain relationship is linear elastic-perfectly plastic. Linear behaviour is defined by elastic shear and bulk modulus. While, plastic behaviour is determined by the angle of internal friction and cohesion of the soil. Shear modulus of sandy shell materials is calculated from the following formula (Kokusho & Esashi, 1986):

$$G_{\max} = 8400\frac{(2.17-e)^2}{1+e}(\sigma'_m)^{0.6} \tag{6}$$

where, G_{max} = maximum (small strain) shear modulus in kPa; e = void ratio; σ'_m = mean effective confining stress in kPa; Poisson's ratio is considered as 0.3 for the shell materials.

Shear modulus of clayey core materials is calculated by the below formula (Hardin & Black, 1968):

$$G_{\max} = 3270\frac{(2.973-e)^2}{1+e}(\sigma'_m)^{0.5} \tag{7}$$

Poisson's ratio for the core materials is taken as 0.45.

Here, the basic elastic-perfectly plastic model is modified in order to better fit with the curves of shear modulus and damping ratio derived from the experimental data. This modified model can predict the seismic behaviour and the associated deformations of earth-fill dams. Masing behaviour is implemented into FLAC via FISH subroutine (Itasca, 2004) in order to represent more accurately the non-linear stress-strain behaviour of soil that follows the actual stress-strain path during cyclic loading. Masing model consists of a backbone curve and several rules that describe the unload-reload behaviour of soil and the cyclic modulus degradation. Backbone curve can be constructed by the modulus reduction curves coupled with the small strain modulus (G_{max}). Unload-reload rules can similarly be formulated to reproduce the hysteretic damping values expected from the standard curves of damping ratio versus shear strain (e.g., Seed et al., 1986; Vucetic & Dobry, 1991). These formulations are described later.

In this study, shear modulus and damping ratio curves, proposed by Seed et al. (1986) for sandy soils and Vucetic & Dobry (1991) for clayey soils, are adopted as the references. The geotechnical properties of earth-fill dam materials, used in the analyses, are presented in Table 1.

Region	γ_{wet} (kN/m^3)	γ_{sat} (kN/m^3)	υ	Porosity (n)	C (kPa)	φ degree	K (cm/s)
Core	20	20.5	0.45	0.41	80	8	10^{-7}
Shell	22	23.0	0.30	0.33	-	40	10^{-2}

Table 1. Soil properties

2.1.2 Boundary conditions

The geotechnical problems can be idealized by assuming that the regions far from the area of interest extend to infinity. The unbounded theoretical models should be truncated to the manageable size by using the artificial boundaries for minimizing the computation time as well as avoiding the outwards propagating waves form reflecting to the model. The viscous boundary, developed by Lysmer & Kuhlemeyer (1969), is used in the current calculations. In this case, independent dashpots are used in the normal and shear directions at the model boundaries, as shown in Fig. 3. During the static analysis, the bottom boundary is fixed in both the horizontal and the vertical directions and the lateral boundaries only in the horizontal direction. In dynamic analysis, when the dam is laid on the foundation (and not on the bedrock), lateral boundaries are changed into free-field boundaries, available in the FLAC library, in order to eliminate the wave reflections from the truncated boundaries.

2.1.3 Element size

Numerical distortion of propagating wave can occur in dynamic analysis as a function of modelling condition. The numerical accuracy of wave transmission is affected by both the frequency content of input wave and the wave speed characteristics of system. Kuhlemeyer & Lysmer (1973) showed that for an accurate representation of wave transmission through the soil model, the spatial element size should be smaller than 1/10 to 1/8 of the wavelength associated with the highest frequency component of input wave i.e.,

$$\Delta L = \lambda / 9 \tag{8}$$

where, λ = wave length associated with the highest frequency component that contains significant energy. Considering the above mentioned criteria, the element size is defined small enough to allow the seismic wave propagation throughout the analysis.

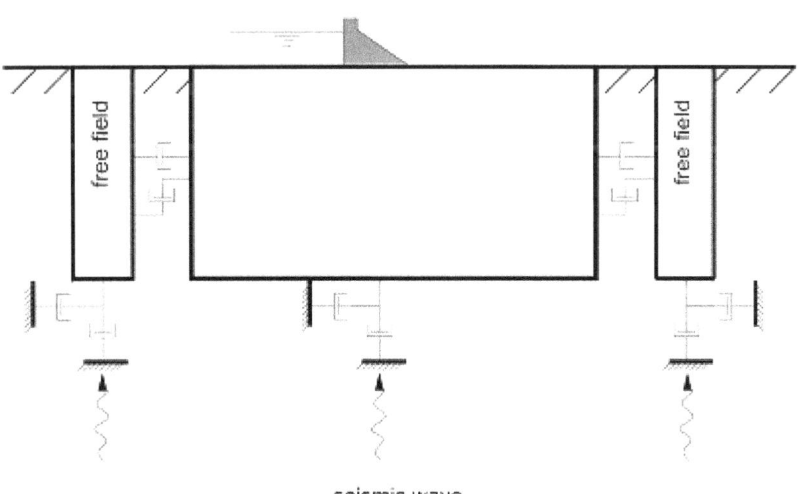

Fig. 3. Free field boundaries in dynamic analysis (Itasca, 2004)

2.1.4 Damping

Material damping in soil is generally because of its viscosity, friction and plasticity development. Indeed, the role of damping in the numerical models is the reproduction of energy losses in the natural systems subjected to dynamic loads. The dynamic damping is provided in the model by Rayleigh damping option available in FLAC. Rayleigh damping was primarily used in analyzing the structures and elastic continua in order to damp the natural oscillation modes of system. Rayleigh damping R_d=5% is used to compensate the energy dissipation in the media (Itasca, 2004). In the dynamic analysis incorporated plasticity constitutive models, considerable amount of energy dissipation can occur during the plastic flow. In the calculations of such cases, minimal percentage of hysteretic damping (e.g. 2%) is considered as well. The dam's natural frequency is determined as the Rayleigh damping parameter by Fourier analysis of its free response, as shown in Fig. 4. The fundamental frequency (f_1 = 1.71 Hz) of the dam with 40 m height is shown in this figure and those of dams with different heights are tabulated in Table 2.

2.1.5 Time step

The governing equations in time should be integrated incrementally for completing the numerical solution. The solution time step should be small enough in order to accurately define the applied dynamic loads and guarantee the stability and convergence of solution. In this regard, time step is about 10^{-6} second in the current FLAC model.

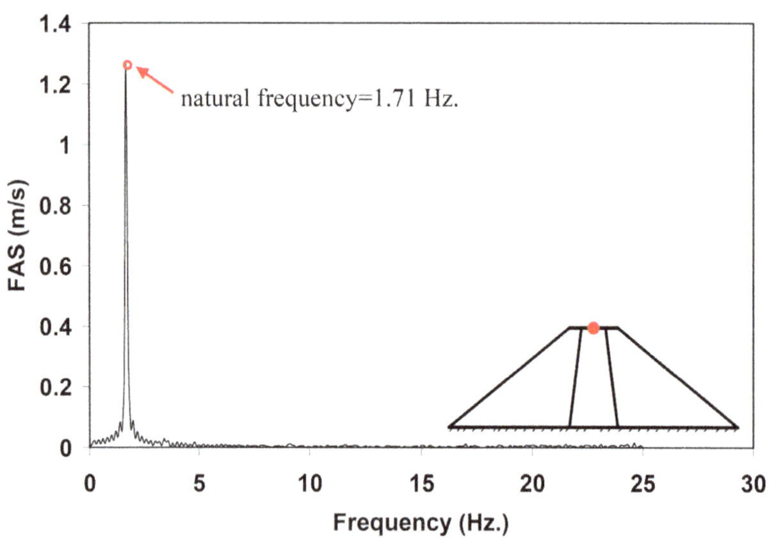

Fig. 4. Fourier amplitude spectrum of free horizontal motion at the dam crest

First mode of vibration	Dam Height (*m*)				
	40	80	120	200	280
f_1 (*Hz.*)	1.7	0.88	0.60	0.36	0.26

Table 2. Fundamental frequency

2.2 Input excitations

Selecting dynamic input motion is an important task in the seismic evaluation processes. In non-linear dynamic analysis, the expected earthquakes should be expressed as a set of ground motion time histories. For more correct evaluation, the input motions which offer an appropriate range of dam responses should be selected in the adaptable time history realizations. This procedure may be intractable due to the number of time-history realizations. However, in reality, the level of earthquake responses, probably achieved by physical system, is limited. Quantifying such responses demands good understanding of the seismic response of the system as well as the ground motion parameters that characterize the damage potential of seismic input (USCOLD, 1999). Different parameters can be employed to identify the severity and damage potential of a certain acceleration time history, assumed as the representative of expected earthquake ground motion; peak ground acceleration value (PGA) is of such kind. The use of this descriptor is intuitively natural since accelerations and resulting inertial forces are directly related by Newton's second law. However, there is no direct relation between PGA and structural response at the dominant natural frequencies of most typical dams. Moreover, large PGA values are not sufficient for generating response conditions which lead to significant damage. Despite these limitations, PGA is still the fundamental parameter used to judge the damage potential of certain acceleration histories. On the other hand, the seismic response of system is strongly affected by the frequency content of earthquake. Therefore, the better characterization of a given input motion can be achieved by using some forms of spectral representations. In particular, using Fourier amplitude spectrum is at the core of earthquake engineering practice. However, such characterizations do not provide direct description of the duration or time variation features of a given input motion.

Based on above, in this chapter, three different real acceleration time histories are selected from a database of earthquake records: Tabas, PGA=0.93g in MCE level; Naghan, PGA=0.72g in MDE level; San Fernando, PGA=0.21g in DBE level. In the dynamic analysis of dams, the scaled records are filtered to the maximum frequency of 10 Hz, transferred to the "inside" bedrock formation by standard de-convolution analysis and applied at the base of numerical model. The information of earthquake records are summarized in Table 3, and their corresponding acceleration time histories and Fourier amplitude spectra are shown in Figs. 5 & 6, respectively.

Earthquake	Station	Date	M	Closest Distance	PGA (g)	PGV	PGD
				(km)		*(cm/s)*	*(cm)*
Tabas	Tabas	1978	7.4	94	0.93	121.4	94.58
Naghan	Naghan	1977	5.4	75	0.72	46.20	61.00
San Fernando	Pasadena	1971	6.6	19	0.21	10.90	2.320

Table 3. Earthquake records data

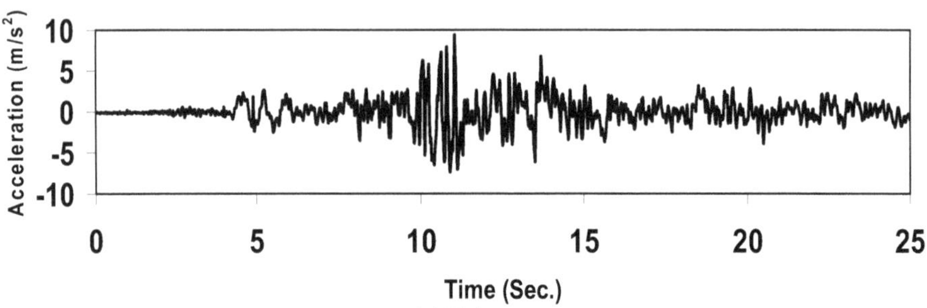

(a) Tabas

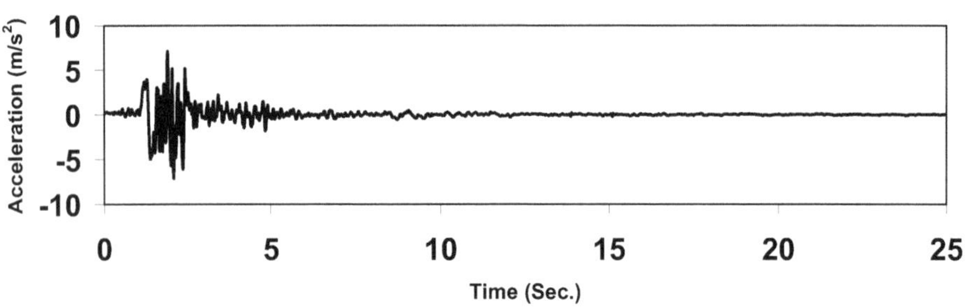

(b) Naghan

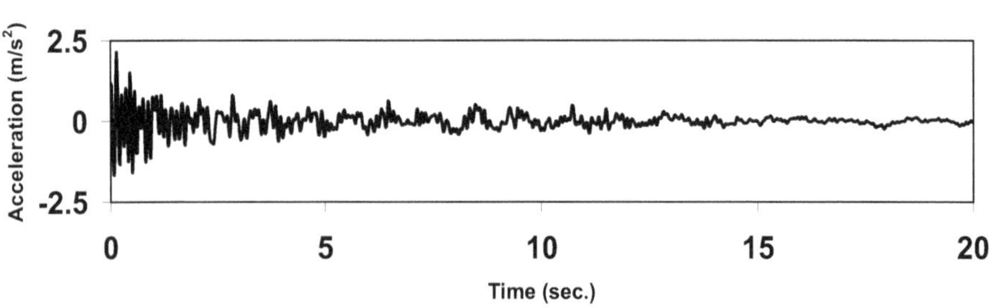

(c) San Fernando

Fig. 5. Input acceleration time histories

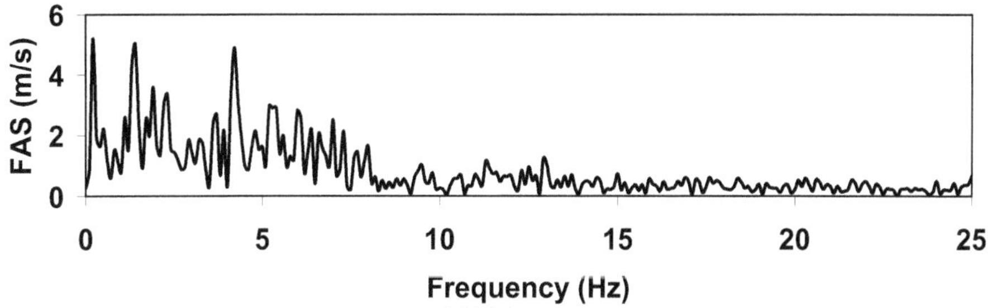

(a) Tabas

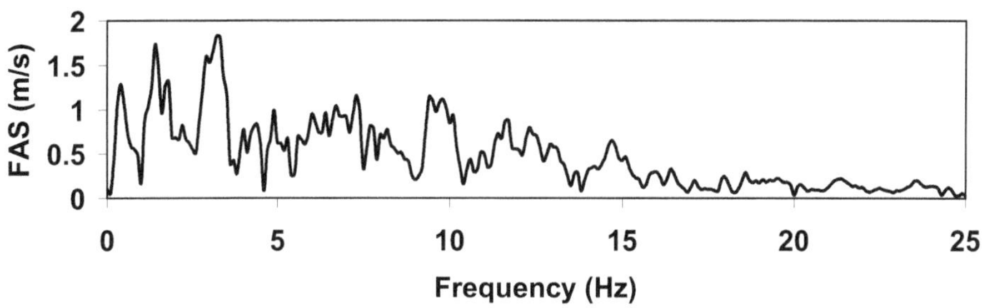

(b) Naghan

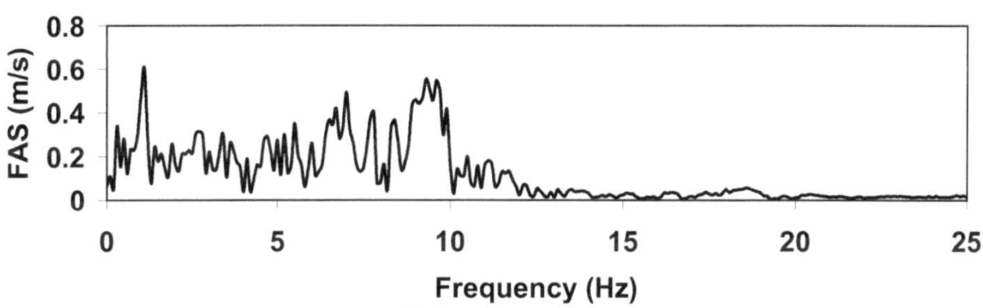

(c) San Fernando

Fig. 6. Fourier amplitude spectra

2.3 Full non-linear dynamic analysis

Equivalent linear analysis is the common method used for evaluating the seismic behaviour of earth structures. In this approach, first, the responses are linearly analyzed using the initial values of damping ratio and shear modulus. Then, the new values of damping ratio and shear modulus are estimated, using maximum value of shear strain and laboratory curves. These values are used for redoing the analysis. This procedure is repeated several times until the material properties show no variation. Therefore, no non-linear effect is directly captured by this method as it assumes linearity during the solution process. Strain-dependent modulus and damping functions are considered roughly in order to approximate some effects of non-linearity (damping and material softening).

In the non-linear analysis, employed in this study, the non-linear stress-strain relationship is directly followed by each zone. Damping ratio and shear modulus of the materials are calculated automatically at different strain levels. The real behaviour of soil, under cyclic loading, is non-linear and hysteretic. Such behaviour can be simulated by Masing model (Masing, 1926), which can model the dynamic behaviour of soil. In this model, the shear behaviour of soil may be explained by a backbone curve as:

$$F_{bb}(\gamma) = \frac{G_{\max}\gamma}{1+(G_{\max}/\tau_{\max})|\gamma|} \tag{9}$$

where, $F_{bb(\gamma)}$ = backbone or skeleton function; γ = shear strain amplitude; G_{max} = initial shear modulus; τ_{max} = maximum shear stress amplitude.

Stress-strain curve follows the backbone curve in the first loading, as shown in Fig. 7(a); however, for explaining the unload-reload process, the above equation should be modified. If load reversal occurs at the point (τ_r, γ_r), stress-strain curve follows the path given by the below formula:

$$\frac{\tau-\tau_r}{2} = F_{bb}\left[\frac{\gamma-\gamma_r}{2}\right] \tag{10}$$

In other words, the shapes of unload-reload curves are similar to that of backbone curve (with the origin shifted to the loading reversal point) except they are enlarged by a factor of 2, as shown in Fig. 7(b). The Eqs. (9) & (10) describe the Masing behaviour (Masing, 1926).

Masing rules seem not to be enough for precise explanation of soil response under general cyclic loading. Finn et al. (1977) developed modified rules to describe the irregular loading. They suggested that unloading and reloading curves follow the concerning two rules. If the new unloading or reloading curve exceeds the last maximum strain and cut the backbone curve, it will follow the backbone curve up to meeting the next returning point, as shown in Fig. 7(c). If a new unloading or reloading curve passes through the previous one, it will follow the former stress-strain curve, as shown in Fig. 7(d). According to this model, the tangent shear modulus can be defined at the points on the backbone and new reloading-unloading curves by the Formulas (11) & (12), respectively, as:

$$G_t = G_{\max} \Big/ \left[1+\frac{G_{\max}|\gamma|}{\tau_{\max}}\right]^2 \tag{11}$$

$$G_t = G_{\max} \Big/ \left[1+\frac{G_{\max}}{2\tau_{\max}}|\gamma-\gamma_r|\right]^2 \tag{12}$$

Based on the results, obtained in this research, the shear stress decreases as the number of load cycles increases; it means that shear stress-strain curves are more inclined. In this study, Masing rules are implemented into FLAC via a series of FISH functions in order to simulate the non-linear stress-strain relationships.

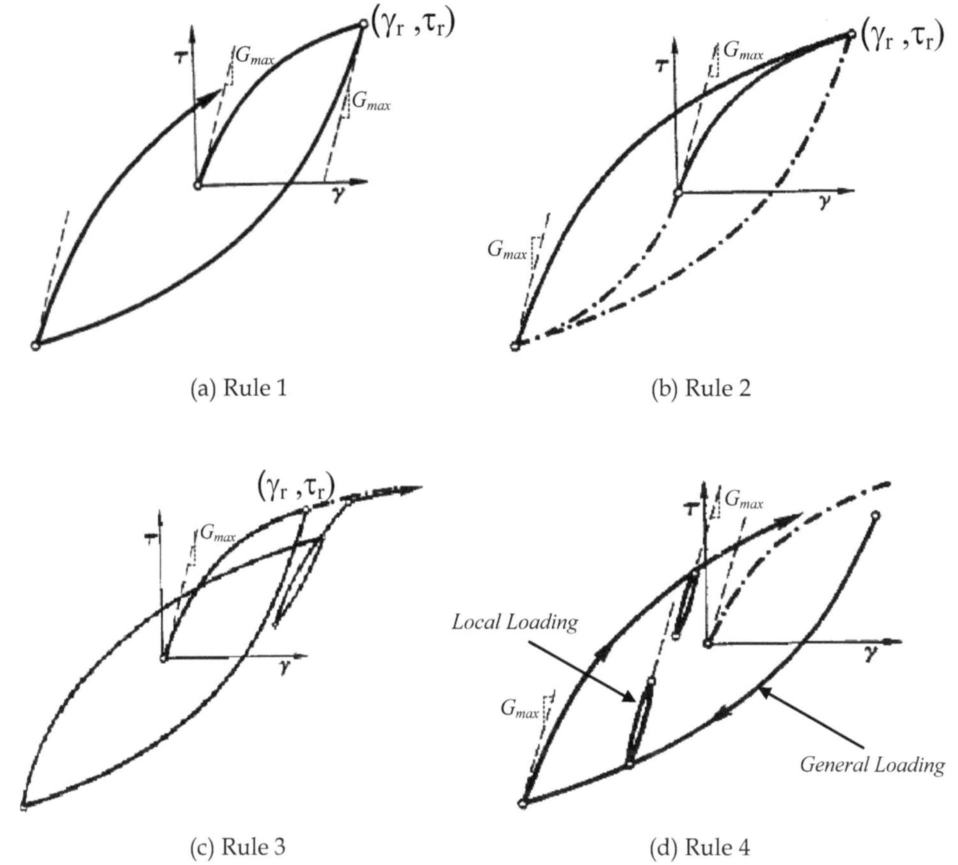

Fig. 7. General patterns of loading, unloading and reloading paths in Masing model

3. Numerical results and discussions

3.1 Validation analysis

In this research, one-zone sample is simulated using the unit cell as shown in Fig. 8, in order to validate the implementation of Masing rules in FLAC program. The one-zone sample consists of sandy soil and a periodic motion is exerted at its base. Vertical loading is established only by gravity and then, the Equilibrium stresses are installed in the soil. The stress/strain loops of the sample are shown in Fig. 9(a) for several cycles. According to the figure, shear modulus decreases as shear strain increases. It seems that the hysteretic model can reasonably handle the multiple nested loops. Energy dissipation and shear stiffness degradation are clearly observed during seismic loading, as shown in Fig. 9(b). Shear modulus reduction curve, obtained in this study, follows well the empirical relation proposed by Seed et al. (1986) and the test data.

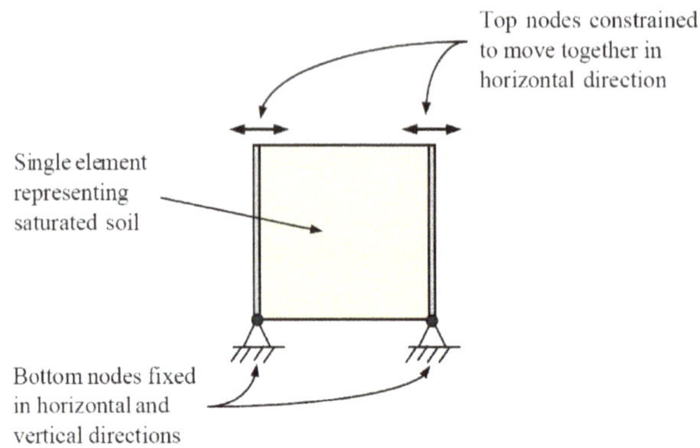

Fig. 8. One-zone model in FLAC for simulating cyclic simple shear test

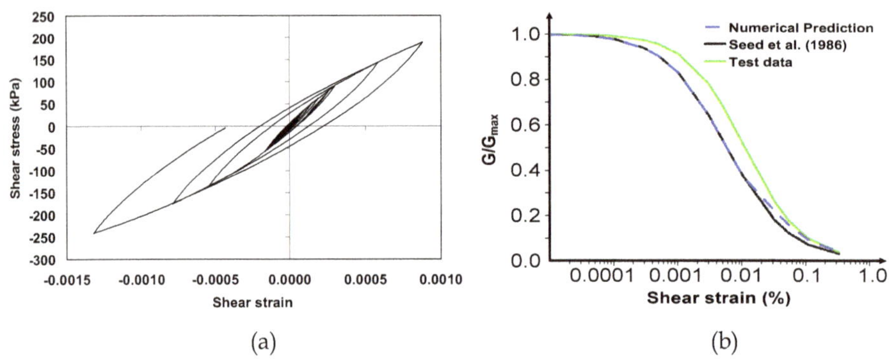

Fig. 9. Simulation results of one-zone sample: (a) hysteresis loop, and (b) comparing the shear modulus reduction curves

The results obtained from the numerical analyses are compared with those of experimental ones in order to evaluate the capability of proposed model. One of the centrifuge tests related to the embankment which was performed in VELACS project (VErification of Liquefaction Analysis using Centrifuge Studies (Arulanandan and Scott, 1993, 1994)) is chosen as a reference. It is attempted to create almost similar conditions for both laboratory model test and numerical model, as shown in Figs. 10(a) & (b), respectively. The numerical results are presented and compared with those obtained from the corresponding centrifuge test data. The computed (numerical simulations) and measured (centrifuge tests) results are shown in Figs. 11(a) & (b). According to these comparisons, the reference numerical model can rationally predict the seismic behaviour of earth-fill dam.

Finally, the model's ability in simulating the seismic behaviour of earth-fill dam during a real earthquake is verified by a real well-documented case history. In this regard, the results of non-linear dynamic analysis of Long Valley (LV) earth-fill dam in California subjected to the 1980 Mammoth Lake earthquake (Griffiths and Prevost, 1988) are presented. Then, the results are compared with the real measurements recorded at the site and also with the results presented by previous authors. LV dam is located in Mammoth Lake area

(California) in close proximity of active faults. The dam is a rolled earth-fill one formed mainly with the impervious zone. The dam has maximum height of 55 m, 182 m length at the crest, and upstream and downstream slopes of 3h/1v. The LV dam was instrumented in the 1970's with a multiple-input-output array; it has 3 accelerometer stations to monitor the boundary conditions, and 5 stations to record the dam response (Fig. 12(a)). Thus, the array comprised a total of 22 accelerometers linked to a common triggering mechanism.

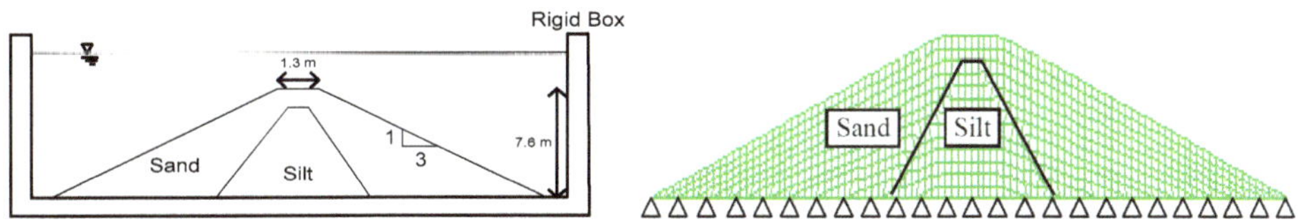

(a) (b)

Fig. 10. Model configuration: (a) schematic sketch of dam in centrifuge container box, and (b) numerical grid constructed in FLAC

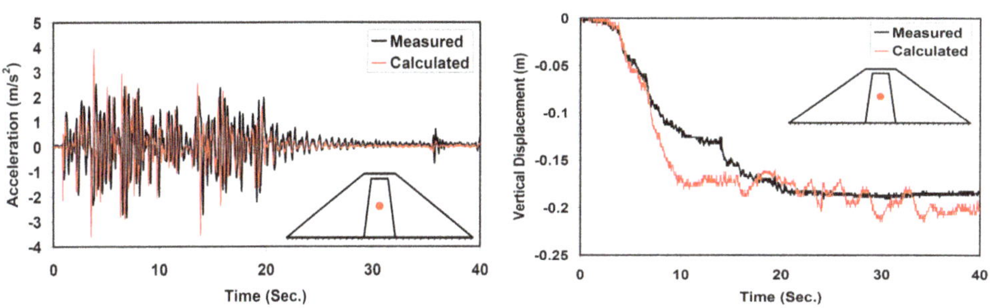

(a) (b)

Fig. 11. Measured time histories versus calculated ones at the middle of dam height: (a) acceleration, and (b) vertical displacement

In May 1980, a series of 6 earthquakes occurred in the Mammoth Lakes area. The magnitudes of these earthquakes were M_L = 4.9 - 6.7, and the induced peak accelerations at the crest centre was 0.5g in the upstream-downstream direction (x direction, as shown in Fig. 13(a)) during the strongest event. Extensive arrays of 22 input-output (excitation-response) accelerations were recorded, providing a valuable information source of the dam seismic responses over a wide range of deformation levels. In this study, the dam is subjected to the input motion, recorded downstream at the outlet during Mammoth Lake earthquakes. The first 12 seconds of the recorded acceleration is used with data point at 0.02 second intervals and the peak acceleration is 0.135g in the upstream-downstream (x) direction and 0.084g in the vertical direction (y).

The cross section of LV dam is shown in Fig. 12(b) and its detailed information is found in (Griffith and Prevost, 1988). The numerical grid constructed in FLAC is presented in Fig. 12(c). The input accelerations are applied in the horizontal and vertical directions of the model base. Free Field boundary conditions are exerted to the lateral boundaries of numerical model.

This research focuses on the computed acceleration at the crest which can be compared directly with the measured values. Previously, LV dam has been analyzed by: Lai & Seed, 1985; Elgamal et al., 1987; Griffiths & Prevost, 1988; Yiagos & Prevost, 1991; Zeghal & Abdel-Ghaffar, 1992; Woodward & Griffiths, 1996. The first natural frequency, obtained in this

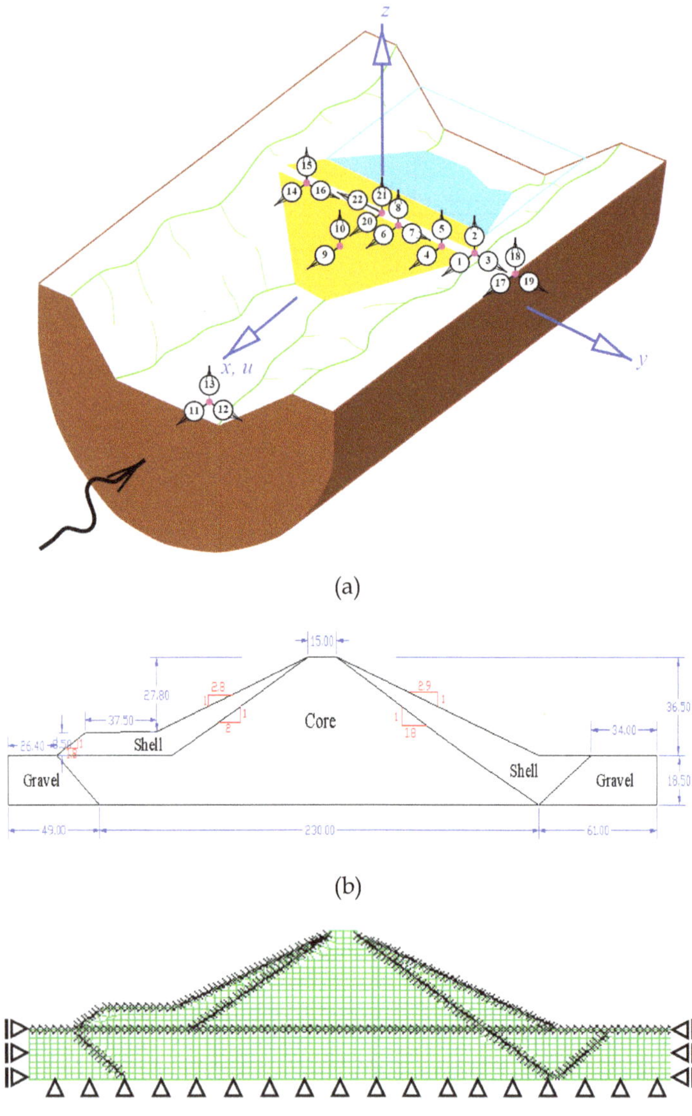

Fig. 12. Long Valley earth-fill dam: (a) schematic view of Long Valley canyon, earth-fill dam and installed instrumentation array acceleration, (b) cross section, and (c) numerical grid for dynamic analysis

study, is presented in Table 4 and compared with the other solutions available in the literature. The results of present study are reasonably in close agreement with those of other relevant numerical investigations. Here, the crest acceleration responses of LV dam are computed and compared with those recorded at the site in both time and frequency domains.

Frequency	Spectral Analysis	2D FE Analysis	3D FE Analysis	Elasto-plastic FE Analysis	Elasto-plastic FE Analysis	**Present Study**
	Griffith & Prevost (1988)	Griffith & Prevost (1988)	Griffith & Prevost (1988)	Yiagos & Prevost (1991)	Woodward & Griffiths (1996)	
f (*Hz.*)	1.85	1.76	1.95	1.987	1.79	**1.71**

Table 4. First natural frequency of Long Valley dam

Fig. 13(a) shows the computed horizontal acceleration of the crest; it indicates that the amplification occurs between the base and the crest. The magnification factor of peak amplitude at the crest is about 5.47 over the peak base amplitude. The crest response, computed in the horizontal direction, is compared with the measured values, as shown in Fig. 13(b); the dashed line corresponds to the computed response there. Excellent overall agreement is achieved between the computed and measured values; however, the computed values show higher amplitudes. The frequency contents of two time records are compared in the form of Fourier amplitude spectra (FAS), as shown in Fig. 13(c). Their peaks are in close agreement although the computed values show rather more energy associated with the fundamental frequency around 1.8 Hz. The frequency content of the up/down stream motion, presented in Fig. 13(c), shows that the energy is concentrated just at the frequencies below 2 Hz.

In the vertical direction, the calculated acceleration shows low agreement with the measured values, as shown in Fig. 14(a). According to this figure, the plots of vertical acceleration are superimposed at the base and crest. This excitation is considerably noisier and less intensive in the vertical direction in compared with that of horizontal one. The maximum accelerations at the crest, recorded in the vertical and horizontal directions are 0.172g and 0.64g, respectively. The computed accelerations in the vertical direction are compared with the measured values in the crest of LV dam, as shown in Fig. 14(b). According to this figure, the computed values have generally lower amplitudes in compared with those of measured values. The Fourier amplitude spectra of these time histories are given in Fig. 14(c). The measured values show a broad band of frequencies, none of which is dominant. The computed values also contain a broad band of frequencies, but with clear peaks in the ranges of 2-3 Hz and 5-6 Hz. It should be mentioned that the time and frequency domain

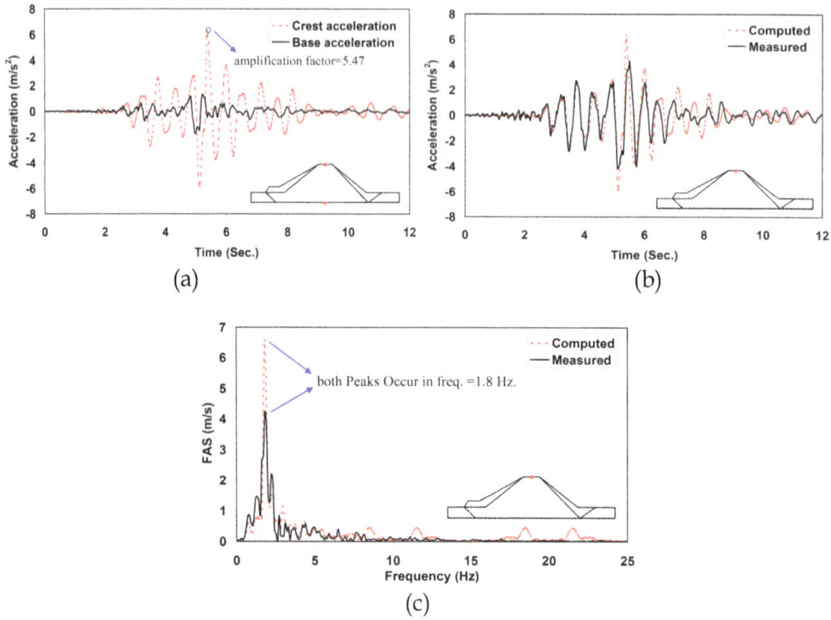

Fig. 13. Comparing computed and measured time histories in horizontal direction: (a) crest and base acceleration, (b) crest acceleration, and (c) Fourier amplitude spectrum

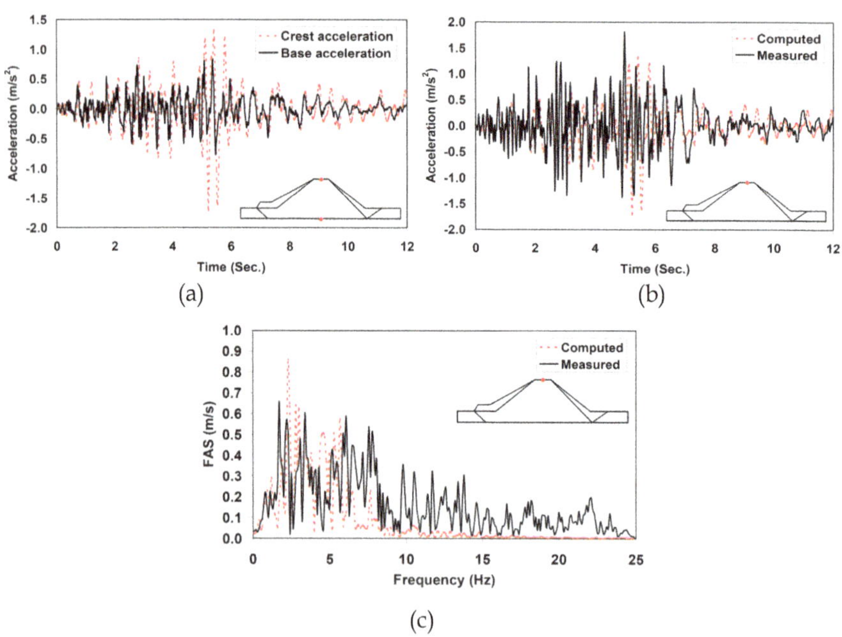

Fig. 14. Comparing computed and measured time histories in vertical direction: (a) crest and base acceleration, (b) crest acceleration, and (c) Fourier amplitude spectrum

results show more appropriate accordance in the horizontal direction in compared with those of the vertical direction.

The results, obtained in the validation analysis of LV dam, in term of crest acceleration are given in Table 5 and compared with the other numerical results presented by previous authors. According to the comparisons, the numerical procedure, presented in this study, can properly capture the fundamental aspects of seismic behaviours of earth-fill dams. As mentioned earlier, the numerical model is then used for parametric studying of hypothetical earth-fill dams due to the satisfactory modelling of validation cases.

	Yiagos & Prevost (1991)	Woodward & Griffiths (1996)	**Present Study**	**Measured values**
Maximum horizontal Acceleration (g)	0.53	0.80	**0.61**	**0.40**
Minimum horizontal Acceleration (g)	-0.65	-0.68	**-0.50**	**-0.50**

Table 5. Comparing the Numerical results obtained for Long Valley dam

3.2 Parametric study

Here, the analyses are carried out to investigate the effects of dam height, input motion characteristics, soil behaviour and strength of shell materials on the seismic behaviour of earth-fill dams. The effects of different earthquakes are studied on the horizontal permanent deformations, permanent shear strains and maximum accelerations, as shown in Fig. 15. The values have been induced at the crests of dams with different heights. The displacements are shown in Fig. 15(a) and the relevant shear strains in Fig. 15(b). It is clear that the shear strain variation is similar to displacement. The horizontal displacements and shear strains in the dam body increase with dam height increasing. The calculated values are much higher in Tabas earthquake and the failure occurs in the dam body. According to Fig. 15(a), the maximum horizontal displacement computed at the crest of dam is about 94 cm at the end of Tabas earthquake. It can be observed in Figs. 15 (a) & (b), that increasing in the input motion energy leads to significant increase of displacements and shear strains. Fig. 15(c) illustrates the coupled effects of dam height and earthquake type on the maximum acceleration induced at the dam crest. According to the figure, the crest acceleration decreases as the dam height increases and no amplification is seen maybe due to more flexible behaviour, larger damping and larger developed plastic zones, observed in higher dams. Therefore, because of these factors, more energy is absorbed in higher dams in compared with that in the shorter ones. It can be seen in Fig. 15 (c) that the accelerations in the dam crest are more reduced in higher dams comparing with the smaller ones. It should be mentioned that PGA of Naghan earthquake (0.72g) is much higher than that of San Fernando earthquake (0.21g). However, the created displacements and shear strains in the dam crest caused by Naghan earthquake are close to those of San Fernando input motion. It can be concluded that using just PGA parameter is not sufficient for evaluating the effect of a certain earthquake time history on the dam response. Therefore, other earthquake parameters such as effective duration, magnitude and frequency content should be considered in the analysis.

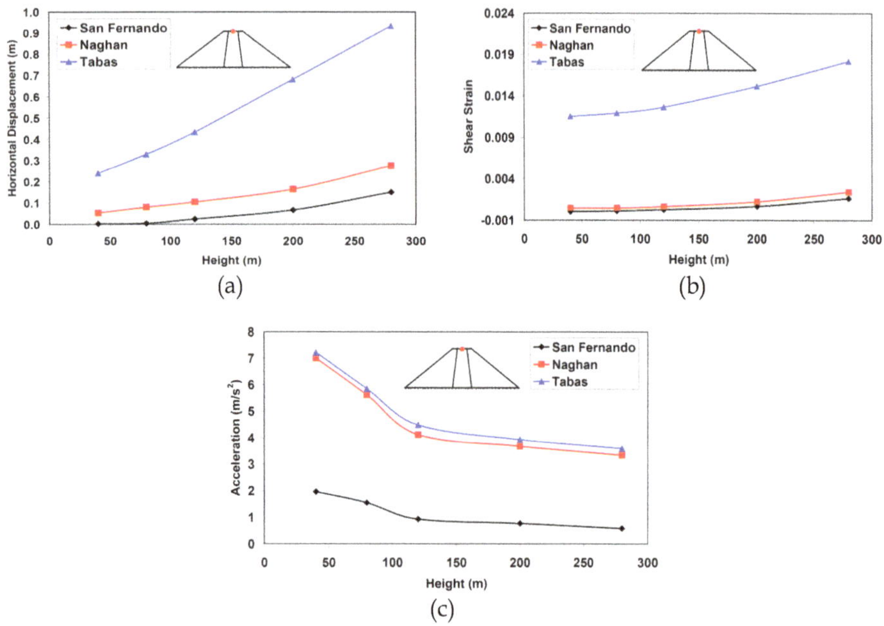

Fig. 15. Computed values of: (a) permanent horizontal displacements, (b) permanent shear strains, and (c) induced maximum accelerations at dam crest versus dam height for different earthquakes

Failure mechanism with permanent shear strain contour in the dam body is shown in Fig. 16, regarding two different heights at the end of Naghan earthquake. The slip surface is much deeper and more obvious in the dam with 280 m height (Fig. 16(b)) in compared with that of 120 m height (Fig. 16(a)).

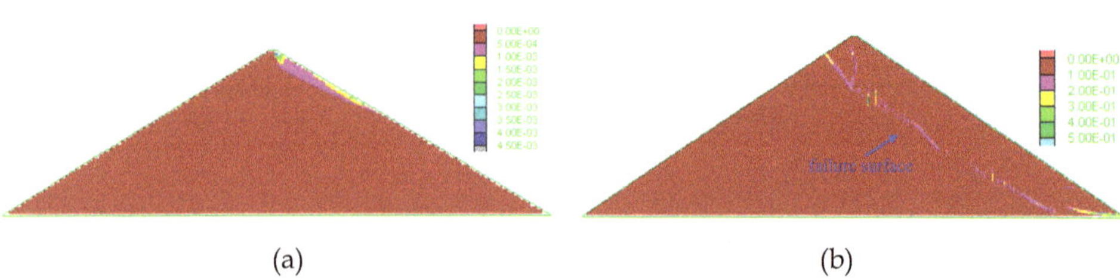

Fig. 16. Failure surface with shear strain contour in dam body at the end of Naghan earthquake: (a) dam height=120 m, and (b) dam height=280 m

A dam with 40 m height is subjected to the mentioned earthquakes and chosen as a reference with two different behaviours, elastic and elastic-perfectly plastic, in order to investigate the effect of soil behaviour on the seismic response of dam body. As it is expected, regarding linear elastic behaviour, smaller displacements and shear strains are observed along the dam height, Figs. 17(a) & (b). However, large amplification occurs especially for the strongest earthquake, Fig. 17(c). It means that plasticity reproduce more energy dissipation during dynamic loading. In such cases, the accelerations are reduced across the dam height and therefore become lower than the base acceleration. According to Fig. 17(a), maximum displacement occurs at about Z/H = 0.88 in linear elastic behaviour, while it happens at the crest of dam in linear elastic-perfectly plastic behaviour. Furthermore, in elastic-perfectly plastic behaviour, the dynamic induced residual (permanent) displacement increases largely in the upper part of dam, especially for Tabas and Naghan earthquakes, confirmed in the previous research works (Ohmachi and Kuwano, 1994; Ozkan et al., 2006). That is why the crest should especially be considered in designing the embankment dams, due to the stronger shaking at the upper parts, for avoiding undesirable deformations. The distribution of shear strain is extremely non-linear along the dam height in the stronger earthquakes, as shown in Fig. 17(b). In the elastic dams, maximum acceleration occurs in the dam crest, as shown in Fig. 17(c). In the acceleration profile of Tabas earthquake, a special increase is seen along the dam centerline at Z/H = 0.38.

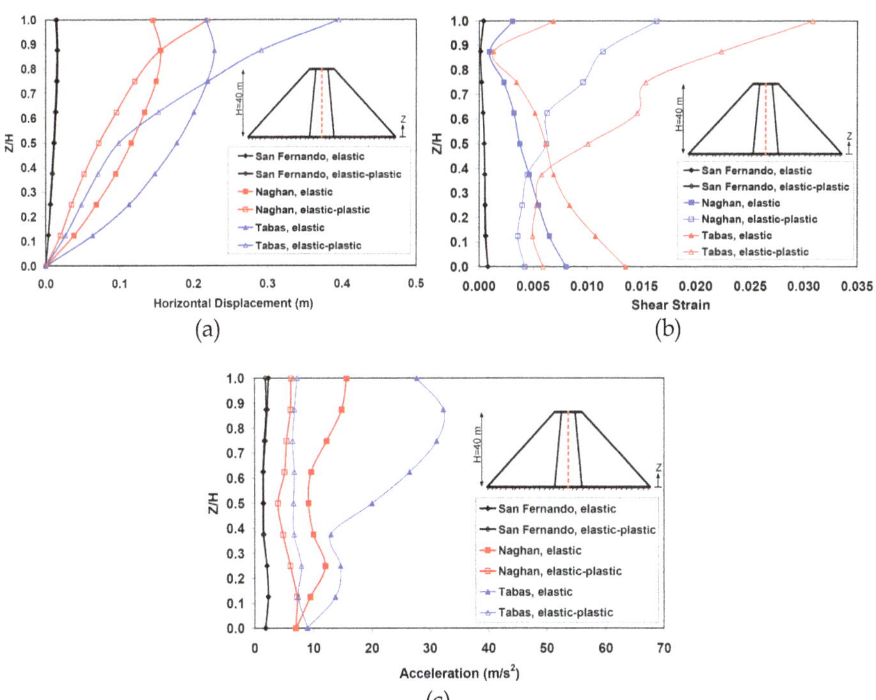

Fig. 17. Computed values of: (a) permanent horizontal displacements, (b) permanent shear strains, and (c) induced maximum accelerations along the dam height for elastic and elastic-plastic behaviours subjected to different earthquakes

Strength of dam materials is an important parameter which can significantly affect the seismic response of dam. In this regard, different friction angles are assumed accompanying with different dam heights, subjected to Naghan earthquake, in order to clarify the above mentioned effect. Fig. 18(a) shows horizontal displacement values versus dam height for different friction angles of shell materials. The variations of shear strains in the crests of dams, with different heights, are shown in Fig. 18(b) for different friction angles of shell materials. As it is expected when the friction angle increases, the horizontal displacement and shear strain in the dam crest decrease. It can be seen in the above figure that the variation of friction angle causes no significant change in the displacement and shear strain regarding $\varphi \leq 40^{\circ}$. However, the variation is more significant in $\varphi = 45^{\circ}$, compared with the lower friction angles; the highest displacement values correspond to $\varphi = 30^{\circ}$. The horizontal displacements computed at the crests of dams with 40 and 120 m heights are about 16 and 13 cm, respectively, and their shear strains are about 3.5^{-3} and 2.5^{-3}, respectively. The maximum acceleration induced at the top of dam decreases as the friction angle decreases or the dam height increases, as shown in Fig. 18(c). Considering larger friction angles for the shell materials (e.g., $\varphi = 45^{\circ}$) leads to about 70% increase in the dynamic amplification. The computed maximum crest acceleration of dam with 40 m height is about 0.89g for $\varphi = 45^{\circ}$, while that of 120 m height is 0.52g. When the dam height deceases, the horizontal displacement and shear strain increase but the acceleration decreases at the crest of dam. All variations are linear for $\varphi = 45^{\circ}$, but for the other friction angles are slightly non-linear, as shown in Fig. 18.

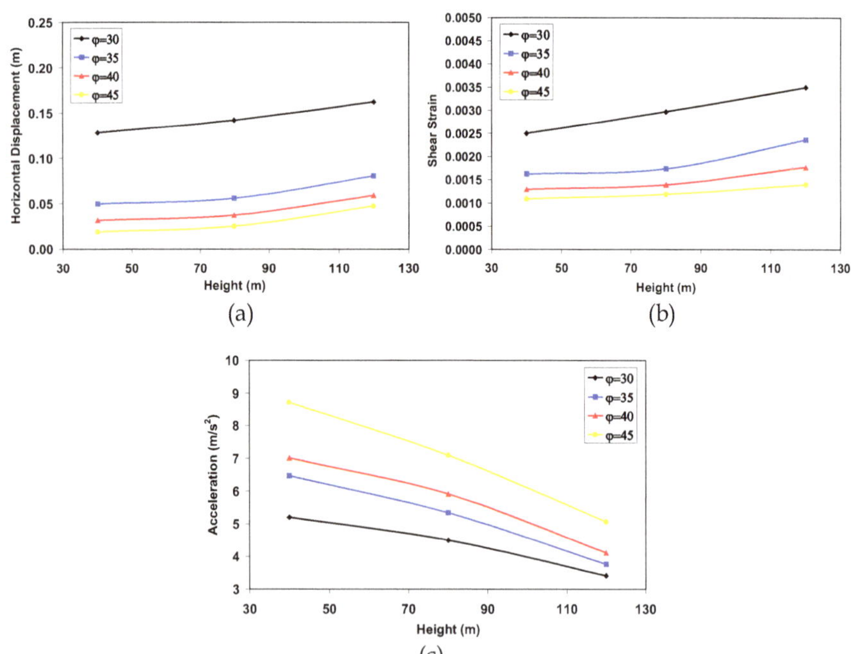

Fig. 18. Computed values of: (a) permanent horizontal displacements, (b) permanent shear strains, and (c) induced maximum accelerations at dam crest versus dam height for different friction angles of shell materials

3.3 Lessons learned

The author experienced several interesting points and noteworthy items during numerical model calibration and numerous dynamic analyses which are listed below:

- In full non-linear dynamic analysis, soil stiffness degradation is automatically taken into account upon constitutive model of soil and just the initial shear modulus is needed as an input parameter. Therefore, it is important to be sure that in the numerical model, the trend of shear modulus decrease and damping ratio increase are in agreement with those of laboratory test results during dynamic loading.
- The poisson's ratio of about 0.5 should not be used in the calculating of bulk modulus of undrained soil layers (such as clay) in the analysis. Otherwise, bulk modulus increases irrationally and the time step of analysis decreases rigorously and consequently the calculation time increases excessively. Therefore, the poisson's ratio should not be more than 0.45 in such cases.
- If a "raw" acceleration record from a site is used as a time history, then FLAC model may exhibit residual displacements once the motion is finished. This arises from the fact that the integral of complete time history may not be zero. Therefore, the process of baseline drift correction should be performed in such cases.
- The input motion should be filtered before being applied to the FLAC grid in order to eliminate all high frequency components form it.
- The stages of construction should be considered in the numerical simulation of earth-fill dams. In the present study, the stages are: initial state of foundation (if any); layer by layer dam replacement; applying the hydrostatic water pressure due to the replacement of dam reservoir; seepage analysis in the dam body; mechanical adjustment to new flow field; and finally dynamic analysis. Regarding the mentioned stages, one is run to equilibrium and then the next stage is started. However, construction sequences have much greater effects on static results than dynamic results.

4. Conclusions

This chapter presents the non-linear seismic behaviour of earth-fill dams using explicit finite difference method. In this regard, a simple elastic-perfectly plastic constitutive model with Mohr-Coulomb failure criterion is used to describe the stress-strain response of the soil. Here, Rayleigh damping is used to promote the level of hysteretic damping during dynamic analysis. Masing rules are implemented into the constitutive model to precisely explain the non-linear soil response under general cyclic loading. The numerical model is then calibrated using centrifuge test data as well as field data. The field data are obtained in real measuring of Long Valley earth-fill dam subjected to the 1980 Mammoth Lake earthquake. The results of dynamic analysis, obtained in this study, are compared with the real measurements of Long Valley dam in terms of accelerations computed at the crest of dam in both time and frequency domains. The proposed numerical model can properly reproduce the overall seismic behaviours of earth-fill dams, their qualities and quantities, under earthquake loading conditions, confirmed by validation analyses. After validation, the effects of dam height, real earthquake loading, soil behaviour and strength of shell materials on the seismic response of earth-fill dams are evaluated through a comprehensive parametric study. The effect of dam height on the non-linear seismic behaviour is particularly focused in this research. The following conclusions are obtained based on the performed parametric study:

- If the dam materials keep their elastic behaviours during dynamic loading, then the horizontal acceleration increases along the dam height (from the base to the top). In this case, the higher dams show larger amplifications, especially if the natural periods of their bodies coincide with the periodical nature of earthquake waves.
- When the dam body shows non-linearity or the materials go towards plastic behaviour during a strong shaking, the attenuation of acceleration waves in the dam body becomes more effective. Consequently, the amplitudes of earthquake accelerations decrease when moving from the base towards the top.
- According to the non-linear elastic-plastic analyses, when the height of the dam increases, then the strongest dynamic loading (Tabas earthquake) induces plasticity in large parts of the dam body. In fact, strong earthquakes are more effective in changing the material behaviour from elastic to plastic condition in comparison with weak earthquakes.
- The higher dams are more flexible than the smaller ones. Consequently, the flexibility affects the shear strains which influence the shear modulus degradation and attenuating coefficient. All these effects are on the trend of weakening the accelerations along the height.
- Soils with less strength (suppose low friction angle) go towards yielding by small amount of dynamic force which cause the attenuation of acceleration along the dam height in the weak materials compared with the strong ones.
- Regarding a dam subjected to the earthquake with lower energy, the dam body behaves as an elastic material. Therefore, the induced seismic accelerations inside the dam body become larger from the base of dam to its top. In this case, small plasticity zones are developed in the dam body and the dam remains safe during dynamic loading.
- Finally, non-linear dynamic analysis shows that plasticity should be considered in the investigation of seismic response of earth-fill dams, because of which the acceleration of the dam crest decreases and the displacements and shear strains of dam body as well as the energy dissipation increase. All these can significantly affect the seismic response of earth-fill dams.

5. References

Abouseeda, H. & Dakoulas, P. (1998). Non-linear Dynamic Earth Dam–foundation Interaction using a BE–FE Method. *Journal of Earthquake Engineering and Structural Dynamics*, Vol.27, pp. 917–936

Adalier, K. & Sharp, M.K. (2004). Embankment Dam on Liquefiable Foundation-Dynamic Behavior and Densification Remediation. *Journal of Geotechnical Geoenvironmental Engineering*, Vol.130, No.11, pp. 1214–1224

Amadio, C.; Fragiacomo, M. & Rajgelj, S. (2003). The Effects of Repeated Earthquake Ground Motions on the Non-linear Response of SDOF Systems. *Earthquake Engineering & Structural Dynamics*, Vol.32, No.2, pp. 291–308

Ambraseys, N.N. (1960). The Seismic Stability of Earth Dams, *Proceedings of the Second World Conference on Earthquake Engineering*, pp. 1345–1363, Tokyo, Japan

Arulanandan, K. & Scott, R.F. (1993). Verification of Numerical Procedures for the Analysis of Soil Liquefaction Problems, *Conference Proceedings of VELACS*, Vol. 1, Balkema, Rotterdam

Arulanandan, K. & Scott, R.F. (1994). Verification of Numerical Procedures for the Analysis of Soil Liquefaction Problems, *Conference Proceedings of VELACS*, Vol. 2, Balkema, Rotterdam

Ashford, S.A.; Boulanger, R.W.; Donahue, J.L. & Stewart J.P. (2011). Geotechnical Quick Report on the Kanto Plain Region during the March 11, 2011, Off Pacific Coast of Tohoku Earthquake, Japan, *Geotechnical Extreme Events Reconnaissance (GEER)*

Basudhar, P.K.; Kameswara Rao, N.S.V.; Bhookya, M. & Dey, A. (2010). 2D FEM Analysis of Earth and Rockfill Dams under Seismic Condition, *Fifth International Conference on Recent Advances in Geotechnical Earthquake Engineering and Soil Dynamics and Symposium in Honor of Professor I.M. Idriss*, San Diego, California

Cascone, E. & Rampello, S. (2003). Decoupled Seismic Analysis of an Earth Dam. *Soil Dynamics and Earthquake Engineering*, Vol.23, pp. 349–365

Chen, M. & Harichandran, R.S. (2001). Response of an Earth Dam to Spatially Varying Earthquake Ground Motion. *Journal of Engineering Mechanics*, Vol.127, pp. 932–939

Eberhard, M.; Baldridge, S.; Marshall, J.; Mooney, W. & Rix, G. (2010). The M_W 7.0 Haiti Earthquake of January 12, 2010: USGS/EERI Advance Reconnaissance Team, *Team Report*, Vol.1

Ebrahimian, B. & Vafaeian, M. (2005). Effects of Dam Height on the Seismic Response of Earth Dam, *Proceedings of 7th International Conference on Civil Engineering*, Tarbiat Modarres University, Tehran, Iran

Ebrahimian, B. (2009). Numerical Modeling of the Seismic Response of an Earth Dam Founded on Liquefiable Soils, *Proceedings of 2nd International Conference on Long Term Behavior of Dams*, E. Bauer, S. Semprich, G. Zenz, (Eds.), pp 610-615, Graz University of Technology, Graz, Austria

Ebrahimian, B. (2011). Numerical Analysis of Nonlinear Dynamic Behavior of Earth Dams. *Frontiers of Architecture and Civil Engineering in China*, Vol.5, No.1, pp. 24–40

Elgamal, A.M.; Abdel-Ghaffar, A.M. & Prevost, J.H. (1987). 2-D Elastoplastic Seismic Shear Response of Earth Dams: Application. *Journal of the Engineering Mechanics ASCE*, Vol.113, No.5, pp. 702-719

Finn, W.D.L.; Lee, K.W. & Martin, G.R. (1977). An Effective Stress Model for Liquefaction. *Journal of Geotechnical Engineering Division ASCE*, Vol.103, No.6, pp. 517-553

Gazetas, G. (1987). Seismic Response of Earth Dams: Some Recent Developments. *Soil Dynamics and Earthquake Engineering*, Vol.6, No.1, pp. 1–48

Griffith, D.V. & Prevost, J.H. (1988). Two- and Three-Dimensional Dynamic Finite Element Analyses of the Long Valley Dam. *Geotechnique*, Vol.38, pp. 367-388

Harder, L.F.; Kelson, K.I.; Kishida, T. & Kayen, R. (2011). Preliminary Observations of the Fujinuma Dam Failure Following the March 11, 2011 Tohoku Offshore Earthquake, Japan, *Geotechnical Extreme Events Reconnaissance (GEER)*

Hardin, B.O. & Black, W.L. (1968). Vibration Modulus of Normally Consolidated Clay. *Journal of Soil Mechanics & Foundation Engineering ASCE*, Vol.84, No.2, pp. 1531-1537

Itasca Consulting Group, Inc. *FLAC (Fast Lagrangian Analysis of Continua)*, Version 4. 2004, Minneapolis, MN

Kokusho, T. & Esashi, Y. (1981). Cyclic Tri-Axial Test on Sands and Coarse Material, *Proceedings of 10th International Conference on Soil Mechanics and Foundation Engineering*, (Quoted by Ishihara 1986)

Krinitzsky, E.L. & Hynes, M.E. (2002). The Bhuj, India, Earthquake: Lessons Learned for Earthquake Safety of Dams on Alluvium. *Engineering Geology*, Vol.66, pp. 163-196

Kuhlmeyer, R.L. & Lysmer, J. (1973). Finite Element Method Accuracy for Wave Propagation Problems. *Journal of the Soil mechanics and Foundation Division ASCE*, Vol.99, No. SM5, pp. 421-427

Lacy, S.J. & Prevost, J.H. (1987). Nonlinear Seismic Response Analysis of Earth Dams. *Soil Dynamics and Earthquake Engineering*, Vol.6, No.1, pp. 48–63

Lai, S.S. & Seed, H.B. (1985). Dynamic Response of Long Valley Dam in the Mammoth Lake Earthquake Series of May 25-27 1980, *Report No. UCB/EERC-85/12, Earthquake Engineering Research Center*

Lysmer, J. & Kuhlmeyer, R.L. (1969). Finite Element Method for Infinite Media. *Journal of Engineering Mechanics ASCE*, Vol.95, No. EM4, pp. 859-877

Masing, G. (1926). Eigenspannungen und Verfestigung Beim Messing, *Proceedings of 2nd International Congress on Applied Mechanics*, Zurich.

Matsumoto, N.; Sasaki, T. & Ohmachi, T. (2011). The 2011 Tohoku Earthquake and Dams, *ICOLD 89th Annual Meeting*, Lucerne, Switzerland

Ming, H.Y. & Li, X.S. (2003). Fully Coupled Analysis of Failure and Remediation of Lower San Fernando Dam. *Journal of Geotechnical and Geoenvironmental Engineering*, Vol.129, No.4, pp. 336–349

Moustafa, A. & Takewaki, I. (2010). Modeling Critical Ground-Motion Sequences for Inelastic Structures. *Advances in Structural Engineering*, Vol. 3, No.4, pp. 665-679

Moustafa, A. (2011). Damage-based Design Earthquake Loads for Single-degree-of-freedom Inelastic Structures. *Journal of Structural Engineering*, Vol.137, No.3, pp. 456-467

Ohmachi, T. & Kuwano, J. (1994). Dynamic Safety of Earth and Rock Fill Dams, *Proceedings of a course*, T. Ohmachi, J. Kuwano, (Eds.), Balkema, Rotterdam

Özkan, M.; Yözyazicioglu, M. & Aksar, U.D. (2006). An Evaluation of Güldürcek Dam Response during 6 June 2000 Orta Earthquake. *Soil Dynamics and Earthquake Engineering*, Vol.26, No.5, pp. 405–419

Özkan, M. (1998). A Review of Considerations on Seismic Safety of Embankments and Earth and Rock-Fill Dams. *Soil Dynamics and Earthquake Engineering*, Vol.17, pp. 439–458

Papalou, A. & Bielak, J. (2004). Nonlinear Seismic Response of Earth Dams with Canyon Interaction. *Journal of Geotechnical and Geoenvironmental Engineering*, Vol.130, No.1, pp. 103–110

Pinto, S.P. (1993). *Soil Dynamics and Geotechnical Earthquake Engineering*. Balkema, Rotterdm

Prevost, J.; Abdel-Ghaffar, A.M. & Lacy, S. (1985). Nonlinear Dynamic Analyses of an Earth Dam. *Journal of Geotechnical Engineering*, Vol.111, No.7, pp. 882–897

Rampello, S.; Cascone, E. & Grosso, N. (2009). Evaluation of the Seismic Response of a Homogeneous Earth Dam. *Soil Dynamics and Earthquake Engineering*, Vol.29, pp. 782–798

Rathje, E.M. (2010). Case History: The Geotechnical Aspects of the Haiti Earthquake, *ISSMGE Bulletin*, Vol.4, Issue 3

Seed, H.B.; Lee, K.L.; Idriss, I.M. & Makdisi, F.I. (1975). The Slides in the San Fernando Dams during the Earthquake of February 9, 1971. *Journal of the Soil mechanics and Foundation Division ASCE*, Vol.101, No. GT7, pp. 651–688

Seed, H.B.; Wong, R.T.; Idriss, I.M. & Tokimatsu, K. (1986). Moduli and Damping Factors for Dynamic Analyses of Cohesionless Soils. *Journal of Geotechnical Engineering ASCE*, Vol.112, No.11, pp. 1016-1032

Sherard, J.I.; Woodward, R.J.; Gizienski, S.F. & Clevencer, W.A. (1963). *Earth and Earth-Rock Dams*. John Wiely and Sons, Inc

Sherard, J.L. (1967). Earthquake Considerations in Earth Dam Design. *Journal of the Soil mechanics and Foundation Division ASCE*, Vol.93, pp. 377–401

Sica, S.; Pagano, L. & Modaressi, A. (2008). Influence of Past Loading History on the Seismic Response of Earth Dams. *Computers and Geotechnics*, Vol.35, pp. 61–85

Siyahi, B. & Arslan, H. (2008). Nonlinear Dynamic Finite Element Simulation of Alibey Earth Dam. *Environmental Geology*, Vol.54, pp. 77–85

Siyahi, B. & Arslan, H. (2008). Earthquake Induced Deformation of Earth Dams. *Bulletin of Engineering Geology and the Environment*, Vol.67, pp. 397–403

Takewaki, I. (2011). Preliminary Report of the 2011 off the Pacific Coast of Tohoku Earthquake. *Journal of Zhejiang University-SCIENCE A (Applied Physics & Engineering)*, Vol.12, No.5, pp. 327-334

USCOLD (US Committee on Large Dams). *Observed performance of dams during earthquakes. Committee on Earthquakes*, 1992, July, Denver, CO

USCOLD (US Committee on Large Dams). *Updated guidelines for selecting seismic parameters for dam projects. Committee on Large Dams*, 1999, Denver, CO

Vucetic, M. & Dobry, R. (1991). Effects of the Soil Plasticity on Cyclic Response. *Journal of Geotechnical Engineering ASCE*, Vol.117, pp. 89–107

Wang, Z.L.; Makdisi, F.I. & Egan, J. (2006). Practical Applications of a Nonlinear Approach to Analysis of Earthquake-Induced Liquefaction and Deformation of Earth Structures. *Soil Dynamics and Earthquake Engineering*, Vol.26, pp. 231–252

Wieland, M. (2008). Large Dams the First Structures Designed Systematically against Earthquakes, *The 14th World Conference on Earthquake Engineering*, October 12-17, 2008, Beijing, China

Woodward, P.K. & Griffiths, D.V. (1996). Non-linear Dynamic Analysis of the Long Valley Dam. *Computer Methods and Advances in Geomechanics*, Vol.11, No.6, pp. 635-644

Yiagos, A.N. & Prevost, J.H. (1991). Tow-phase Elasto-plastic Seismic Response of Earth Dams: Applications. *Soil Dynamics and Earthquake Engineering*, Vol.10, No.7, pp. 371-381

Zeghal, M. & Abdel-Ghaffar, A.M. (1992). Analysis of Behavior of Earth Dam Using Stron-Motion Earthquake Records. *Journal of Geotechnical Engineering ASCE,* Vol.118, No.2, pp. 266-277

8

Selection of the Appropriate Methodology for Earthquake Safety Assessment of Dam Structures

Hasan Tosun[1] and Evren Seyrek[2]
[1]*Eskisehir Osmangazi University, Civil Engineering Department, Eskisehir*
[2]*Dumlupinar University, Civil Engineering Department, Kutahya*
Turkey

1. Introduction

Important lessons which reflect the seismic performance of dams under large earthquakes are available in the literature, (1971, San Fernando earthquake; 1985, Mexico earthquake; 1999, Kocaeli earthquake; 2001, Bhuj earthquake; 2008, Wenchuan earthquake). Seismic behavior of dams subjected to these severe earthquakes shows that earthquake safety of dams is an important phenomenon in dam engineering and requires a more comprehensive seismic studies.

Dams built on the site with high seismicity have a high-risk potential for downstream life and property. Active faults near dam sites can cause to damaging deformation of the embankment. In general, strong ground shaking can result instability of the dam and strength loss of foundations. (Seed et al., 1969; Seed et al., 1975; Jansen, 1988; Castro et al.,1985). In the last decade, large earthquakes have killed many thousands of people and caused economic devastation, commonly as a result of building failures in seismic events. Therefore, meaningful seismic parameters are needed to perform a satisfactory evaluation of dam structure (Tosun, 2002).

ICOLD (1989) stated that safety concerns for embankment dams subjected to earthquakes involve either the loss of stability due to a loss of strength of the embankment of foundation materials or excessive deformations such as slumping, settlement, cracking and planer or rotational slope failures. To obtain preliminary information about seismic parameters, the simplified procedures can be used. If the materials used in embankment are not susceptible to loss of strength and the hazard and risk ratings are low, the simplified analyses are entirely sufficient to define the seismic evaluation parameters. The safety concerns for concrete dams subjected to earthquakes involve evaluation of the overall stability of the structure, such as verifying its ability to resist induced lateral forces and moments and preventing excessive cracking of the concrete. For analyzing the loads, different procedures are performed. In the simplified analyses, peak ground motion parameters and response spectra are sufficient to define the seismic evaluation parameters. It is suggested the finite element method to be used for analyzing of most dams in high risk or hazard class.

The seismic hazard study at a dam site basically depends on the seismicity of a region, the types of structures involved and the consequences of failure. FEMA (2005) states that the design and evaluation of dams for earthquake loading should be based on a comparable level of study and analyses for each phase of the study including seismo-tectonic, geological, geotechnical and structural investigations. The level of study should reflect both the criticality of the structure and the complexity of the analysis procedures. Basic seismic studies for earthquake safety assessment of dams generally rely on existing seismological studies, available site data and simplified methods of design or evaluation developed for similar projects or structures. In other words, these studies use preliminary values of the ground motions obtained from existing studies, a simplified structural analysis and a general assessment of soil liquefaction and deformation. Detailed seismic studies involve the use of site-specific studies in evaluating the earthquake hazard and dynamic analyses for determining the response of project features to seismic loading. Detailed geological studies should define the seismic tectonic province, characterize the site, and investigate all faults that can be the source of ground shaking at the dam site.

Extensive field exploration and testing programs are necessary to perform earthquake safety assessment. The earthquake history, earthquake recurrence relationship and strong motion records should be defined by seismological investigation (Erdik et al, 1985). Structural investigations should consider all relevant factors that affect the seismic hazard at the specific site and the actual dynamic performance of the structure. Geotechnical studies should relieve the types and spatial distribution of foundation and embankment materials and the engineering properties of soil and rock, liquefaction potential of the foundation and embankment soils, stability of natural and artificial slopes and estimation of deformations. The final results of all studies should be used as a basis for making design or evaluation decisions and for designing remedial measures.

2. Selection of design and safety evaluation earthquake

The selection of site-dependent seismic input is an important stage for determining the safety evaluation of dam structures. The earthquakes have been specified by different terms such as the Operating Basis Earthquake (OBE), the Maximum Credible Earthquake (MDE) and the Safety Evaluation Earthquake (SEE). These earthquakes have been defined by separate organization with different value. FEMA (2005) has meaningfully defined those earthquakes in recent. Their short definitions are given below as based on this guideline of FEMA.

2.1 Operating Basis Earthquake (OBE)

This earthquake is defined for the ground motions at the site to be expected to occur within the service life of the project. FEMA (2005) states that the associated performance requirement is that the project function with little or no damage and without interruption of function. The OBE means to protect against economic losses from damage or loss of service. Consequently, the return period of the OBE can be based on economic considerations.

2.2 Maximum Credible Earthquake (MCE)

According to FEMA (2005), this is the largest earthquake magnitude that could occur along a recognized fault or within a particular seismo-tectonic province or source area under the current tectonic framework. The loading resulting from MCE can be exceeded for probabilistic methods, which is discussed later on, for high return period faults close in, such as North Anatolian Fault in Turkey and San Andreas Fault in USA.

2.3 Maximum Design Earthquake (MDE) or Safety Evaluation Earthquake (SEE)

The MDE is the earthquake that produces the maximum level of ground motion for which a structure is to be designed, while SEE is defined just for safety evaluation. These earthquakes may be considered as earthquakes which are equal to the MCE or to a design earthquake less than the MCE. As based on the FEMA (2005), the associated performance requirement for the MDE or SEE is that the project performs without catastrophic failure, such as uncontrolled release of a reservoir, although significant damage or economic loss may be tolerated.

Earthquake ground motions at a particular site are estimated through a seismic hazard evaluation. The geologic and seismologic inputs needed for completing a seismic hazard evaluation. Two different methods are widely used for describing earthquake ground motions for seismic design. These are the deterministic seismic hazard analysis (DSHA) and the probabilistic seismic hazard analysis (PSHA). According to Kramer (1996), DSHA considers a seismic scenario that includes a four-step process. This procedure gives rational solutions for large dams because it provides a straightforward framework for evaluation of the worst situation. The DSHA procedure is outlined as below and introduced schematically in Figure 1.

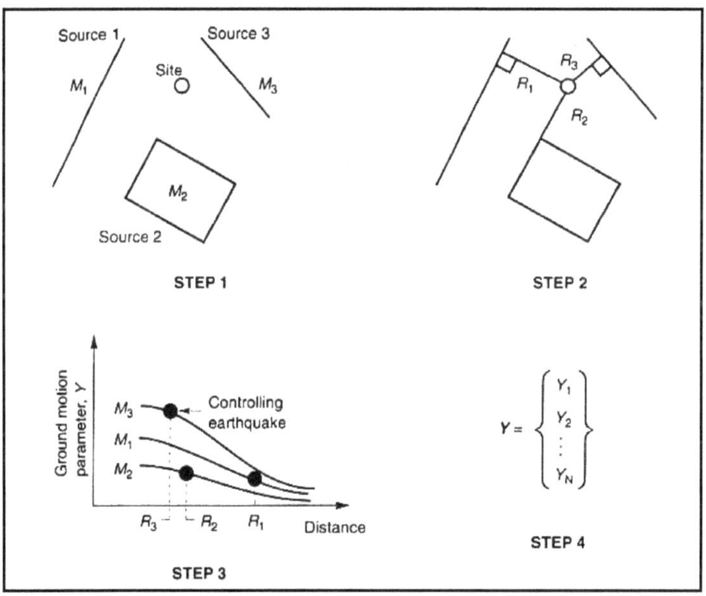

Fig. 1. Four steps of deterministic seismic hazard analysis (Kramer, 1996)

1. Identification and characterization of all earthquake sources capable of producing significant ground motion at the site.

2. Selection of a source-to-site distance parameter for each source zone.
3. Selection of the controlling earthquake.
4. Definition of hazard at the site in terms of the ground motions produced at the site by controlling earthquake.

PSHA has allowed the uncertainties in the size, location and rate of recurrence of earthquakes, as well as in the variation of ground motion characteristics with earthquake size and location, to be explicitly considered in the evaluation of seismic hazards (Figure 2). This method is generally similar with DSHA and outlined as follows:

1. Identification and characterization of earthquake sources.
2. Characterization of seismic activity and temporal distribution of earthquake recurrence.
3. Determination of ground motion with use of predictive relationships.
4. Prediction of ground motion parameters

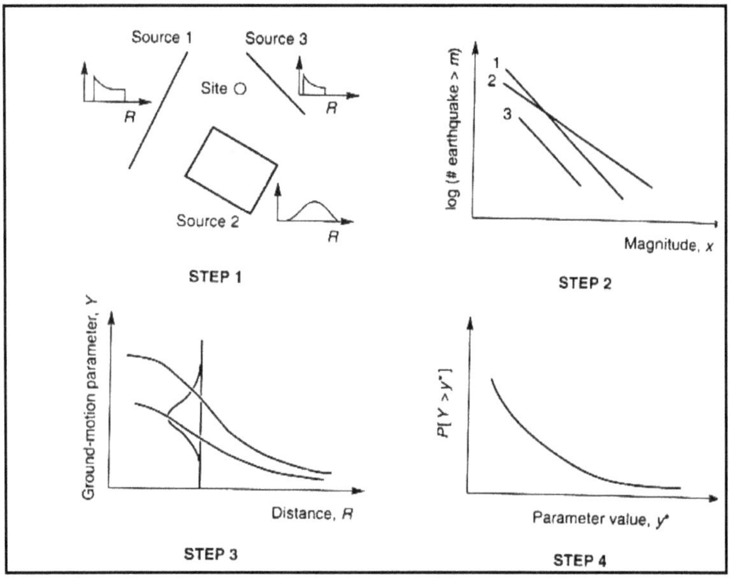

Fig. 2. Four steps of probabilistic seismic hazard analysis (Kramer, 1996)

DSHA and PSHA have been performed by different researchers in specific sites according to their project goals and importance (Chandler et al. 2001; Fat-Helbary and Tealb 2002; Al-Homoud, 2003; Ardeleanu et al. 2005; Simeonova et al. 2006; Orhan et al. 2007; Nakajima et al. 2007; Tosun and Seyrek, 2006; Tosun et al. 2007; Seyrek and Tosun, 2011).

Both probabilistic and deterministic methods have a role in hazard and risk analyses performed for decision-making purposes. One method may have priority over the other, depending on the seismic environment and the scope of the project (McGuire 2001). McGuire (2001) claims that the analysis of a specific site usually requires a probabilistic approach, but a deterministic check on the resulting decision is appropriate. Generally seismic sources contribute to the seismic hazard and risk at a site, and the integration of these through a probabilistic analysis provides the most insight.

Orozova and Suhadolc (1999) express that all the earthquakes with different magnitudes and distance influences the seismic hazard at a site and PSHA correctly reflects the actual knowledge of seismicity. Another advantage of probabilistic approach is that offers a rational framework for risk management by considering the frequency or probability of exceedance of the ground motion against which a structure or facility is designed (Bommer and Abrahamson 2006).

A severe criticism of PSHA came from Castanos and Lomnitz (2002), who consider that the problem with PSHA is that its data are inadequate and its logic is defective. They suggest that the deterministic procedures especially when coupled with engineering judgment to be more reliable and more scientific.

3. Methods of analysis

In this section, seismic hazard analysis results for two different methodology presented in previous section will be discussed by using acceleration values of dam site locations subjected to different seismicity and geological setting.

For the seismic hazard analysis of the dams in Turkey, all possible seismic sources were identified and their potential was evaluated in detail, as based on the guidelines given by Fraser and Howard (2002). Various seismic source models and active fault maps have been previously reported in Turkey (Yücemen, 1982; Erdik et al. 1985; Erdik et al. 1999; Şaroğlu et al. 1992; Kayabalı and Akın, 2003; Ulusay et al. 2004). These seismic-source models have been modified taking advantage of recent neotectonic and seismic data for Turkey (Figure 3). The data about 20th century instrumentally recorded earthquakes for Turkey and vicinity

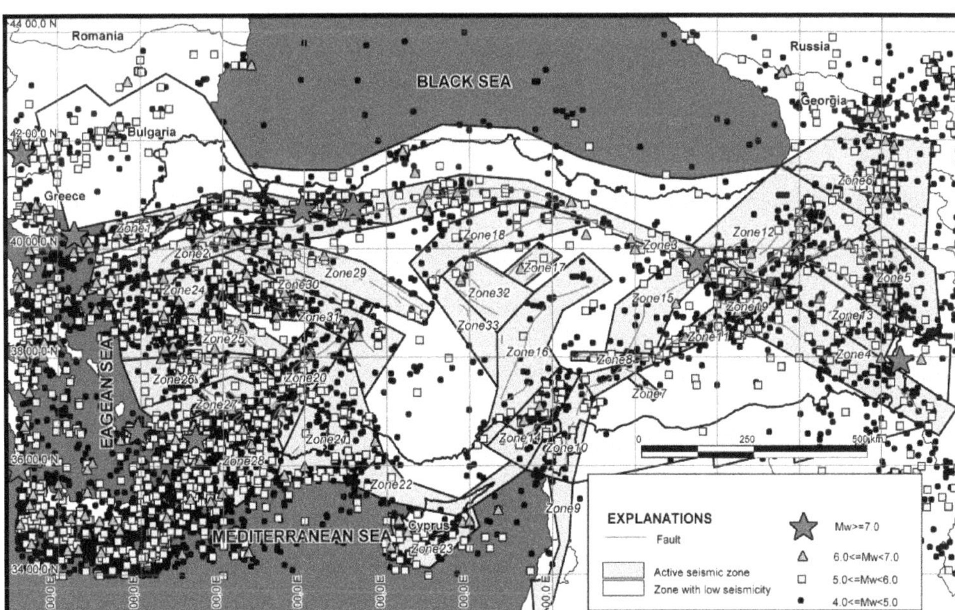

Fig. 3. Seismo-tectonic map of Turkey

were collected by the Bogazici University Kandilli Observatory and Earthquake Research Institute, National Earthquake Monitoring Center. It should be noted that moment magnitude scale is used for hazard calculations.

To research the effect of seismic hazard models on the results, Altinkaya, Ayhanlar, Boztepe, Hatap ve Kilickaya dam sites are considered (Figure 4). Three of them are located in Yesilirmak basin and others are in Kizilirmak basin. The neotectonics of the region around the dams is governed by four major elements: (1) the North Anatolian Fault (NAF), which is the main structural feature in the basin, (2) the Ezinepazari Fault, which lies in south-west direction as a secondary feature of NAF, (3) Ecemis Fault, which starts from Mediterranean Sea at the south and has approximately same direction as Ezinepazari Fault. (4) Shear zone including secondary faults at the central part of basin (Figure 4).

The North Anatolian Fault is one of the best-known strike-slip faults in the world, because of its significant seismic activity and well developed surface features. It is approximately 1,500 km long and extends from eastern Turkey at the east to Greece at the west. Its width ranges from a single zone of a few hundred meters to multiple shear zones of 40 km. This fault produces very large earthquakes, which have resulted in the death of one thousand people and severe structural damages. The Ezinepazari fault which is a secondary branch of NAF has approximately 260 km length and extends to the central part of Anatolia in south-west direction. Its width can be defined by a single zone of a few hundred meters. This zone also comprises the Bala fault, which recently generated the earthquake near Ankara city, and Merzifon fault, which is known as the source of large earthquakes occurred in the past. The Ecemis fault, presently called as Central Anatolian Fault Zone, is also a strike-slip fault, which is located at the southern part of basin. It is composed of several Fault segments and its width ranges from 2 to 15 km. Kocyigit and Beyhan (1997) stated that this fault connects to the North Anatolian Fault and extends to south as far as the Hellenic trench, reaching up to several hundred kilometers in length. The shear zone, which is located from the central part to the south-west of Kızılırmak basin, contains structural features such as Kırsehir, Gumuskent, Delice, Akpinar, and Bala faults. Seyrek and Tosun (2011) remark that this zone has low seismic activity, although an earthquake with moderate magnitude occurred in 1938.

During the analysis process, seismic zones and earthquakes within the area having a radius of 100 km around the dam site were considered. Deterministic and probabilistic seismic hazard analyses were performed by the computer program DAMHA, which was developed at the Earthquake Research Center, Eskişehir Osmangazi University. DAMHA is capable of performing the deterministic and probabilistic seismic hazard analyses on digitized tectonic map of Turkey. This program includes databases for earthquakes, faults, area seismic sources and attenuation equations (Seyrek, 2009). This program calculates the probabilistic hazard for three hazard level as OBE, MDE and SEE. These levels correspond to the return periods of 144, 475 and 2475 years respectively.

For horizontal peak ground acceleration calculation, eight separate predictive relationships (Campbell, 1981; Boore et al. 1993; Campbell and Bozorgnia, 1994; Ambraseys et al. 1995; Boore et al. 1997; Gülkan and Kalkan, 2002; Kalkan and Gülkan, 2004; Ambraseys et al. 2005) were considered. It is clear that the use of several attenuation laws can result in more reliable evaluation than a single relationship (ICOLD 1989). It was noted that Gülkan and

Kalkan (2002) and Ambraseys et al. 2005 were derived from Turkey earthquakes. Other attenuation relationships used for this study were selected owing to the similarities between the mechanisms of North Anatolia Fault Zone and San Andreas Fault.

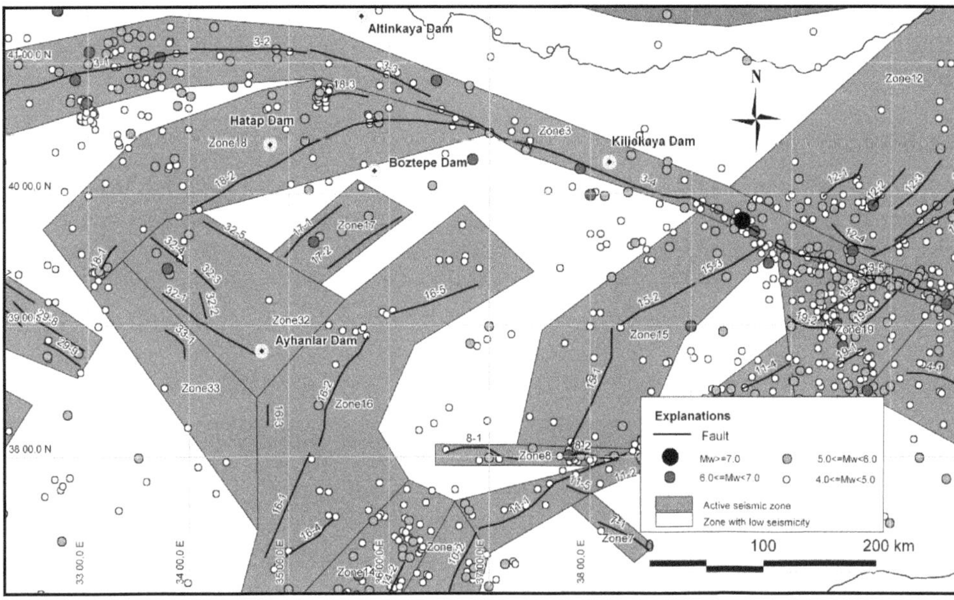

Fig. 4. Locations of dams on seismo-tectonic map

One of the most important parameters of source seismicity is the size (magnitude) of the maximum earthquake. The general assumption is that one-third to half of the total length of the fault would break when it generates the maximum earthquake (Mark, 1977). In this study, for each source the maximum earthquake magnitude was determined using the empirical relationships proposed by Wells and Coppersmith (1994) as given in Table 1. Where M_w is the earthquake moment magnitude and L is the fault rupture length in km.

Fault type	Equation
Strike slip	$M_w = 5.16+1.12\log L$
Reverse	$M_w = 5.00+1.22\log L$
Normal	$M_w = 4.86+1.32\log L$
All	$M_w = 5.08+1.16\log L$

Table 1. Relationship between earthquake magnitude (M_w) and rupture length (L)

Once the maximum earthquake magnitude is determined for each seismic source, a linear regression is performed to estimate the coefficients of Gutenberg-Richter (1944) relationship using the computer program DAMHA. Orhan et al. (2007) express that the records which have a magnitude equal to or greater than 4.0 is more credible in Turkish earthquake catalogue.

4. Analyses and results

For the seismic hazard analyses of the dam sites, a detailed study was performed. Local geological features and seismic history referred in previous section were used to quantify the rate of seismic activity in the basin. To present the effect of methodology on seismic hazard results, five dam sites are chosen. Below, details of analysis results will be given for each dam site.

4.1 Altinkaya dam

Altinkaya dam with a storage capacity of 5763 hm^3 is also located on a Kizilirmak river (Figure 5). It was designed as rockfill dam and its construction was finished in 1988. Its height from river bed is 195 meter. It produces electricity of 1632 GWh per year with an installed capacity of 700 MW. Altinkaya dam is one the most important dam projects in Kizilirmak basin.

Fig. 5. View of the Altinkaya dam

As a result of detailed evaluation, two seismic sources are considered for hazard calculations. Seismic parameters used for hazard assessment are given in Table 2.

Results of deterministic and probabilistic analyses are given in Table 3. It should be noted that each PGA value introduced for a dam site in these tables represent the average of those obtained from eight different attenuation relationships discussed in previous section. Total seismic hazard curve of Altinkaya dam site is presented in Figure 6. PGA value for SEE level is 0.25 g.

Zone no	Fault no	Fault name	Fault type*	M_{max}	a**	b**
3	3-1	North Anatolian Fault Zone Segment	SS*	7.7	5.172	0.684
	3-2	North Anatolian Fault Zone Segment	SS	7.4		
	3-3	North Anatolian Fault Zone Segment	SS	7.4		
	3-4	North Anatolian Fault Zone Segment	SS	7.9		
	3-5	North Anatolian Fault Zone Segment	SS	7.6		
18	18-1	Bala Fault	N	6.7	5.457	0.878
	18-2	Ezinepazari Fault	SS	7.9		
	18-3	Merzifon Fault	SS	6.9		

* SS : Strike slip N: Normal
**from Gutenberg-Richter (1944) law

Table 2. Hazard parameters of seismic sources for Altinkaya dam site

Critical Zone	Critical Segment	Closest distance (km)	PGA (g)				
			DSHA		PSHA		
			50th percentile	84th percentile	OBE	MDE	SEE
3	3-3	39.8	0.11	0.19	0.13	0.18	0.25

Table 3. DSHA and PSHA results for Altinkaya dam site

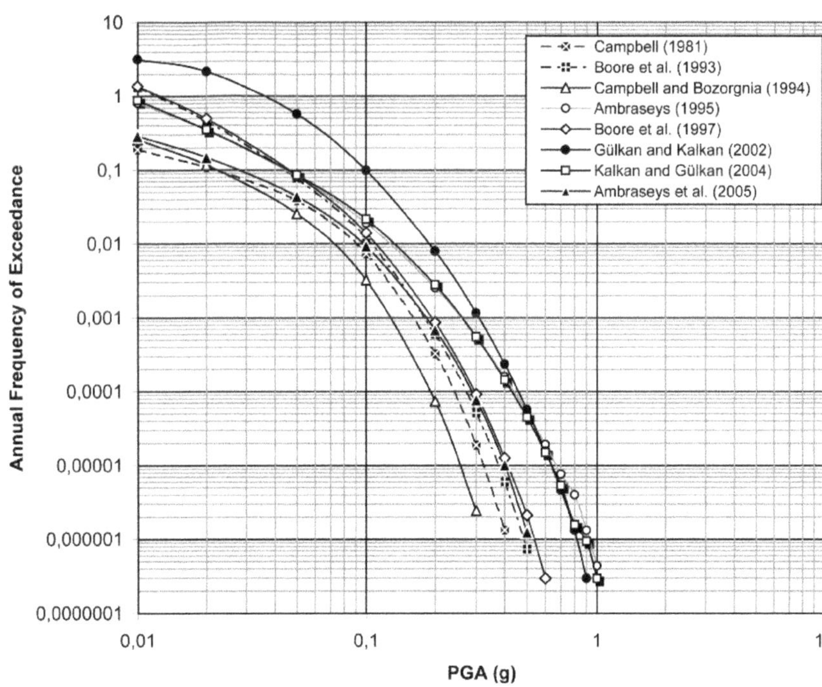

Fig. 6. Seismic hazard curve for Altinkaya dam site

4.2 Kilickaya dam

Kilickaya dam is one of the large dams located in Yesilirmak basin and was constructed on Kelkit river for energy and flood control purposes. It was designed as rockfill dam and finished in 1990. When the reservoir is at normal capacity, the facility impounds 1400 hm^3 of water with a reservoir surface area of 64 km^2. It has a height of 103 m from river bed. It produces the electricity with an installed capacity of 124 MW (Figure 7).

Fig. 7. View of the Kilickaya dam

North Anatolian Fault Zone and Malatya Ovacik Fault Zone are taken into consideration for hazard calculations. Seismic parameters used for hazard assessment are given in Table 4.

Zone no	Fault no	Fault name	Fault type*	M_{max}	a**	b**
3	3-1	North Anatolian Fault Zone Segment	SS*	7.7	5.172	0.684
	3-2	North Anatolian Fault Zone Segment	SS	7.4		
	3-3	North Anatolian Fault Zone Segment	SS	7.4		
	3-4	North Anatolian Fault Zone Segment	SS	7.9		
	3-5	North Anatolian Fault Zone Segment	SS	7.6		
15	15-1	Malatya Ovacik Fault Zone	SS	7.4	4.084	0.626
	15-2	Malatya Ovacik Fault Zone	SS	7.2		
	15-3	Malatya Ovacik Fault Zone	SS	7.2		

* SS : Strike slip
**from Gutenberg-Richter (1944) law

Table 4. Hazard parameters of seismic sources for Altinkaya dam site

The analyses show that the most critical zone is North Anatolian Fault Zone and closest distance from this seismic source to dam site is 11.0 km. PGA value for 84th percentile is greater than SEE level (Table 5). Total seismic hazard curve of Altinkaya dam site is presented in Figure 8. PGA value for 84th percentile is greater than SEE level.

Critical Zone	Critical Segment	Closest distance (km)	PGA (g)				
			DSHA		PSHA		
			50th percentile	84th percentile	OBE	MDE	SEE
3	3-4	11.0	0.37	0.60	0.17	0.26	0.40

Table 5. DSHA and PSHA results for Kilickaya dam site

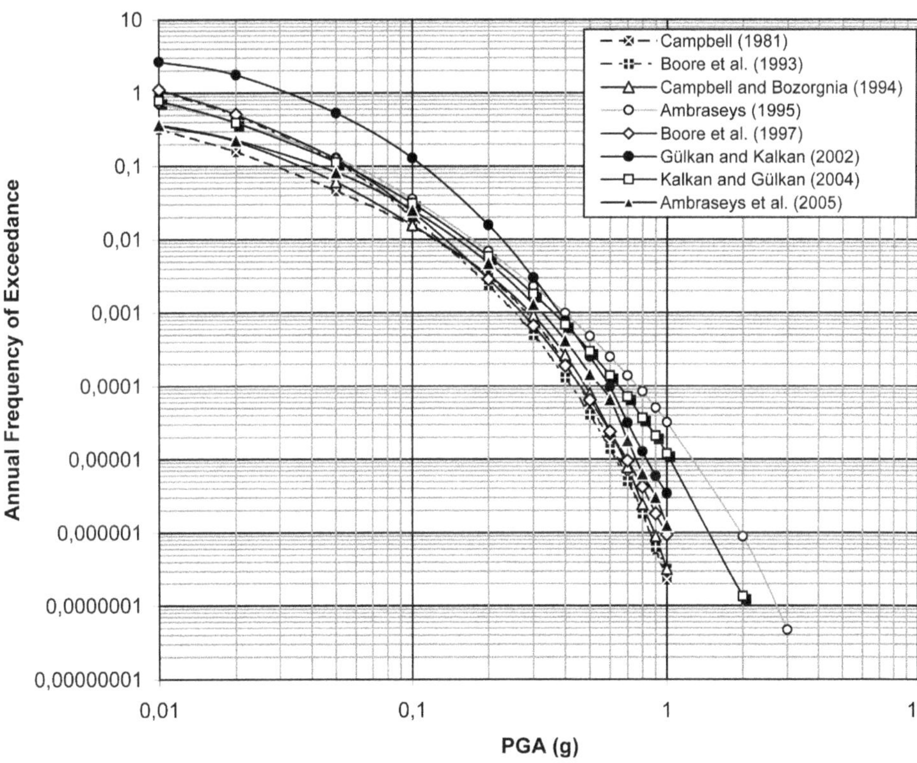

Fig. 8. Seismic hazard curve for Kilickaya dam site

4.3 Hatap dam

Hatap dam with a body volume of 1.25 hm³ was designed as rockfill dam with a clay core. The construction of the dam was finished in 2009. It is located on the Hatap creek and has

42-m height from river bed (Figure 9). Main purposes are irrigation and water supply. When the reservoir is at normal capacity, the facility impounds 11.6 hm^3 of water with a reservoir surface area of 1.02 km^2.

The project site are affected from source zone 3, 17, 18 and 32. Seismic parameters used for hazard assessment are given in Table 6.

Zone no	Fault no	Fault name	Fault type*	M_{max}	a**	b**
3	3-1	North Anatolian Fault Zone Segment	SS*	7.7	5.172	0.684
	3-2	North Anatolian Fault Zone Segment	SS	7.4		
	3-3	North Anatolian Fault Zone Segment	SS	7.4		
	3-4	North Anatolian Fault Zone Segment	SS	7.9		
	3-5	North Anatolian Fault Zone Segment	SS	7.6		
17	17-1	Sorgun Fault	SS	7.2	2.655	0.413
	17-2	Sarikaya Akdagmadeni Fault	SS	7.3		
18	18-1	Bala Fault	N	6.7	5.457	0.878
	18-2	Ezinepazari Fault	SS	7.9		
	18-3	Merzifon Fault	SS	6.9		
32	32-1	Gumuskent Fault	SS	7.4	3.054	0.476
	32-2	Kirsehir Fault	SS	6.6		
	32-3	Akpinar-Kirsehir Fault Zone	SS	6.7		
	32-4	Akpinar-Kirsehir Fault Zone	SS	6.9		
	32-5	Delice(Yerkoy) Fault	SS	7.2		

*SS: Strike slip N: Normal U: Unknown
**from Gutenberg-Richter (1944) law

Table 6. Hazard parameters of seismic sources for Hatap dam site

Seismic hazard analysis results are presented in Table 7. The critical segment is Ezinepazari Fault with a closest distance of 17.7m from dam site. Figure 10 gives the probabilistic hazard curve based on different attenuation relationships. Average PGA values are 0.16, 0.24 and 0.37 g respectively.

Fig. 9. View of the Hatap dam

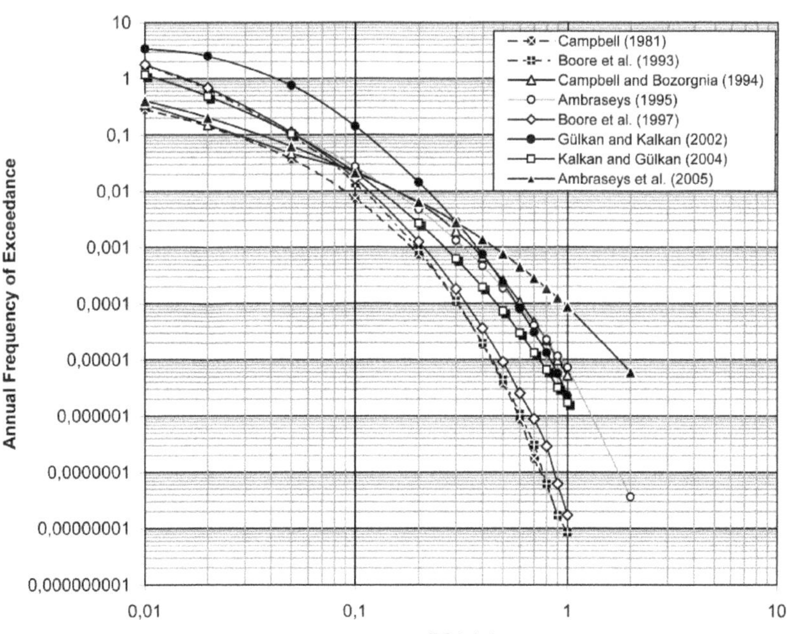

Fig. 10. Seismic hazard curve for Hatap dam site

Critical Zone	Critical Segment	Closest distance (km)	PGA (g)				
			DSHA		PSHA		
			50th percentile	84th percentile	OBE	MDE	SEE
18	18-2	17.7	0.27	0.45	0.16	0.24	0.37

Table 7. DSHA and PSHA results for Hatap dam site

4.4 Ayhanlar dam

Ayhanlar dam is located on Kiziloz creek with a storage capacity of 21.8 hm^3 (Figure 11). It was designed as earthfill dam and its construction was finished in 2003. Its body volume is 1.2 hm^3 and height from river bed is 35 m. Main purpose is to perform irrigation.

Fig. 11. View of the Ayhanlar dam

From seismo-tectonic map of the investigation area, four zones with eleven segment are included into the seismic hazard analyses. Details of the seismic hazard parameters of each source are given in Table 8.

PGA values of dam site are given on the basis of deterministic and probabilistic approach (Table 9). The critical segment is Gumuskent Fault with a magnitude of 7.4. Figure 12 presents the probabilistic hazard results by means of seismic hazard curve. PGA values for return period of 475 and 2475 years are 0.16 and 0.30g respectively.

Zone no	Fault no	Fault name	Fault type**	M_{max}	a***	b***
16	16-1	Ecemis Fault Zone Segment	SS	7.4	4.805	0.814
	16-2	Ecemis Fault Zone Segment	SS	7.4		
	16-3	Derinkuyu Fault	N	6.5		
	16-4	Karsanti-Karaisali Fault Zone	SS	7.2		
	16-5	Deliler Fault	SS	7.3		
17	17-1	Sorgun Fault	SS	7.2	2.655	0.413
	17-2	Sarikaya Akdagmadeni Fault	SS	7.3		
32	32-1	Gumuskent Fault	SS	7.4	3.054	0.476
	32-2	Kirsehir Fault	SS	6.6		
	32-3	Akpinar-Kirsehir Fault Zone	SS	6.7		
	32-4	Akpinar-Kirsehir Fault Zone	SS	6.9		
	32-5	Delice(Yerkoy) Fault	SS	7.2		
33	33-1	NA*	U	6.8	4.112	0.846

* NA: Non available
**SS: Strike slip N: Normal U: Unknown
***from Gutenberg-Richter (1944) law

Table 8. Hazard parameters of seismic sources for Hatap dam site

Critical Zone	Critical Segment	Closest distance (km)	PGA (g)				
			DSHA		PSHA		
			50th percentile	84th percentile	OBE	MDE	SEE
32	32-1	6.9	0.38	0.64	0.09	0.16	0.30

Table 9. DSHA and PSHA results for Hatap dam site

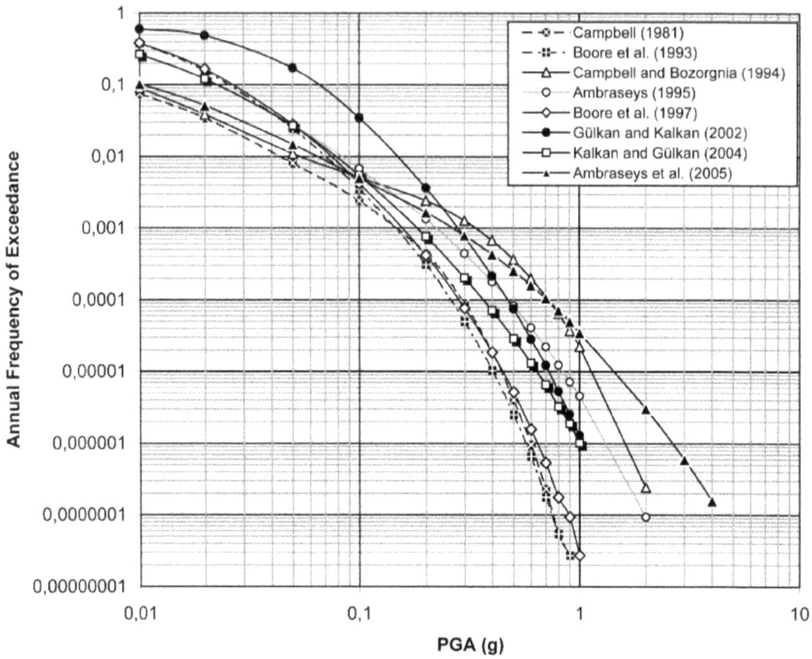

Fig. 12. Seismic hazard curve for Ayhanlar dam site

4.5 Boztepe dam

Boztepe dam with a storage capacity of 14.2 hm³ is also located on a Boztepe creek (Figure 13). It was designed as earthfill dam and its construction was finished in 1984. Its height from river bed is 27 meter. Main aim of this dam is irrigation.

Fig. 13. View of the Boztepe dam

Five seismic source zone are included into the analyses. Seismic parameters used for hazard assessment are given in Table 10.

Zone no	Fault no	Fault name	Fault type*	M_{max}	a**	b**
8	8-1	Elbistan Fault	SS	6.9	5.050	0.831
	8-2	Surgu Fault	SS	6.6		
10	10-1	East Anatolian Fault Zone Segment	SS	7.0	5.585	0.925
	10-2	East Anatolian Fault Zone Segment	SS	6.8		
11	11-1	East Anatolian Fault Zone Segment	SS	7.0	5.882	0.948
	11-2	East Anatolian Fault Zone Segment	SS	6.6		
	11-3	East Anatolian Fault Zone Segment	SS	7.0		
	11-4	East Anatolian Fault Zone Segment	SS	6.7		
	11-5	Tut Fault	SS	6.3		
14	14-1	Karatas-Osmaniye Fault Zone	SS	6.8	5.280	0.818
	14-2	Karatas-Osmaniye Fault Zone	SS	6.5		
16	16-1	Ecemis Fault Zone Segment	SS	7.4	4.805	0.814
	16-2	Ecemis Fault Zone Segment	SS	7.4		
	16-3	Derinkuyu Fault	N	6.5		
	16-4	Karsanti-Karaisali Fault Zone	SS	7.2		
	16-5	Deliler Fault	SS	7.3		

*SS: Strike slip N: Normal U: Unknown
***from Gutenberg-Richter (1944) law

Table 10. Hazard parameters of seismic sources for Boztepe dam site

Results of deterministic and probabilistic analyses are given in Table 11. Average PGA values from eight different attenuation relationships are seen in this table. Total seismic hazard curve of Boztepe dam site is presented in Figure 14. PGA value for SEE level is 0.24 g.

Critical Zone	Critical Segment	Closest distance (km)	PGA (g)				
			DSHA		PSHA		
			50th percentile	84th percentile	OBE	MDE	SEE
10	10-2	27.6	0.11	0.19	0.13	0.18	0.24

Table 11. DSHA and PSHA results for Boztepe dam site

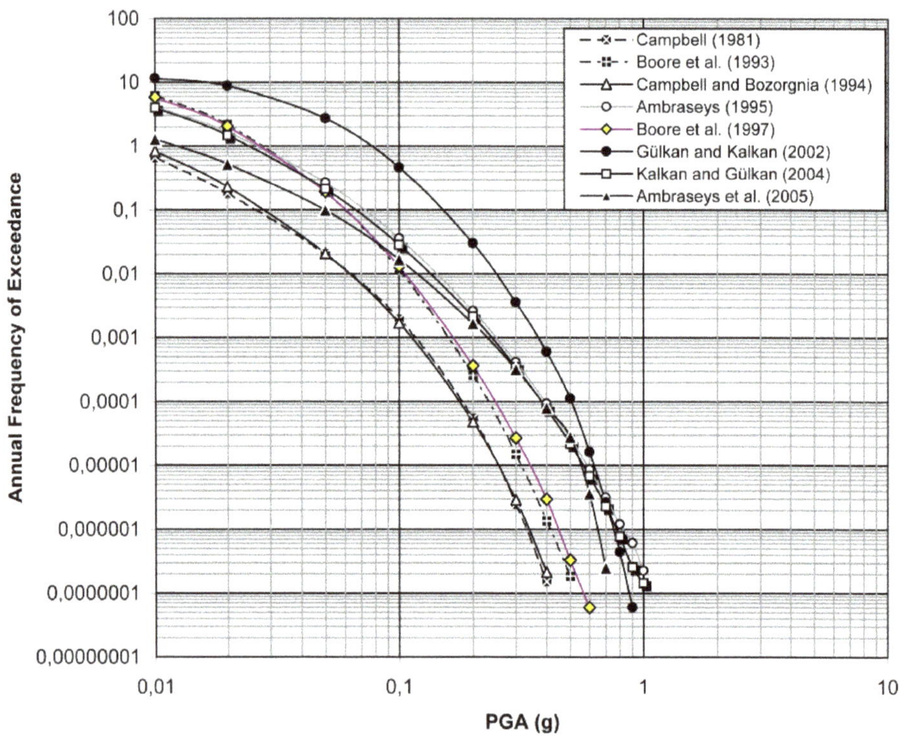

Fig. 14. Seismic hazard curve for Boztepe dam site

5. Discussions

Turkey, which has at least 1200 large dams, is one of the most seismically active regions in the world and major earthquakes with the potential of threading life and property occur frequently here. It is obvious that many of the large dams planned in Turkey are located in zones of moderate-high seismicity. Thus it is very important that dams to be resistant to the strong earthquakes. Many of guidelines about earthquake safety of dams promise different methodology for dam site locations with different seismicity.

The seismic hazard of a dam site is based on the peak ground acceleration. This value derived from the defined design earthquake produces the main seismic loads. For preliminary study, the existing map of seismic zones can be used to estimate the seismic hazard of a dam site. However, authors believe that the detailed seismic hazard analyses should be performed for safety evaluation of existing dams. Because, the main requirement of an earthquake–resistant design of a dam is to protect public safety and property. Therefore, seismic criteria and analysis parameters for dams should be selected more conservatively than for conventional structures since the failures are more disastrous.

To reveal the effects of methodologies on seismic hazard results, five dam sites are chosen and repetitive analyses are performed. Deterministic and probabilistic methods give different results for each dam sites. It is obvious that deterministic method (84th percentile) gives the maximum value for dams within a near seismic source zone. For the dams which are not close to the seismic sources, deterministic PGA value is lower than SEE level. But it is necessary to perform a lot of examination for locations with low to high seismicity. Another important effect on these differences is sensitiveness of attenuation equations to distance and earthquake magnitude. For further study, analysis results should be compared for each attenuation equation.

Authors state that seismic performance of dams within the near source zone must be reevaluated in detail. Large dams which have high-risk class should be evaluated with highest priority as a part of the National Dam Safety Program. Both DSHA and PSHA must be performed for these dams.

6. References

Al-Homoud, A., 2003, Evaluation of strong motion acceleration for embankment dam design considering local seismotectonics, Natural Hazards, 29, p: 37-56.

Ambraseys, N.N., 1995, The prediction of earthquake peak ground acceleration in Europe,Earthquake Engineering and Structural Dynamics, Vol. 24, 467-490.

Ambraseys, N.N., Douglas, J., Karma, S.K. and Smit, P.M., 2005, Equations for the estimation of strong ground motions from shallow crustral earthquakes using data from Europe and the Middle East: horizontal peak ground acceleration and spectral acceleration, Bulletin of Earthquake Engineering, 3, 1-53.

Ardeleanu, L., Leydecker, G., Bonje,r K., Busche, H., Kaise,r D., Schmitt, T., 2005, Probabilistic seismic hazard map for Romania as a basis for a new building code. Natural Hazards Earth Syst Sci 5:679–684.

Bommer, JJ, Abrahamson, N., 2006, Why do modern probabilistic seismic hazard analyses often lead to increased hazard estimates? Bull Seismol Soc Am 96(6):1967–1977.

Boore, D.M., Joyner, W.B. and Fumal, T. E., 1993, Estimation of response spectra and peak accelerations from Western North American Earthquakes: An interim report, U.S. Geol. Surv. Open-File Rept. 93-509, 72 p.

Boore, D.M., Joyner, W.B. and Fumal, T.E., 1997, Equations for estimating horizontal response spectra and peak acceleration from Western North American earthquakes: A summary of recent work, Seis. Res. Let., 68, 128-153.

Campbell, K.W., 1981, Near-source attenuation of peak horizontal acceleration. Bulletin Seism. Soc. Am., vol. 71, N.6, 2039-2070.

Campbell, K.W. and Bozorgnia, Y., 1994, Near-source attenuation of peak horizontal acceleration from worldwide accelerograms recorded from 1957 to 1993, Proc. Fifth U.S. National Conference on Earthquake Engineering 3, Chicago, Illinois, 283-292.

Castanos, H. and Lomnitz, C., 2002, PSHA: is it science?, Engineering Geology, 66, 315-317.

Castro, G., Poulos, S.J., Leathers, F. 1985, Re-examination of Slide of Lower San Fernando Dam. ASCE, Journal of Geotechnical Engineering, V.111.

Chandler, A.M., Chan, L.S. and Lam, N.T.K., 2001, Deterministic seismic hazard parameters and engineering risk implications fort he Hong Kong region, Journal of Asian Earth Sciences, 20, 59-72.

Erdik, M., Doyuran, V., Gülkan, P., Akkaş, N. 1985, Evaluation of Earthquake Hazard in Turkey with Statistical Approach. Middle East Technical University Earthquake Engineering Research Center, Ankara, 116 p, (in Turkish).

Erdik, M., Biro, Y.A., Onur, T., Sesetyan, K. and Birgoren, G., 1999, Assessment of earthquake hazard in Turkey and neighboring rgions, Annali Di Geofisica, Vol. 42, No. 6, 1125-1138.

Fat-Helbary, R., Tealb, A.A., 2002, A study of seismicity and earthquake hazard at the proposed Kalabsha dam site, Aswan, Egypt, Natural Hazards, 25, p: 117-133.

FEMA 2005, Federal Guidelines for Dam Safety-Earthquake Analyses and Design of Dams.

Fraser, W.A. and Howard, J.K., 2002, Guidelines for use of the consequence-hazard matrix and selection of ground motion parameters, Technical Publication, Depatment of Water Resources, Division of Safety of Dam, 7p.

Gülkan, P. and Kalkan, E., 2002, Attenuation modeling of recent earthquakes in Turkey, Journal of Seismology, 6, No. 3, 397-409.

ICOLD 1989, Selecting Parameters for Large Dams-Guidelines and Recommendations. ICOLD Committee on Seismic Aspects of Large Dams, Bulletin 72.

Jansen, R.B. (ed) 1988, Advanced Dam Engineering for Design, Construction and Rehabilitation. Van Noswtrand Reinhold, New York, 884p.

Kalkan, E., Gülkan, P., 2004, Site-dependent spectra derived from ground motion records in Turkey. Earthquake Spectra 20(4):1111–1138.

Kayabalı, K. and Akın, M., 2003, Seismic hazard map of Turkey using the deterministic approach, Engineering Geology, 69, 127-137.

Kocyigit, A., Beyhan, A., 1997, A new intracontinental transcurrent structure-the central anatolian fault zone, Turkey. Tectonophysics 284:317–336.

Kramer, S.L., 1996, Geotechnical earthquake engineering, Prentice Hall, New Jersey, 653 p.

Mark, R.K., 1977. Application of linear statistical models of earthquake magnitude versus fault length in estimating maximum expectable earthquakes. Geology 5, 464- 466.

McGuire, R.K., 2001, Deterministic vs. probabilistic earthquake hazards and risks, Soil Dynamics and Earthquake Engineering, 21, 377-384.

Nakajima, M., Choi, I., Ohtori, Y. And Choun, Y., 2007, Evaluation of seismic hazard curves and scenario earthquakes for Korean sites based on probabilistic seismic hazard analysis, Nuclear Engineering and Design, 237, 277-288.

Orhan, A., Seyrek, E. and Tosun, H., 2007, A probabilistic approach for earthquake hazard assessment of the Province of Eskisehir, Turkey, Nat. Hazards Earth Syst. Sci., 7, 607–614.

Orozova, I.M. and Suhadolc, P., 1999, A deterministic-probabilistic approach for seismic hazard assessment, Tectonophysics, 312, 191-202.

Seed, H.B., Lee, K.L., and Idriss, I.M. 1969, Analysis of Sheffield Dam failure. Journal of Soil Mechanics and Foundations-ASCE, 95(SM6), 1453–1490.

Seed, H.B., Lee, K.L., Idriss, I.M., and Makdisi, F.I. 1975, The slides in the San Fernando Dams during the Earthquake of February 9, 1971, J. Geotech. Eng.-ASCE, 101(GT7), 651–688.

Simeonova, S.D., Solakov, D.E., Leydecker, G., Busche, H., Schmitt, T., Kaiser, D., 2006, Probabilistic seismic hazard map for Bulgaria as a basis for a new building code. Nat Hazard Earth Syst Sci 6:881–887.

Seyrek, E., 2009, Numerical solution models for seismic hazard analysis of dam sites and a case study. PhD thesis, Eskisehir Osmangazi University

Seyrek, E., Tosun, H., 2011, Deterministic approach to the seismic hazard of dam sites in Kizilirmak Basin, Turkey, Natural Hazards, 14 p. (DOI 10.1007/s11069-011-9795-7, Published online 01 April 2011)

Saroglu, F., Emre, O., Kuscu, I., 1992, Active Fault Map of Turkey. General Directorate of Mineral Research and Exploration, Ankara, Turkey

Tosun, H. 2002, Earthquake-Resistant Design for Embankment Dams. Publication of General Directorate of State Hydraulic Works, Ankara, 208 p.(in Turkish).

Tosun, H., Seyrek, E., 2006, "Seismic studies on Turkish dams", International Water Power&Dam Construction, February 2006, 20-23.

Tosun, H., Zorluer, İ., Orhan, A., Seyrek, E., Türköz, M., Savaş, H., 2007, "Seismic hazard and total risk analyses for large dams in Euphrates Basin in Turkey", Engineering Geology, Volume 89, Issues 1-2 , 12 January 2007, 155-170.

Ulusay, R., Tuncay, E., Sönmez, H. and Gökçeoğlu, C., 2004, An attenuation relationship based on Turkish syrong motion data and iso-acceleration map of Turkey, Engineering Geology, 74, 265-291.

Wells, D.L. and Coppersmith, K.J., 1994, New empirical relationships among magnitude, rupture length, rupture width, rupture area, and surface displacement, Bulletin of the Seismological Society of America, Vol. 84, No. 4, 974-1002.

Yücemen, S., 1982, Seismic Risk Analysis, Publication of Middle East Technical University, Ankara (in Turkish)

Recent Landslide Damming Events and Their Hazard Mitigation Strategies

Ahsan Sattar and Kazuo Konagai
Institute of Industrial Science, University of Tokyo
Japan

1. Introduction

Earthquakes often strike the vulnerable parts of our society. A 7.0-magnitude earthquake shattered Port-au-Prince, the capital of Haiti. Based on historical records in the Port-au-Prince area, the last large earthquake happened in 1770. Therefore, throughout its history, the capital of the country had never experienced any deadly earthquake, but long-lasting political violence. Being the poorest country in the Americas as per the Human Development Index, due attention has never been paid to seismic preparedness, causing the catastrophic devastation with 316,000 casualties to occur (Reuters, 12 January 2010). Even in Japan with advanced technologies in disaster preparedness, the earthquake and tsunami of March 11th 2011 revealed vulnerabilities of disaster prevention schemes/ systems given the tsunami heights exceeding the determined design heights. These devastations often cause long lasting problems. In Haiti 2 million people are managing to get by on food rations. In Japan, huge amount of tsunami debris are just piled up in temporary damp yards with the fear that they might have been radiation-contaminated, thus discouraging all attempts for quick rehabilitations.

One of the most serious and costliest post earthquake problems is the stabilization of earthquake-induced landslide and debris masses. Shaking from the January 12, 2010, Haiti earthquake caused devastating structural damage and triggered many landslides that blocked roads, dammed rivers and streams, and threatened infrastructure in many parts of Haiti. About 4000 to 5000 landslides were estimated by the USGS (Randall at el., 2011). Many of the landslides blocked stream drainages and impounded lakes. In Japan, it has never been given a front-page coverage that total 4 billion JPY is spent annually even to this date to stabilize unstable debris mass of total 200 million m^3 blocking the upper reaches of the 50km long Joganji River in Tateyama Caldera; the debris mass of total 400 million m^3 at that time was originated in the 1858 Hietsu Earthquake with the estimated magnitude of 7 to 7.1.

A landslide mass that blocks water and impounds a lake at its upstream is referred to as a "landslide dam". Landslide damming events are very common around the world. They pose hazards to the population at both their upstream and downstream areas because of inundations due to lake impoundment and slope stability problems associated with such impoundment, and the risk of downstream flooding, respectively. Researches on landslide dams have been focused mostly on their distribution and failure potential (e.g, Costa & Schuster 1988, Ermini & Casagli 2003), and less attention has been given to their emergency treatments. A retrogressive review of the emergency response and mitigation treatment for recent landslide damming events around the world highlights the problems and uncertainty involved in mitigating the risk from landslide dams.

The simplest and the most commonly used mitigation measure has been the excavation of a spillway which is normally constructed at the lowest point at the crest of a landslide deposit. The excavation of the spillway has been proven to be the best engineering technique for mitigating risks posed by landslide dams, and its application can be traced about 500 years back (Bonnard, 2006). Sometimes blasting is also done to help excavating the channel; however blasting is seldom done for complete removal of large landslide masses. Drainage of the dammed lakes by means of pumps and siphons is a relatively less common treatment and is limited to small lakes. Tunnels or diversion channels are also rarely adopted as mitigation measures owing to the cost and time required.

Fig. 1. True-color image of Tangjiashan landslide dam and dammed lake captured by NASA's Earth Observing-1 (EO-1) (a) on 3rd June 2008 prior to overtopping and (b) on 8th June after overtopping but before the breaching of dam

A retrogressive review of the emergency response and mitigation strategies for recent landslide damming events around the world is presented in this chapter. Four landslide dams are presented for this purpose, consisting of the Tangjiashan landslide dam which was triggered by the 2008 Wenchuan earthquake in China, the Higashi Takezawa landslide dam that was triggered by the 2004 Mid-Nigata earthquake in Japan, the Attabad landslide that was formed in 2010 in Hunza Pakistan, and the Hattian Bala landslide dam that was formed by the 2005 Kashmir earthquake in Pakistan. A brief description of the mitigation measures adopted for the dams is given and failures of Tangjiashan landslide dam and Hattian landslide dam are also explained. A detailed description of the geomorphologic changes resulting from the Hattian landslide dam is also provided to elucidate long lasting hazards that landslide dams pose to their surrounding areas.

2. Tangjiashan landslide dam

The Ms 8.0 Wenchuan earthquake of China occurred on 12 May 2008 that have reportedly created 256 landslide dams most of which were formed along the Yingxiu–Beichuan (Lonmenshan) fault, which was dislocated over its 270km stretch in the earthquake (Qiang at el., 2009). The Tangjiashan landslide dam (N31° 50' 39.54", E104° 25' 45.93") is located in northern part of the fault zone, and impounded a lake on the Jiangjian River (Fig. 1). This lake behind the 124 m high landslide dam had a maximum storage capacity of 320 million m^3 (Liu at el., 2010). The dam posed a great threat of flooding to the 1.2 million people living along the lower reach of the river, therefore required immediate mitigation measures.

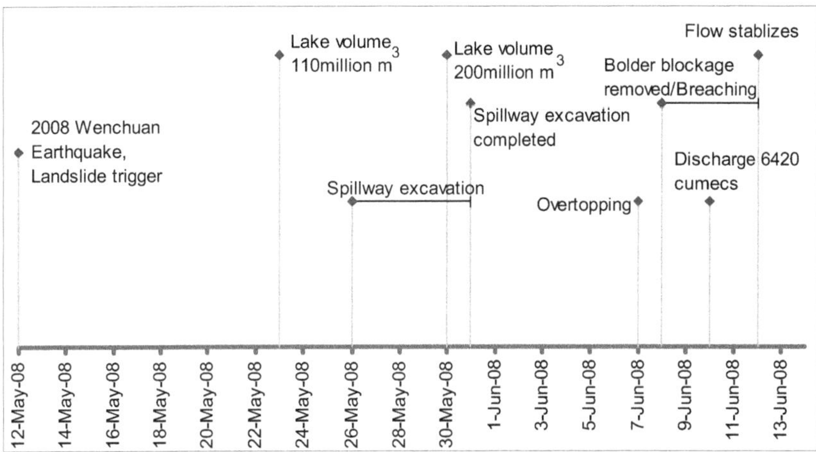

Fig. 2. Timeline of Tangjiashan landslide dam emergency response

2.1 Emergency response and mitigation measures

The water level of the dammed lake initially rose at a rate of 1.2 m/day under the average flow rate of 80-90m^3/s, and later on the rate decreased to 0.5m/day (Liu at el., 2010). Six towns and two cities are located downstream of the dam that were threatened to be inundated by the dam-break-caused flood. The dam break studies showed that a catastrophic flood would occur if no mitigation measures were adopted. Till 23rd of May the

Fig. 3. Higashi Takezawa landslide dam and the dammed lake (Satellite imagery from Google Map)

total volume of the lake had reached 110 million m^3 and then 200 million m^3 till the end of May, 2008. The figure 2 presents the timeline of the Tangjiashan landslide dam emergency response. The lake hypsometric curve showed that if the lake level was kept at 740.3 m, the storage capacity of the lake would be reduced to 238 million m^3 instead of 320 million m^3 corresponding to the natural potential overtopping level of 755.0 m (Liu at el., 2010). Accordingly the civil works were carried out to excavate a spillway to minimize the potential dam breach damage. It took 7 days work from May 26 to 31, to excavate a 13 m deep and 8 m wide channel covering 475 m in length. The primary intension for the excavation of the spillway was to reduce the total volume of the lake and meanwhile to create a controlled flood to drain the lake. About 275,000 people evacuated from the downstream areas for the expected flood.

2.2 Retrospective review of breaching event

The lake level reached 740 m, and the water started overflowing through the channel on June 7 (Qiang at el., 2009, Liu at el., 2010). The breaching process did not initiated immediately; instead the water kept rising for two days. A huge block slowed the channel erosion which was later removed by blasting on 8th June. However, no rapid flow occurred till 10th June when the water level reached 742 m. Then the outflow increased rapidly, and the channel walls and bed eroded rapidly. The outflow discharge peaked at 6420 m^3/s on June 10, and the erosion stopped on 12th June. Downstream areas were inundated where the people were evacuated before the breaching. The debris mass from the breached dam were transported and deposited along the downstream reaches. The emergency work for Tangjiashan landslide dam was a successful case history of response where the early identification of the hazard had made it possible to map out necessary measures for the area.

3. Higashi Takezawa landslide dam

The M6.8 Mid-Niigata Prefecture earthquake of 23rd October 2004 triggered numerous landslides and formed more than 50 landslide dams along the Imokawa River and its tributaries. Among these landslide dams the most crucial dam was the Higashi Takezawa

Fig. 4. Emergency works at Higashi Takezawa landsldide dam. Photo taken on 20.12.2004 by Takeji & Tadashi, 2005

landslide dam (N37° 18' 16.41", E138° 54' 16.13"), the largest landslide dam triggered by the earthquake (Fig. 3). The landslide originated from the left bank of the Imokawa River at Higashi Takezawa district. The landslide mass was about 350 m long and had volume up to 1 million m^3 (Sassa, 2005). The blockage started to impound a lake and soon the water started to inundate the Kogomo village. Due to the possible danger of overtopping and successive breaching of the dam emergency measure were taken to prevent the failure.

3.1 Emergency response and mitigation measures

Due to the danger of possible failure of the dammed lake, it was necessary to reduce the total capacity of the lake. Two possible mitigation measures were considered to cope with the flood hazard that the landslide dam posed. One of the possible measures was to remove the blockage to the original river level to drain the lake completely. This option would favor people living both upstream and downstream reaches of Imogawa River. However, there was a concern that removing only the toe part of the landslide mass would reactivate the movement of the remaining landslide mass. Removing the entire landslide mass would not be considered as a cost-effective solution. Therefore, the landslide mass was planned to be consolidated with the combination of diversion pipes and spillway. This meant that in exchange for abandonment of the upstream Kogomo village, the relevant authorities have secured the lower reach of the dam. Inhabitants of Kogomo Village were offered alternate sites for their new dwellings.

The following steps have been taken for stabilizing the landslide mass: twelve pumps were installed first to lower and keep the lake water level 11.5 m below the overflow point. In addition, the outflow level was raised by 3 m with sand bags at the outlet. Keeping in view the pumping power for this temporary system, diversion pipes were installed in case water overtopped the landslide mass (Fig. 4). Finally, a 280m long concrete-faced spillway was constructed with its cross-section large enough for snow melts.

Closed circuit televisions (CCTV) were installed to monitor the conditions in the catchment in case of emergency. Other monitoring activities included observations of cracks, installation of the water level gauges and monitoring the surface displacement of the collapsed slope.

4. Attabad landslide dam

On 4th January 2010 a landslide (36° 18′ 50″N, 74° 49' 13″E) occurred at Attabad in Hunza, Gilgit Baltistan. The debris mass dammed the Hunza river and formed a lake which was 120 m deep at the saddle point of the debris mass. At about 3 km downstream of the landslide dam, there is a site of an older landslide that blocked a 30 km long lake that formed in 1958 and eventually breached and caused a catastrophic flood. Prior to the 2010 Attabad landslide, cracks were being observed by the villagers at Attabad, which were being progressively lengthened. The area was declared as high hazard area by the local authorities, and evacuations of the affected slopes was being implemented by the end of 2009.

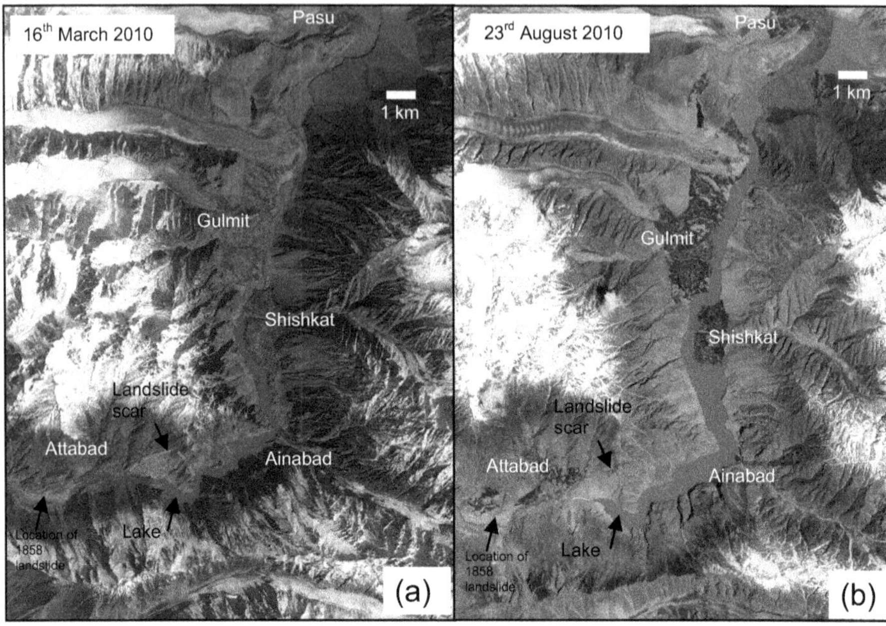

Fig. 5. The true-color image of the Attabad landslide dam (36° 18′ 50″N, 74° 49' 13″E) and lake captured by Advanced Land Imager (ALI) on NASA's Earth Observing-1 (EO-1) satellite, on (a) 16th March 2010 and (b) 23rd March 2010

4.1 Emergency response and mitigation measures

A lake immediately started to develop at upstream of the debris mass. The impoundment created huge problems for the 25,000 people living at upstream of the landslide dam. Mitigation studies were conducted (Petley, 2010) and as an emergency repose a 30 m deep spillway was planned to be excavated to reduce the maximum possible volume of the lake. Seepage points emerged progressively at six different locations till April 2010. Meanwhile, closed circuit television, (CCTV) system was installed to issue a warning in case of much feared seepage failure. The excavation of the spillway started on 30th January 2010. The surface silty-clay deposit created problems during the excavation works as the construction machinery were bogged down with the muddy conditions at the site, and the work was stopped when the excavation had reached 15 m deep across the crest till 16th May 2010. The water started overflowing on 29th May 2010. At the time of overtopping the lake reached a length of 21 km and had an estimated volume of 450 x 10^6 m^3. Fig 5a and 4b shows the true-color image of the Attabad landslide dam and lake captured by Advanced Land Imager (ALI) on NASA's Earth Observing-1 (EO-1) satellite, on 16th March 2010 and 23rd August 2010 respectively. About 25,000 people were cut off from the downstream population and are still suffering at the moment of writing.

4.2 Post overtopping situation

The Attabad landslide dam survived the much feared failure upon overtopping. The dam has emerged as a permanent geological feature that has caused a 23 km stretch of the Karakorum highway and about 232 houses to be inundated (NDMA) and isolated the communities at its upstream reaches. Moreover the longevity of the dam is coming under question. Erosion of the spillway occurred at the time of overtopping that continued for several days which later on was followed by a slow down-cutting of the channel. The channel erosion was stopped by the presence of large boulders embedded in the channel bed, which prevented the catastrophic failure, while slowed the drainage of the lake. The dam survived the high discharge of glacier melting of 2010. Blasting option was considered

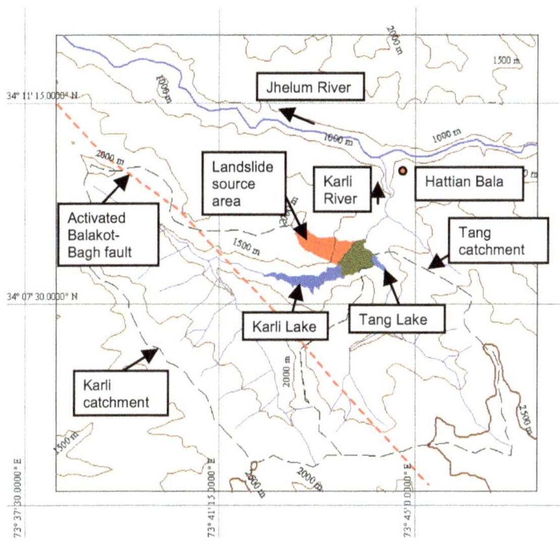

Fig. 6. Contour map of Hattian landside, dammed lakes with their catchments and downstream areas.

Fig. 7. Front view of the Hattian landslide mass, showing the downstream face erosion.

to widen the spillway in late June, 2010 which was conducted at small scale and was unsuccessful (Pamir Times). As the water level of the lake dropped, attempts were made to lower the level of the spillway in January 2011, which was eventually unsuccessful.

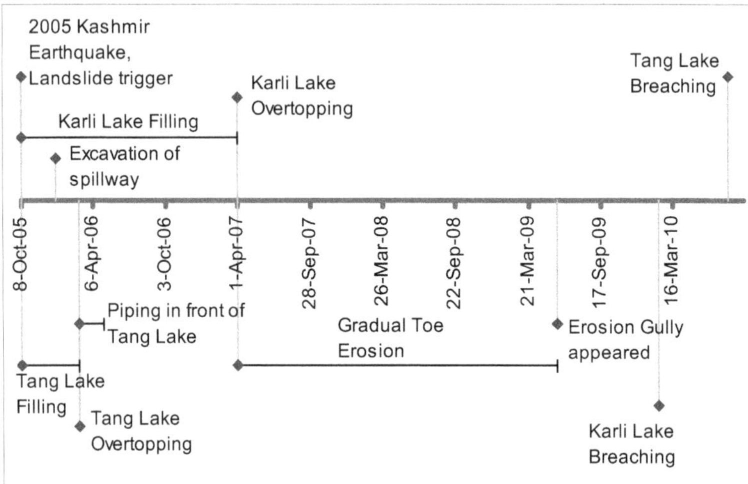

Fig. 8. Timeline of post formation features and major events at Hattian Bala landslide dam

Possible failures will include those due to down cutting erosion, seepage or piping failure, earthquake induced dam collapse, landsliding into the lake, and the glacial lake outburst flood (GLOF). The concerns are rising as the water level of the lake started increasing in June-July 2011 because of the glacier melting. The dam is intact at the moment of writing; however the stability of the dam is again threatened as the monsoon of 2011 approaches.

5. Hattian Bala landslide dam

On the 8th October 2005 earthquake (epicenter 34°29′35″ N, 73°37′44″ E, focal depth 26 km, 7.6 Mw (USGS)) caused widespread destruction in northern areas of Pakistan and Pakistan-administered Kashmir, killing more than 86,000 people. The earthquake triggered a large landslide mass near the south-eastern end of the Balakot-Bagh fault, the fault that caused the earthquake (Fig. 6). The slopes around the landslide source area have been a site of frequent landslide activities predating the earthquake. The 65 million m^3 landslide destroyed a village and killed about 1000 people (about 1.1% of total earthquake casualties) making it one of the most devastating recorded historical landslide events (Dunning at el., 2007). The landslide blocked two tributaries of the Karli River and created two lakes, Karli Lake (large lake) with a maximum natural capacity of 86 million m^3, and a Tang lake (small lake) with maximum natural capacity of 5 million m^3. Immediately the water started to be impounded in the two lakes behind the landslide mass, and thus a threat of dam breach flooding was eminent.

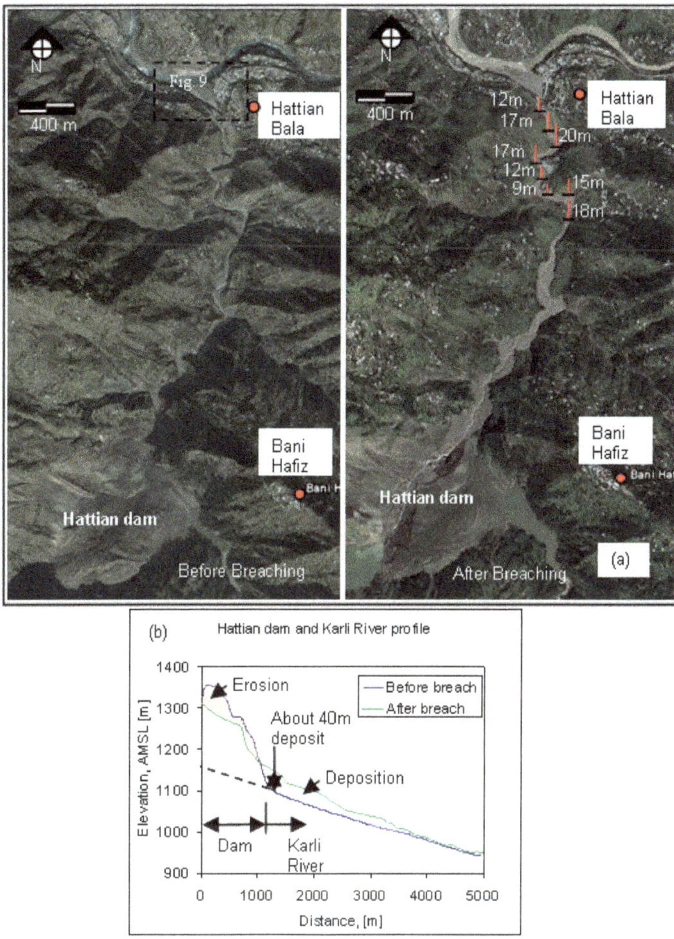

Fig. 9. (a) Satellite imageries of Karli River before and after breaching of Hattian dam (Google Earth), along with the maximum flow depths measured along the Karli River after failure of the dam (b) Profile of Hattian dam and Karli River before and after breaching of the Hattian dam

5.1 Emergency response and mitigation measures

As the water level in the lakes rose, the concern about the stability of the landslide dam was increasing. Mitigation studies were conducted and based on the recommendations the relevant authorities decided to excavate a spillway to reduce the maximum possible volume of the lake. The seepage through the landslide mass, the water levels and inflows to the lakes were continuously monitored as part of emergency response (WAPDA, 2007). A 450 m long spillway was excavated for the Karli Lake and relatively short 130 m spillway was excavated for the Tang Lake. The excavation of the spillway considerably reduced the total natural storage capacity of the Karli Lake from 86 million m^3 to 62 million m^3. The Tang Lake filled up in February 2006 and started overflowing. A timeline of various events occurring at Hattian Bala landslide dam after its formation are presented in figure 8.

The debris mass developed seepage erosion and the water level of the Tang Lake gradually reduced by 7-8 m over a period of two months (May-June 2006) and later on reached a state of equilibrium between inflow and seepage discharges, keeping the lake volume at approximately 3.5 million m^3. The Karli Lake filled up in April, 2007 and started to overflow. The dam survived the overtopping and hence less attention was drawn towards the monitoring of the dam as the spillway was performing well and no immediate threat was imminent. On the basis of the comparison between the volumes of the debris mass with the volume of the water stored in the lake the dam was thought to be stable by many experts. Therefore, less attention was drawn towards the monitoring of the dam site. However, after overtopping gradual erosion of the dam body started to occur due to the overflowing and seepage water.

5.2 Breaching and breach inflected morphological changes

Gradual toe erosion was initially observed, which was then followed by a rapid head cutting erosion of the downstream face of the dam (Fig. 7). On 9th February 2010, the water of Karli Lake breached the debris dam after five days of incessant rain, and the breaching drained about 36 million m^3 of water from the lake. According to the back analysis conducted by Konagai, at el., (2011) the peak outflow was about 5500 m^3/s. A moderate 32 mm/day rainfall was estimated on the day prior to the day of breaching. About 7.78 million m^3 of soil was estimated to have been eroded during the breaching event. The flushed debris mass was deposited along the Karli River. Fig. 9 shows the satellite image and profile of the Karli River before and after the breaching. A major portion of the eroded material is deposited over the approximately 2km downstream stretch from the dam along Karli River and less significantly beyond the stretch. The flow along the Karli River reached the depth up to 20 m. The debris deposit was about 40 m thick near the toe of the dam.

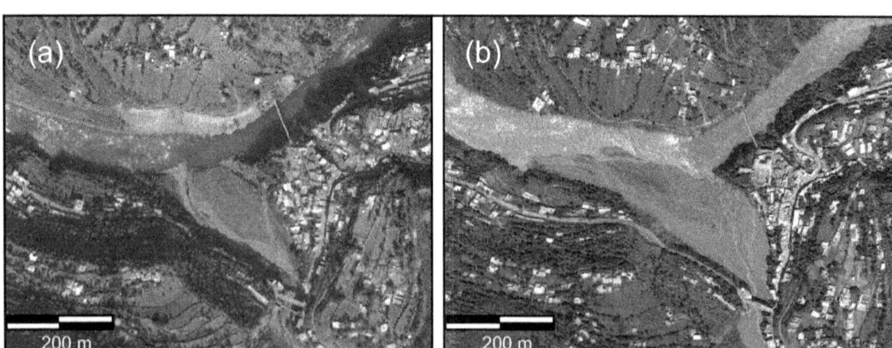

Fig. 10. (a) Satellite image of the junction of the Karli River with the Jhelum River before and after the breaching of Hattian dam (Google earth)

The dam break flood and debris destroyed one bridge near the junction of the Karli River with the Jhelum River and damaged about 24 houses as shown in Fig. 10. However major damage from the dam breach occurred as a landslide on the right bank of the Karli lake that was triggered by emptying of lake water. The figure 11 shows the satellite image of the right bank landslide after the breaching event, along with the contour map of the landslide and longitudinal profile of the landslide. This landslide destroyed 174 houses forcing about 1000 residents to evacuate.

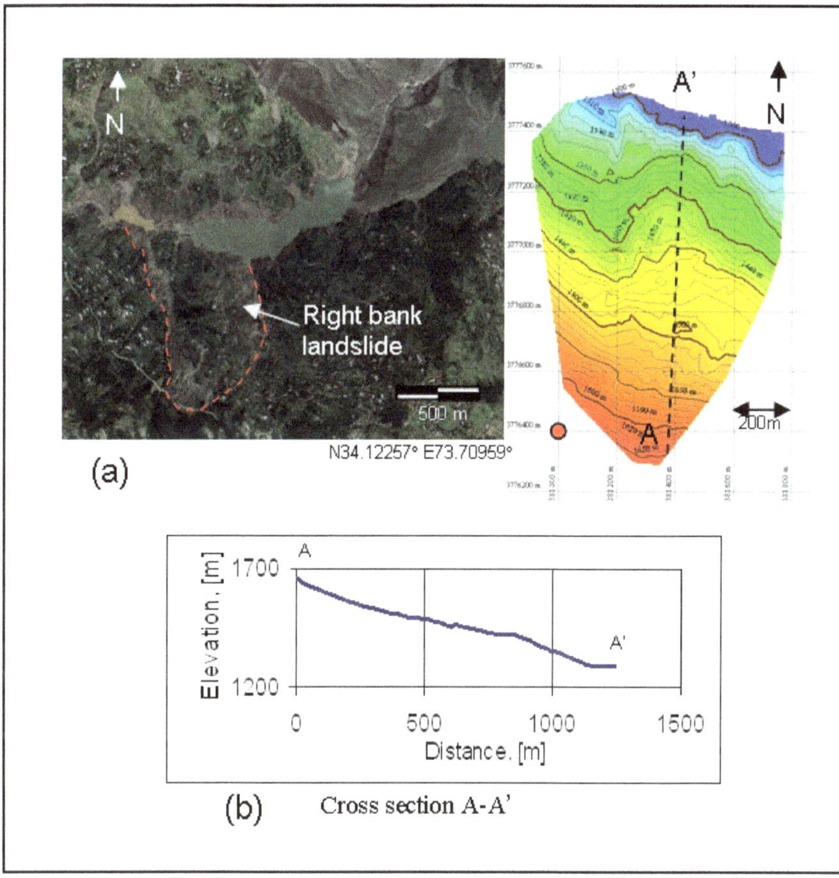

Fig. 11. (a) Satellite image of the Hattian landslide dam, Karli lake and the right bank landslide after the breaching event, along with the contour map of the landslide, (b) the longitudinal profile of the right bank slide. Modified from Konagai at el., 2011

The Tang Lake breached during the monsoon rains (July-August) of 2010. In the breaching process, the water from Tang Lake eroded the north-eastern end of the landslide lobe, which had been covered with boulders segregated up at the time of deposit, and approximately 15 m depth of the lake water was drained leaving 1.9 million m^3 water in the lake (Fig. 12).

Fig. 12. Tang Lake after breaching (Photo taken on 6th Dec-2010). From Konagai at el., 2011

6. Conclusion

A review of emergency responses and mitigation strategies adopted for the landslide dams formed in the recent earthquakes shows the problems being faced in the risk management of the hazards posed by the landslide dams. Excavation of the spillway seems to be the most common and effective mitigation measure. Meanwhile the time seems to be the most crucial factor during the emergency response, as the rise of water level inundates more area at its upstream and increase the lake volume. Higashi Takezawa landslide dam presented an example where a combination of different mitigation techniques such as pumps, siphons, diversion channels, and spillway, were possible for the relatively small landslide mass which was easily accessible. For the large landslide masses at remote locations, spillway remains as the most suitable mitigation measure.

It is also observed that the failure potential of landslide dammed lakes is not a well understood phenomenon. The Attabad landslide dam which was expected to fail upon overtopping due to the material composition of the dam body survived and continues to pose threats. On the other hand the Hattian landslide dam that seemed to be relatively stable and expected to remain intact, after showing some signs of instability failed as a result of moderate rainfall. The risk assessment for landslide dam mostly considers the expected inundation from dam breach, which might not be sufficient. From the case study of Hattian landslide it is observed that the damage resulting from the slope instability as a result of drawdown can be equally severe as the damage resulting from dam breach flood.

The available response time varies from case to case, as in the case of Hattian dam it took over one year for the lake to fill, while Tangjiashan landslide dammed lake provided less then a month for the response before the lake filled. The consideration of the recent landslide dams and their emergency treatment leads to the conclusion that for potentially large landslide dams, the early identification of possible situation is essential to organize appropriate preventive actions in due time. A detailed knowledge of the past events like those mentioned above will always be helpful determining the optimum actions to be adopted considering the fact that the significant bases in this field are lacking.

7. Acknowledgment

This paper summarizes one of the outcomes of the MEXT Research Project, "Scientific surveys for long-lasting geotechnical problems caused by large earthquakes and their implementations for rational rehabilitation strategies", Konagai K. Leader of the project, 2008 Grant-in-aid for scientific research (A) No. 20254003, Ministry of Education, Culture, Sports, Science and Technology (MEXT).

8. References

Bonnard C. (2006). *Technical and Human aspects of historic rockslide dammed lakes and landslide dam breaches,* Italian journal of engineering geology and environment, Special issue 1 doi: 10.4408/IJEGE.2006-01.S-03

Costa J. E. and Schuster R. L. (1988). *The formation and failure of natural dams,* GSA Bulletin; July 1988; v. 100; no. 7; p. 1054-1068; DOI: 10.1130/0016-7606(1988)100<1054:TFAFON>2.3.CO;2

Dunning S. A., Mitchell W. A., Rosser N. J., Petley D. N. (2007). *The Hattian Bala rock avalanche and associated landslides triggered by the Kashmir Earthquake of 8th October 2005,* Eng Geol 93(3–4):130–144

Ermini L., Casagli N. (2003). *Prediction of the behaviour of landslide dams using a geomorphological dimensionless index,* Earth Surface Processes and Landforms, Volume 28, Issue 1, pages 31–47, January 2003 DOI: 10.1002/esp.424

Geological Survey of Pakistan (GPS). (2010). *Post disaster threats in Attabad and adjacent areas, Hunza valley, Gilgit-Baltistan,* Information release no 893, April 2010

Konagai K., Ahsan S. (2011). *Partial breaching of Hattian Bala landslide dam formed in the 8th October 2005 Kashmir earthquake, Pakistan,* Landslides Journal (in review)

Liu N., Chen Z., Zhang J., Lin W., Chen W. and Xu W. (2010). *Draining the Tangjiashan Barrier Lake,* DOI: 10.1061/ ASCE┐ HY.1943-7900.0000241, Journal of Hydraulic Engineering, Vol. 136, No. 11, November 1, 2010

Pamir Times. *Voices of the Mountain Communities,* community news and views blog of Gilgit - Baltistan. http://pamirtimes.net/

Petley D. (2010). *The landslide at Attabad in Hunza, Gilgit/Baltistan: current situation and hazard management needs.* Report prepared for Focus Humanitarian Assistance, Pakistan, based upon a rapid field assessment on 26th February - 4th March 2010

Qiang X., Xuan-Mei F., Run-Qiu H., Cee V. (2009). *Westen Landslide dams triggered by the Wenchuan Earthquake, Sichuan Province, south west China.* Bull Eng Geol Environ (2009) 68:373–386 DOI 10.1007/s10064-009-0214-1

Randall W. Jibson and Edwin L. Harp (2011). *Field Reconnaissance Report of Landslides Triggered by the January 12, 2010, Haiti Earthquake* Open-File Report 2011–102 U.S. Department of the Interior U.S. Geological Survey

Sassa K. (2005). *Landslide disasters triggered by the 2004 Mid-Niigata Prefecture earthquake in Japan Landslides,* (2005) 2: 135–142. DOI: 10.1007/s10346-005-0054-4

Takeji Kokusho and Tadashi Hara (2005): Photographs of Landslides triggered in the 2004 Chuets Earthquake, Earthquake damage in active-folding areas: Creation of a comprehensive data archive and suggestions for its application to remedial measures for civil-infrastructure systems (Project leader: Kazuo Konagai), Research and Development program for Resolving Critical Issues, Special Coordination Funds for Promoting Science and Technology, Ministry of Education, Culture, Sports, Science and Technology, 2005-2007.

WAPDA Report. (2007). *Study of Hattian Ballah landslide, Potential hazards of land sliding and mitigation measures at Hattian Ballah and other earthquake hit areas,* National Engineering Services Pakistan limited (NESPAK) and Geological Survey of Pakistan (GSP), Water and Power Development Authority (WAPDA), 2007, Government of Pakistan

10

Rate-Dependent Nonlinear Seismic Response Analysis of Concrete Arch Dam

Xiao Shiyun
Faculty of Infrastructure Engineering, Dalian University of Technology, Dalian, China

1. Introduction

In civil engineering, all kinds of concrete structures inevitably encounter some form of dynamic loading during their lifetime. For example, bridges and tall buildings encounter wind loading, dams suffer from hydrodynamic pressure loading, ocean platforms encounter the impact of ocean waves, and all kinds of structures may suffer from earthquake loading. Because of their unpredictability and destructive capacity, these kinds of loadings always become important factors in controlling structural design.

The concrete is a typical rate-dependent material: its strength, stiffness, and ductility (or brittleness) are affected by loading rates. Researches on the rate dependency of concrete started in 1917 by the Abrams' dynamic compressive experiment (Abrams 1917). Jones (1936) investigated the relationship between compressive strengths of concrete and loading rates. Their experiments gave the conclusion that the compressive strengths of concrete increased with loading rates. Numerous tests have been carried out to investigate the response of concrete to rapid loading. Watstein (1953) observed that the compressive strengths of concrete increased an average of 80 percent when the strain rate increased from the static loading rate $10^{-6}s^{-1}$ to $10s^{-1}$. Based on the results of his experiments, Norris (1959) proposed an empirical formula and predicted that the compressive strengths were increased up to 33%, 24%, and 17% greater than the static strengths when the strain rates were $3s^{-1}$, $0.3s^{-1}$, and $0.1s^{-1}$, respectively. Atchley (1967) reported that the dynamic compressive strength increased from 25% to 38%. Experimental results from Hughes (1972) illustrated that the compressive strengths of concrete increased 90% more than the static strength on the impact loading.

Although researchers are not in complete agreement with which strain rates cause the increase in strength to be significant, it is generally accepted that a definite increase in the uniaxial compressive strength of concrete correlates with the increase of strain rates. However, confusion also has arisen in regard to the increase in magnitude of dynamic strengths. Some experimental results (Abrams 1917; Jones 1936; Watstein 1953; Rush 1960; Atchley 1967; Spooner 1972; Hughes 1972; Sparks 1973; Dilger 1984) showed that an increase of 30 percent more than the static strength of concrete, and even up to 80 percent, is possible. Others, such as Moore (1934), Evans (1942) and Dhir (1972), indicated that the increase in the strength of concrete was less than 20 percent and was not influenced by the rate of loading. Bischoff (1991) reviewed and analyzed the dynamic compressive experiments of concrete and deduced that the confusion about the increased magnitude of dynamic strengths arose because many factors, such as concrete quality, aggregate, age, curing and moisture conditions, influenced the behavior during the rapid loading.

The reported dynamic tensile tests of concrete in literature are more difficult to perform and the results are few. Birkimer (1971) conducted two sets of dynamic tensile tests using plain concrete cylinders. In the first set, the dynamic strength at the strain rate of $20s^{-1}$ was between 17.2 MPa and 22.1 MPa, whereas the static tensile strength was 3.4 MPa at the quasi-static strain rate of $0.57\times10^{-6}s^{-1}$. In the second set, the concrete dynamic strength was between 15.4 MPa and 27.6 MPa. Zielinski (1981) studied the behavior of concrete subjected to the uniaxial impact tensile loading and found that the ratios of impact and static tensile strengths were between 1.33 and 2.34 for various concrete mixes. Oh (1987) presented a realistic nonlinear stress-strain model that could describe the dynamic tensile behavior of concrete. An equation was proposed to predict the increase of tensile strengths resulting from an increase of strain rate. Tedesco (1991) conducted the direct tension tests of plain concrete specimens on a split-Hopkinson pressure bar to investigate the effects of increasing strain rate on the tensile strength of concrete. Rossi (1994) made an experimental study of rate effects on the behaviors of concrete under tensile stress to investigate the effect of the water/cement ratio on the tensile strength enhancement. In addition, an analysis of the physical mechanisms was developed to investigate how the Stefan effect, the cracking process, and the inertia forces participated together in the dynamic behavior of a specimen subjected to a uniaxial tensile test (Rossi, 1996). Cadoni (2001) studied the effect of strain rate on the tensile behavior of concrete at different relative humidity levels. Malvar (1998) reviewed the extant data characterizing the effects of strain rate on the tensile strength of concrete and compared the DIF formulation with that recommended by the European CEB. Finally, an alternative formulation was proposed based on the experimental data.

Many high arch dams have been built and will be built in areas of China with high seismic activity. Some of them will reach 300 meters in height. For researchers and engineers, the significant concern is paid on the safety of these structures against earthquake shocks. During the past two or three decades, many sophisticated computer programs are developed and used for numerical analysis of the arch dams. Our ability to analyze mathematical models of dam structures subjected to earthquake ground motions has been improved dramatically. Nevertheless, the current design practice in the seismic design of arch dams has to be based on the linear elastic assumption. The key property that determines the capacity of arch dams to withstand earthquakes is the tensile strength of concrete. However, the design criterion for the tensile stress remained a problem at issue. A widely accepted standard has not been available. The conventional design practice accounts for the rate sensitivity by means of drastic simplifying assumptions. That is, in all cases, the allowable stresses of an arch dam under earthquake load are increased by, such as a Chinese Standard (2001), 30% of the value specified for static case. Similarly, the dynamic modulus of elasticity is assigned 30% higher than its static value. Raphael (1984) carried out the dynamic test of concrete from dam cores and reported an average dynamic-static splitting tensile strength ratio of 1.45, and an average dynamic-static compressive strength ratio of 1.31 for the same loading rate ranges. Harris (2000) performed laboratory tests on concrete cores drilled from dams and tested at strain rates that simulated dynamic and static loading conditions. Results indicated that dynamic-static strength ratios were greater than those for both the tensile and compressive strength tests. Thereby, it is improper that the same increments of strengths and elastic modulus of concrete at different strain rates are adopted during the process of analyzing the seismic response of dams.

Few researchers considered the effect of strain rates on dynamic responses of arch dams because there was a lack of rate-dependent dynamic constitutive models of concrete. Cervera (1996) developed a rate-dependent isotropic damage model for the numerical analysis of concrete dams. The application of the proposed model to the seismic analysis of a large gravity concrete dam showed that the tensile peak strength of concrete could be increased up to 50 percent for the range of strain rates that appear in a structural safety analysis of a dam subjected to severe seismic actions. Lee (1998) developed a plastic-damage model for the concrete subjected to cyclic loading using concepts of fracture-energy-based damage and stiffness degradation. The rate-dependent regularization was used to guarantee a unique converged solution for softening regions. No effect for the rate-dependency on the stress distribution has been involved. Chen (2004) proposed a rate-dependent damage constitutive model for massive concrete by introducing rate-dependant plastic damage variables as internal variables. The nonlinear seismic responses of arch dams were computed using this model and the results were compared with the results given by the corresponding rate-independent damage model. It showed that the distribution of strain rates not only influenced the vibration modes of dam but also had significant effects on the dynamic damage of arch dams. Li (2005) analyzed the seismic response of a high arch dam, in which a rate-dependent damage constitutive model of concrete was considered and the nonlinear contact of joints was simulated by direct stiffness method based on the Lagrange multiplier. The study showed that the nonlinear concrete model had great effects on the dynamic opening of the contraction joints caused by the nonlinear softening and cracking. Bai (2006) established a rate-dependent damage constitutive model for simulating the mechanical behaviors of concrete by introducing the effect of strain rate into the damage tensor. The model was applied to analyze the seismic overload response of a typical concrete gravity dam. Results indicated that the distribution of strain rate caused by seismic loading varied at the dam surface and significantly affected the dynamic response.

The effect of strain rates on dynamic behaviors of concrete is an important aspect in the evaluation of the seismic responses of concrete structures. To evaluate the seismic behaviors of concrete structures, the dynamic experiment and the dynamic constitutive model of concrete are necessary. The main objective of this study is, based on the results of the dynamic uniaxial tensile and compressive experiments on the concrete, to establish the dynamic constitutive model of concrete and study the effect of strain rates on dynamic responses of concrete dams.

2. Dynamic experiments of concrete

2.1 Dynamic uniaxial tensile experiment of concrete

2.1.1 Tensile specimen

A concrete mix with proportions, by weight, of cement: water: gravel: sand content = 1.00:0.75:4.09:2.56 was used in the study. The employed cement is 425 Portland cement, the fine aggregate is general river sand, and the coarse aggregate is crushed rock. Specimens were cast in steel moulds and cured in moisture condition for 7 days, then they were naturally cured at 20±3 Celsius degree temperature in the laboratory. Fifty dumbbell-shaped specimens were cast for the tensile experiment, as shown in **Fig.1**. The specimens of dumbbell shape ensure that the specimens were destroyed at the middle of specimen in tension.

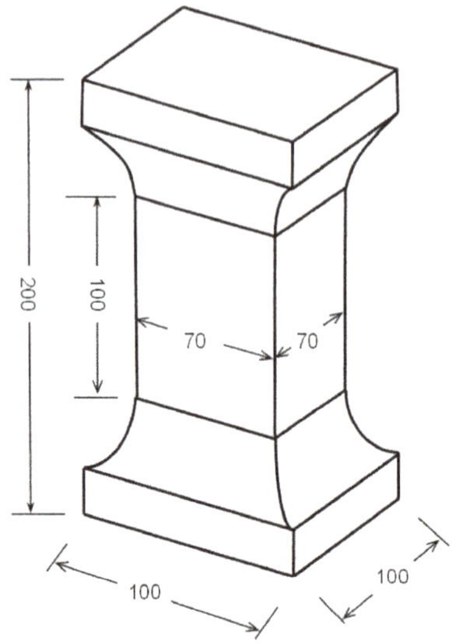

Fig. 1. Tensile specimen

2.1.2 Tensile test loading system and measuring system

The tensile dynamic test was carried out using the 1000-kN servo fatigue testing machine at the State Key Laboratory of Coastal and Offshore, Dalian University of Technology. During the process of experiments, the loading sign is sent by the control center, and then it is transferred to the servo fatigue testing machine. The magnitude and frequencies of loading are controlled by the control center.

During the tensile test, the specimen was adhered to two steel plates by the constructional glue. The bottom steel plate was fixed to the base with bolts and the upper steel plate was connected to the load cell with the load transducer, as shown in **Fig. 2**. In order to increase the stiffness of the loading system, a frame was formed by four steel bars connecting the load cell to the base with bolts.

Fig. 2. Tensile test setup

The data acquisition processor is 32-channel. Vertical and lateral strains of specimens were measured by four pairs of decussate strain gauges adhered on the four sides of specimens. Vertical and lateral displacements were measured by two opposite Linear Variable Displacement Transducers (LVDT) fixed to two opposite sides of the specimen. The load was measured by the load transducer fixed to the specimen. All measured signals were transmitted to the data acquisition and processing system of the computer through a specially allocated amplifier.

2.1.3 Analysis of the tensile experimental results

2.1.3.1 Stain-rate influence on uniaxial strength of concrete

The dynamic tensile strengths of 15 specimens under different strain rates of $10^{-5}s^{-1}$, $10^{-4}s^{-1}$, $10^{-3}s^{-1}$and $10^{-2}s^{-1}$ are shown in **table 1**. It shows that the uniaxial tensile strengths of concrete increase with the increasing of strain rates. Compared to the quasi-static tensile strength of concrete at the strain rate of $10^{-5}s^{-1}$, the dynamic tensile strengths of concrete at strain rates of $10^{-4}s^{-1}$, $10^{-3}s^{-1}$ and $10^{-2}s^{-1}$ increase 6%, 10% and 18%, respectively.

Strain rate / s^{-1}	Tensile Strengths / MPa			
	10^{-5}	10^{-4}	10^{-3}	10^{-2}
1	1.388	1.435	1.549	1.639
2	1.582	1.480	1.502	1.753
3	1.469	1.604	1.696	1.731
4	1.355	1.606	1.629	
average	1.449	1.531	1.594	1.696

Table 1. Dynamic strength of concrete in tension

According to the references, the increases in strengths follow a linear-logarithmic relationship with the increases in strain rates. By test results the linear-logarithmic relationship between the tensile strength enhancement with the strain rate enhancement, is given by

$$f_t / f_{ts} = 1.0 + 0.057 \log\left(\dot{\varepsilon}_t / \dot{\varepsilon}_{ts}\right) \tag{1}$$

Fig.3 shows the relationship between the dynamic tensile strength and the static tensile strength of concrete at different strain rates of $10^{-5}s^{-1}$, $10^{-4}s^{-1}$, $10^{-3}s^{-1}$and $10^{-2}s^{-1}$.

2.1.3.2 Stain-rate influence on elastic modulus

The stress-strain curves of concrete in tension at different strain-rate loading are illustrated in **Fig.4**. It is clear that during different strain rate loading the slope of curves is linear at the beginning of loading, indicating that the initial tangent modulus of concrete is independent of strain rate.

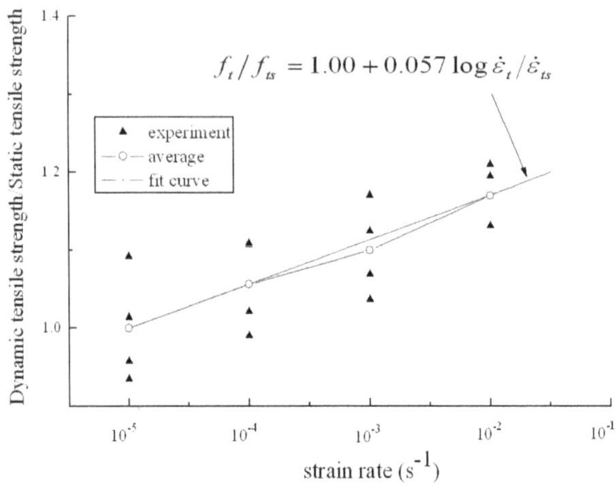

Fig. 3. The relationship between the dynamic tensile strength and the static tensile strength

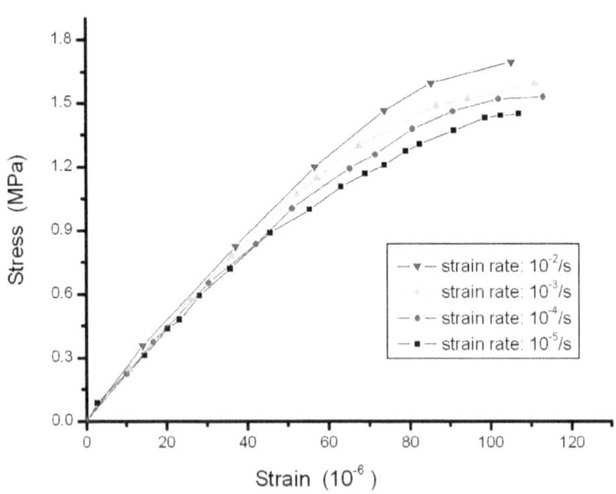

Fig. 4. The stress-strain curves of concrete in tension

2.1.3.3 Stain-rate influence on critical strain

Table 2 illustrates the critical strain of concrete in tension obtained from the present test. **Table 2** implies that in the range of strain rate from $10^{-5}s^{-1}$ to $10^{-2}s^{-1}$, the effect of strain rate on the critical strain value of concrete in tension is little, if any.

Strain rate	$10^{-5}s^{-1}$	$10^{-4}\ s^{-1}$	$10^{-3}\ s^{-1}$	$10^{-2}\ s^{-1}$
1	106	106	122	102
2	127	130	98	114
3	113	112	108	123
4	102	114	110	
Average	112	116	110	113

Table 2. The critical strain of concrete in tension

2.1.3.4 Stain-rate influence on Poisson's ratio

Table 3 shows the change in Poisson's ratio of concrete in tension and compression. It is found that the maximum of Poisson's ratio is 0.20 but the minimum is 0.13. And it is concluded that Poisson's ratio isn't obviously dependent on the loading rate according to the average of results.

Strain rate	$10^{-5}s^{-1}$	$10^{-4}\ s^{-1}$	$10^{-3}\ s^{-1}$	$10^{-2}\ s^{-1}$
1	0.16	0.17	0.14	0.15
2	0.14	0.16	0.14	0.19
3	0.19	0.18	0.15	0.13
4	0.14	0.13	0.16	
Average	0.16	0.16	0.15	0.16

Table 3. The Poisson's ratio of concrete in tension

2.2 Dynamic uniaxial compressive experiment of concrete

2.2.1 Compressive specimen

Similar to the tensile specimen, the concrete mix with proportions, by weight, of cement: water: gravel: sand content was still 1.00:0.75:4.09:2.56. The employed cement is 425 Portland cement, the fine aggregate is general river sand, and the coarse aggregate is crushed rock. Specimens were cast in steel moulds and cured in moisture condition for 7 days, then they were naturally cured at 20±3 Celsius degree temperature in the laboratory. Fifty cuboid specimens with 100×100×300mm were cast for the compressive experiment, as shown in **Fig.5**.

Fig. 5. Compressive specimen

2.2.2 Compressive test loading system and measuring system

The compressive dynamic test was also carried out using the 1000-kN servo fatigue testing machine at the State Key Laboratory of Coastal and Offshore, Dalian University of Technology. As for the compressive test, the cuboid specimen was placed vertically on the circular steel plate connecting the base with the load transducer, as shown in **Fig. 6**. The measure system is same as the tensile test.

Fig. 6. Compressive test setup

2.2.3 Analysis of the compressive experimental results

2.2.3.1 Stain-rate influence on uniaxial strength of concrete

Table 4 gives the dynamic compressive strengths of concrete at different strain rates. It can be concluded that the uniaxial compressive strengths of concrete increase with the increasing of strain rate. Compared to the quasi-static compressive strength of concrete at the strain rate of $10^{-5}s^{-1}$, the dynamic compressive strengths of concrete at strain rate of $10^{-4}s^{-1}$, $10^{-3}s^{-1}$, $10^{-2}s^{-1}$ and $10^{-1}s^{-1}$ increase 4.8%, 9.0%, 12.0% and 15.6%, respectively.

Strain rate / s^{-1}	Compressive strengths / MPa				
	10^{-5}	10^{-4}	10^{-3}	10^{-2}	10^{-1}
1	21.89	22.09	24.63	24.04	26.19
2	22.03	24.40	23.18	25.97	24.94
3	20.67	21.72	22.60	22.70	26.72
4	23.35	23.93	25.44	25.76	23.76
average	22.00	23.03	23.96	24.62	25.40

Table 4. Dynamic strengths of concrete in compression

Similar to the dynamic tensile strength, the linear-logarithmic relationship between the compressive strength enhancement with the strain rate enhancement, is also given by

$$f_c/f_{cs} = 1.0 + 0.040\log\left(\dot{\varepsilon}_c/\dot{\varepsilon}_{cs}\right) \tag{2}$$

Fig.7 gives the relationship between the dynamic compressive strength and the static compressive strengths of concrete.

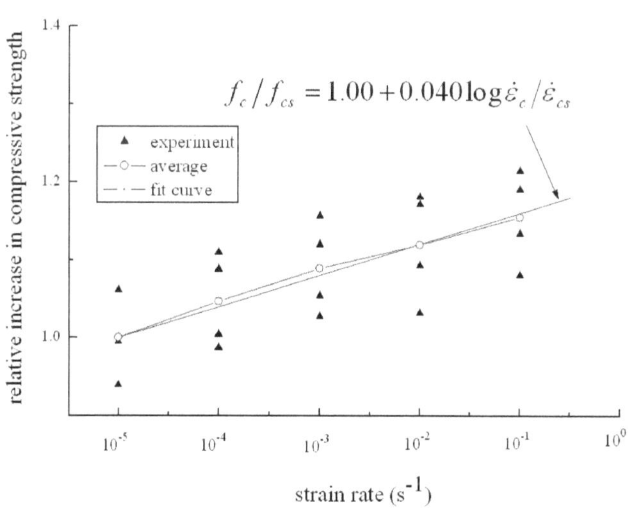

Fig. 7. The relationship between the dynamic and static compressive strength

2.2.3.2 Stain-rate influence on elastic modulus

The stress-strain curves of concrete in compression at different strain-rate loading are plotted in **Fig.8**. The initial tangent modulus of concrete in compression slightly increased as the strain rate increased. The initial tangent moduli of concrete at strain rate of $10^{-4}s^{-1}$, $10^{-3}s^{-1}$, $10^{-2}s^{-1}$ and $10^{-1}s^{-1}$ increase to 1.3×10^{4}MPa, 1.38×10^{4}MPa, 1.48×10^{4}MPa and 1.60×10^{4}MPa from 1.23×10^{4}MPa at strain rate of $10^{-5}s^{-1}$, respectively.

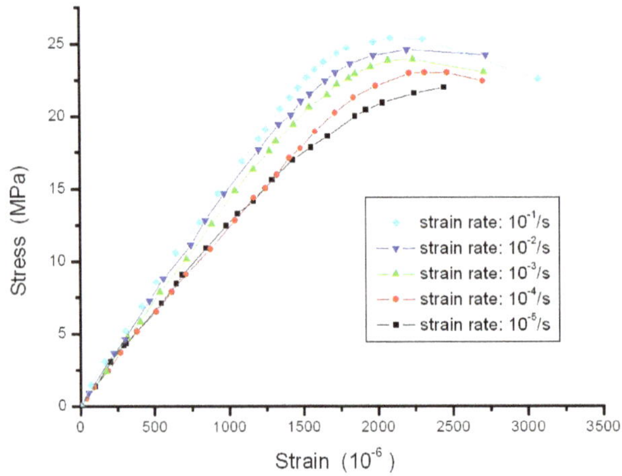

Fig. 8. The stress-strain curves of concrete in compression

2.2.3.3 Stain-rate influence on critical strain

Table 5 gives the results of the critical compressive strain of concrete obtained from the present test. It shows that a slight decrease in the critical compressive strain value was observed as the strain rate was increased.

Strain rate	$10^{-5}s^{-1}$	$10^{-4}\ s^{-1}$	$10^{-3}\ s^{-1}$	$10^{-2}\ s^{-1}$	$10^{-1}\ s^{-1}$
1	2524	2164	2480	2382	2030
2	2713	2153	2244	2172	2114
3	2180	2633	2083	2231	2234
4	2353	2304	2137	2003	1965
Average	2443	2314	2236	2197	2086

Table 5. The critical strain of concrete in compression

2.2.3.4 Stain-rate influence on Poisson's ratio

Table 6 shows the change in Poisson's ratio of concrete in tension and compression. It is found that the maximum of Poisson's ratio is 0.20 but the minimum is 0.13. And it is concluded that Poisson's ratio isn't obviously dependent on the loading rate according to the average of results.

Strain rate	$10^{-5}s^{-1}$	$10^{-4}\ s^{-1}$	$10^{-3}\ s^{-1}$	$10^{-2}\ s^{-1}$	$10^{-1}\ s^{-1}$
1	0.18	0.14	0.20	0.14	0.15
2	0.13	0.19	0.18	0.18	0.17
3	0.15	0.16	0.14	0.17	0.17
4	0.17	0.15	0.17	0.20	0.15
Average	0.16	0.16	0.17	0.17	0.16

Table 6. The Poisson's ratio of concrete in compression

3. Rate-dependent constitutive model of concrete

3.1 Consistency viscoplastic model theory

The consistency viscoplastic model can be seen as an extension of the classic elasto-plastic model to account for the rate-dependent behavior of materials. This model, which uses the Von Mises yield surface, was applied by Wang (1997) to analyze metal. With this method, the consistency viscoplastic Hoffman model of concrete was modified by Winnicki (2001). In this model, during viscoplastic flow, the actual stress state must remain on the yield surface, and the consistency condition is imposed.

The viscoplastic yield function can be expressed as

$$F(\sigma_{ij},\kappa,\dot{\kappa})=0 \quad \text{for } \dot{\lambda}>0 \tag{3}$$

In uniaxial tension and compression, equation (3) can be expressed as

$$\left.\begin{aligned} F_c &= F(\sigma_{ij},\kappa_c,\dot{\kappa}_c) \\ F_t &= F(\sigma_{ij},\kappa_t,\dot{\kappa}_t) \end{aligned}\right\} \tag{4}$$

It is difficult to establish biaxial or triaxial constitution relations because of the lack of biaxial dynamic experiment results for the concrete. For simplicity, the biaxial dynamic behavior of concrete is assumed to be the same as the uniaxial dynamic behavior with an increasing factor K_{bc} for the strength such that:

$$f_{bc}=K_{bc}f(\sigma_{ij},\kappa_c,\dot{\kappa}_c) \tag{5}$$

On an arbitrary stress state, it is assumed that:

$$\kappa_c=\varphi_c(\sigma_{ij})\kappa \quad \kappa_t=\varphi_t(\sigma_{ij})\kappa \tag{6}$$

Functions $\varphi_t(\sigma_{ij})$ and $\varphi_c(\sigma_{ij})$ should be chosen in such a way that, for loading processes with dominant tensile stress states, $\varphi_t(\sigma_{ij})=1$ and $\varphi_c(\sigma_{ij})=0$; similarly, for loading processes with dominant compressive stress states $\varphi_t(\sigma_{ij})=0$ and $\varphi_c(\sigma_{ij})=1$, and for loading processes with tensile-compressive stress states $0<\varphi_t(\sigma_{ij})<1$ and $0<\varphi_c(\sigma_{ij})<1$. These functions must satisfy the condition of $\varphi_t(\sigma_{ij})+\varphi_c(\sigma_{ij})=1$. Consequently, weight functions $\varphi_t(\sigma_{ij})$ and $\varphi_c(\sigma_{ij})$ can be achieved expediently.

Using such conditions, the yield function can be expressed as

$$F(\sigma_{ij},\kappa,\dot{\kappa})=F(\sigma_{ij},\kappa_c,\dot{\kappa}_c,\kappa_t,\dot{\kappa}_t)=0 \tag{7}$$

At the same time, the viscoplastic consistency condition must be satisfied so that:

$$\frac{\partial F}{\partial \sigma_{ij}}d\sigma_{ij}+\frac{\partial F}{\partial \kappa_c}d\kappa_c+\frac{\partial F}{\partial \dot{\kappa}_c}d\dot{\kappa}_c+\frac{\partial F}{\partial \kappa_t}d\kappa_t+\frac{\partial F}{\partial \dot{\kappa}_t}d\dot{\kappa}_t=0 \tag{8}$$

The effects of κ_t and $\dot{\kappa}_t$ on f_t are assumed to be independent and the instantaneous tensile strength f_t is formulated in a very general way as follows

$$f_t = f_{ts}H_t(\kappa_t)R_t(\dot{\kappa}_t) \tag{9}$$

Similarly, the instantaneous compressive strength is computed as

$$f_c = f_{cs}H_c(\kappa_c)R_c(\dot{\kappa}_c) \tag{10}$$

where $H_t(\kappa_t)$, $R_t(\dot{\kappa}_t)$, $H_c(\kappa_c)$ and $R_c(\dot{\kappa}_c)$ are the assumption functions achieved by experiment. Here, κ is adopted as $\kappa=\overline{\varepsilon}^{VP}=\sqrt{2\varepsilon_{ij}^{vp}\varepsilon_{ij}^{vp}/3}$.

3.2 Consistency viscoplastic William-Warnke three-parameter model of concrete (xiao, 2010)

In this study, the yield surface function is assumed to be the same as the failure surface function of concrete so that the yield function of the William-Warnke three-parameter model can be expressed as

$$F(\sigma_m,\tau_m,\theta)=\frac{1}{\rho}\sigma_m+\frac{1}{r(\theta)}\tau_m-f_c=0 \tag{11}$$

where two parameters ρ and $r(\theta)$ are chosen by three conditions: 1) uniaxial tensile strength f_t, 2) uniaxial compressive strength f_c, and 3) biaxial compressive strength $f_{bc}=k_{bc}f_c$.

Defining $m_{ij}=\dfrac{\partial F}{\partial \sigma_{ij}}$, one can achieve:

$$\begin{aligned} m_{ij}&=\frac{\partial F}{\partial \sigma_{ij}}=\frac{\partial F}{\partial \sigma_m}\cdot\frac{\partial \sigma_m}{\partial \sigma_{ij}}+\frac{\partial F}{\partial \tau_m}\cdot\frac{\partial \tau_m}{\partial \sigma_{ij}}+\frac{\partial F}{\partial r}\cdot\frac{\partial r}{\partial \sigma_{ij}}\\ &=\frac{1}{\rho}\cdot\frac{1}{3}\delta_{ij}+\frac{1}{r}\cdot\frac{1}{5\tau_m}s_{ij}-\frac{\tau_m}{r^2}\frac{\partial r}{\partial \sigma_{ij}} \end{aligned} \tag{12}$$

where $\delta_{ij}=\partial I_1/\partial\sigma_{ij}$ and $s_{ij}=\partial J_2/\partial\sigma_{ij}$.

For the sake of simplicity, defining $A=\dfrac{1}{3\rho}$, $B=\dfrac{1}{5r\tau_m}$ and $C=-\dfrac{\tau_m}{r^2}$ and substituting them into equation (11), one obtains:

$$m_{ij} = \frac{\partial F}{\partial \sigma_{ij}} = A\delta_{ij} + Bs_{ij} + C\frac{\partial r}{\partial \sigma_{ij}} = A\delta_{ij} + Bs_{ij} + C\frac{\partial r}{\partial \theta}\frac{\partial \theta}{\partial \sigma_{ij}} \tag{13}$$

Defining $r = \frac{u}{v}$ and taking the partial derivative with respect to the stress tensor σ_{ij}, one obtains:

$$\frac{\partial r}{\partial \theta} = \frac{\partial (u/v)}{\partial \theta} = \frac{v \cdot du/d\theta - u \cdot dv/d\theta}{v^2} = D \tag{14}$$

where

$$u = 2r_c(r_c^2 - r_t^2)\cos\theta + r_c(2r_t - r_c)[4(r_c^2 - r_t^2)\cos^2\theta + 5r_t^2 - 4r_c r_t]^{1/2}$$

$$v = 4(r_c^2 - r_t^2)\cos^2\theta + (r_c - 2r_t)^2$$

$$\frac{du}{d\theta} = 2r_c(r_t^2 - r_c^2)\sin\theta + \frac{4r_c(2r_t - r_c)(r_t^2 - r_c^2)\sin\theta\cos\theta}{[4(r_c^2 - r_t^2)\cos^2\theta + 5r_t^2 - 4r_c r_t]^{1/2}}$$

$$\frac{dv}{d\theta} = 8(r_t^2 - r_c^2)\sin\theta\cos\theta$$

According to the William-Warnke three-parameter model, the relationship between θ and σ_{ij} is given as follows:

$$\cos 3\theta = \frac{3\sqrt{3}}{2} \cdot \frac{J_3}{J_2^{3/2}} \tag{15}$$

By taking the partial derivative of equation (11) with respect to the stress tensor σ_{ij}, the following expression can be obtained as:

$$\begin{aligned} \frac{\partial \theta}{\partial \sigma_{ij}} &= \frac{\partial \theta}{\partial J_2}\frac{\partial J_2}{\partial \sigma_{ij}} + \frac{\partial \theta}{\partial J_3}\frac{\partial J_3}{\partial \sigma_{ij}} = \frac{3\sqrt{3}}{4\sin 3\theta} \cdot \frac{J_3}{J_2^{5/2}} \cdot \frac{\partial J_2}{\partial \sigma_{ij}} + \frac{-\sqrt{3}}{2\sin 3\theta} \cdot \frac{1}{J_2^{3/2}} \cdot \frac{\partial J_3}{\partial \sigma_{ij}} \\ &= \frac{3\sqrt{3}}{4\sin 3\theta} \cdot \frac{J_3}{J_2^{5/2}} \cdot s_{ij} - \frac{\sqrt{3}}{2\sin 3\theta} \cdot \frac{1}{J_2^{3/2}} \cdot t_{ij} \end{aligned} \tag{16}$$

where $t_{ij} = \frac{\partial J_3}{\partial \sigma_{ij}}$ and t_{ij} has the behavior such as:

(1) $t_{ij}\delta_{ij} = 0$ (2) $t_{ij}s_{ij} = 3J_3$ (3) $t_{ij}t_{ij} = \frac{2}{3}J_2^2$

Simply, defining $E = \frac{3\sqrt{3}}{4\sin 3\theta} \cdot \frac{J_3}{J_2^{5/2}}$ and $U = -\frac{\sqrt{3}}{2\sin 3\theta} \cdot \frac{1}{J_2^{3/2}}$, equation (16) can be expressed as

$$\frac{\partial \theta}{\partial \sigma_{ij}} = E \cdot s_{ij} + U \cdot t_{ij} \tag{17}$$

Substituting equations (10) and (13) into equation (9), m_{ij} is simplified as:

$$\begin{aligned} m_{ij} &= \frac{\partial F}{\partial \sigma_{ij}} = A \cdot \delta_{ij} + B \cdot s_{ij} + CD(E \cdot s_{ij} + U \cdot t_{ij}) \\ &= A \cdot \delta_{ij} + (B + CDE) \cdot s_{ij} + CDU \cdot t_{ij} \\ &= \alpha \delta_{ij} + \beta s_{ij} + \gamma t_{ij} \end{aligned} \tag{18}$$

where $\alpha = A$, $\beta = B + CDE$ and $\gamma = CDU$.

In equation (5), $\frac{\partial F}{\partial \kappa_c}$, $\frac{\partial F}{\partial \dot{\kappa}_c}$, $\frac{\partial F}{\partial \kappa_t}$ and $\frac{\partial F}{\partial \dot{\kappa}_t}$ could be expressed as

$$\frac{\partial F}{\partial \kappa_c} = \frac{\partial F}{\partial f_c}\frac{\partial f_c}{\partial \kappa_c} = (\frac{\partial F}{\partial \rho}\frac{\partial \rho}{\partial f_c} + \frac{\partial F}{\partial r}\frac{\partial r}{\partial f_c})\frac{\partial f_c}{\partial \kappa_c}$$

$$\frac{\partial F}{\partial \dot{\kappa}_c} = \frac{\partial F}{\partial f_c}\frac{\partial f_c}{\partial \dot{\kappa}_c} = (\frac{\partial F}{\partial \rho}\frac{\partial \rho}{\partial f_c} + \frac{\partial F}{\partial r}\frac{\partial r}{\partial f_c})\frac{\partial f_c}{\partial \dot{\kappa}_c}$$

$$\frac{\partial F}{\partial \kappa_t} = \frac{\partial F}{\partial f_t}\frac{\partial f_t}{\partial \kappa_t} = (\frac{\partial F}{\partial \rho}\frac{\partial \rho}{\partial f_t} + \frac{\partial F}{\partial r}\frac{\partial r}{\partial f_t})\frac{\partial f_t}{\partial \kappa_t}$$

$$\frac{\partial F}{\partial \dot{\kappa}_t} = \frac{\partial F}{\partial f_t}\frac{\partial f_t}{\partial \dot{\kappa}_t} = (\frac{\partial F}{\partial \rho}\frac{\partial \rho}{\partial f_t} + \frac{\partial F}{\partial r}\frac{\partial r}{\partial f_t})\frac{\partial f_t}{\partial \dot{\kappa}_t}$$

where the curves of $\frac{\partial f_c}{\partial \kappa_c}$, $\frac{\partial f_c}{\partial \dot{\kappa}_c}$, $\frac{\partial f_t}{\partial \kappa_t}$ and $\frac{\partial f_t}{\partial \dot{\kappa}_t}$ can be achieved by uniaxial compressive and tensile tests of concrete.

Based on the associated plastic flow rule, the viscoplastic strain is defined as

$$d\varepsilon_{ij}^{vp} = d\lambda \frac{\partial F}{\partial \sigma_{ij}} = d\lambda m_{ij} \tag{19}$$

The invariable can be expressed as

$$d\kappa = \sqrt{\frac{2}{3} d\varepsilon_{ij}^{vp} d\varepsilon_{ij}^{vp}} = \sqrt{\frac{2}{3} m_{ij} m_{ij}} d\lambda = g(\sigma_{ij}) d\lambda \tag{20}$$

Using these functions, the consistency equation (4) can be expressed as

$$m_{ij}d\sigma_{ij} + hd\lambda + sd\dot{\lambda} = 0 \tag{21}$$

where

$$h = \frac{\partial F}{\partial \lambda} = a_c h_c(\kappa_c)R_c(\dot{\kappa}_c) + a_t h_t(\kappa_t)R_t(\dot{\kappa}_t)$$

$$s = \frac{\partial F}{\partial \dot{\lambda}} = a_c H_c(\kappa_c)r_c(\dot{\kappa}_c) + a_t H_t(\kappa_t)r_t(\dot{\kappa}_t)$$

In which

$$a_t = \frac{\partial F}{\partial f_t} f_{ts}\varphi_t(\sigma_{ij})g(\sigma_{ij}) \quad a_c = \frac{\partial F}{\partial f_c} f_{cs}\varphi_c(\sigma_{ij})g(\sigma_{ij})$$

$$h_t(\kappa_t) = \frac{\partial H_t(\kappa_t)}{\partial \kappa_t} \quad h_c(\kappa_c) = \frac{\partial H_c(\kappa_c)}{\partial \kappa_c}$$

$$r_t(\dot{\kappa}_t) = \frac{\partial R_t(\dot{\kappa}_t)}{\partial \dot{\kappa}_t} \quad r_c(\dot{\kappa}_c) = \frac{\partial R_c(\dot{\kappa}_c)}{\partial \dot{\kappa}_c}$$

Dynamic tensile and compressive tests were carried out to investigate the effect of strain rates on the dynamic tensile and compressive behaviors of concrete (Xiao, 2008). Test results indicate that the tensile and compressive strengths of concrete increase with the increase of the loading rate. The initial tangential modulus and the critical strain of concrete in tension are independent of strain rate, but those in compression slightly increased with the strain rate. Poisson's ratio of concrete in both tension and compression is not obviously dependent on loading rate.

Based on the experimental data, the functions $H_t(\kappa_t)$ and $R_t(\dot{\kappa}_t)$ are given in **Fig. 9(a)** and **Fig. 9(b)**, where the plotted curves are the fitting curves for the later calculation.

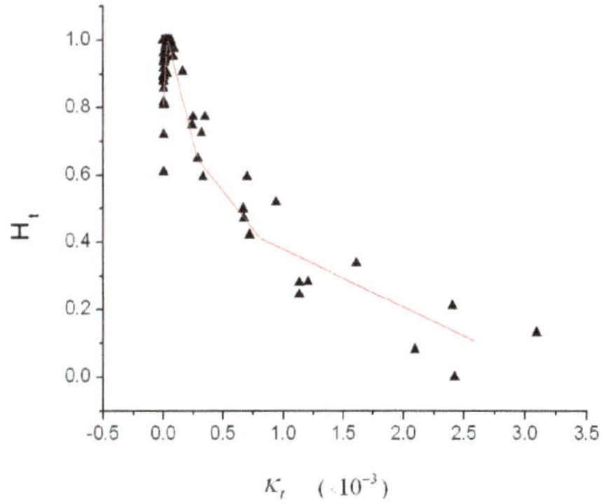

(a) Function of $H_t(\kappa_t)$

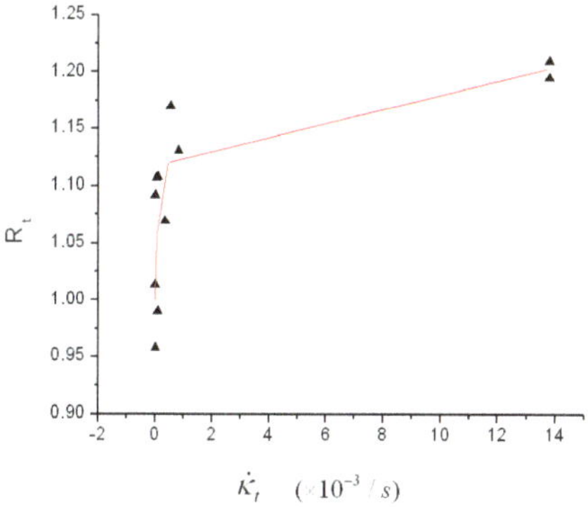

(b) Function of $R_t(\dot{\kappa}_t)$

Fig. 9. Function of $H_t(\kappa_t)$ and $R_t(\dot{\kappa}_t)$

Similarly, the functions of $H_c(\kappa_c)$ and $R_c(\dot{\kappa}_c)$ are given in **Fig. 10(a)** and **Fig. 10(b)**.

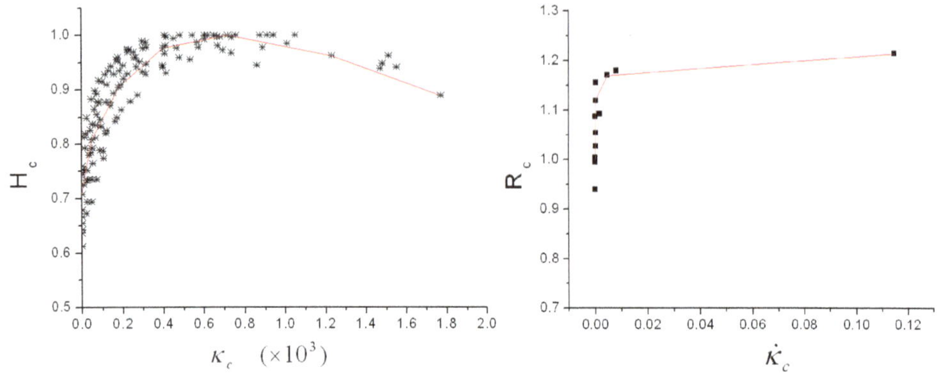

(a) Function of $H_c(\kappa_c)$ (b) Function of $R_c(\dot{\kappa}_c)$

Fig. 10. Function of $H_c(\kappa_c)$ and $R_c(\dot{\kappa}_c)$

3.3 Euler return mapping algorithm

At the time t, the stress σ_{ij}^t, the invariable κ^t and the rate of invariable $\dot{\kappa}^t$ should satisfy the yield condition:

$$F(\sigma_{ij}^t, \kappa^t, \dot{\kappa}^t) = 0 \tag{22}$$

At the time $t + \Delta t$, the stress can be written as

$$\sigma_{ij}^{t+\Delta t} = \sigma_{ij}^t + \Delta\sigma_{ij} = \sigma_{ij}^t + D_{ijkl}^{ep}(\sigma_{ij}, \Delta\lambda)\Delta\varepsilon_{kl} \tag{23}$$

where $D_{ijkl}^{ep}(\sigma_{ij}, \Delta\lambda)$ is the tangent module of the consistency viscoplastic model.

Also, the stress $\sigma_{ij}^{t+\Delta t}$, the invariable $\kappa^{t+\Delta t}$ and the rate of invariable $\dot{\kappa}^{t+\Delta t}$ at the time $t + \Delta t$ should satisfy the yield condition:

$$F(\sigma_{ij}^{t+\Delta t}, \kappa^{t+\Delta t}, \dot{\kappa}^{t+\Delta t}) = 0 \tag{24}$$

At the short time increment, the assumption is an approximation for $\dot{\lambda}$ as

$$\dot{\lambda} = \frac{\lambda^{t+\Delta t}}{\Delta t} \tag{25}$$

Thus, the internal parameter $\kappa^{t+\Delta t}$ and its rate $\dot{\kappa}^{t+\Delta t}$ can be expressed by

$$\dot{\kappa}^{t+\Delta t} = \dot{\lambda}^{t+\Delta t} g(\sigma_{ij}^{t+\Delta t}) = \frac{\Delta\lambda^{t+\Delta t}}{\Delta t} g(\sigma_{ij}^{t+\Delta t}) \tag{26}$$

$$\kappa^{t+\Delta t} = \kappa^{t} + \Delta\lambda^{t+\Delta t} g(\sigma_{ij}^{t+\Delta t}) \tag{27}$$

Consequently, the yield condition at the time $t+\Delta t$ can be formulated using the above expression as in the classic rate-independent plasticity as follows:

$$F(\sigma_{ij}^{t+\Delta t}, \Delta\lambda^{t+\Delta t}) = 0 \tag{28}$$

During the kth iteration at the time $t+\Delta t$, the stress σ_{ij}^{k} and plastic multiplier $\Delta\lambda^{k}$ might not satisfy the yield condition:

$$F^{k}(\sigma_{ij}^{k}, \Delta\lambda^{k}) \neq 0 \tag{29}$$

Generally, the stress σ_{ij}^{k} does not reflect the real stress $\sigma_{ij}^{t+\Delta t}$ at the end of the given time $t+\Delta t$. So, the residual stress can be expressed as

$$r_{ij}^{k}(\sigma_{ij}^{k}, \Delta\lambda^{k}) = \sigma_{ij}^{k} - \sigma_{ij}^{t+\Delta t} = \sigma_{ij}^{k} - \sigma_{ij}^{t} - D_{ijkl}^{ep}(\sigma_{ij}, \Delta\lambda)\Delta\varepsilon_{kl} \tag{30}$$

During the $(k+1)$th iteration at the time $t+\Delta t$, the yield function value F^{k+1} and the residual stress r_{ij}^{k+1} can be achieved from the truncated Taylor's series expansion of the yield F and the residual stress r_{ij} about position k, and then set to zero:

$$F^{k+1} = F^{k} + \frac{\partial F^{k}}{\partial \sigma_{ij}}\delta\sigma_{ij}^{k} + \frac{\partial F^{k}}{\partial \Delta\lambda}\delta\Delta\lambda^{k} = 0 \tag{31}$$

$$r_{ij}^{k+1} = r_{ij}^{k} + \frac{\partial r^{k}}{\partial \sigma_{ij}}\delta\sigma_{ij}^{k} + \frac{\partial r^{k}}{\partial \Delta\lambda}\delta\Delta\lambda^{k} = 0 \tag{32}$$

Equations (31) and (32) represent a set of linear equations for the iterative stress update $\delta\sigma_{ij}^{k}$ and the iterative viscoplastic multiplier update $\delta\Delta\lambda^{k}$.

Defining

$$[\mathbf{A}] = \begin{bmatrix} \dfrac{\partial F^{k}}{\partial \Delta\lambda} & \dfrac{\partial F^{k}}{\partial \sigma_{ij}} \\ \dfrac{\partial r^{k}}{\partial \Delta\lambda} & \dfrac{\partial r^{k}}{\partial \sigma_{ij}} \end{bmatrix}$$

$$\{\mathbf{b}\} = \begin{Bmatrix} -F^{k} \\ -r_{ij}^{k} \end{Bmatrix}$$

equations (31) and (32) can be expressed as

$$[\mathbf{A}]\begin{Bmatrix} \delta\Delta\lambda^k \\ \delta\sigma_{ij}^k \end{Bmatrix} = \{\mathbf{b}\} \tag{33}$$

where

$$\frac{\partial F^k}{\partial \Delta\lambda} = \frac{\partial F^k}{\partial \lambda} + \frac{\partial F^k}{\partial \dot{\lambda}}\frac{d\dot{\lambda}}{d\Delta\lambda} = \frac{\partial F^k}{\partial \lambda} + \frac{\partial F^k}{\partial \dot{\lambda}}\frac{1}{\Delta t} = h^k + \frac{s^k}{\Delta t}$$

$$\frac{\partial F}{\partial \sigma_{ij}} = m_{ij} = \alpha\delta_{ij} + \beta s_{ij} + \gamma t_{ij}$$

$$\frac{\partial r}{\partial \Delta\lambda} = -\frac{\partial (D_{ijkl}^{ep}\Delta\varepsilon_{kl})}{\partial \Delta\lambda}$$

$$\frac{\partial r_{ij}}{\partial \sigma_{kl}} = I_{ijkl} - \frac{\partial}{\partial \sigma_{kl}}(D_{ijpq}^{ep}\Delta\varepsilon_{pq})$$

where I_{ijkl} was the identity tensor of the fourth order.

Taking into account the symmetry of the stress tensor, the iterative stress update $\delta\sigma_{ij}^k$ and the iterative viscoplastic multiplier update $\delta\Delta\lambda^k$ are achieved by solving the set of linear equations. The iteration process continues until the norms $|F|$ and $|r_{ij}|$ reaches to reasonably small ammount. The final values of $\Delta\lambda$ and $\Delta\sigma_{ij}$ at the end of the time step are obtained by a summation process:

$$\Delta\lambda = \sum_{k=1}^{N} \delta\Delta\lambda^k \tag{34}$$

$$\Delta\sigma_{ij} = \sum_{k=1}^{N} \delta\sigma_{ij}^k \tag{35}$$

where N is the total number of iterations.

3.4 Tangent module of the consistency viscoplastic model

According Hooke's law, the stress change can be written as

$$d\sigma_{ij} = D_{ijkl}^e d\varepsilon_{kl}^e = D_{ijkl}^e (d\varepsilon_{kl} - d\varepsilon_{kl}^{vp}) = D_{ijkl}^e d\varepsilon_{kl} - D_{ijkl}^e d\varepsilon_{kl}^{vp} \tag{36}$$

Substituting the equation (36) into the equation (8), the consistency viscoplastic condition is expressed by

$$\frac{\partial F}{\partial \sigma_{ij}}(D_{ijkl}^e d\varepsilon_{kl} - D_{ijkl}^e d\varepsilon_{kl}^{vp}) + \frac{\partial F}{\partial \kappa}d\kappa + \frac{\partial F}{\partial \dot{\kappa}}d\dot{\kappa} = 0 \tag{37}$$

According to the flow law, the viscoplastic strain change is given by the following:

$$d\varepsilon_{ij}^{vp} = d\lambda \frac{\partial Q}{\partial \sigma_{ij}} \tag{38}$$

Substituting it into equation (37), one obtains

$$(\frac{\partial F}{\partial \sigma_{ij}})D_{ijkl}^{e} d\varepsilon_{ij} - d\lambda(\frac{\partial F}{\partial \sigma_{ij}})D_{ijkl}^{e}(\frac{\partial Q}{\partial \sigma_{ij}}) + \frac{\partial F}{\partial \kappa}d\kappa + \frac{\partial F}{\partial \dot{\kappa}}d\dot{\kappa} = 0 \tag{39}$$

Therefore, the viscoplastic multiplier is expressed as

$$d\lambda = \frac{(\frac{\partial F}{\partial \sigma_{ij}})D_{ijkl}^{e} d\varepsilon_{ij}}{(\frac{\partial F}{\partial \sigma_{ij}})D_{ijkl}^{e}(\frac{\partial Q}{\partial \sigma_{kl}}) - \frac{1}{d\lambda}\frac{\partial F}{\partial \kappa}d\kappa - \frac{1}{d\lambda}\frac{\partial F}{\partial \dot{\kappa}}d\dot{\kappa}} \tag{40}$$

Substituting it into equation (36), the following formula is obtained as

$$\begin{aligned}
d\sigma_{ij} &= D_{ijkl}^{e} d\varepsilon_{kl} - D_{ijkl}^{e} d\varepsilon_{kl}^{vp} \\
&= D_{ijkl}^{e} d\varepsilon_{kl} - D_{ijkl}^{e} d\lambda \frac{\partial Q}{\partial \sigma_{kl}} \\
&= D_{ijkl}^{e} d\varepsilon_{kl} - D_{ijkl}^{e} \frac{\frac{\partial F}{\partial \sigma_{mn}} D_{mnpq}^{e} d\varepsilon_{pq}}{\frac{\partial F}{\partial \sigma_{mn}} D_{mnpq}^{e} \frac{\partial Q}{\partial \sigma_{pq}} - \frac{1}{d\lambda}\frac{\partial F}{\partial \kappa}d\kappa - \frac{1}{d\lambda}\frac{\partial F}{\partial \dot{\kappa}}d\dot{\kappa}} \frac{\partial Q}{\partial \sigma_{kl}} \\
&= D_{ijkl}^{e} d\varepsilon_{kl} - \frac{D_{ijkl}^{e} \frac{\partial F}{\partial \sigma_{mn}} D_{mnpq}^{e} \frac{\partial Q}{\partial \sigma_{kl}} \delta_{kp}\delta_{lq}}{\frac{\partial F}{\partial \sigma_{mn}} D_{mnpq}^{e} \frac{\partial Q}{\partial \sigma_{pq}} - \frac{1}{d\lambda}\frac{\partial F}{\partial \kappa}d\kappa - \frac{1}{d\lambda}\frac{\partial F}{\partial \dot{\kappa}}d\dot{\kappa}} d\varepsilon_{kl} \\
&= (D_{ijkl}^{e} + D_{ijkl}^{vp}) d\boldsymbol{\varepsilon}_{kl} = D_{ijkl}^{ep} d\varepsilon_{kl}
\end{aligned} \tag{41}$$

where D_{ijkl}^{ep} is the tangent module of the consistency viscoplastic model.

$$\begin{aligned}
D_{ijkl}^{ep} &= D_{ijkl}^{e} + D_{ijkl}^{vp} \\
&= D_{ijkl}^{e} - \frac{D_{ijkl}^{e} \frac{\partial F}{\partial \sigma_{mn}} D_{mnpq}^{e} \frac{\partial Q}{\partial \sigma_{kl}} \delta_{kp}\delta_{lq}}{\frac{\partial F}{\partial \sigma_{mn}} D_{mnpq}^{e} \frac{\partial Q}{\partial \sigma_{pq}} - \frac{1}{d\lambda}\frac{\partial F}{\partial \kappa}d\kappa - \frac{1}{d\lambda}\frac{\partial F}{\partial \dot{\kappa}}d\dot{\kappa}}
\end{aligned} \tag{42}$$

Therefore,

$$D_{ijkl}^{vp} = -\frac{D_{ijkl}^{e}\frac{\partial F}{\partial \sigma_{mn}} D_{mnpq}^{e}\frac{\partial Q}{\partial \sigma_{kl}}\delta_{kp}\delta_{lq}}{\frac{\partial F}{\partial \sigma_{mn}} D_{mnpq}^{e}\frac{\partial Q}{\partial \sigma_{pq}} - \frac{1}{d\lambda}\frac{\partial F}{\partial \kappa}d\kappa - \frac{1}{d\lambda}\frac{\partial F}{\partial \dot{\kappa}}d\dot{\kappa}} \tag{43}$$

The yield function of the classic William-Warnke three-parameter model is expressed as

$$F(\sigma_m, \tau_m, \theta) = \frac{1}{\rho}\sigma_m + \frac{1}{r(\theta)}\tau_m - f_c = 0 \tag{44}$$

The divergence of the yield function is written as

$$\frac{\partial F}{\partial \sigma_{ij}} = \alpha\delta_{ij} + \beta s_{ij} + \gamma t_{ij} \tag{45}$$

To the relative flow criteria, the flow law is the same as the yield function of the model:

$$Q(\sigma_{ij}, \kappa, \dot{\kappa}) = F(\sigma_{ij}, \kappa, \dot{\kappa}) \tag{46}$$

Then, the tangent module of the consistency viscoplastic model can be written as follows

$$\begin{aligned} D_{ijkl}^{vp} &= -\frac{D_{ijkl}^{e}\frac{\partial F}{\partial \sigma_{mn}} D_{mnpq}^{e}\frac{\partial Q}{\partial \sigma_{kl}}\delta_{kp}\delta_{lq}}{\frac{\partial F}{\partial \sigma_{mn}} D_{mnpq}^{e}\frac{\partial Q}{\partial \sigma_{pq}} - \frac{1}{d\lambda}\frac{\partial F}{\partial \kappa}d\kappa - \frac{1}{d\lambda}\frac{\partial F}{\partial \dot{\kappa}}d\dot{\kappa}} \\ &= -\frac{D_{ijkl}^{e}(\alpha\delta_{mn} + \beta s_{mn} + \gamma t_{mn})D_{mnpq}^{e}(\alpha\delta_{kl} + \beta s_{kl} + \gamma t_{kl})\delta_{kp}\delta_{lq}}{(\alpha\delta_{mn} + \beta s_{mn} + \gamma t_{mn})D_{mnpq}^{e}(\alpha\delta_{pq} + \beta s_{pq} + \gamma t_{pq}) - h - \frac{s}{\Delta t}} \\ &= -\frac{(3K\alpha\delta_{ij} + 2G\beta s_{ij} + 2G\gamma t_{ij})(3K\alpha\delta_{kl} + 2G\beta s_{kl} + 2G\gamma t_{kl})}{9K\alpha^2 + 4G\beta^2 J_2 + \frac{4}{3}G\gamma^2 J_2^2 + 12G\beta\gamma J_3 - h - \frac{s}{\Delta t}} \end{aligned} \tag{47}$$

Defining

$$P_0 = \sqrt{9K\alpha^2 + 4G\beta^2 J_2 + \frac{4}{3}G\gamma^2 J_2^2 + 12G\beta\gamma J_3 - h - \frac{s}{\Delta t}}$$

$$P_1 = 3K\alpha / P_0$$

$$P_2 = 2G\beta / P_0$$

$$P_3 = 2G\gamma / P_0$$

Therefore, the equation (47) can be expressed as

$$D_{ijkl}^{vp} = -(P_1\delta_{ij} + P_2 s_{ij} + P_3 t_{ij})(P_1\delta_{kl} + P_2 s_{kl} + P_3 t_{kl}) \tag{48}$$

The tangent module of the consistency viscoplastic model is obtained as

$$\begin{aligned} D_{ijkl}^{ep} &= D_{ijkl}^{e} + D_{ijkl}^{vp} \\ &= D_{ijkl}^{e} - (P_1\delta_{ij} + P_2 s_{ij} + P_3 t_{ij})(P_1\delta_{kl} + P_2 s_{kl} + P_3 t_{kl}) \end{aligned} \tag{49}$$

3.5 Comparisons with experimental data

The stress-strain curves of concrete for the uniaxial tension at the strain rate 10^{-3}/s, shown in **Fig. 11(a)**, are calculated and compared with experimental results. **Fig. 11(b)** shows the stress-strain curves of concrete for the uniaxial tension at the strain rate $10^{-5}s^{-1}$, $10^{-4}s^{-1}$, $10^{-3}s^{-1}$ and $10^{-2}s^{-1}$nd the comparisons with the experimental data. Similarly, the stress-strain curves of concrete for the uniaxial compression at the strain rate $10^{-2}s^{-1}$, shown in **Fig. 12(a)**, are calculated and compared with the experimental results. **Fig. 12(b)** illustrates the stress-strain curves of concrete for the uniaxial compression at the strain rate $110^{-5}s^{-1}$, $10^{-4}s^{-1}$, $10^{-3}s^{-1}$, $10^{-2}s^{-1}$ and $10^{-1}s^{-1}$ and the comparisons with the experimental data.

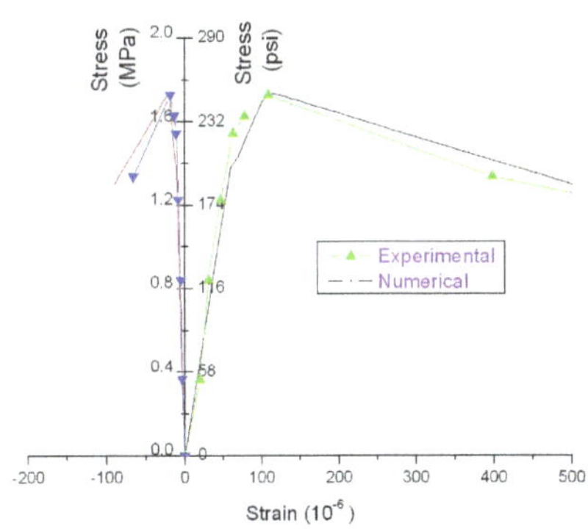

(a) strain rate 10^{-3}/s

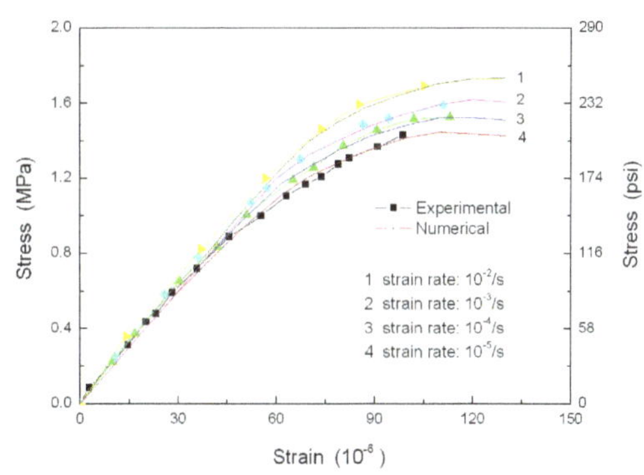

(b) different strain rates

Fig. 11. Model and tensile test results

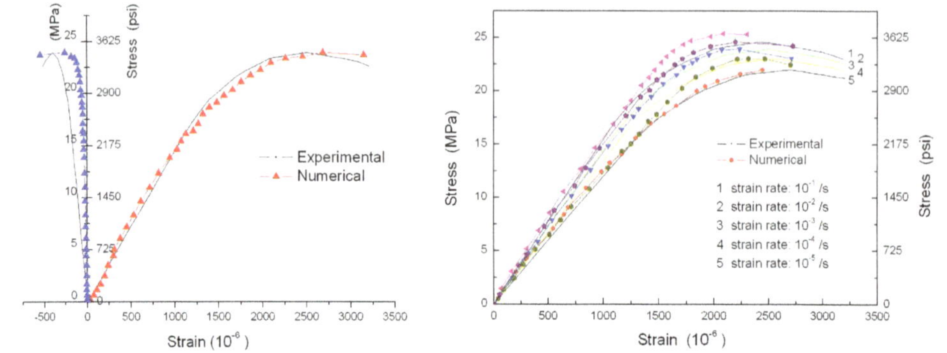

(a) strain rate 10^{-2}/s (b) different strain rates

Fig. 12. Model and compressive test results

The consistency viscoplastic model is modified from the classic William-Warnke three-parameter model of concrete and has the advantages and disadvantages of the William-Warnke three-parameter model. It may directly simulate the dynamic behaviors of concrete and it is simple and easy to calculate. Thus, the proposed model is good for analyzing the dynamic responses of concrete structures.

3.6 Numerical example

In order to study the effect of strain rate on the dynamic response of concrete structures, the dynamic response of a simple-supporting beam with dimensions 8mx1mx1m is analyzed with this model. **Fig. 13** shows the discretized beam and calculated elements adopting three-dimension eight-node equivalent parameter elements. An impact loading is imposed on the midpoint of beam and **Fig. 14** depicts the loading history. Dynamic response is analyzed with the ADNFEM program compiled by the authors. The material properties are as follows: the elastic modulus of concrete is 1.6×10^4MPa, the Poisson's ratio is 0.17, the mass density is 2.4×10^3kg/m^3, the static compressive strength is 22MPa and the static tensile strength is 2.2MPa.

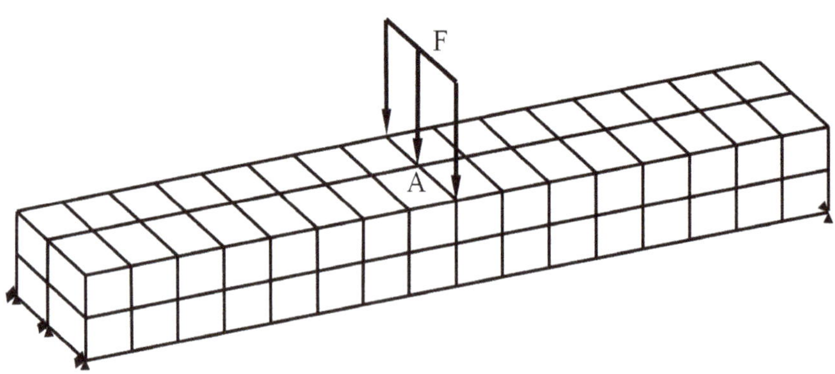

Fig. 13. Structure and meshes of the beam

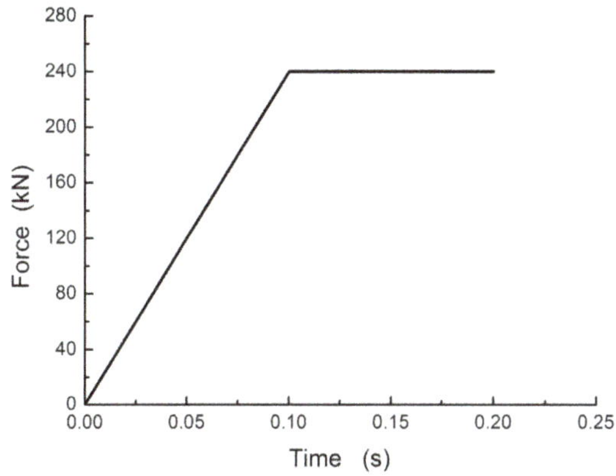

Fig. 14. Time-dependent curve of loading

Fig. 15 illustrates the displacement of midpoint A with time. Three curves shown in the figure denote the vertical displacement of the beam at model I (linear elastic model), model II (rate-dependent William-Warnke three-parameter model), and model III (rate-independent William-Warnke three-parameter model), respectively. **Fig. 15** shows that, at the beginning of loading, the beam is at the elastic state and three displacement curves are the same. When the stress of the beam reaches to the initial yield stress, the displacement curves of three models separate, and when time is 0.1 second, the loadings of three models reach to their maximums, but the displacements of the three models do not reach to their maximums at the same time. The vertical displacement of model I reaches to its maximum 1.50mm at time 0.114 second, but that of model II reaches to its maximum 1.62mm at time 0.110 second, and that of model III reaches to its maximum 1.90mm at time 0.117 second. It is clearly shown that displacement of the beam changes greatly after considering the effect of strain rate. The displacement of the beam with the model II decreases with 14.7 percent compared with that of the beam with model III but increases with 8.0 percent compared with that of the beam with model I.

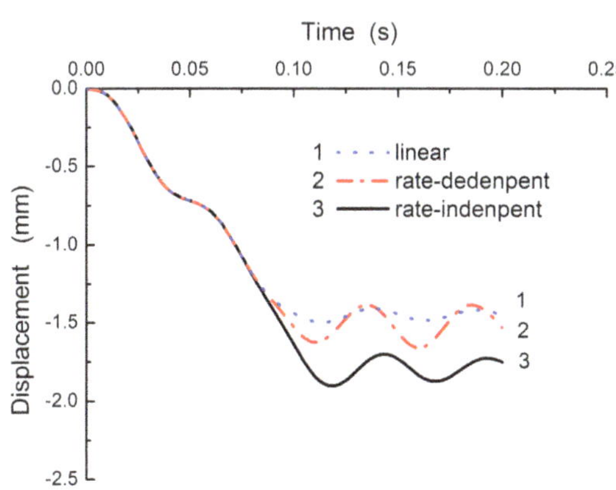

Fig. 15. Displacement of point A

At the same time, the stresses of the beam with the three models differ greatly from each other. **Table 7** lists the maximums of the tensile and compressive principal stress of the beam with the three models. It can be seen clearly that principal stresses of the beam with different models vary greatly. The tensile principal stress of the beam with model II increases with 13.10 percent compared with that of the beam with model III, but the compressive principal stress decreases with 11.6 percent. Compared with model I, the tensile principal stress of the beam with model II decreases with 16.80 percent, but the compressive principal stress increases with 19.1 percent. Similarly, the stress distribution of the beam changes greatly. **Fig. 16(a)**, **(b)** and **(c)** show the stress distribution of the beam with the three models respectively when the displacement is maximum. The tensile stress distribution figure is shown above and the compressive stress distribution figure is illustrated below. It can be seen clearly from these figures that the stress magnitudes and distributions of the beam change greatly with the different models. Consequently, it can be seen that the dynamic response of the concrete beam, the displacement, and the stress magnitude and distribution, change greatly after considering the effect of strain rate.

model	The first principal stress of point B MPa	The third principal stress of point A MPa
model I	1.97	-2.04
model II	1.64	-2.43
model III	1.45	-2.75

Table 7. Maximal principal stress of beam

4. Seismic response of arch dam

4.1 Model and parameters of arch dam

In order to illustrate the effect of the rate dependency on the dynamic structural response, a 278m high arch dam in China subjected to earthquake excitation is analyzed by the proposed model. The dam and the foundation are discretized into 450 and 1,040 three-dimensional isoparametric 8-node elements, respectively. **Fig. 17** shows the discretized dam-foundation system.

The material properties are as follows: for the dam body, the elastic module is 2.4×10^4MPa, the Poisson's ratio is 0.17, the density is 2.4×10^3kg/m^3, the static compressive strength is 30MPa, and the static tensile strength is 3MPa; for the foundation rock, the elastic module is 1.6×10^4MPa, the Poisson's ratio is 0.25, and the density is 2.0×10^3kg/m^3. The five lowest vibration frequencies of the dam in care of full reservoir are: f_1 =0.997Hz, f_2 =1.004Hz, f_3 =1.450Hz, f_4 =1.497Hz and f_5 =1.542Hz. An assumption of massless foundation is introduced to simplify the dam-foundation interaction analysis, although more rigorous interaction effects can be included.

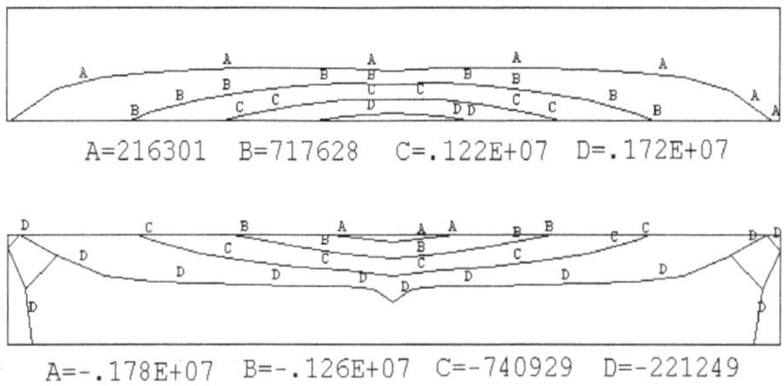

(a) model I

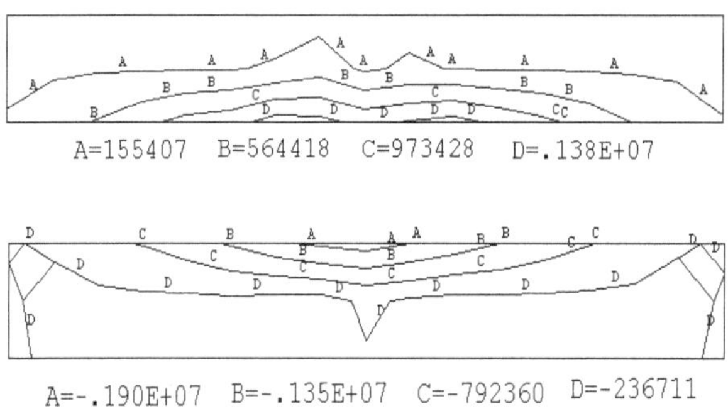

(b) model II

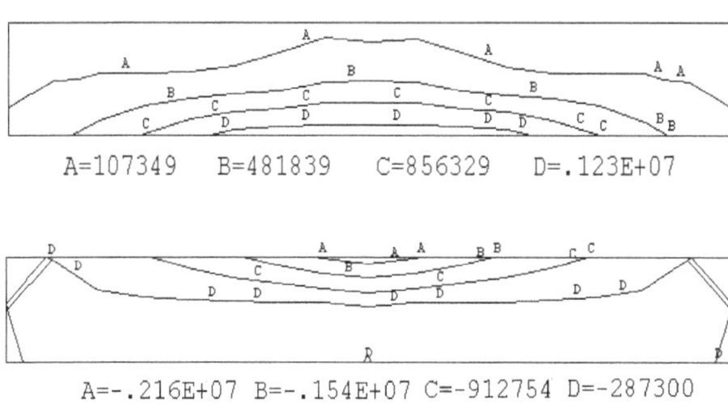

(c) model III

Fig. 16. Distribution of principal stress of beam, Pa

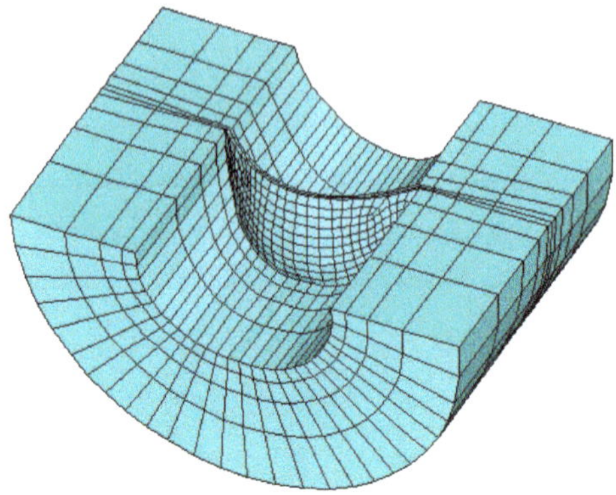

Fig. 17. Geometry and mesh of arch dam

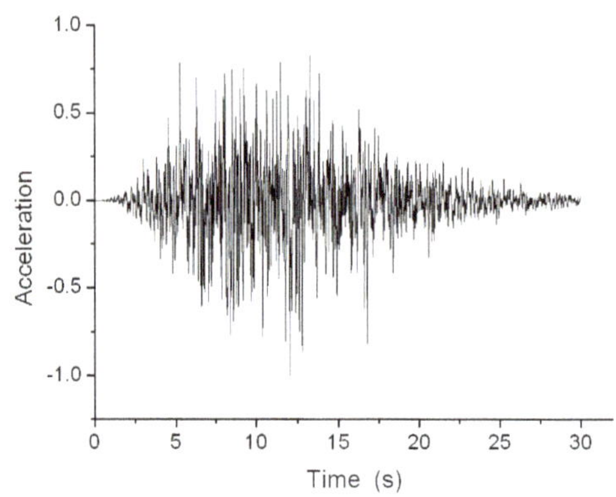

Fig. 18. Time history of earthquake input

Three-dimensional earthquake waves are used as the input. The design earthquake acceleration peak is 0.321 *gal* . **Fig. 18** shows the typical artificial unitary acceleragram that met the requirements of the Chinese Specifications for Seismic Design of Hydraulic Structures.

4.2 Design of the ADNFEM Program

The ADNFEM (Arch Dam Nonlinear Finite Element Method) program compiled by the authors is used to calculate the dynamic response of arch dams in the rate-dependent constitutive model. The program is validated by ANSYS in the rate-independent model when the strain rate was zero in the rate-dependent constitutive model.

The dynamic response of arch dams includes two parts: static analysis and seismic response analysis. The static analysis of arch dams is carried out to calculate the initial stress and strain state of the seismic response of arch dams. In order to form the Rayleigh damping matrix, the model of arch dams is calculated during the process of seismic response analysis. The processes of static analysis and seismic response analysis are listed as **Table 8** and **Table 9**.

1. Input initial data and form one-dimension bandwidth memory vector and constrain condition vector;
2. Input the constrain condition;
3. Set the loading step $I = 1$;
4. Form the Ith static load vector $\mathbf{f}$;
5. Judge elastic-plasticity state of every element and calculate the elastic-plasticity impact factor;
6. Form the tangent stiffness matrix of structures $\mathbf{K}$;
7. Calculate the elastic-plasticity predicted state;
8. Set the iterative $J = 1$ of the Ith loading step;
9. Solve for the iterative displacement $\delta\mathbf{u} = \mathbf{K}^{-1}\mathbf{f}$;
10. Update incremental displacements $\Delta\mathbf{u} = \Delta\mathbf{u} + \delta\mathbf{u}$;
11. For every Gaussian integration point:
    a. Compute the incremental strains $\Delta\boldsymbol{\varepsilon}$
    b. Compute the elastic-plastically predicted stress state $\boldsymbol{\sigma}^{ep} = \boldsymbol{\sigma}_t + \mathbf{D}^{ep}\Delta\boldsymbol{\varepsilon}$
    c. If $\boldsymbol{\sigma}^{ep}$ violates the yield criterion, perform the return mapping for the consistency model
12. Check for convergence, if not, $J \leftarrow J+1$, go to step 9;
13. Update the loading step $I \leftarrow I+1$, go to step 4;
14. Output the static analysis results.

Table 8. Summary of static analysis

1. Form the mass matrix $\mathbf{M}$ and initial stiffness matrix $\mathbf{K}$;
2. Calculate the first two frequencies ω_1 and ω_2 with the subspace iterative method;

Input structural damping ratio ξ and calculate the Rayleigh damping factor $\alpha_1 = \dfrac{2\omega_1\omega_2\xi}{\omega_1+\omega_2}$, $\alpha_2 = \dfrac{2\xi}{\omega_1+\omega_2}$;

3. Form the damping matrix $\mathbf{C} = \alpha_1\mathbf{M} + \alpha_2\mathbf{K}$;
4. Input the earthquake wave acceleration $\mathbf{a}_g$;
5. Initialize the parameters of the Newmark-β method β_N and γ_N;
6. Set the loading step $I = 1$;
7. Form the Ith earthquake load vector $\mathbf{f}_g^{t+\Delta t} = -\mathbf{M}\mathbf{a}_g^{t+\Delta t}$;
8. Initialize the displacements, velocities, and accelerations;

$$\Delta\mathbf{u} = 0\,,\ \delta\mathbf{u} = 0\,;$$

$$\mathbf{a}_{t+\Delta t}^0 = -\frac{1}{\beta_N \Delta t}\mathbf{v}_t - (\frac{1}{2\beta_N} - 1)\mathbf{a}_t\,,\ \mathbf{v}_{t+\Delta t}^0 = (1-\frac{\gamma_N}{\beta_N})\mathbf{v}_t + (1-\frac{\gamma_N}{2\beta_N})\Delta t\mathbf{a}_t$$

9. Set the iterative $J=1$ of the Ith loading step;
10. Update velocities and accelerations

$$\mathbf{a}_{t+\Delta t}^{J}=\frac{1}{\beta_N \Delta t^2}\delta\mathbf{u}+\mathbf{a}_{t+\Delta t}^{J-1},\ \mathbf{v}_{t+\Delta t}^{J}=\frac{1}{\beta_N \Delta t}\delta\mathbf{u}+\mathbf{v}_{t+\Delta t}^{J-1};$$

11. Judge elastic-plasticity state of every element and calculate the elastic-plasticity impact factor;
12. Form the tangent stiffness matrix of structures $\mathbf{K}$;
13. Compute the equivalent stiffness matrix and equivalent force vector;

$$\overline{\mathbf{K}}=\mathbf{K}+\frac{1}{\beta_N \Delta t^2}\mathbf{M}+\frac{\gamma_N}{\beta_N \Delta t}\mathbf{C};$$

$$\overline{\mathbf{f}}=\mathbf{f}_g^t+\mathbf{M}(\frac{1}{\beta_N \Delta t}\mathbf{v}_{t+\Delta t}^J+\frac{1}{2\beta_N}\mathbf{a}_{t+\Delta t}^J)+\mathbf{C}\{\frac{\gamma_N}{\beta_N}\mathbf{v}_{t+\Delta t}^J+(\frac{\gamma_N}{2\beta_N}-1)\mathbf{a}_{t+\Delta t}^J\};$$

14. Solve for the iterative displacement $\delta\mathbf{u}=\overline{\mathbf{K}}^{-1}\overline{\mathbf{f}}$;
15. Update incremental displacements $\Delta\mathbf{u}=\Delta\mathbf{u}+\delta\mathbf{u}$;
16. For every Gaussian integration point:
a. Compute the incremental strains $\Delta\boldsymbol{\varepsilon}$
b. Compute the elastic-plastically predicted stress state $\boldsymbol{\sigma}^{ep}=\boldsymbol{\sigma}_t+\mathbf{D}^{ep}\Delta\boldsymbol{\varepsilon}$
c. If $\boldsymbol{\sigma}^{ep}$ violates the yield criterion, perform the return mapping for the consistency model
17. Check for convergence, if not, $J \leftarrow J+1$, go to step 11;
18. Update the loading step $I \leftarrow I+1$, go to step 8;
19. Output the seismic response analysis results.

Table 9. Summary of dynamic analysis

4.3 Seismic response of arch dam

4.3.1 Stresses in arch dams

The dynamic response analyses of the arch dam are performed with three models: model I (linear elastic model), model II (rate-dependent William-Warnke three-parameter model), and model III (rate-independent William-Warnke three-parameter model). The maximum values of the first and the third principle stresses in the dam are shown in **Table 10**. **Fig. 19** shows the distributions of the third principle stresses obtained from the three models. It is seen that, in all three cases, the maximum compressive stress is the same and appeared at the bottom of the upstream face; the material remaines working in the elastic range. While, for the maximum tensile stress there is the marked difference among the calculated results of the three models. Owing to the plasticity of concrete, the maximal values of the first principal stresses of model II and III decrease with 37.7% and 44.5%, respectively, compared with model I. Because the dynamic tensile strength of concrete increases with the increase of strain rates, the maximal values of the first principal stresses of model II, taking into account the effect of strain rates, increase with 12.2% compared with model III. **Fig. 20**, **Fig. 21** and **Fig. 22** show the distributions of the first principle stress obtained from the three models.

model	The first principal stress MPa	The third principal stress MPa
model I	5.17	-12.50
model II	3.22	-12.50
model III	2.87	-12.50

Table 10. Maximal principal stress of arch dam

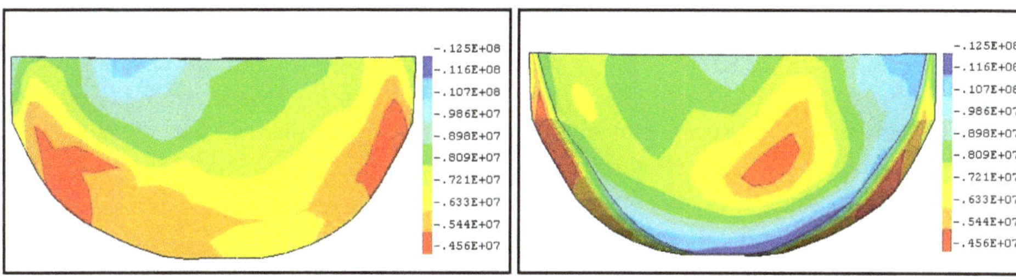

(a) upstream face (b) downstream face

Fig. 19. Distribution of the third principle stress (model I, II, and III), Mpa

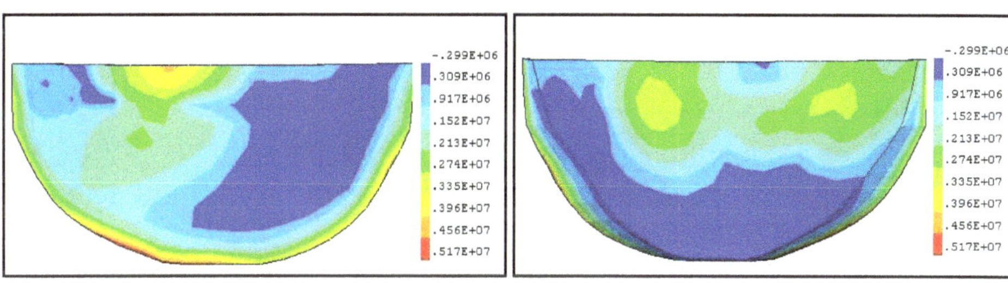

(a) upstream face (b) downstream face

Fig. 20. Distribution of the first principle stress (model I), Mpa

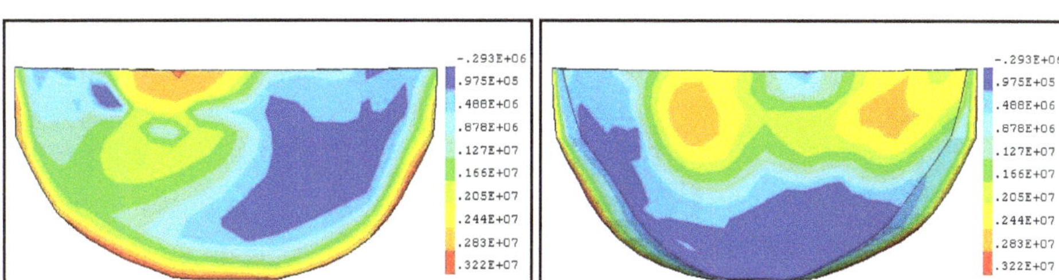

(a) upstream face (b) downstream face

Fig. 21. Distribution of the first principle stresses (model II), Mpa

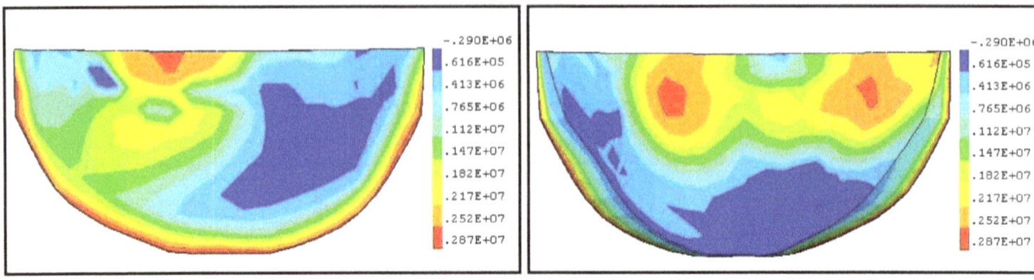

(a) upstream face (b) downstream face

Fig. 22. Distribution of the first principle stresses (model III), MPa

4.3.2 Strain and strain rate of arch dam

The maximal equivalent strain of concrete in three cases are the same because the compressive strain of concrete plays a more important role in the equivalent strain of concrete in the dominant compressive stress states, although the tensile strain is important in the dominant tensile stress states, but the values are smaller than those for the compressive strain. **Fig. 23** shows the distributions of the equivalent strain of the arch dam from the three models. It is clear that the maximal equivalent strain is 4.75×10^{-4}, and it appears at the bottom of the downstream face. Similarly, the maximal equivalent strain rates in the three cases are the same, and **Fig. 24** shows the distributions of the equivalent strain rates of the arch dam from the three models. The maximal equivalent strain rate is up to $3.47\times10^{-2}s^{-1}$ and it also appears at the bottom of the downstream face.

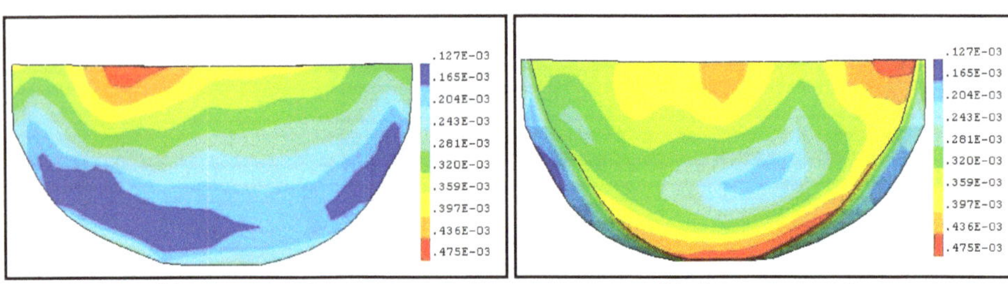

(a) upstream face (b) downstream face

Fig. 23. Distribution of the equivalent strain (model I, II, and III)

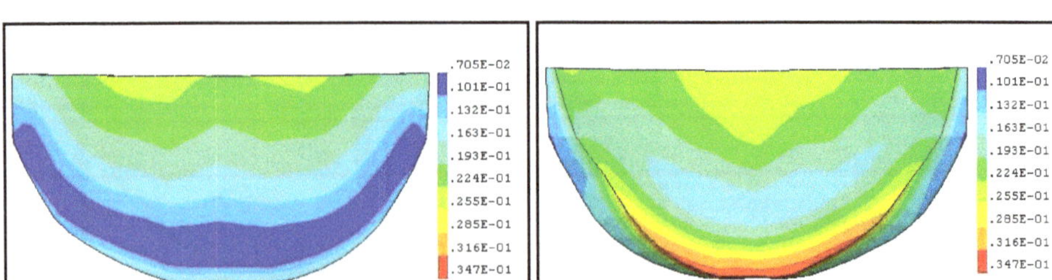

(a) upstream face (b) downstream face

Fig. 24. Distribution of the equivalent strain rate (model I, II, and III), s^{-1}

4.3.3 Plastic strain and plastic strain rate of arch dam

The equivalent viscoplastic strain of concrete appears only on the tensile zones of the arch dam. **Fig. 25** and **Fig. 26** show the distributions of the maximal equivalent viscoplastic strains obtained from model II and model III, respectively. It is shown that the maximal equivalent viscoplastic strains appear on the bottom of the upstream face and that the strain rates has little effect. **Fig.27** and **Fig.28** show the distributions of the maximal equivalent viscoplastic strain rates obtained from model II and model III, respectively. Similarly, the maximal equivalent viscoplastic strain rate appears on the bottom of the upstream face but it decreases with 17.5% after taking into account the effect of strain rates.

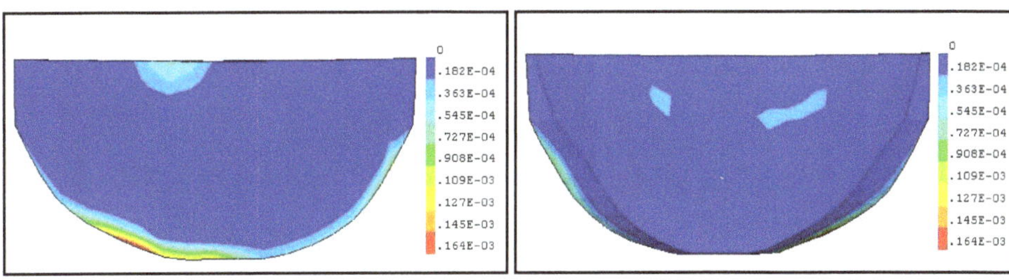

(a) upstream face (b) downstream face

Fig. 25. Distribution of the equivalent viscoplastic strain (model II)

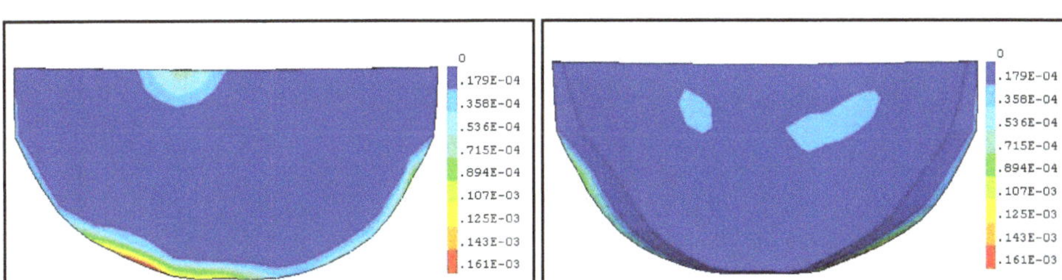

(a) upstream face (b) downstream face

Fig. 26. Distribution of the equivalent viscoplastic strain (model III)

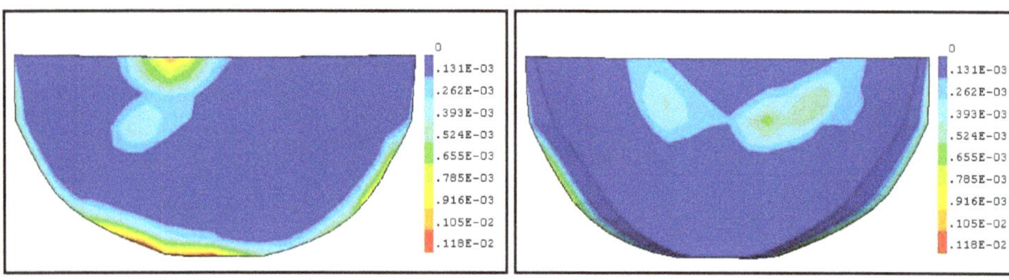

(a) upstream face (b) downstream face

Fig. 27. Distribution of the equivalent viscoplastic strain rate (model II), s^{-1}

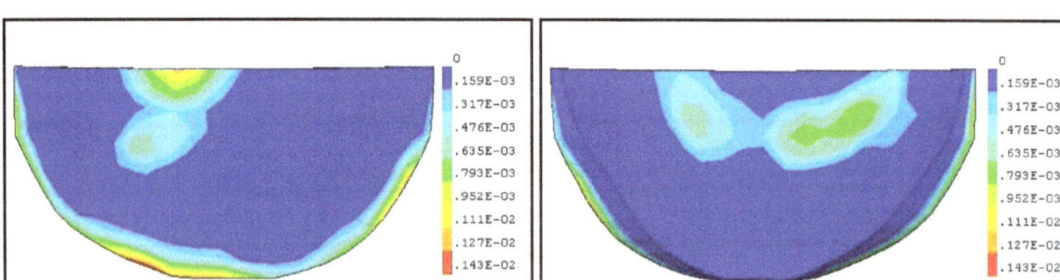

(a) upstream face (b) downstream face

Fig. 28. Distribution of the equivalent viscoplastic strain

5. Conclusions

Based on the numerical results obtained, the following conclusions are drawn:

1. Results indicate that the tensile and compressive strengths of concrete increase with increase of the rate of loading. The initial tangent modulus and the critical strain of concrete are independent of strain rate in tension but slightly increased as the strain rate increased in compression. Poisson's ratio of concrete isn't obviously dependent on the loading rate both in tension and in compression.
2. Comparisons between the models and experimental data show that the consistency model may simulate directly the uniaxial dynamic behaviors of concrete. The dynamic responses of a simple-supporting beam show that dynamic responses of concrete beam, the displacement, the stress magnitude and distribution, change greatly after considering the effect of strain rate.
3. In all three cases, the maximum values of the compressive stress are the same and the concrete remains working in the elastic range. The maximal values of the first principal stresses of the arch dam, taking into account the effect of strain rates, increase with 12.2% because the dynamic tensile strength of concrete increases with the increase of strain rates. There are no effects of strain rates on the maximal equivalent strain and the maximal equivalent strain rate of the arch dam. The strain rates have little effect on the maximal equivalent viscoplastic strain, while the maximal equivalent viscoplastic strain rate decreases with 17.5% after taking into account the effect of strain rates.

6. Acknowledgements

This study was founded by the National Science Foundation of China under grant No.50408030 and No. 51121005, the College Innovation Group of Liaoning Province under grant No: 2008T223 and the Fundamental Research Funds for the Central Universities under grant No: DUT10LK29.

7. Symbol notation

Symbol	Description
f_t	dynamic tensile strength of concrete
f_{ts}	static tensile strength of concrete
f_c	dynamic compressive strength of concrete
f_{ts}	static compressive strength of concrete
$\dot{\varepsilon}_t$	dynamic tensile strain rate
$\dot{\varepsilon}_{ts}$	quasi-static tensile strain rate
$\dot{\varepsilon}_c$	dynamic compressive strain rate
$\dot{\varepsilon}_{cs}$	quasi-static compressive strain rate
E	elastic modulus of concrete
K	volume module of concrete
G	shear module of concrete
μ	Poisson's ratio of concrete

σ_{ij}	the stress tensor
s_{ij}	the deviatoric stress tensor
r_{ij}	the residual stress tensor
σ_m	the average stress
τ_m	the average shear stress
ε_{ij}	the strain tensor
ε_{ij}^{vp}	the viscoplastic strain tensor
κ	internal variable
$\dot{\kappa}$	rate of internal variable
λ	the viscoplastic multiplier
φ_t	weight function of tensile invariable
φ_c	weight function of compressive invariable
F	yield function
Q	plastic flow function
I_1	the first invariant of stress tensor
J_2	the second invariant of deviatoric stress tensor
J_3	the third invariant of deviatoric stress tensor
D_{ijkl}^{e}	elastic stiffness of concrete
D_{ijkl}^{vp}	viscoplastic stiffness of concrete

8. References

Abrams DA (1917), "Effect of Rate of Application of Load on the Compressive Strength of Concrete," *ASTM Journal,* 17: 364-377.

Atchley BL and Furr HL (1967), "Strength and Energy-absorption Capabilities of Plain Concrete under Dynamic and Static Loading," *ACI Journal,* 64: 745-756.

Bai WF, Chen JY and Zhong H (2006), "Seismic Overload Response Analysis of Concrete Gravity Dam Based on Strain Rate-dependent Damage Model," *Journal of Hydraulic Engineering,* 37(7): 820-826. (in Chinese)

Birkimer DL and Lindemann R (1971), "Dynamic Tensile Strength of Concrete Materials," *ACI Journal Proceedings,* 1: 47-49.

Bischoff PH and Perry SH (1991), "Compressive Behaviour of Concrete at High Strain Rates," *Materials and Structures,* 24(144): 425-450.

Cadoni E, Labibes K and Albertini C (2001), "Strain-rate Effect on the Tensile Behaviour of Concrete at Different Relative Humidity Levels," *Materials and Structures,* 34(235): 21-26.

Cervera M, Oliver J and Manzoli O (1996), "A Rate-dependent Isotropic Damage Model for the Seismic Analysis of Concrete Dams," *Earthquake Engineering and Structural Dynamics,* 25: 987-1010.

Chen JY, Hu ZQ and Lin G (2004), "Seismic Analysis of Arch Dam with Joints Based on a New Strain-rate-dependant Plastic Damage Model," *Chinese Journal of Computational Mechanics,* 21(1): 45-49. (in Chinese)

Chinese Standard (2001), *Specifications for Seismic Design of Hydraulic Structures,* Beijing: Chinese Electric Power Press. (in Chinese)

Dhir RK and Sangha CM (1972), "A Study of Relationships between Time, Strength, Deformation and Fracture of Plain Concrete," *Magazine of Concrete Research,* 24(81): 197-208.

Dilger WH, Koch R and Kowalczyk R (1984), "Ductility of Plain and Confined Concrete under Different Strain Rates," *ACI Journal,* 81(1): 73-81.

Evans RH (1942), "Effect of Rate of Loading on the Mechanical Properties of Some Materials," *Journal of Institute of Civil Engineers,* 18(7-8): 96-306.

Harris DW, Mohorovic CE and Dolen TP (2000), "Dynamic Properties of Mass Concrete Obtained from Dam Cores," *ACI Material Journal,* 97(3): 290-296.

Hughes BP and Gregory R (1972), "Concrete Subjected to High Rates of Loading in Compression," *Magazine of Concrete Research,* 24(78): 25-36.

Jones PG and Richart FE (1936), "The Effect of Testing Speed on Strength and Elastic Properties of Concrete," *ASTM Journal,* 36: 380-392.

Lee J and Fenves GL (1998), "A Plastic-damage Concrete Model for Earthquake Analysis of Dams," *Earthquake Engineering and Structural Dynamics,* 27(9): 937-956.

Li J, Chen JY and Bai WF (2005), "Application of Strain Rate-dependent Damage Model in Analysis of Dynamic Opening of High Arch Dam Transverse Joints," *Journal of Hydraulic Engineering,* 36(7): 870-875. (in Chinese)

Malvar LJ and Ross CA (1998), "Review of Strain Rate Effects for Concrete in Tension," *ACI Materials Journal,* 95(6): 735-739.

Moore OL (1934), "Report of Working Committee of Committee C-1 on Plastic Mortar Tests for Portland Cement," *Proceedings of the American Society for Testing Materials,* 34: 322.

Norris CH, Hansen RJ, Holley MJ, Biggs JM, Namyet S and Minami JK (1959), *Structural Design for Dynamic Loads,* New York: McGraw-Hill Book Co. Inc.

Oh BH (1987), "Behavior of Concrete under Dynamic Tensile Loads," *ACI Materials Journal,* 84(1): 8-13.

Raphael JM (1984), "Tensile Strength of Concrete," *ACI Journal,* 87(17): 158-165.

Rossi P and Toutlemonde F (1996), "Effect of Loading Rate on the Tensile Behaviour of Concrete: Description of the Physical Mechanisms," *Materials and Structures,* 29(186): 116-118.

Rossi P, Van M and Jan GM (1994), "Effect of Loading Rate on the Strength of Concrete Subjected to Uniaxial Tension," *Materials and Structures,* 27(169): 260-264.

Rush H (1960), "Research towards a General Flexural Theory for Structural Concrete," *ACI Journal,* 57(1): 1-28.

Spark PR and Menzies JB (1973), "The Effect of Rate Loading upon the Static and Fatigue Strengths of Plain Concrete in Compression," *Magazine of Concrete Research,* 25(83): 83-80.

Spooner DC (1972), "Stress-stress-time Relationships for Concrete," *Magazine of Concrete Research,* 24(81): 197-208.

Tedesco JW, Ross CA, McGill PB and O'Neil BP (1991), "Numerical Analysis of High Strain-rate Concrete Direct Tension Tests," *Computers and Structures,* 40(2): 313-327.

Wang W (1997), *Stationary and Propagative Instabilities in Metals – a Computational Point of View,* PhD dissertation, TU Delft, Netherlands.

Watstein D (1953), "Effect of Straining Rate on the Compressive Strength and Elastic Properties of Concrete," *ACI Journal,* 49: 729-744.

Winnicke A, Pearce CJ and Bicanic N (2001), "Viscoplastic Hoffman Consistency Model for Concrete," *Computers and Structures,* 79(1): 7-19.

Xiao SY, Li HN and Lin G (2008), "Dynamic Behaviour and Constitutive Model of Concrete at Different Strain Rates," *Magazine of Concrete Research,* 60(4): 271-278.

Xiao SY, Li HN and Lin G (2010), "3-D Consistency Dynamic Constitutive Model of Concrete," *Earthquake Engineering and Engineering Vibration,* 9(2): 233- 246.

Zielinski AJ, Reinhardt HW and Körmeling HA (1981), "Experiments on Concrete under Uniaxial Impact Tensile Loading," *Materials and Structures,* 14(80): 103-112.

11

Seismic Response Analysis and Protection of Underground Monumental Structures – The Catacombs of Kom EL-Shoqafa, Alexandria, Egypt

Sayed Hemeda
Conservation Department, Faculty of Archaeology, Cairo University
Egypt

1. Introduction

Alexandria is located in eastern part of the Mediterranean Basin (Northern Egypt) and it is a place of great historical and religious interest. Numerous Catacombs and cemeteries for Greek-roman were erected in Greek-roman and Christian era has been found. They represent actually a large complex of an underground necropolis. The aim of the present study is the investigation and documentation of the existing stability conditions of the site of Catacombs in order to define the instability problems to interpret the pathology and to propose the best retrofitting procedure.

On 28 September 1900 the ground on the Hill of potsherds (Kom El- Shoqafa) spontaneously opened, and a donkey disappeared into the crevasse. The unfortunate beast had inadvertently discovered one of Alexandria's most important archaeological sites, the principal hypogeum of a funerary complex dating from the end of the first century of the Christian era and still in use at the beginning of the fourth (Empereur,2003). It had been known for some time that this area held antique tombs, since the hill has being extensively quarried to provide building materials for a fast -expanding modern Alexandria. Much had already been destroyed, though certain archaeologists of the late 19th century had been able to record other tombs that were subsequently to disappear. These reports have descriptions and drawings, which show that the complex that can visit today was part of a vast necropolis, traces of which must still exist under the foundations of the neighboring buildings.

The method of construction of these underground monuments and accesses was mainly rock cutting and carving. The Ancient Egyptian did have a large experience in cutting and digging in rocks. They used simple hand tools and with the experience they got in treatment of rock either soft (e.g. limestone and sandstone) or hard (e.g. granite, basalt, quartzite), they brought up the large number of monuments that we have discovered until now or that which may be revealed in the future. It is hard to think nowadays how they did such works with their simple tools—comparing them with the resources we have now available either mechanical, electrical or other forms of developed equipment. The tools available at that time—as already mentioned—did not exceed those made from hard stone, wood, copper, iron and bronze. We can hardly imagine that all these monuments, pyramids, temples and underground tombs were constructed by hand and by these simple tools.

The stability conditions of the historical monuments are of crucial interest, especially in regions like the Mediterranean Basin and particularly Alexandria, Egypt, where the seismotectonic and weathering regime is active and the geological structure is complex .Phenomena like settlement and slope movements as well as earthquakes and tectonic activity contribute to the damage of the historical buildings. The ground water activity is also an important factor, especially in cases underground monuments the environmental factor is also necessary to be taken in mind, when different protection measures are decided to apply (hemeda, et al, 2007).

The earthquakes affected a large variety of structural systems. The severity of damage is a function of the structural type, quality of work manship, material, and local soil conditions. Observations of the damaged areas close to the epicenter of the Dahshour earthquake (Cairo, 1992) indicated that they can be rated VII on the Modified Mercalli intensity (MMI) scale with an estimated peak ground acceleration of about 10% g, (Badawi and Mourad, 1994).

The damages which occurred can be classified as:

- STRUCTURAL; resulting in damage to load carrying components. Foundations, Bearing walls, columns, etc.
- NON STRUCTURAL; producing damage to architectural or functional components, such as partitions, suspended ceiling parapets.

Structural damage

Old load bearing unreinforced Masonry walls, constructed of solid blocks, had a large number of failures and collapse. The extent of damage varied from minor cracks to complete collapse of walls.

- Failures were due to inplane cyclic shear cracking. Out-of -plan e instability, and impact with adjacent buildings (pounding)
- Masonry walls that were acting as a shear walls showed flexural failures between windows, combined with some diagonal shear cracking.
- In spite of damage to wall- bearing structures, recently constructed reinforced concrete buildings did not suffer severe damage, except the complete collapse of a residential building in Cairo.

Non structural damage

- - Different types of cracks, shear and bending were observed in partitions.
- - Poorly built parapets were either dislocated or collapsed due to relative motion.
- - Collapsed many masonry parapets and facades

Geotechnical effects of the earthquake

The soil in the area (Cairo and its districts) is silty to fine-sandy soil that makes up the valley of the Nile. When this type of soil is water-saturated and strongly shaken, it is highly susceptible to the liquefaction phenomenon.

- The soil can quickly convert from solid material into liquid which has no strength to support structures that may rest its surface on when subjected to prolonged shaking.
- The potential for widespread liquefaction is very high on the Nile flood plain, especially in irrigated fields. A highway west of the Nile sunk as much as105m because the agricultural land adjacent to the pavement had liquefied over a few hectares

- When the strength of the soil beneath the structure vanished (due to the excess pore pressure) during the earthquake, the structures settled as the liquefied soil escaped latterly into the field.
- Also, many zones of liquefaction in the agricultural fields were reported when revealed on the ground by sand boils.

The instability problems and depreciation phenomena of subterranean monuments in Alexandria is not likely to be dominated by gravity fall or sliding on structural features, other factors such as excessively high rock stress, creep effect, poor geotechnical properties of rock structures, weathering and /or swelling rock and excessive groundwater pressure or flow, seismic loading as well as utter lack of preservation become important and can be evaluated by means of a classification of rock quality

The aim of the analysis carried out in this research is to investigate the safety margins of the underground monuments, under their present conditions, against unfavorable environmental (i.e. weathering and high underground water table), utter lack of preservation, geotechnical and extreme seismic conditions.

Underground structures safety analysis is performed using the finite element (FE) method. The research presents a comprehensive study for the underground monuments safety analysis. The safety analysis includes not only a failure analysis but the effect of weathering specially the underground water on the differential settlement will be investigated. The commercial FE package Plaxis (Karstunen et al, 2006) is used for conducting stress, as well as settlement analysis. PLAXIS is a finite element program developed for numerical analysis of geotechnical and underground structures (plaxis manual, 2002).

To compute the deformation of these underground monuments as realistically as possible, an advanced nonlinear elastoplastic material model needs to be utilized in PLAXIS which is capable of utilizing such advanced material models. Mohr's–Coulomb model is used for deformation and consolidation analysis in this study. The consolidation analysis is performed using PLAXIS utilizing Biot's consolidation theory in 2D (Biot, 1941) and the nonlinear material behavior is taken into account as mentioned before.

Also in this research, we attempt to construct and analyze a three-dimensional (3D) finite element model (FEM) of the central rotunda in catacombs of Kom El-Shoqafa with its six supporting rock piers excavated in sandy oolitic limestone deposit, using the FLAC 3D code.

For the seismic analysis, we have modeled the complex catacomb assuming an equivalent plane strain model and applying the Plaxis b.v. 8 with different seismic scenarios, corresponding to the seismotectonic features of Alexandria. Advanced soil-rock elastoplastic modeling has been used. Extensive time domain parametric analysis were performed in order to examine the response of the catacombs subjected to seismic motions with different amplitudes of ground motion and different frequency content. (Kalamata in Greece, 1986, Erzincan in Turkey, 1992, Aqaba in Egypt, 1995). The analysis takes into account the complex behavior of the structure with the aim to determine the threshold peak ground acceleration (PGA) and the corresponding developed stresses, which should remain lower than the actual strength of different elements composing the catacombs.

2. The objectives of this study

The Catacombs in Alexandria, Egypt from the Greek-Roman era represent cultural heritage of outstanding universal values. They suffer weathering - aging as well as multiple geotechnical and earthquake problems. A pilot study has been carried out for on the Catacombs of Kom El-Shoqafa in order (a) to define the pathology and the causes (b) to investigates the safety assessment of the structure under static and seismic loads. (c) To assess the global risk due to several factors and (d) to define the appropriate retrofitting

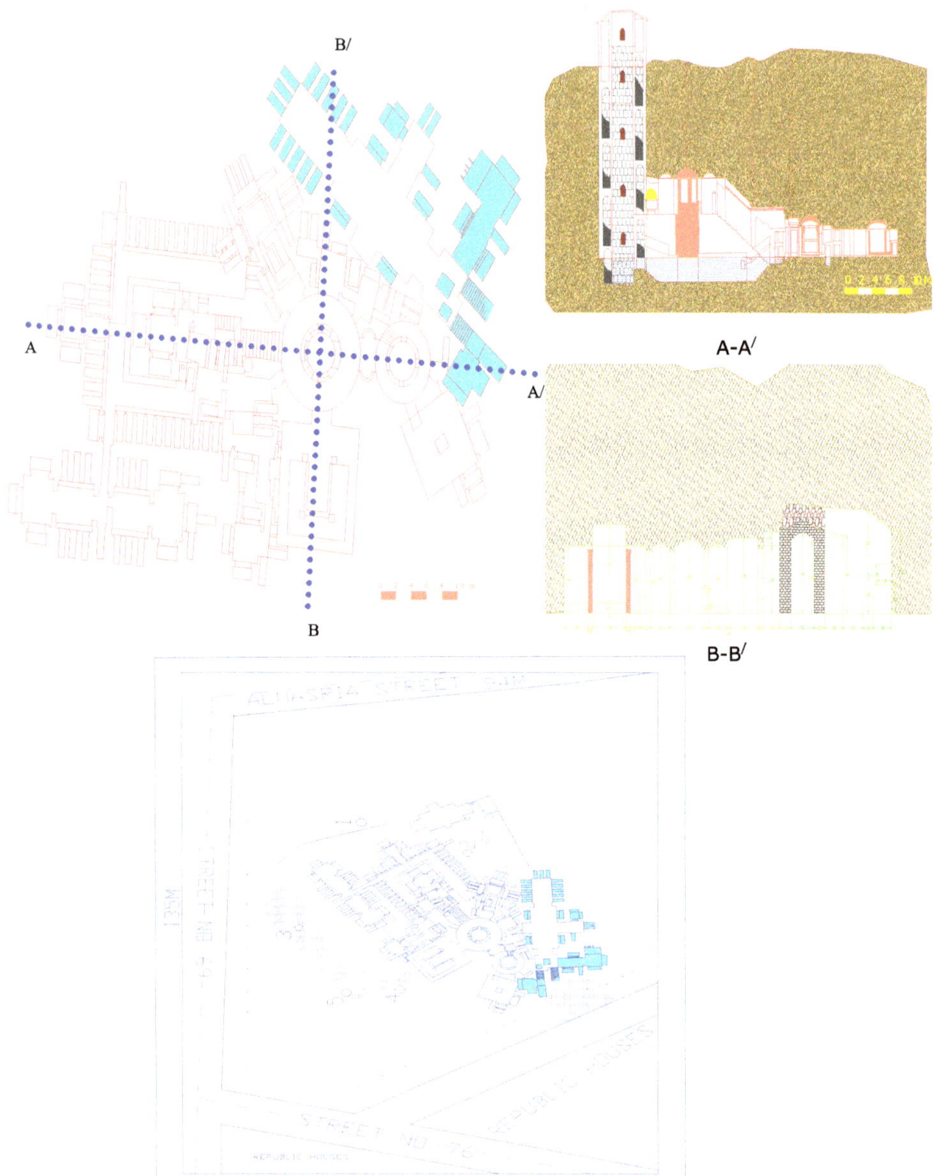

Fig. 1. Plan view and the two main cross- sections of Catacombs of Kom El-Shoqafa. (a) cross-sections and (b) plan view.

Fig. 2. Present state of the catacombs of Kom El-Shoqafa..

techniques to be applied. In the paper a general outline of the various tests, surveys and analyses is presented, highlighting the most important issues related to the static and seismic stability of underground structures like Catacombs in particularly unfavorable geotechnical and environmental conditions.

Parameters	Symbol and unit	Rock material	Brick (old restoration)
Rock unit weight	γ_{unsat} $kN/_{m^3}$	18	17
Rock unit weight	γ_{sat} $kN/_{m^3}$	22	20
Young's modulus	E_{ref} kN/m^2	2.270 x10^6	1.350 x10^6
Shear modulus	G_{ref} kN/m^2	8.867 x10^5	5.400 x10^5
Oedometer modulus	E_{oed} kN/m^2	2.902 x10^6	1.620 x10^6
Poisson's ratio	ν (nu)	0.28	0.25
Cohesion	c_{ref} kN/m^2	500	450
Friction angle	φ^O	35	31
Shear velocity	Vs m/s	715	557.9
Longitudinal wave velocity	Vp m/s	1293	9663
Uniaxial compressive strength	UCS kN/m^2	2400	1100
Bending strength	σy kN/m^2	560	200
Shear strength	T_f kN/m^2	364	-
Dilatancy	Ψ (o)	1°	0

Table 1. Material properties of rock and other construction materials used for the stability analysis of the Catacombs of Kom El-Shoqafa.

3. Preliminary 2D static analysis

In the initial static analysis, the excavation is modeled by assuming non-linear soil / rock behavior and the Mohr coulomb failure criterion. The following parameters are used: $\varphi=36^o$, c=500 kN/m^2, E=2.270E+06 KN/m^2, v=0.28, Vs=715 m/sec for the rock material, and $\varphi=31^o$, c=500 kN/m^2, E=1.350E+06 KN/m^2, v=0.25, Vs=550 m/sec for the modern brick support walls and piers. Figure 3 presents the total displacements of the soft rock excavations of catacombs of Kom El-Shoqafa, for two typical cross sections, 1 and 2, for the first and second floor, respectively. The results from the preliminary static analysis indicate that the ground displacements were small (of the order of a few millimetres; 1.13 mm); some rock pillars are under relatively high compression stresses. The calculated effective peak principal compressive stresses on supporting rock pillar 1 is about 1.42 MPa.

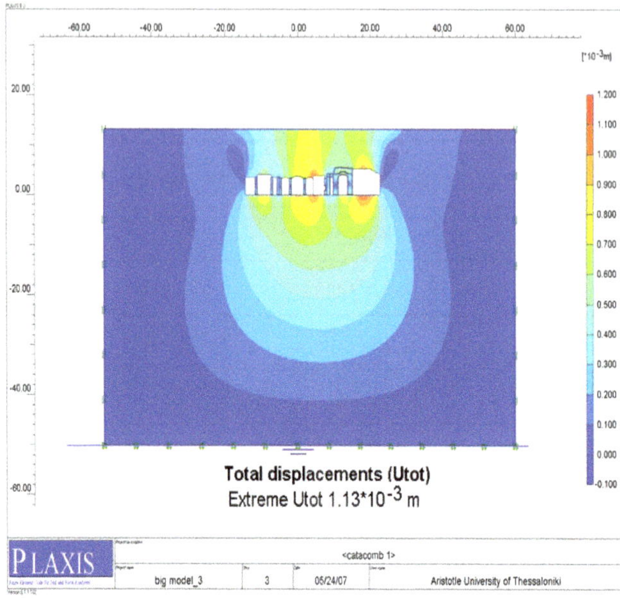

First floor

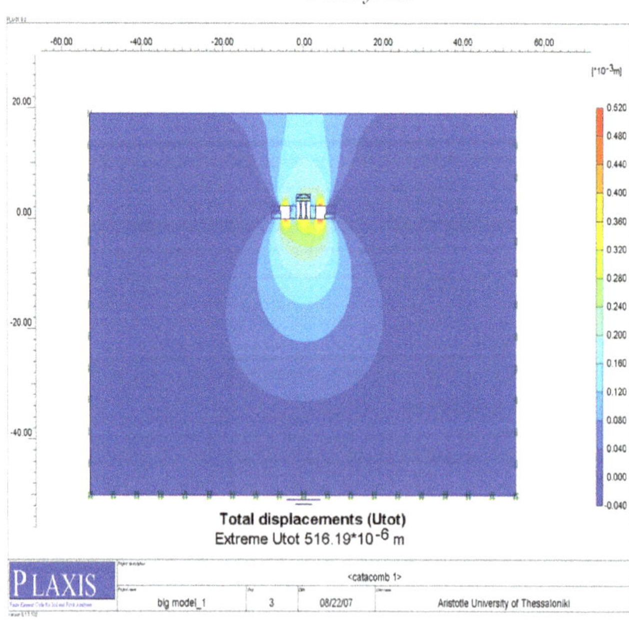

Second floor

Fig. 3. Total vertical static displacements at the catacombs of Kom El-Shoqafa for the first and second floor.

4. Preliminary 3D static analysis

An attempt has been made to construct a three-dimensional (3D) numerical model for the central Rotunda in the catacombs of Kom El-Shoqafa with its six supporting rock pillars excavated in sandy oolitic limestone deposit. The objective of the 3D analyses is to evaluate the stress state in the pillars taking into account the 3D geometry. The 3D effects issue is considered on a basic engineering approach in the subsequent sections. The various simulations described herein are conducted using the FLAC 3D code (Itasca, 2007).

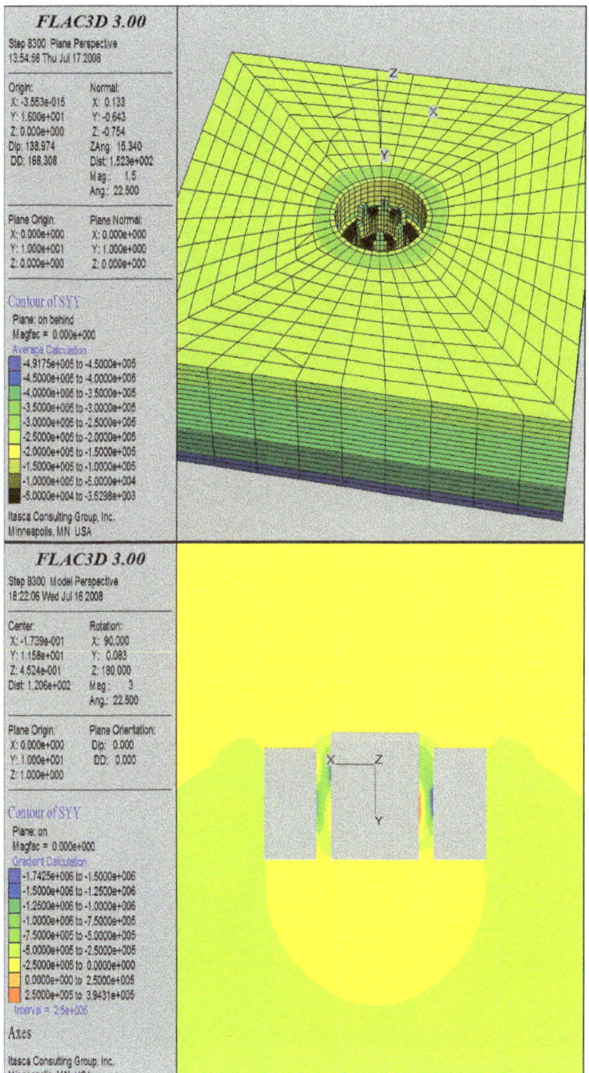

Fig. 4. Contour lines of vertical effective stresses σ_{yy} through the rotunda. The maximum effective compressive stresses on pillar_1 = -1.74*10^3 (×) kN/m^2.

The results from the 3D static analysis indicate that the ground displacements above the rotunda (catacombs) small. The maximum total (vertical) displacements range between 2.6 mm and 3 mm in the whole domain. The peak horizontal displacement is about 1.0 mm. Some rock piers are under relatively high compression stresses. The calculated peak effective principal vertical compressive stresses on supporting rock pillar 1 is 1.74×10^3 KN/m^2 and the calculated peak effective principal tensile stress is about 200 KN/m^2. The factor of safety of the rock pillar 1 is 1.47, (note that the acceptable safety factor for the underground structures is > 1.6 in static state). Also the overstress state is beyond the elastic regime (limit of domain) (Hemeda, 2008).

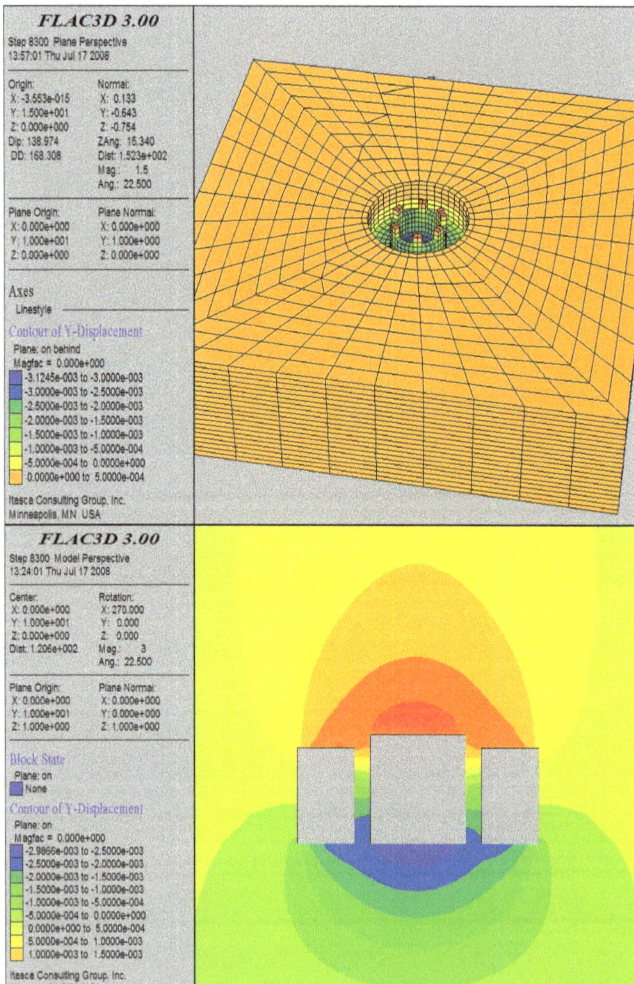

Fig. 5. Results from the 3D analysis indicating that the maximum ground displacements above the rotunda are relatively small (of the order of 2.6 mm), and the maximum horizontal displacement is about 1 mm.

5. Seismic response analysis

In the present study, we have selected three reference earthquakes to be used as input motions to the time-history analysis of the monumental structure studied, namely: (i) Aqaba, Egypt, 1995, (ii) Erzincan, Turkey, 1992 and (iii) Kalamata, Greece, 1986. The time histories (Figure 6) of these earthquakes representing different seismotectonic settings and frequency content, were scaled to three peak ground acceleration values equal to 0.08g, 0.16g, and 0.24g (g is acceleration due to gravity), respectively. The design acceleration in Alexandria according to the Egypt seismic code is 0.08g (ECP-203). The structures have been analyzed under single horizontal ground acceleration and not under the three ground acceleration.

5.1 Seismicity of Egypt

Egypt lies at the north east of the African tectonic plate whose borders are the Red Sea and the Jordan Valley. Another major tectonic feature is the African Rift Valley which extends from the Red Sea down through Ethiopia towards South and East Africa. The contemporary geological setting of the region is mainly governed by active extensional tectonics taking place within the Gulf of Suez Rift. For over 20 years, geologically young faulting has been known to exist both along the margins and within the Gulf of Suez (Said, 1990).

In the 15th century, an Egyptian scholar listed more than 30 earthquake events that happened from AD 796 to 1500. The Dahshour earthquake was the latest in a long history of earthquakes that occasionally cause damage to buildings in Cairo and surrounding areas in northeastern Egypt. On 7 August 1847, an earthquake with an estimated magnitude of 6.8 occurred about 100 km southwest of Cairo, near E1-Fayoum. Three thousand houses and 42 mosques were destroyed in E1- Fayoum and damage was extensive in Cairo (Kebeasy et al., 1976).

Two recent earthquakes occurred within the last three decades. The first, having a magnitude 7.0, occurred in the northern part of the Red Sea on 31 March 1969. It caused landslides, rock falls, and fissures, in addition to damage to a power station and some hotels in a city nearby (Maamoun et al., 1984). The second earthquake occurred on 14 November 1981, in Aswan, 1000 km south of Cairo, along the Nile River. It measured 5.6 on the Richter scale, in an area that was considered a seismic. Some scientists attribute this earthquake to the construction of a dam along the river, resulting in an artificial reservoir, approximately 300 km long, and having a maximum capacity of 164000 million cubic meters (Bolt, 1988).

5.2 Seismicity of Alexandria

Alexandria during his long history has suffered important seismic damages from near and distant sources earthquakes. As the under-water archeological remains in Abou Kir bay strongly support, either local or remote earthquakes (El-Sayed, 2004) destroyed the city. Seismogenic zones such as the Red Sea, Gulf of Aqaba-Dead Sea Hellenic Arc, Suez-Cairo-Alexandria, Eastern-Mediterranean-Cairo-Faiyoum and the Egyptian coastal area may all affect the city. However, the seismic hazard of the city has not been fully defined.

Alexandria is located approximately 300 to 600 km from three known active plate boundaries, namely: the Red-Sea, the Gulf of Aqaba and the Hellenic Arc (Mckenzie 1970, 1972; Sestini, 1984; Mesherf, 1990). The interaction among these three plate boundaries created major fault zones in Egypt as: (1) Eastern- Mediterranean Cairo Faiyoum fault zone (Neev, 1975; Sestini, 1984; Mesherf, 1990), (2) Suez-Cairo-Alexandria fault zone (Kebeasy,

1990). These fault zones are very close to the city of Alexandria. Moreover, deep seismic sounding reveals that there are minor faults such as Rosetta fault which trend a few kilometers from Alexandria city (Hussein & Abd Allah, 2001).

Because of this complex tectonic setting, many earthquakes occurred approximately Alexandria, both in recent and historical time (Ambraseys et al., 1994; Maamoun et al., 1984). The spatial distribution of the earthquakes epicenters shows that there are areas of very intense (e.g., plate boundaries) and others of low (e.g., offshore area) activities. For those of intense seismicity, there are a considerable number of focal mechanisms that allow us to understand its geodynamic behaviors (Mckenzie, 1970; and Rotstein and Kafka, 1982; Sestini, 1984; CMT database). Controversy, the number and quality of the focal mechanisms available for those of low seismic activity (like the Egyptian coastal zone) are not enough to have the clear understanding for the tectonic setting (Sestini, 1984; Kebeasy, 1990; Mesherf 1990).

Because of this complex stress regime, damaging earthquakes had occurred in the vicinity of Alexandria (Ambraseys et al., 1998; Kebeasy, 1990; Maamoun et al., 1984). Some of these damaging earthquakes are apparently missing. As an example, the damage in Menouthis and Herakleion is more likely caused by an earthquake (which is not known yet). This is supported by: (1) the collapsed columns are falling down in the same direction NE-SW, (2) the presence of coins and jewelry suggest a sudden collapse, (3) the sharp sand grains in the bottom of Abou kir Bay reflecting active tectonic environment, and (4) the Cairo- Alexandria fault system (NE-SW) is passing by the area and recently generated frequent moderate earthquakes (GEOTIMES, 2000). This does not exclude the possibilities of land subsidence as suggested in the STANFORD REPORT (2000).

5.3 Seismic input

In the present study, we have selected three reference earthquakes. (i) Aqaba, Egypt, 1995 (ii) Erzincan, Turkey, 1992 and (iii) Kalamata, Greece, 1986. The time histories (Figure.6) of these earthquakes representing different seismtectonic settings and frequency content were scaled to three peak ground acceleration values equal to 0.08g, 0.16g,and 0.24g respectively. And they are used as input motions at the bedrock. The design acceleration in Alexandria according to the Egypt code is 0.08g.

We believe that with the advances in computational methods it is now possible to predict with reasonable accuracy the seismic demands on these geometrically complex monuments. Specially, computer modeling and simulations are very useful tools for identifying regions of stress concentration where only non-invasive techniques are allowed. Accurate quantification of stresses are also useful for understanding the direction of cracks propagation and for quantifying the seismic demands on whatever new materials may be introduced in the retrofit program. Three earthquakes were chosen:

Aqaba: 22/11/1995, M=7.1, Ml=6.2: Station Eilat. Distance (km): Closest to fault rupture (93.8).

Erzincan:13/3/1992, Mw=6.9, Ms=6.8.Station:95 Erzincan, Rrup=2km, Re=1km.

Kalamata:13/9/1986.17:24:35, Ms=5.8 .Mw=5.9, Ml=5.5. Station: Old telecommunication building, Re=10 km.

22 November 1995, an major earthquake (Aqaba earthquake) with a magnitude of 6.2 on the local scale ML, and a moment magnitude of MW D 7:1, (PDE bulletin, 1996), struck the

shorelines cities of the Gulf of Aqaba, such as Aqaba (Jordan), Eilat (Israel), Hagel (Saudi Arabia) and Nuweiba (Egypt), Damage occurred in many parts northeastern Egypt as far as Cairo. This major event was followed by 2089 earthquakes ranging in magnitude from 2 to 5.5 on local magnitude (ML) recorded and/or analyzed by the Jordan Seismological Observatory (JSO bulletin, 1998). This seismic swarm activity began on 22 November 1995 and continues at least until the end of December 1997, (EID AL -TARAZI, 2000).

The Kalamata earthquake was recorded on hard ground at a distance of about 9 km from the epicenter and its magnitude was Ms=6.2. The record samples the near field strong motion that caused considerable damage to the buildings of the city of Kalamata. The duration of the strong motion is about 6 sec and the maximum accelerations are 0.24g in the N-S direction and 0.27g in the E-W direction. The corresponding peak velocities are 32.0 and 23.5 cm/s, respectively.

The 13 March 1992 Erzincan earthquake, M=6.8, occurred in the eastern half of the Erzincan basin. The largest aftershock took place near Pülümür on 15 March 1992. No clear surface breaks were observed, although teleseismic studies suggested that it was a strike-slip earthquake striking parallel to the North Anatolian fault, with a focus of approximately 10±2 km depth, 30 km rupture length, 95 cm of slip, and a 1.16×1026 dyn.cm seismic moment. The aftershock distribution concentrated at an area of the intersection between the North Anatolian fault and the Ovacik fault. These results indicate that the previously suggested seismic gap along the North Anatolian fault, east of Erzincan, remains unruptured (AYKUT et al 1993).

The criterion for this choice was their different frequency content, as they will give information about the response of these structures in different period ranges Figure (6) shows one of the horizontal components of acceleration for each record. The records were retrieved from PEER and ESMD online database.

In order to estimate the threshold PGA values to collapse, a set of parametric analysis was carried out. These structures were subjected to increasing level of horizontal accelerations.

Figures (7) through (13) depict the main results of the analysis. In case of the Aqaba earthquake, it is clear that a great part of seismic energy is dissipated by the upper parts of catacombs (ground surface) even for small values of PGA. Kalamata and Erzincan input motions give much lower displacements values (Arias' intensity and total duration will explain the difference in responses from the three ground accelerations). The maximum horizontal displacement at the top of catacombs for Aqaba earthquake at PGA = 0.24g earthquake scenario was u_x = 7.95 cm, while the peak vertical effective principal stress was 4190 kN/m^2. For the Erzincan and Kalamata earthquakes, the respective values were 3400 kN/m^2 and 3580 kN/m^2, respectively. Moreover, the maximum horizontal displacement at the top of the catacombs was 2.34 cm and 2.01 cm for Erzincan and Kalamata earthquakes respectively. The maximum vertical displacement at the top of catacombs, are 3 mm, 2.6 mm, and 5.3 mm for Kalamata, Erzincan, and Aqaba earthquakes, respectively.

Given the value of the static strength estimated in the laboratory (UCS=2.5 MPa), the seismic analysis of the catacombs complex proves that the supporting rock piers and columns, which are the most vulnerable parts of the whole complex, are rather safe for PGA values lower than 0.24g in case of the Kalamata and Erzincan earthquakes and PGA =0.12g for the Aqaba seismic scenario (see Figure 13a, b).

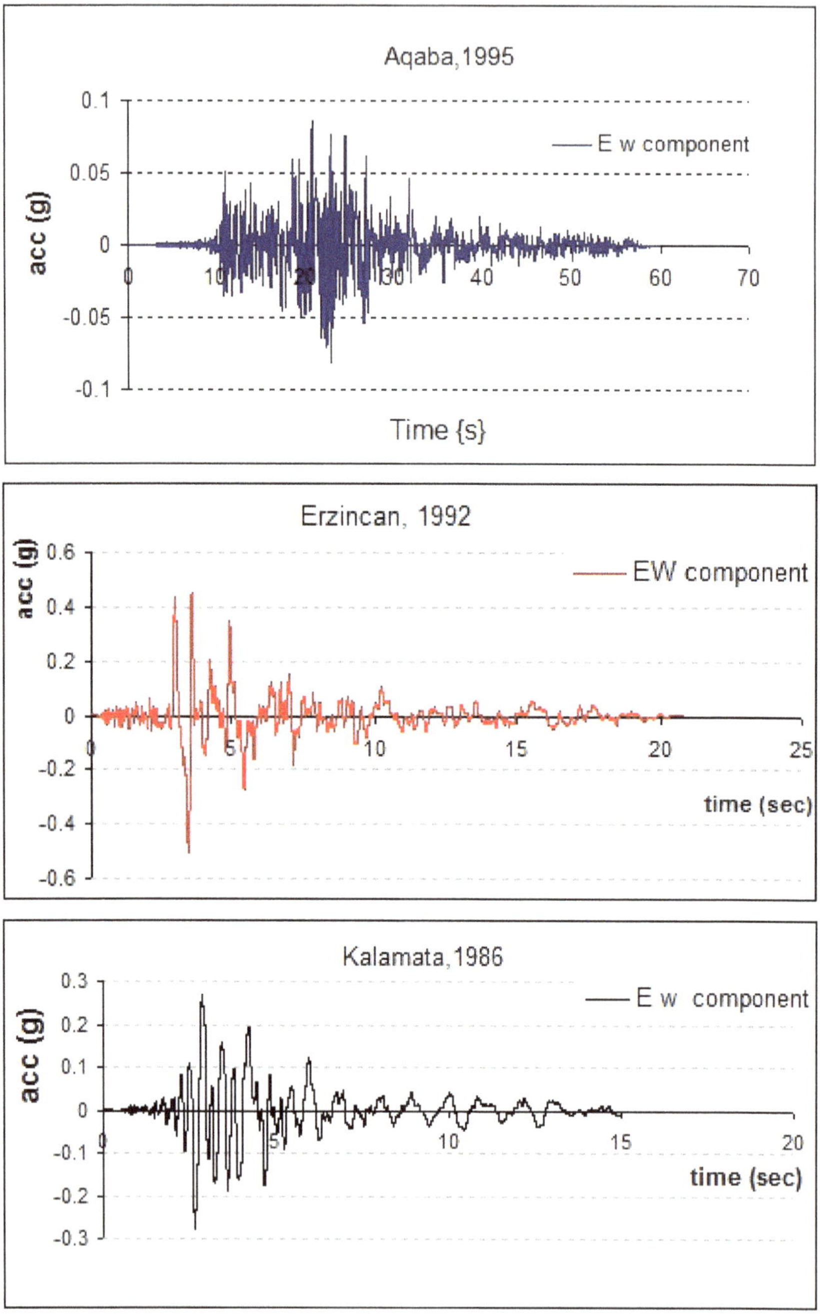

Fig. 6. Seismic excitations at the bedrock for the reference ground motions (acceleration-time history). (a) Aqaba, 1995 (b) Erzincan 1992, and (c) Kalamata, 1986 earthquakes.

For larger earthquakes, which are most likely to happen in the region, the seismic stability of the catacombs is not satisfied and it is necessary to proceed to specific retrofitting works to upgrade their seismic performance. The maximum relative horizontal displacements of the top and the base of the rock piers are of the order of 3 to 5 mm (Figure 10). Considering that the induced seismic ground deformations are better correlated with the intensity of damages in underground structures, the seismic design of the catacombs must be based on these kinematic forces.

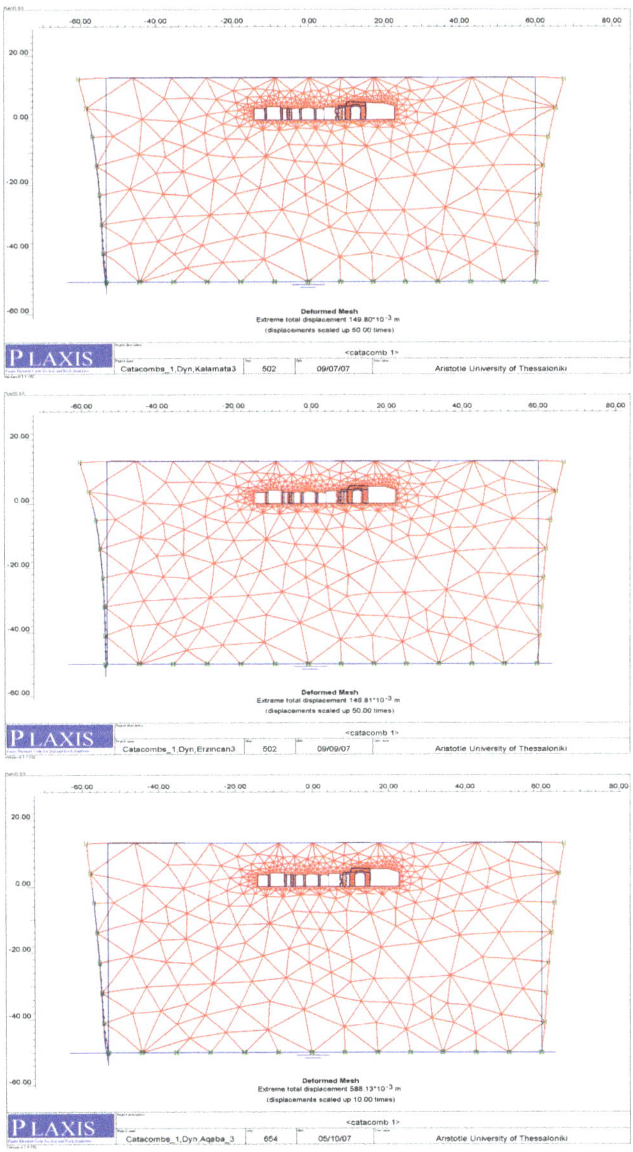

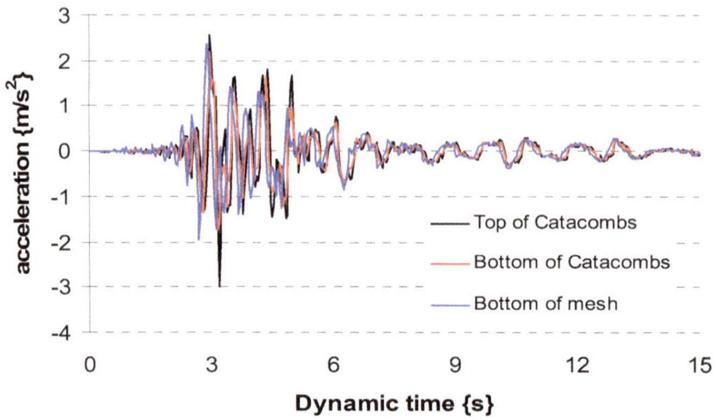
acceleration {m/s²}
3
2
1
0
-1
-2
-3
-4
0
3
6
9
12
15
Dynamic time {s}
Top of Catacombs
Bottom of Catacombs
Bottom of mesh

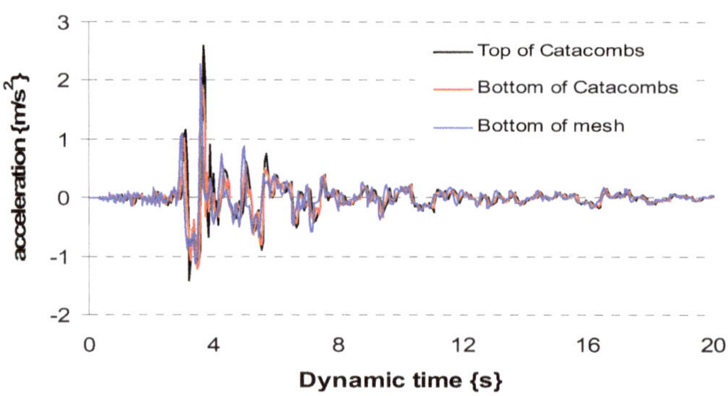
acceleration {m/s²}
3
2
1
0
-1
-2
0
4
8
12
16
20
Dynamic time {s}
Top of Catacombs
Bottom of Catacombs
Bottom of mesh

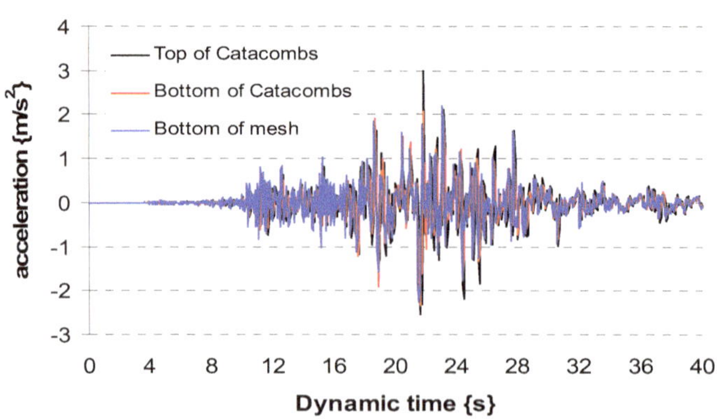
acceleration {m/s²}
4
3
2
1
0
-1
-2
-3
0
4
8
12
16
20
24
28
32
36
40
Dynamic time {s}
Top of Catacombs
Bottom of Catacombs
Bottom of mesh

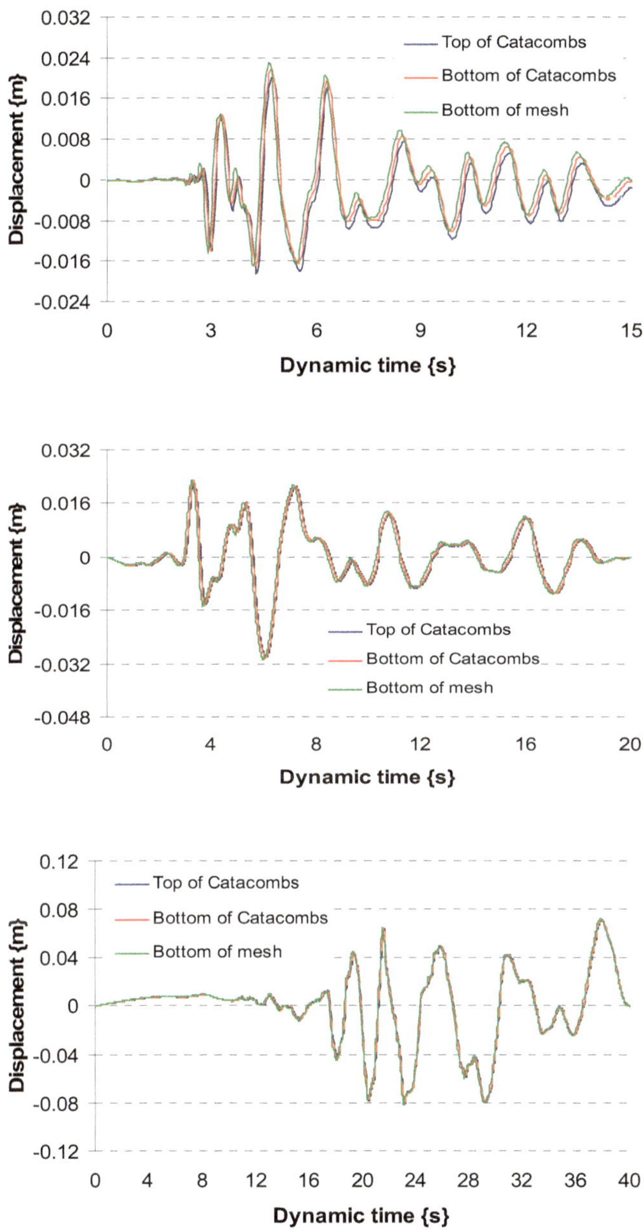

Fig. 7. Deformed meshes and peak total displacement, and acceleration, horizontal displacements-time histories, (a) Kalamata , (b) Erzincan , and (c) Aqaba eartquakes. Input motion PGA = 0.24g. Catacombs of Kom El-Shoqafa, a typical cross section 1, first floor.

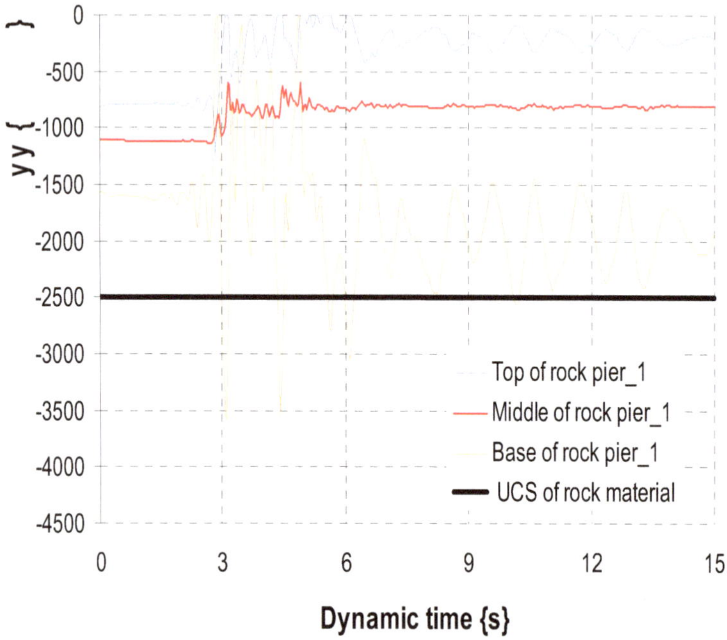
0
-500
-1000
-1500
-2000
-2500
-3000
-3500
-4000
-4500
0
3
6
9
12
15
Top of rock pier_1
Middle of rock pier_1
Base of rock pier_1
UCS of rock material
Dynamic time {s}

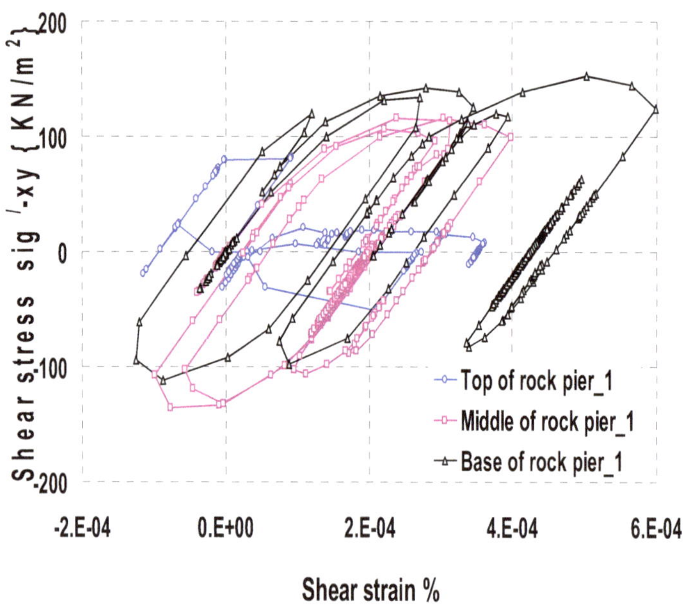
Shear stress sig -xy { KN/m 2 }
200
100
0
-100
-200
-2.E-04
0.E+00
2.E-04
4.E-04
6.E-04
Top of rock pier_1
Middle of rock pier_1
Base of rock pier_1
Shear strain %

a)

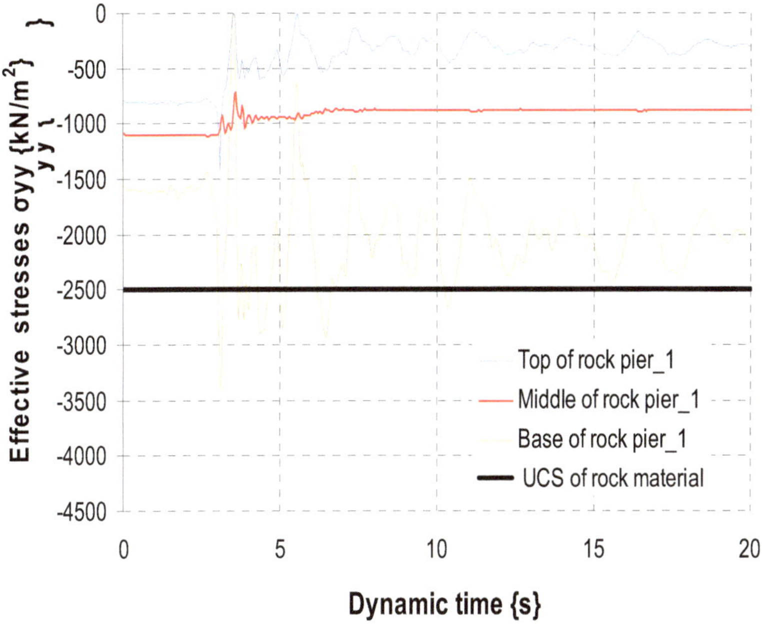
Effective stresses σyy {kN/m2}
Dynamic time {s}
Top of rock pier_1
Middle of rock pier_1
Base of rock pier_1
UCS of rock material

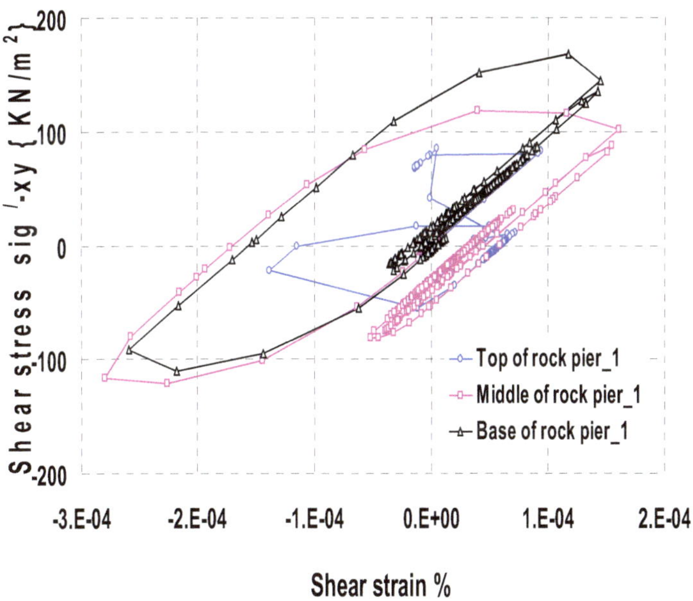
Shear stress sig -xy {KN/m2}
Shear strain %
Top of rock pier_1
Middle of rock pier_1
Base of rock pier_1

b)

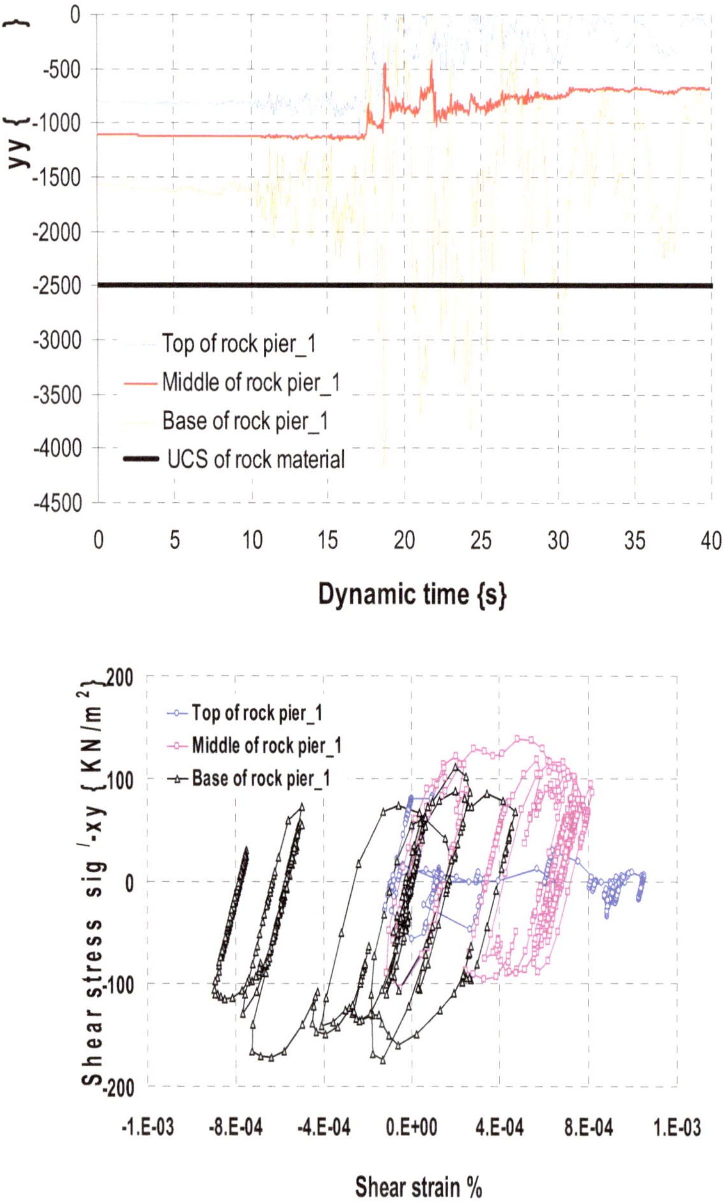

c)

Fig. 8. Effective vertical compressive stresses σ/YY - time histories and shear stress versus shear strain for the most critical rock pier 1 (Figure 9 at the right side of rotunda). (a) Kalamata (b) Erzincan (c) Aqaba earthquakes. PGA value = 0.24g.

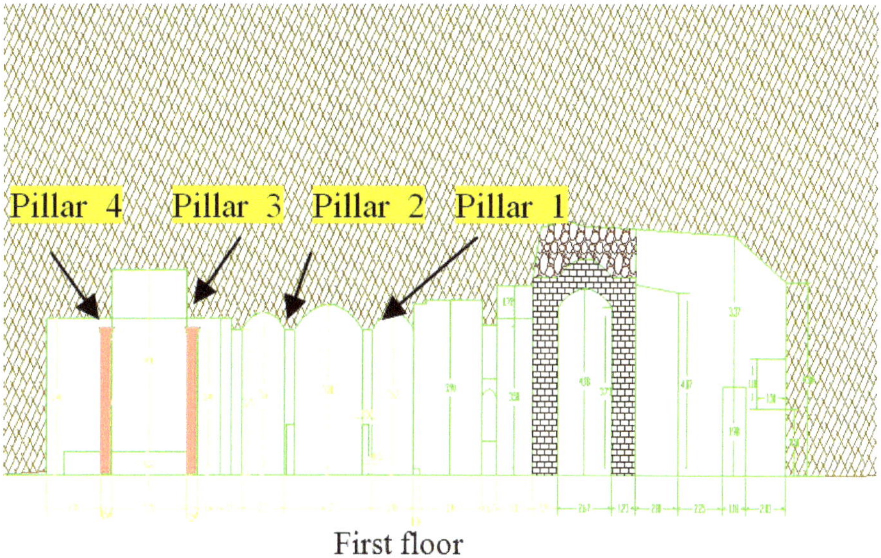

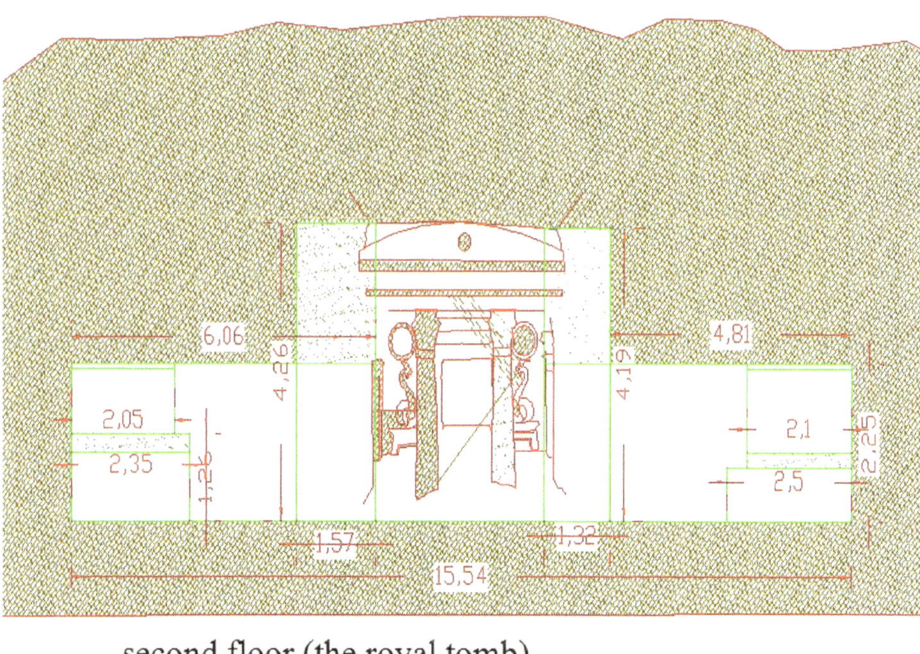

Fig. 9. The typical cross sections 1and 2, which have been used in the seismic analysis.

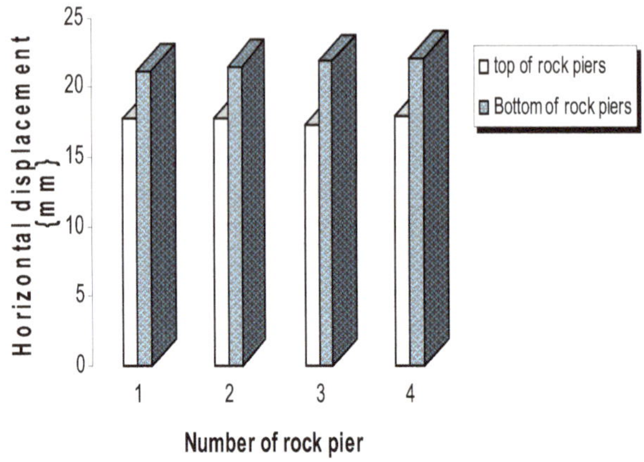
25
20
15
10
5
0
Horizontal displacement {mm}
1
2
3
4
Number of rock pier
top of rock piers
Bottom of rock piers

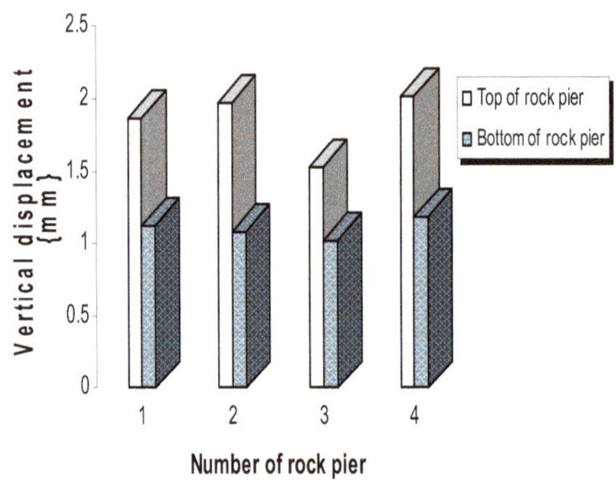
2.5
2
1.5
1
0.5
0
Vertical displacement {mm}
1
2
3
4
Number of rock pier
Top of rock pier
Bottom of rock pier

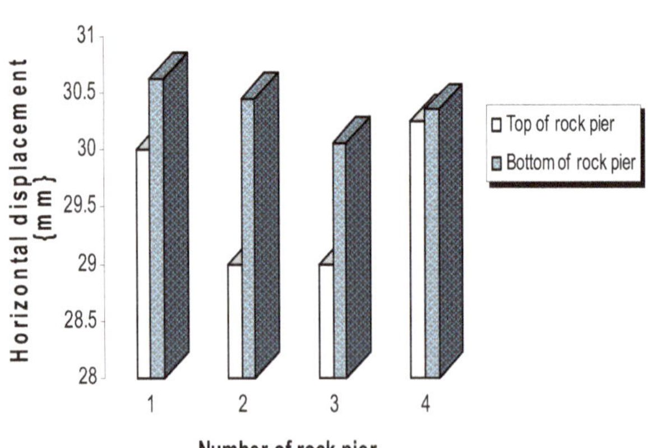
31
30.5
30
29.5
29
28.5
28
Horizontal displacement {mm}
1
2
3
4
Number of rock pier
Top of rock pier
Bottom of rock pier

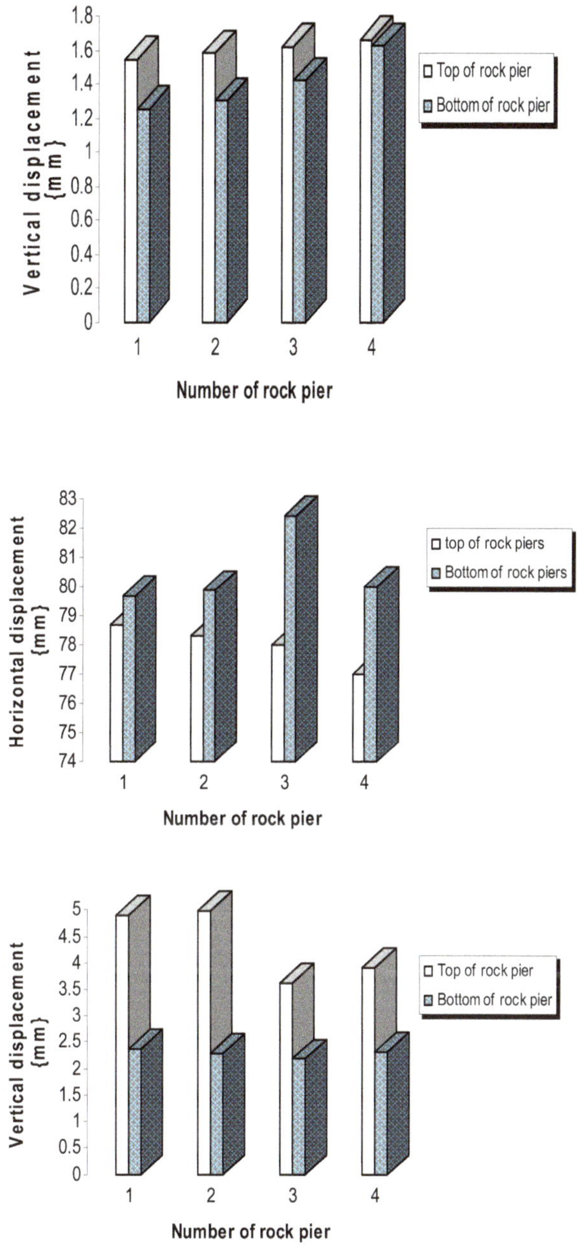

Fig. 10. Differential horizontal and vertical displacements on the top and base of the rock piers in the catacomb complex (a) Kalamata at time (t) =4.60 sec, (b) Erzincan at (t) = 7.09 sec, (c) Aqaba at time (t) = 23.22 sec. PGA = 0.24g.

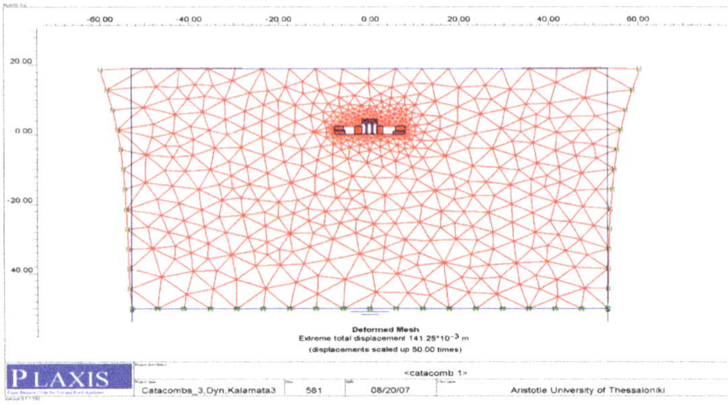
-60.00
-40.00
-20.00
0.00
20.00
40.00
60.00
20.00
0.00
-20.00
-40.00
Deformed Mesh
Extreme total displacement 141.25*10^-3 m
(displacements scaled up 50.00 times)
PLAXIS
<catacomb 1>
Catacombs_3.Dyn.Kalamata3
581
08/20/07
Aristotle University of Thessaloniki

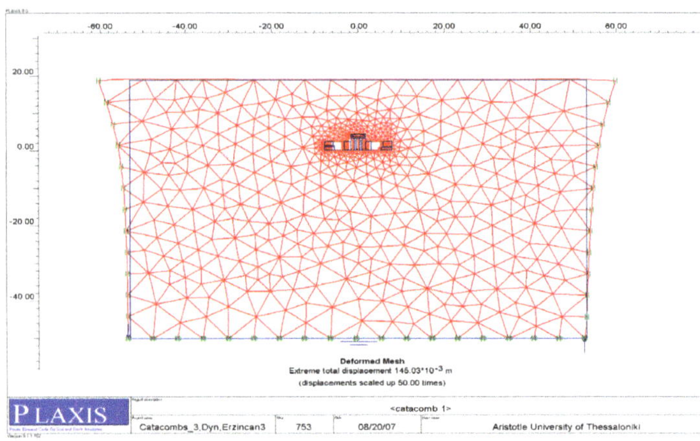
-60.00
-40.00
-20.00
0.00
20.00
40.00
60.00
20.00
0.00
-20.00
-40.00
Deformed Mesh
Extreme total displacement 145.03*10^-3 m
(displacements scaled up 50.00 times)
PLAXIS
<catacomb 1>
Catacombs_3.Dyn.Erzincan3
753
08/20/07
Aristotle University of Thessaloniki

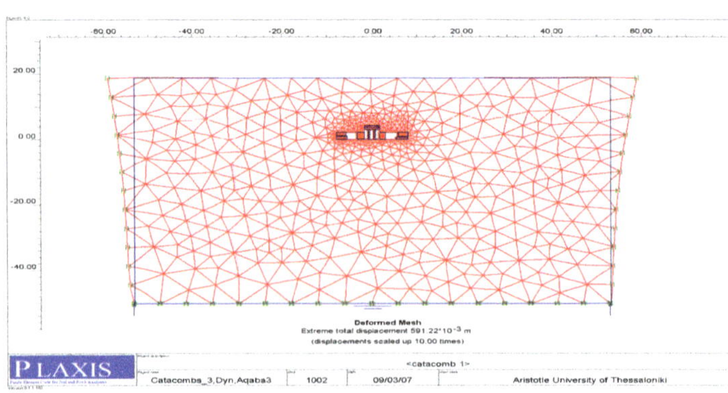
-60.00
-40.00
-20.00
0.00
20.00
40.00
60.00
20.00
0.00
-20.00
-40.00
Deformed Mesh
Extreme total displacement 591.22*10^-3 m
(displacements scaled up 10.00 times)
PLAXIS
<catacomb 1>
Catacombs_3.Dyn.Aqaba3
1002
09/03/07
Aristotle University of Thessaloniki

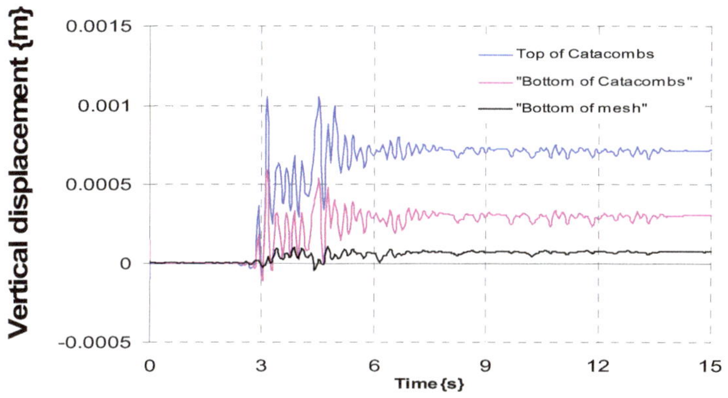
Vertical displacement {m}
0.0015
0.001
0.0005
0
-0.0005
0
3
6
9
12
15
Time {s}
Top of Catacombs
"Bottom of Catacombs"
"Bottom of mesh"

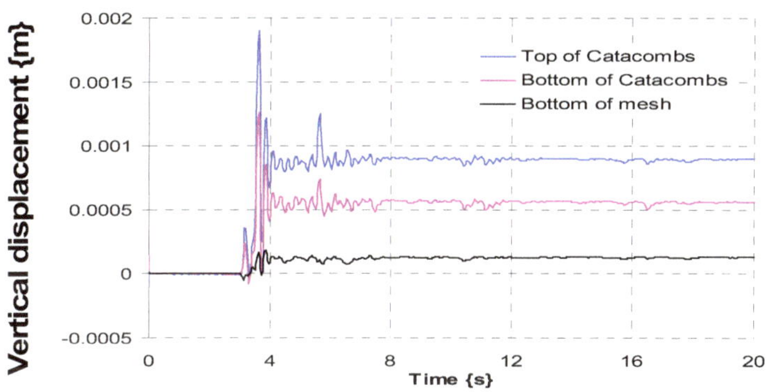
Vertical displacement {m}
0.002
0.0015
0.001
0.0005
0
-0.0005
0
4
8
12
16
20
Time {s}
Top of Catacombs
Bottom of Catacombs
Bottom of mesh

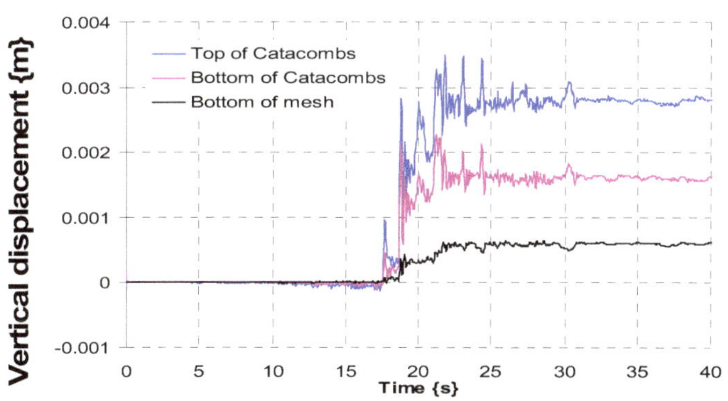
Vertical displacement {m}
0.004
0.003
0.002
0.001
0
-0.001
0
5
10
15
20
25
30
35
40
Time {s}
Top of Catacombs
Bottom of Catacombs
Bottom of mesh

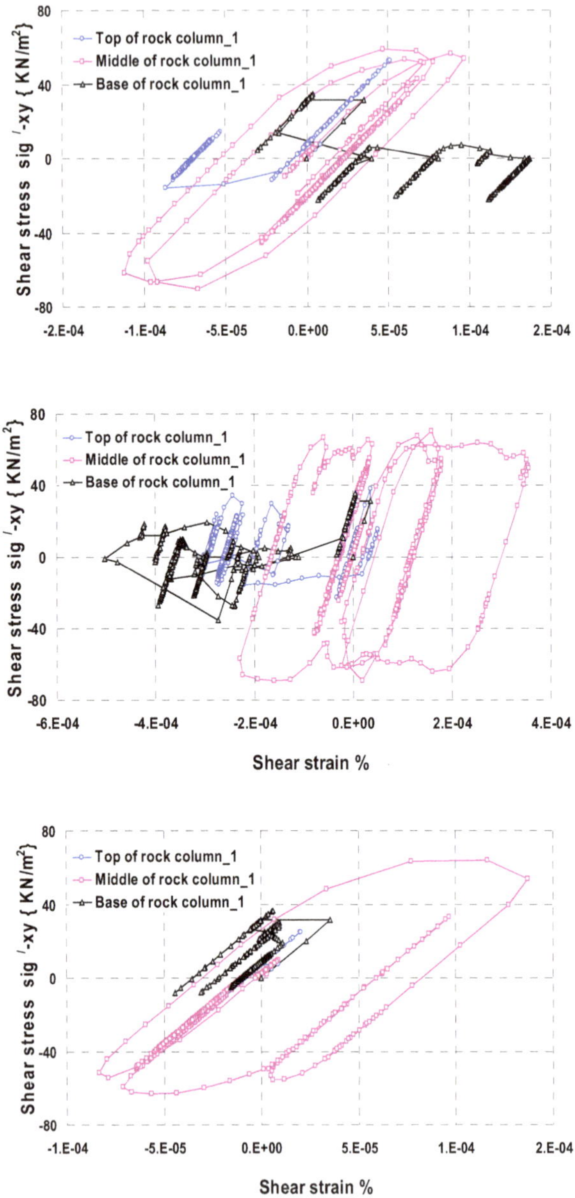

Fig. 11. Deformed meshes, vertical displacement-time history, shear stresses-shear strain, (a) Kalamata , (b) Erzincan , and (c) Aqaba earthquakes. Input motion PGA = 0.24g. Catacombs of Kom El-Shoqafa, a typical cross section 2, second floor, the royal tomb.

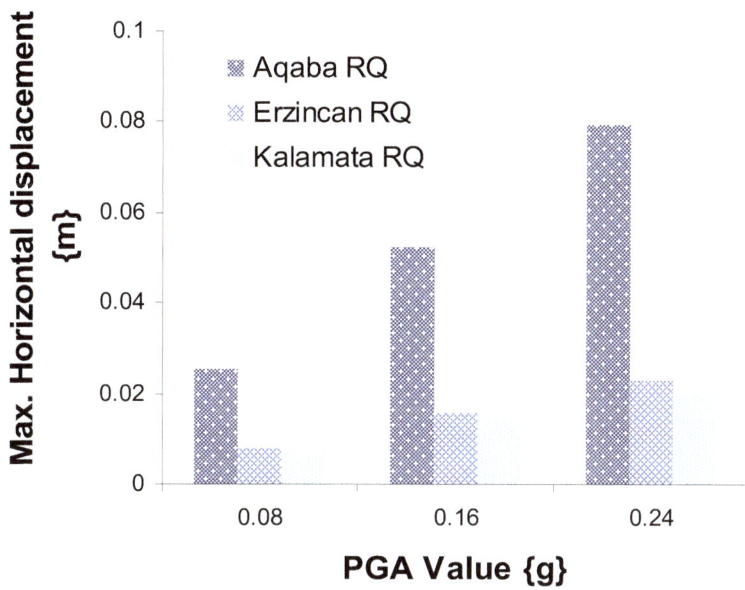

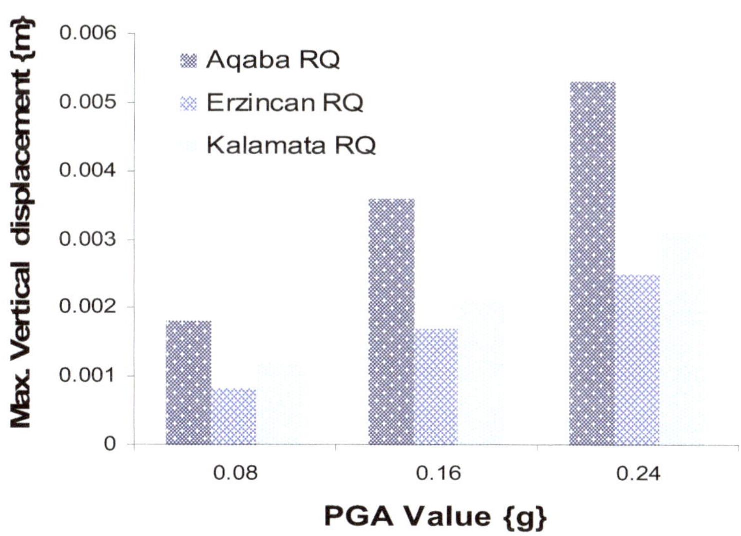

Fig. 12. (a) Maximum horizontal displacements, (b) Maximum vertical displacements on the top of Catacombs, for Aqaba, Erzincan, and Kalamata earthquakes, scaled to several values of PGA.

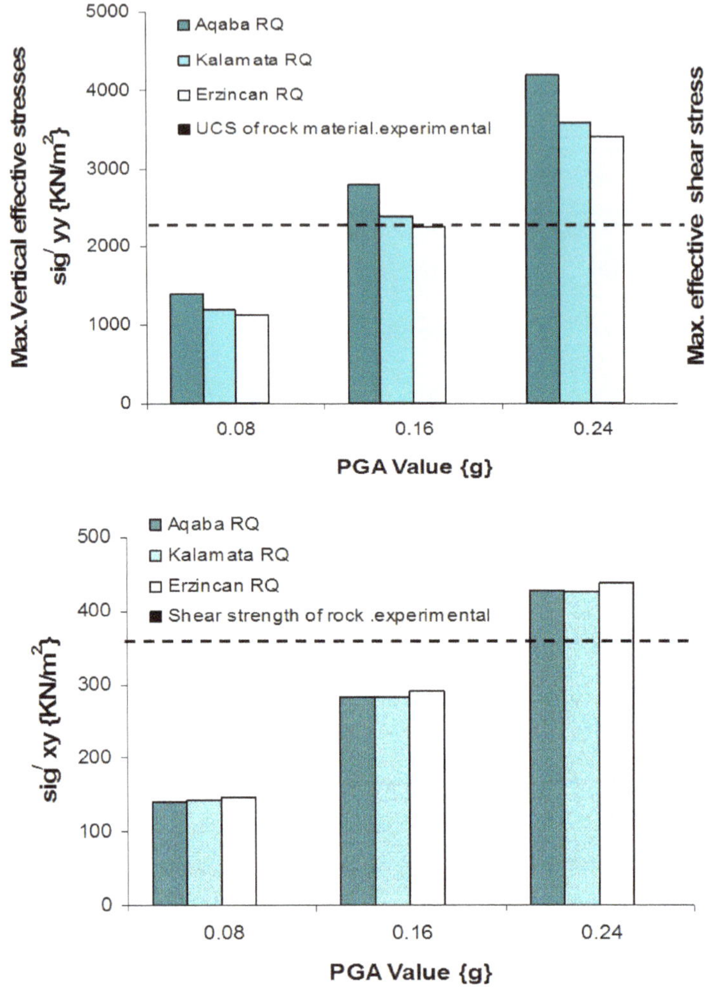

Fig. 13. (a) Maximum vertical effective compressive stresses σ/yy, (b) Maximum effective shear stresses σ/xy on the base of rock pier_1, for Aqaba, Erzincan, and Kalamata earthquakes, scaled to several values of PGA.

6. Polymer anti-seismic piling to protect the catacombs against strong earthquakes

We employed a newly developed polymer seismic isolation method, which has been employed successfully to retrofitting the Nakagawa underground station in Yokohama city (Japan), to protect the underground structures of catacombs of Kom El-Shoqafa against strong earthquakes with PGA>0.24g. The polymer seismic isolation method outlined in Figure 14 presents certain advantages compared to other conventional anti-seismic methods (such as the steel jacket method).This method is apriority suitable for the seismic protection of the underground monuments, because the application of this technique is non-destructive and do not involve any change in the original materials of the monuments; in addition we do not employ new materials or constructions to the monumental initial structure.

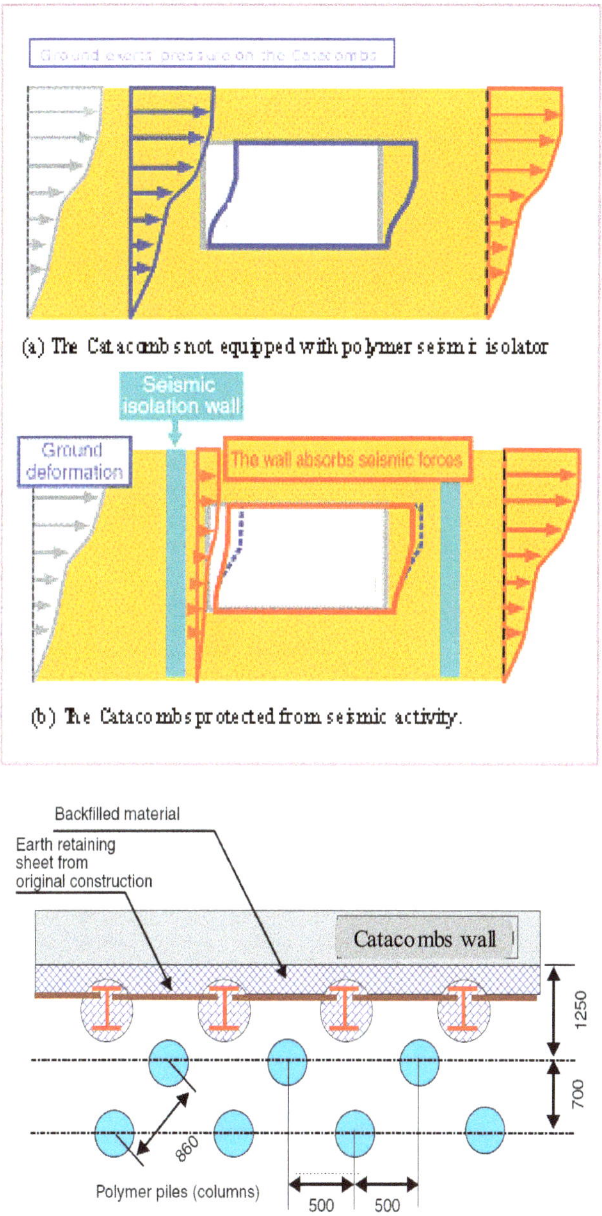

Fig. 14. The polymer seismic isolation method to protect the catacombs against strong earthquakes, (conceptual diagram).

The polymer seismic isolation method' involves the construction of polymer walls on both sides of underground structure in order to reduce the seismic actions transmitted from the surrounding ground to the structure. The stiffness of the polymer material should be about 1/10 to 1/100 that of the surrounding ground. This method is not intended to prevent or control the seismic ground deformation itself, but to isolate structures from seismic forces transmitted from the surrounding ground (Hemeda, 2008).

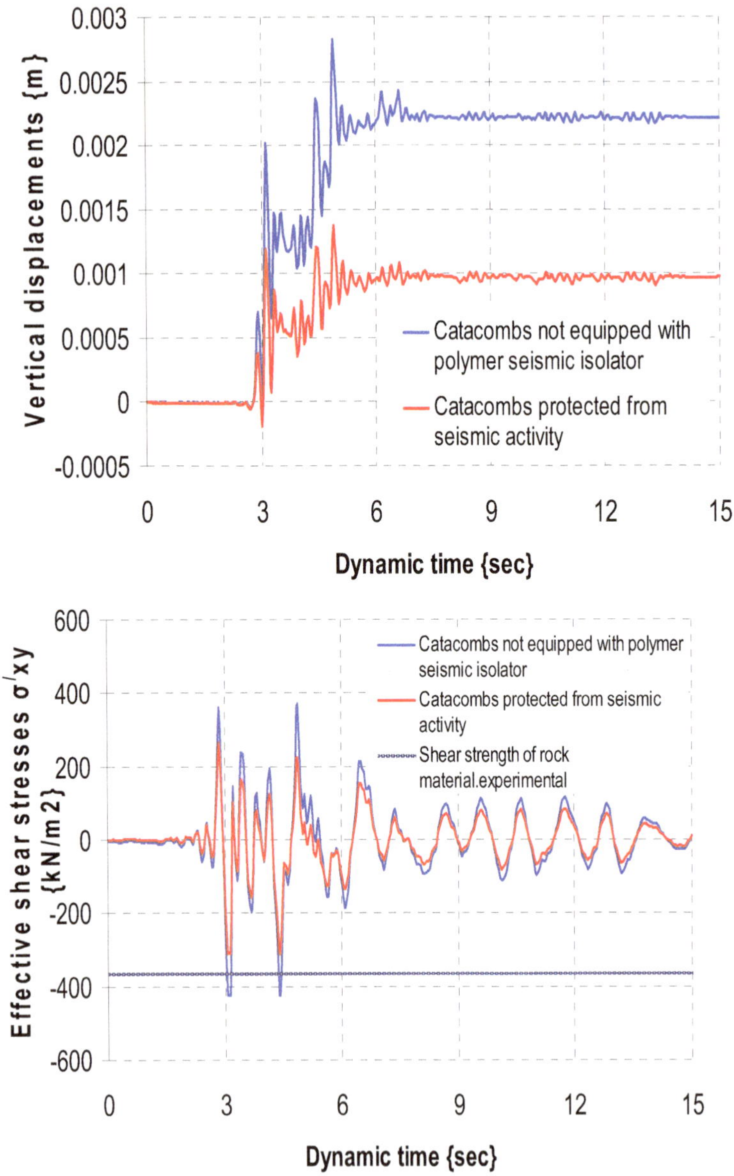

a)

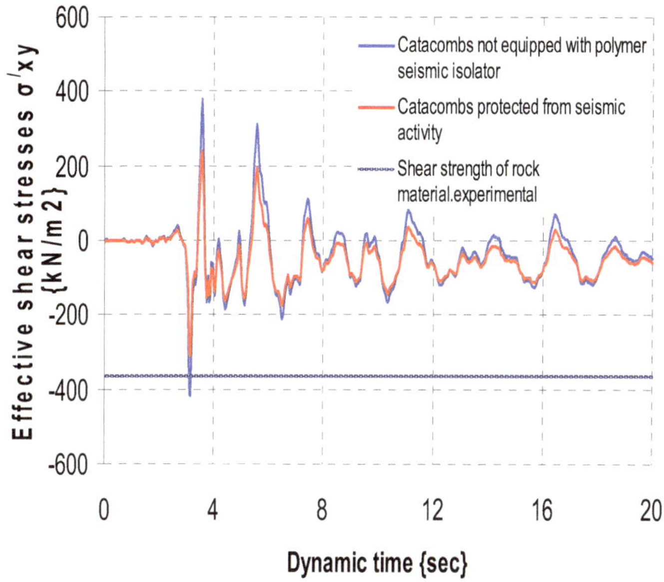
600
400
200
0
-200
-400
-600
0
4
8
12
16
20
Effective shear stresses σ'xy {kN/m2}
Dynamic time {sec}
Catacombs not equipped with polymer seismic isolator
Catacombs protected from seismic activity
Shear strength of rock material.experimental

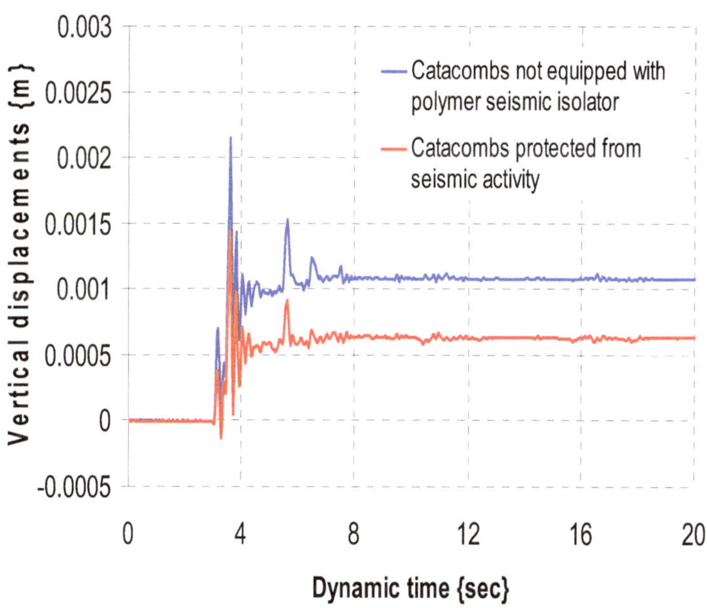
0.003
0.0025
0.002
0.0015
0.001
0.0005
0
-0.0005
0
4
8
12
16
20
Vertical displacements {m}
Dynamic time {sec}
Catacombs not equipped with polymer seismic isolator
Catacombs protected from seismic activity

b)

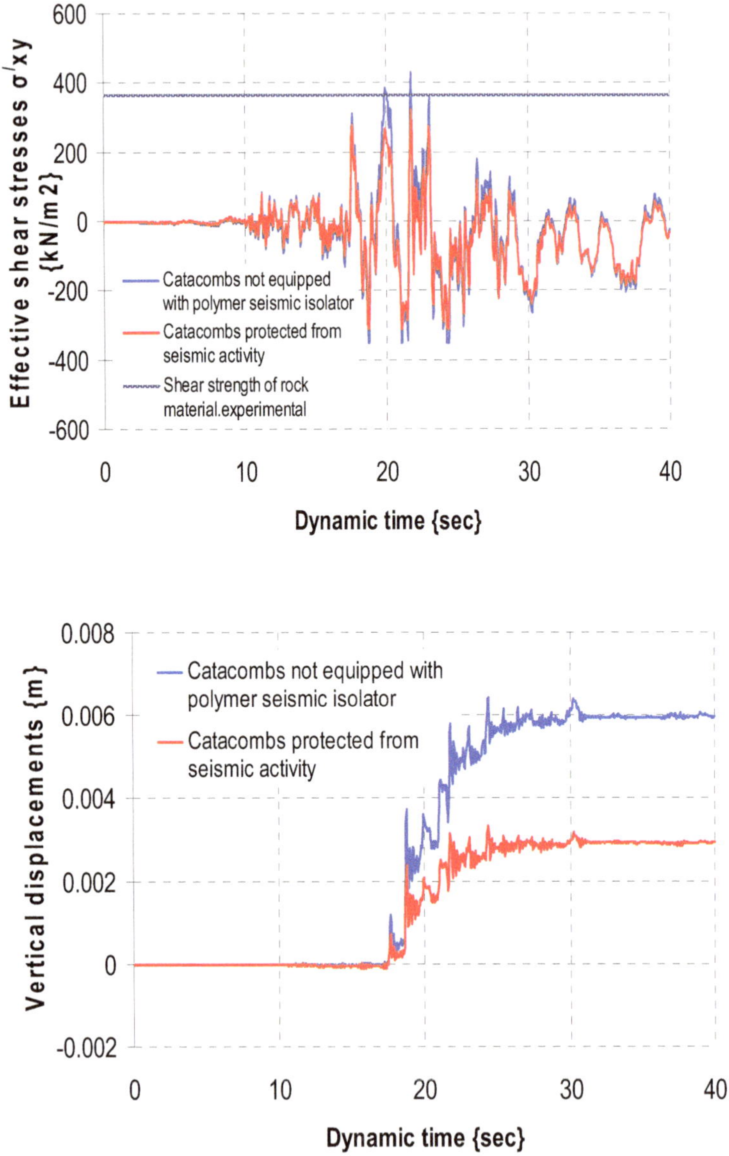

c)

Fig. 15. Vertical displacements, Effective shear stresses $\sigma/_{xy}$ - time histories on the base of rock pier 1 before and after the installation of polymer anti-seismic slurry walls. (Figure 9 at the right side of rotunda). (a) Kalamata (b) Erzincan (c) Aqaba earthquakes. PGA = 0.24g.

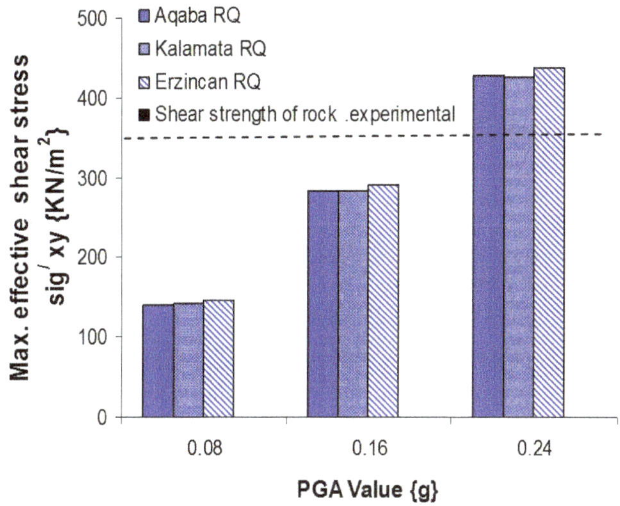

a)

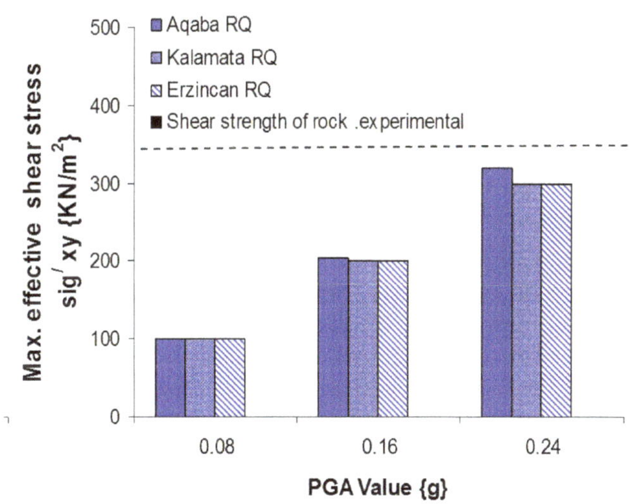

b)

Fig. 16. Maximum effective shear stresses sig/xy on the base of rock pier_1, (a) Initial model (b) After the installation of polymer anti-seismic piling. For Aqaba, Erzincan, and Kalamata earthquakes, scaled to several values of PGA.

With the installation of polymer anti-seismic slurry walls, the shear forces on the rock pillars, which are the most vulnerable parts inside the catacombs, reduce by up to 50% for the three earthquakes scenarios, (see Figure 15). The effective shear stresses on the pillars are also reduced considerably: in particular from 410 kN/m^2 to 250 kN/m^2 in the case of

Kalamata earthquake for PGA = 0.24g, from 400 kN/m^2 to 280 kN/m^2 in the case of Erzincan earthquake, and from 430 kN/m^2 to 300 kN/m^2 in the cases of Aqaba earthquake. The decrease of the computed acceleration values was also obvious in all the three earthquakes scenarios; for example the horizontal acceleration on the top of the catacombs decreases from 3 m/s^2 to 2 m/s^2 in the case of Aqaba earthquake and from 2.6 m/s^2 to 2.3 m/s^2 in the case of Erzincan earthquake. The horizontal displacements on the top of catacombs decrease from 24 mm to 20 mm in the case of Kalamata earthquake and from 25 mm to 21 mm in the case of Erzincan earthquake, while in the case of Aqaba earthquake, for PGA =0.24g, the displacement reduction is lower; from 80 mm to 78 mm. The vertical displacements at the top of catacombs decrease from 2.8 mm to 1.0 mm in the case of Kalamata earthquake, from 2.4 mm to 1.5 mm in the case of Erzincan earthquake, and from 6 mm to 2.5 mm in the case of Aqaba earthquake.

From the above short presentation and discussion it is obvious, that the seismic stability of the catacombs has been upgraded after the installation of the polymer anti-seismic slurry walls in the perimeter of an underground monument. The relative deformations are reduced considerably and the developed seismic shear forces on the sidewalls can be easily controlled within acceptable safety margins even for major earthquakes.

It is expected that the above-mentioned polymer seismic isolation method may contribute effectively to the improvement of the seismic safety margin of underground monuments, without employing other retrofitting techniques, which may modify the architectural and archeological principles of the preserved monuments.

7. Conclusions

Considering all other affecting factors (aging, weathering, multiple geotechnical and seismic instability problems and the specific geometry of the complex it has been shown that the low rock strength affects seriously the safety of the catacombs both under static and seismic loading conditions.

The results from the 2D-3D static analysis indicate that the ground displacements above the catacombs are small (the maximum total vertical displacements are of the order of 2.6 mm to 3 mm in the whole domain, and the peak horizontal displacement is 1.0 mm). Some rock piers exhibit relatively high compression stresses. The calculated peak effective principal vertical compressive stress on the pillar 1 is $-1.74^{*} \times 10^3$ KN/m^2. The calculated peak effective principal tensile stress is 200 KN/m^2 and the factor of safety of the rock pillar 1 is 1.47, which is not adequate, where the acceptable safety factor must be > 1.6. Also the overstress state is beyond the elastic limit. It is so then damage does occur. You can read more on damage assessment and future work that can be done using damage indices (journal of structural engineering 137(3): 456-567: damage-based design earthquake loads for sdof inelastic structures).

The seismic analysis of these underground monumental structures for three seismic scenarios of different PGA values, proved that for PGA > 0.10g, which is a rather low value considering the seismic activity and the past seismic history of the city, there are some critical supporting parts of these catacombs structures (i.e. rock piers and columns) that are not safe, and in general, the catacombs need considerable strengthening.

We also presented some preliminary results applying a polymer seismic isolation method which has been proved to contribute effectively in reducing the induced inertial forces on this type of underground monuments.

8. Acknowledgment

The author wish to express his deep acknowledgement to Professor Pitilakis.K, Dr Bakasis. I, Civil Engineering Department, AUTH, Greece for his support in performing part of the experimental and numerical analysis.

9. References

Biot, M.A. (1941). General theory of three- dimensional consolidation. *International Journal of Applied Physics* Vol.12, pp155–164.

Badawi, H. S. (1991). High earthquake risk buildings in Egypt. International. Sympos. Geophysical Hazards in Developing Countries and their Environmental Impacts, Perugia, Italy, August 1991.

Badawi, H. S and Mourad,Sh. (1994) .Observations from the 12 October 1992 Dahshour Earthquake in Egypt. Natural Hazards 10: 261-274, 1994.

Bolt, B. A. (1988).Earthquakes. W. H. Freeman, New York, pp. 147-150.

El-Sayed, A & Korrat, I & Hussein, M. (2004). Seismicity and seismic hazard in Alexandria (Egypt) and its surrounding. *International Journal of pure appl.geophys.*Vol.161 (2004), pp1003-1019.

Hemeda, S., Pitilakis, K., Bandis, S., Papayianni, I., and Gamal M.(2007). The Underground monuments in Alexandria, Egypt. Proc. 4th International Conference on Earthquake Geotechnical Engineering, Thessaloniki, Greece, 25-27 June, pp. 715-738.

Hemeda, S. (2008). An integrated approach for the pathology assessment and protection of underground monuments in seismic regions. Application to some Greek-Roman monuments in Alexandria, Egypt, PhD Thesis, Aristotle University of Thessaloniki, Greece.

Pitilakis, K., Hemeda, S., and Bandis, S. (2009). Geotechnical investigation and seismic analysis of underground monuments in Alexandria, Egypt. 17th International Conference on Soil Mechanics & Geotechnical Engineering, Alexandria, Egypt, October 5-9.

ITASCA Consulting Group. (2007). Inc, 3 Dimensional distinct element code (3D DEC).

Karstunen, M & Wiltrafsky, C & Krenn, H & Scharinger, F. (2006). Modeling the behavior of an embankment on soft clay with different constitutive models. Vol. 30, Issue 10, 25 August, pp 953–982.

Kebeasy, R. M. and Albert, R. N. (1976). Investigation of the seismo-tectonic nature of the Middle East region, Helwan Institute of Astronomy and Geophysics Bulletin, No. 126.

Maamoun, M., Megahed, A., and Allam, A. (1984). Seismicity of Egypt, Helwan Institute of Astronomy and Geophysics Bulletin, Vol. IV, Ser. B, pp. 109-162.

Ministry of Development, New Communities, Housing and Utilities: 1989, Egyptian Code of Practice for Design and Execution of Reinforced Concrete Constructions.

Moustafa,A. (2009). Egypt seismicity and national seismic network, Proceedings of the thirteenth Ain Shams Conference on geotechnical and Structural Engineering, 27-29 December 2009, Cairo, pp 30-39 .

PLAXIS Manual. (2002). Finite element code for soil and rock analysis published and distributed by AA Balkema Publishers, Nederland's Comput. Geotech. 32(5):326–339.

Said, R. (ed.). (1990). The Geology of Egypt, Balkema, Rotterdam.

12

Seismic Potential Improvement of Road Embankment

Ken-ichi Tokida
Osaka University
Japan

1. Introduction

Roads or highways are very important and fundamental infrastructures for the activities of life and production, and their safety are required to be secured against the various disasters such as rainfalls and earthquakes etc. The road networks are composed of many types of attached structures such as bridges, embankments, slopes, tunnels etc. During earthquakes, the abovementioned structures have been damaged in the past earthquakes and will be possible to be damaged in the near future.

For reducing the damage potential of the road networks, the seismic design and/or counter-measures for bridges have been conducted prior to other structures in the last about 40 years in Japan because bridges are considered to be difficult to retrieve when they will be damaged by earthquakes. On the other hand, road embankments have not been reinforced at all against seismic disasters because it is believed that embankments can be retrieved easily.

However, severe damages of road embankments have been occurred and it has been clarified that the performance of road networks can be lost severely for a long time, for example, 2weeks by the damages of road embankments because it needs a long time to retrieve numbers of damaged embankments and the ones located at the mountainous area.

As mentioned above, road embankments are also necessary to be reinforced against seismic ground motion, in other words, the seismic potential of road embankments should be improved against seismic ground motion during future earthquakes.

However the performance-based design will be very reasonable and effective for improving the seismic design of road embankments in the future, the design concepts have not been clarified yet and necessary to be established as soon as possible, similarly with bridges.

In this chapter, based on the typical damages of road embankments induced by recent earthquakes in Japan, the macroscopic assessment method for seismic potential of road embankments and the new concepts for the performance-based design of road embankments from the engineering view points are proposed and their applications are discussed through large-scaled field tests, centrifuge tests and analytical simulations. The following concerns summarizes the results of the recent research and development initiated by Osaka University in Japan from 2004 to the present through Tokida et al. (2007), Tokida et al. (2008), Tatta et al. (2009), Tokida et al.(2009) and Tokida et al. (2010).

2. Damage of road embankment in past earthquakes in Japan

During the recent decade in Japan, severe damages of many road embankments occurred during the earthquakes such as the 2004 Mid-Niigata Prefecture Earthquake, the 2007 Noto Peninsula Earthquake, the 2009 Suruga Bay Earthquake and the 2011 Off the Pacific Coast of Tohoku Earthquake. The typical characteristics of damages of road embankments and the several important lessons obtained can be indicated as follows from the engineering view points.

2.1 The 2004 Mid-Niigata prefecture earthquake

The 2004 Mid-Niigata Prefecture Earthquake is an inland earthquake and occurred at the mountainous area in October 23, 2004. The middle-class magnitude of this earthquake is 6.8 and the shallow depth of epicenter is 13 km.

Fig. 1 shows the time history just after the main shock occurrence for opening Kanetsu Highway and Hokuriku Highway which were stopped mainly by the damages relating to not the bridges but the embankments to the emergency vehicles such as ambulances etc. and the public cars. It can be seen it took about 13 days to open the highways for the personal vehicles under the speed regulation. As mentioned above, we can realize the importance of seismic design and reinforcement for road embankments in the future.

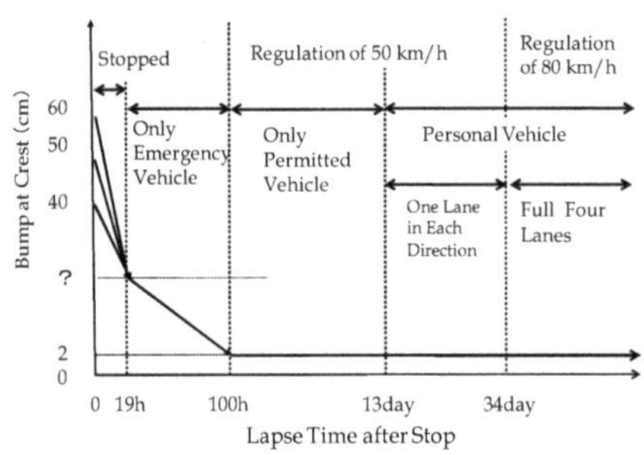

Fig. 1. Recovery of traffic performance after stopping caused by earthquake

Item 2 on the status of drainage system can be selected because the damage risk relates to the level of moisture in the embankment. For example, the assessment rank and each relating score are assumed as follows.

The typical examples of severe or large-scaled damages and slight or small-scaled damages of road embankments can be shown as Photo. 1 and Photo. 2, respectively. The former indicates three examples, i.e. (a) very severely damaged site reported by Higashi Nippon Highway Co. Ltd. where all 4 lanes were slid perfectly, (b) severely damaged site reported by Hokuriku Regional Bureau of Ministry of Land and Transportation where all 2 lanes were slid perfectly and (c) a little severely damaged site also reported by Hokuriku Regional Bureau where almost 2 lanes were slid partially and each traffic performance was lost perfectly. While, the latter indicates two examples, i.e. (a) partially slid site also reported by Hokuriku Regional Bureau where the sliding occurred only at the shoulder of the embankment and was retrieved easily with large-scaled sand bags as shown in Photo. 2, and (b) cracked site reported by Higashi Nippon Highway Co. Ltd. where only the shoulder was cracked a little and dump trucks can run without urgent retrieval, then their loss of traffic performance were small and the road could be used just after the earthquake by small-scale and short-time restoration or not.

(a) 4 lanes slid perfectly (b) 2 lanes slid perfectly (c) Almost 2 lanes slid

Photo 1. Large-scaled damage of road embankment without traffic performance

(a) Partial sliding and urgent retrieval with sand bags (b) Slight crack at shoulder

Photo 2. Small-scaled damage of road embankment with a certain traffic performance

The above two types of damage levels of road embankments give us the important lessons to consider the fundamental concepts for the performance-based design of road embankments in the future which are shown in Chapter 3.

2.2 The 2007 Noto peninsula earthquake

The 2007 Noto Peninsula Earthquake is one type of inland earthquakes and occurred near the coast of Noto Peninsula in March 25, 2007. The middle-class magnitude of this earthquake is 6.9 and the shallow depth of epicenter is 11 km.

Photo.3 reported by Ishikawa Prefecture shows four examples of the slid road embankments which locate at Noto Toll Road at the mountainous area of Noto Peninsula. These photos indicate (a) severe damaged site where 2 lanes were slid perfectly, (b) severe damaged site where almost 2 lanes were slid, (c) severe damaged site where slight 2 lanes were slid and (d) severe damaged site where one side of 2 lanes slid and each traffic performance was lost perfectly.

The severe sliding damages such as Photo. 3 which resulted in the perfect stopping of traffic performance occurred at 13 sites located along the road of about 48.2 km. The site shown in Photo. 3(c) is selected to apply the analytical method named Newmark Method to simulate the sliding plane and its displacement in Chapter 5.

In Photo. 4 reported by Ishikawa Prefecture, Photo. 4(a) shows the severe damaged on ramp close to the toll gate. Most damaged embankments were retrieved urgently and permanently with use of the geotextile which can be shown in Photos. 4(b) and 4(c), respectively. Noto Toll Road was opened to the public urgently in about one month later after the earthquake and the permanent retrieval of all damaged embankments was completed in about five months later after the urgent retrieval.

(a) 2 lanes slid perfectly (b) Almost 2 lanes slid perfectly

(c) A little more 1 lane slid (d) 1 lane slid/1 lane remain

Photo 3. Large-scaled damage of road embankment without traffic performance

(a) Sliding of entrance lane (b) Urgent retrieval (c) Permanent retrieval

Photo 4. Damage and seismic retrieval with geotextile

It can be indicated that the seismic design with the seismic coefficient of 0.2 was applied to reinforce the retrieved embankments strongly more than that before the earthquake for the future earthquakes. Such seismic retrieval for the road embankments at Noto Toll Road may be the first case in Japan and can be referred for other embankments in the future.

2.3 The 2009 Suruga bay earthquake

The 2009 Suruga Bay Earthquake is a plate-slid earthquake and occurred at Suruga bay neighboring to Pacific Ocean in August 11, 2009. The middle-class magnitude of this earthquake is 6.6 and the depth of epicenter is 20 km.

In this earthquake, only one embankment slid at Tohmei Highway which is the most important trunk road between Tokyo and Nagoya as shown in Photo. 5 reported by Naka Nippon Highway Co. Ltd. However the sliding occurred partially at the shoulder and slope as shown in Photo. 5(a) and (b), the traffic performance was lost for about 5 days and the economical influences and so on were very large for Japan. However the urgent and extended retrieval was executed, it took 5 days to recover the limited traffic performance for the personal vehicles as shown in Photo. 5(c). The permanent retrieval ended on June 10, 2010, i.e. in about ten months and the traffic performance was recovered perfectly.

(a) Overview of sliding (b) Shoulder and slope slid (c) Urgent retrieval

Photo 5. Partially damaged site without traffic performance and urgent retrieval

We can get the important lesson from this experience that the only one damaged site warns the more severe damages of embankments by the very large earthquakes named Tokai Earthquake and/or Tokai & Nankai Earthquake whose occurrence has been indicated around this site in the near future in Japan.

2.4 The 2011 off the pacific coast of Tohoku earthquake

The 2011 Off the Pacific Coast of Tohoku Earthquake is a typical plate-slid earthquake and occurred about 130 km off the main island of Japan in March 11, 2011. The very large-class magnitude of the earthquake is 9.0 and the depth of epicenter is 24 km. The very wide area where post earthquakes occurred and/or are occurring at present, i.e. the scale of moved plate by the main-shock and after-shocks is assumed to be 200 km multiplied by 500 km. The outline on the charactoristics of the earthquake and the observed ground motion can be referred the prompt reports by the Japan Meteorological Agency (JMA) and the National Research Institute for Earth Science and Disaster Prevention (NIED), respectively.

However the damages by tsunami occurred widely and severely, other damages of civil engineering structures such as road bridges, road embankments, river dykes and so on were not so remarkable because the epicenter of the main-shock is rather far and the damages occurred along the coasts may be washed away by tsunami attacks. The most severely damaged embankment of all ones at Tohoku Highway and Joban Highway which are the main trunk routes to connect Tokyo with Tohoku Region, where the earthquake occurred can be introduced as shown in Photo. 6 reported by Kanto Branch Office of Higashi Nippon Highway Co. Ltd. As shown in Photo. 6(a), the one side of 2 lanes in one direction slid. However the thickness of the soft ground beneath the embankment body is about 3.6m, the type of the failure is a toe sliding failure and a little small failure. The work flow of the urgent retrieval for this embankment can be shown as Photo. 6(b), 6(c) and 6(d), respectively. As shown in Photo. 7 under urgent retrieval work, the height and the damaged length of the embankment are about 8m and 150m, respectively. The traffic performance for the personal vehicles was recovered urgently on March 21, 2011. Then it takes about 10 days to complete the urgent retrieval and open the highway to the public.

(a) Crushed (b) Sliding occurrence (b) Urgent retrieval with gravel and soil (c) Pavement

Photo 6. One side of two lanes damaged and urgent retrieval

Photo 7. Under urgent retrieving

Some lessons concerning to road embankments against tsunami attacks can be obtained as follows. Photo. 8 is a newly constructed river dyke along the channel which locates in Miyagi Prefecture facing the Pacific Ocean where the earthquake occurred. As shown in the Photo. 8, the river dyke which height and width at the crest are about 4m and 6m, respectively is paved and can be used as a road. Then the lessons obtained from the damages of this dyke can be applied for the future design against tsunami disasters. Photo. 8(a) shows the condition of the front side of dyke slope against tsunami attacks which came from the left side of the Photo. 8(a). It can be seen that almost all of the front sides were not damaged, i.e. the front side may be strong against tsunami attacks which height is assumed to be about 10m around this site and then the overflown water depth at the dyke crest can be estimated to be about 4m. Photo. 8(b) shows the condition at the dyke crest where the asphalt pavements are stripped partially, but the road function was not lost so much. Then it can be indicated that the asphalt pavement at the crest is very effective against the overflowed water by tsunami. Photo. 8(c) shows the conditions of the back side of dyke slope washed away by tsunami. It can be known that the back side slope is eroded near the surface and the toe ground is eroded severely and trenched deeply. However the back side slope is fallen down partially, the minimum stability of the dyke can be maintained to keep the performance of the dyke and/or road. However the above lessons may be limited to this dyke, they become very important hints for the design against tsunami attacks.

(a) Front slope non-damaged (b) Crest stripped partially (c) Back slope and toe eroded

Photo 8. Dyke overflown by tsunami

(a) Opposite side from sea (b) Sea side

Photo 9. Comparison of damages at both sides of highway embankment

Photo. 9 shows another example from which we can get some lessons on the performance of the road embankments against tsunami attacks. Photos. 9(a) and 9(b) show the landscapes after tsunami attacks at the opposite side and the front side of the road embankment against tsunami attacks, respectively. However the road embankment was expected to perform the barrier function against tsunami attacks, the performance may be small as shown in Photo. 9(a) because the box culvert and the elevated bridge permit tsunami flow to go through.

From this example, it can be known that the road embankment may be effective as a barrier against tsunami when the open space such as box culvert etc. can be constructed carefully considering tsunami attacks.

3. Macroscopic assessment for seismic potential of road embankments

It is necessary to select the objective road embankments with high seismic potential for the further detail investigations and the economical countermeasures against future earthquakes. Based on the damages in the past earthquakes as shown in Chapter 2, the assessment items and ranks for elemental functions traversing damage of embankments caused by the sliding failure can be obtained and a macroscopic assessment method to estimate the seismic potential of the road embankments can be proposed as follows.

3.1 Factors related to damage potential

Macroscopic assessment methods of seismic and disaster prevention functions are discussed and proposed by Tokida et al. (2007) based on the on-site investigations as for the damages of expressways and national highways in the 2004 Mid Niigata Prefecture Earthquake and the 2007 Noto Peninsula Earthquake.

Typical 8 items such as Item 1 to Item 8 to assess the structural vulnerability of each road embankment are selected and the estimation weight between each item and their detail ranks and their added scores to estimate the risk potential quantitatively can be recommended as follows.

Item 1 on the topographical conditions below embankments can be selected because the damage risk relates to the original terrain condition under the embankment such as ridge slope or stream. For example, the assessment rank and each relating score are assumed as shown in Fig. 2.

Rank-1 with score = 0 in case of embankment located on large ridge slope
Rank-2 with score = 1 in case of embankment located on small ridge slope
Rank-3 with score = 2 in case of embankment located on slope in parallel

Rank-4 with score = 4 in case of embankment located on small stream
Rank-5 with score = 5 in case of embankment located on large stream

The relative weight for Item 1 between 8 items is assumed to be 2, then the maximum value of the score S1 for Item 1 is 10 multiplying 2 by 5.

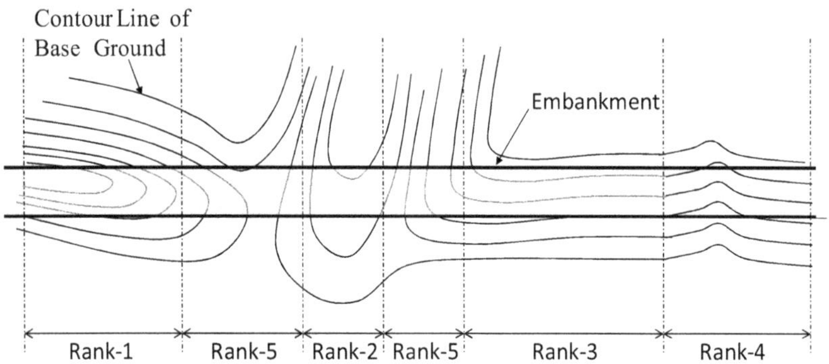

Fig. 2. Identification of topographical ranks for Item 1

Rank-1 with score = 0 in case of embankment where the whole methods of A to E are applied.
Rank-2 with score = 1 in case of embankment where both A and B methods are applied.
Rank-3 with score = 2 in case of embankment where the C method is applied.
Rank-4 with score = 4 in case of embankment where both D and E methods are applied.
Rank-5 with score = 5 in case of embankment where no drain method is applied.

The above drain methods of A to E are identified as follows.

A is a method to drain the underground water from the mountain side with pipes or ditches laid on the base ground of embankments.
B is a method to set the horizontal or vertical drain materials laid inside of embankments.
C is a method to drain the surface water from the mountain side with pipes and ditches laid inside of the embankments.
D is a method to drain the crest surface water with side ditches.
E is a method to drain the slope surface water by side ditch.

The relative weight for Item 2 between 8 items is assumed to be 1, then the maximum value of the score S2 for Item 2 is 5 multiplying 1 by 5.

Item 3 on the soil properties and quality of construction can be selected because the damage risk relates to the embankment soil properties or the quality of workmanship used during construction, such as compaction.

Rank-1 with score = 0 in case of embankment where the embankment material is enough well, and good compaction or improvement is done.
Rank-2 with score = 3 in case of embankment where the embankment material is normal and construction is normal.
Rank-3 with score = 5 in case of embankment where the embankment material is not good and the construction condition is not good.

The relative weight for Item 3 between 8 items is assumed to be 3, then the maximum value of the score S3 for Item 3 is 15 multiplying 3 by 5.

Item 4 on the traverse structure of embankment can be selected because the damage risk relates to the traverse structure of the embankment, including the incline of the foundation ground, the configuration of the embankment, as shown in Fig. 3.

Rank-1 with score = 0 in case of embankment on flat base ground.
Rank-2 with score = 1 in case of embankment on ups and downs base or ridge slope or one side cut and other side fill at the inclination base ground.
Rank-3 with score = 2 in case of embankment on one side fill at the inclination base ground where the embankment edge is flat.
Rank-4 with score = 3 in case of embankment on one side fill at the inclination base ground where the embankment edge slopes.

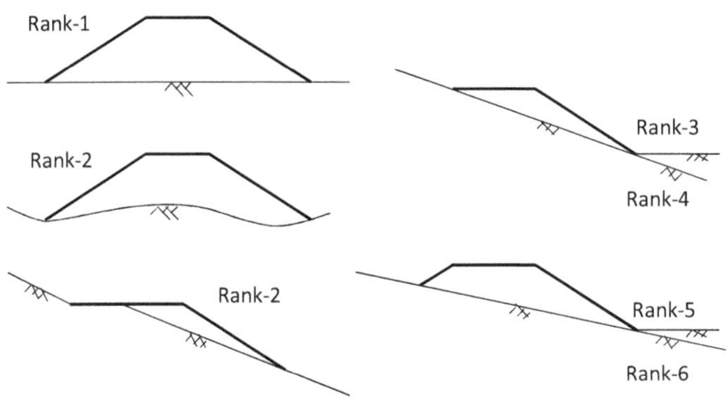

Fig. 3. Identification of traverse structures for Item 4

Rank-5 with score = 4 in case of embankment on both side fills at the inclination base ground where the embankment edge is flat.
Rank-6 with score = 5 in case of embankment on both side fills at the inclination base ground where the embankment edge slopes.

The relative weight for Item 4 between 8 items is assumed to be 2, then the maximum value of rthe score S4 for Item 4 is 10 multiplying 2 by 5.

Item 5 on the height of embankment can be selected because the occurrence and scale of damage risk relate to the embankment height as follows.

Rank-1 with score = 0 in case of embankment which height is 4 m or less.
Rank-2 with score = 2 in case of embankment which height is 8 m or less exceeding 4 m.
Rank-3 with score = 4 in case of embankment which height is 20 m or less exceeding 8 m.
Rank-4 with score = 5 in case of embankment which height is 20 m excess.

The relative weight for Item 5 between 8 items is assumed to be 3, then the maximum value of the score S5 for Item 5 is 15 multiplying 3 by 5.

Item 6 on the inclination of embankment slope can be selected because the stability of embankment relates to the slope inclination as follows.

Rank-1 with score = 0 in case of embankment which slope inclination is below the standard one.
Rank-2 with score = 5 in case of embankment which slope inclination exceeds the standard one.

The relative weight for Item 6 between 8 items is assumed to be 1, then the maximum value of the score S6 for Item 6 is 5 multiplying 1 by 5.

Item 7 on the conditions reinforcing the embankment can be selected because the seismic and/or a certain improvements are effective to reduce damage of embankments as follows.

Rank-1 with score = 0 in case of embankment reinforced based on the seismic design fully.
Rank-2 with score = 2 in case of embankment where the retaining wall is applied.
Rank-3 with score = 4 in case of embankment where the attached structures such as gabions are set.
Rank-4 with score = 5 in case of embankment where the countermeasures are not done.

The relative weight for Item 7 between 8 items is assumed to be 3, then the maximum value of the score S7 for Item 7 is 15 multiplying 3 by 5.

Item 8 on the transformation history and/or present state of embankments can be selected because the past and/or present transformation indicates the sign and warning for the future damage occurrence. Four sub-items are selected to judge the risk of embankments considering the past and/or present conditions as follows.

Sub-Item-1 on the existence level of water spring on the slope and/or toe of embankment can be selected because the water conditions inside of embankments are most important. Each score of 0 to 5 are set considering the level of water spring and the relative weight is assumed to be 2, then the maximum value of the score S8-1 for Sub-Item 1 is 10 multiplying 2 by 5.

Sub-Item-2 on the transformation level of embankment crest such as cracks can be selected because the transformation is one of signs to be failed. Each score of 0 to 5 are set considering the level of transformation and the relative weight is assumed to be 0.6, then the maximum value of the score S8-2 for Sub-Item 2 is 3 multiplying 0.6 by 5.

Sub-Item-3 on the transformation level of embankment slope such as cracks can be selected because the transformation is one of signs to be failed. Each score of 0 to 5 are set considering the level of transformation and the relative weight is assumed to be 0.4, then the maximum value of the score S8-3 for Sub-Item 3 is 2 multiplying 0.4 by 5.

Sub-Item-4 on the present performance of drainage system can be selected because the conditions of drainage system relate to the water conditions inside of embankments. Each score of 0 to 5 are set considering the level of transformation and the relative weight is assumed to be 2, then the maximum value of the score S8-4 for Sub-Item 4 is 10 multiplying 2 by 5.

The maximum value of the score S8 for Item 8 added each score of the above four sub-items gives 25. As a summary, the total score named SRP added from Item 1 to Item 8 which means the score of risk potential can be obtained in the Eq. (1) and scores and weight for each item and rank are summarized in Table 1.

$$SRP=S1+S2+S3+S4+S5+S6+S7+S8 \tag{1}$$

Item	Rank-1	Rank-2	Rank-3	Rank-4	Rank-5	Rank-6	Weight	Max. of Score
Item 1	0	1	2	4	5	-	2	10
Item 2	0	1	2	4	5	-	1	5
Item 3	0	3	5	-	-	-	3	15
Item 4	0	1	2	3	4	5	2	10
Item 5	0	2	4	5	-	-	3	15
Item 6	0	5	-	-	-	-	1	5
Item 7	0	2	4	5	-	-	3	15
Item 8 Sub-Item 1	0 - 5						2	10
Item 8 Sub-Item 2	0 - 5						0.6	3
Item 8 Sub-Item 3	0 - 5						0.4	2
Item 8 Sub-Item 4	0 - 5						2	10
SRP								100

Table 1. Relation between item, rank, weight and scores

The value of SRP gives 0 to 100 and is named the evaluation value of seismic risk potential of an embankment at a certain site, i.e. means that when the value is larger, the damage potential can be indicated to become larger. Basing on the proposed assessment method with use of SRP, the seismic risk potential of a road embankment can be estimated easily, macroscopically and practically.

3.2 Application of macroscopic assessment method

The example of application of the macroscopic assessment method with the value of SRP can be obtained as shown in Fig. 4 where 13 damaged and 19 non-damaged embankments in the 2007 Noto Peninsula Earthquake are selected and compared each other.

The average score μ and standard deviation σ are 60.7 and 8.2, respectively on the 19 non-damaged embankments. Then the score range of $\mu \pm \sigma$ is from 52.5 to 68.9. The average score and standard deviation are 75.2 and 4.8, respectively on the 13 damaged embankments and then the score range of $\mu \pm \sigma$ is from 70.4 to 80.0.

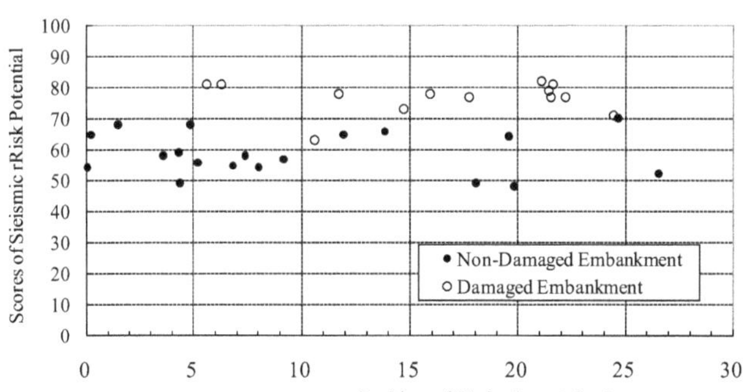

Fig. 4. Scores of seismic risk potential between non-damaged and damaged embankments in the 2007 Noto Peninsula Earthquake

Based on the results as shown in Fig. 4, the following lessons can be drawn to use the evaluation value of seismic risk potential practically.

0≦ SRP < 60 : Seismic Potential of road embankment is very low or low.
60≦ SRP < 70 : Seismic Potential of road embankment is a little high or high.
70≦ SRP ≦100 : Seismic potential of road embankment is high or very high.

4. Performance-based design concept for road embankment

Recently the new design concept, i.e., performance-based design is necessary for reasonable and economical design of road embankments against severe earthquakes. In this Chapter, four design concepts considering the seismic performance of road embankments are proposed, and the criterion to judge the damage level of settlements or sliding displacements is introduced.

4.1 Concepts based on performance-based design

Comparing the severe damage of road embankment shown in Photo. 1(b) with the slight damage shown in Photo. 2(a), it can be hinted to control the position of the sliding plane occurred not to reach to the traffic lane, then to keep the traffic performance at a certain level just after an earthquake. This fundamental design concept was proposed by Tokida et al. (2007) and named "Sliding Failure Control Method". The control of the sliding failure aims to change the positon of sliding plane and/or to decrease the sliding displacement occurred during earthquakes not to disturb the traffic performance perfectly. Furthermore, as shown in Fig. 5, four specific design concepts such as "Crest Reinforcement Structure", "Barrier Reinforcement Structure", "Artificial Reinforcement Structure" and "Toe Reinforcement Structure" can be proposed to achieve the fundamental design concept. The performance of each reinforcement structure is as follows.

Crest Reinforcement Structure aims to reinforce the crest of embankment partially with use of geotexstile etc. and not to occur the sliding plane and decrease the sliding displacement on the top of the embankment, then maintain the traffic performance enough just after an earthquake. Toe Reinforcement Structure aims to reinforce the toe of embankment to control the sliding plane and decrease the sliding displacement. Artificial Reinforcement Structure aims to control the sliding plane by improving the embankment body stiffly as shown in Fig. 5 to control the sliding plane not to reach the traffic lane. Barrier Reinforcement Structure aims to control the position of the sliding plane with use of sheet piles etc. and not to occur at the traffic lane, then maintain the traffic performance enough just after an earthquake.

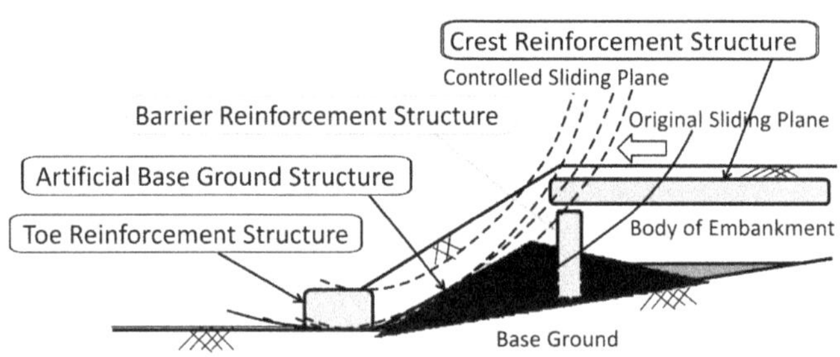

Fig. 5. Typical concepts proposed for performance-based design of road embankment

As mentioned above, these four reinforcing structures are constructed partially and then economically.

4.2 Performance level for seismic design of road embankment

For the performance-based design of road embankments, the performance level as the purpose of the seismic design is necessary to be set. Because the performance level for road embankments has not been clarified in Japan, Tokida et al. (2007) conducted the driving tests to investigate the relation between the step height and the driving velocity through the steps to know the limit height of steps occurred at the pavement surface during earthquakes as shown in Fig. 6(a).

The test results are summarized in Fig. 6(b) which shows the limit height of steps able to drive through without repair, relating to some conditions of the small or large vehicle size, the emergency or personal vehicle and the velocity of vehicles driving through the steps such as almost stop (0 to 10 km/h), slow down (15 to 20km/h) and normal (for example, 60 to 80km/h).

Based on the driving test results such as in Fig.6(b), the performance level considering traffic performance just after an earthquake can be proposed as shown Table 2. The performance levels are classified into four types as follows.

(a) Driving test

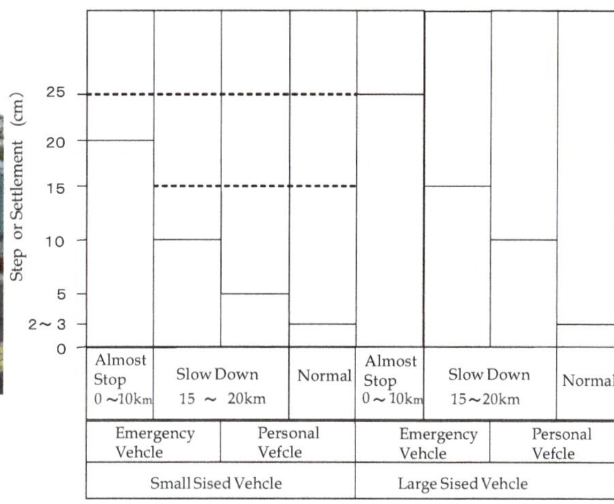

(b) Relations between step and driving velocity

Fig. 6. Driving tests and results

Performance Level	Traffic Performance Just After EQ.	Factors to Estimate Damage			
		Step of Road Surface	Sliding Failure	Settlement at Crest	
				Transverse	Longitudinal
1	Maintained Normally	Step Height Smaller than 2~3 cm	Not Slid	Not Settled	Not Settled
2	Decreased a Little and Recovered Easily	Step Height Larger than 2~3 cm and Smaller than 20~25 cm	Slid at Shoulder or Slope Partially	Settled Small and Uniformly	Settled Small and Uniformly
3	Decreased and Difficult a Little to Recover	Step Height Larger than 20~25 cm and Smaller than 50 cm	Slid at One Traffic Lane Partially	Settled Small and Not Uniformly	Settled Small and Not Uniformly
4	Decreased and Difficult to Recover	Step Height Larger than 50 cm	Slid at Two Traffic Lanes Totally	Settled Largely and Not Uniformly	Settled Largely and Not Uniformly

Table 2. Performance level considering traffic performance just after earthquake

Level 1: The traffic performance can be maintained normally just after an earthquake, because the steps occurred are smaller than 2 to 3 cm.

Level 2: The traffic performance is decreased a little just an earthquake, but can be recovered rapidly and easily, because the steps occurred are smaller than 20 to 25 cm where the vehicles can be go through the steps as shown in Fig. 6 (a).

Level 3: The traffic performance is decreased just an earthquake and difficult a little to recover soon, because the steps occurred are larger than 20 to 25 cm and smaller than 50cm.

Level 4: The traffic performance is decreased just an earthquake and it takes long time to recover, because the steps occurred are larger than 50 cm.

As shown above, Table 2 can be applied to set the objective performance level of the sliding failure of road embankments in the design of both new embankments and existing ones in the future.

5. Analytical estimation of sliding failure of road embankment

It is necessary to estimate the damage level of road embankments for the performance-based design. In this chapter, the typical analytical method named Newmark Method (Tateyama et al., 1998) modified the Original Newmark Method (Newmark, 1965) for calculating the sliding displacement of road embankments during earthquakes is shown and its application is discussed based on the damaged examples of road embankments in the past earthquakes shown in Chapter 2.

Fig. 7 is the example appling the Newmark Method to the embankment damaged during the Noto Peninsula Earthquake in 2007. The ground motion inputted at the bottom of the embankment model shown in Fig. 7(b) is revised the maximum acceleration from the ground acceleration of 580gal in the East-West direction shown in Fig. 7(a) which was observed at Togi City set by National Research Institute for Earth Science and Disaster Prevention (NIED) to the one of 500gal. As known in Fig. 7(b), the position of the sliding plane measured after the earthquake is almost similar to the analytical one. Furthermore, the measured sliding displacement at the crest of the embankment is 10.9m and similar to the 9.7m measured.

In this case study, the sliding failure such as the position of the sliding plane and the sliding displacement estimated with use of the Newmark Method is well estimated analytically.

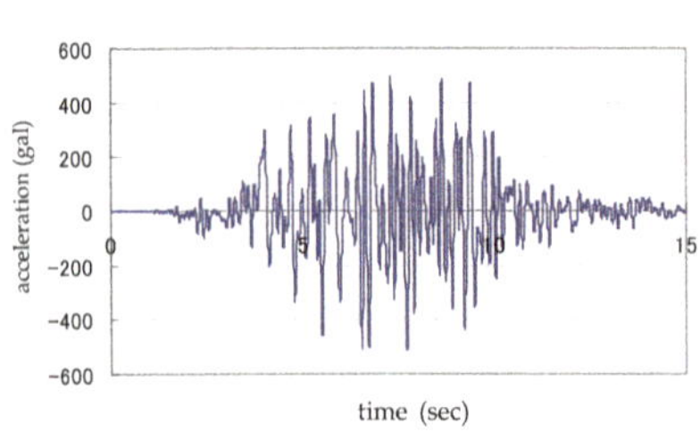

(a) Observed E-W ground motion at Togi

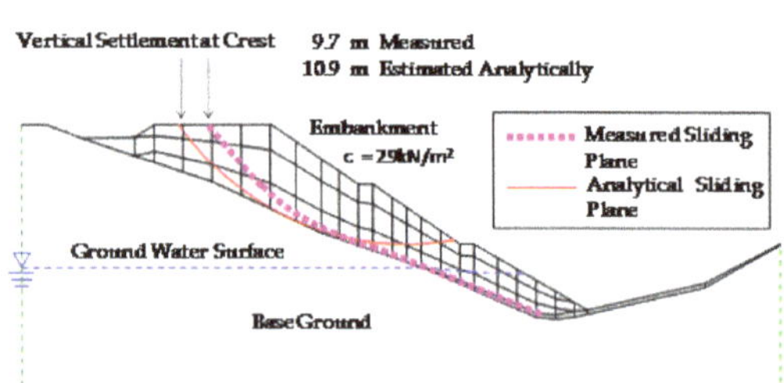

(b) Measured and analytical sliding failure

Fig. 7. Example of application of Newmark Method

6. Performance-based countermeasures for road embankment

Several seismic countermeasures are proposed concretely and their applications are discussed. In the case of the Crest Reinforcement Structure, the crest of road embankments is reinforced with use of geosynthetics and its application is verified based on the static large-scaled model test, the dynamic centrifuge test and the static and dynamic analyses quantitatively and/or qualitatively. In case of the Toe Reinforcement Structure, the toe of road embankments is reinforced by large-scaled gabions and its application is verified based on analyses qualitatively. In case of the Artificial Base-ground Structure, the body of road embankments is improved partially to construct the artificial base ground where the sliding plane is assumed not to be occurred and its application is verified based on analyses qualitatively. In case of the Barrier Reinforcement Structure, the barrier structures such as sheet piles and underground walls may be constructed practically and its application is verified based on the centrifuge test quantitatively.

6.1 Crest reinforcement structure with geosynthetics

Fig. 8 shows the example of the Crest Reinforcement Structure with use of the geosynthetics. This structure can be applied to both existing and newly constructed embankments. In the case of the existing embankment, the sheets of geosynthetics are installed at the overall crest of the embankment as shown in Fig. 8 based on the excavation and the re-pavement, or around the shoulder of the embankment partially. The reinforcing work is necessary to be paid attention to reduce the effects on the traffic performance as little as possible. On the other hand, in the case of newly construction, the sheets of geosynthetics can be easily set inside of the embankment. The practical applications of the crest reinforcement structure with use of geosynthetics have been discussed and clarified as follows.

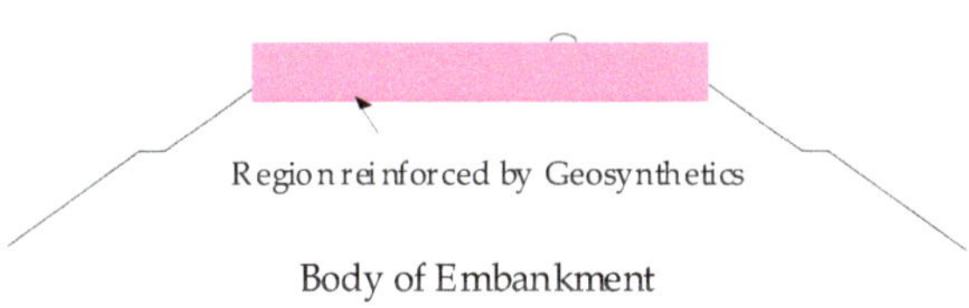

(a) Reinforced with geosynthetics

(b) Geosynthetics

Fig. 8. Example of crest reinforcement structure

6.1.1 Static large-scaled model test

The large-scaled embankment model is constructed in extension of 3.6m, depth of 1.8m and height of 4.1m, which is surrounded by both wooden panels and sandbags shown Fig. 9(a). As shown in Fig. 9(b), the model is divided into 4 reinforcement patterns. In each pattern, the overburden surface layer with the thickness of 50cm are set at the crest of the embankment model. Pattern 1 is the fundamental model without reinforcement. In cases of Pattern 2 and Pattern 3, 5 layers with the thickness of 30cm in length of 1m and 2m, respectively are reinforced with geosynthetics connected by metal fittings each other. In case of Pattern 4, the embankment body is lapped with the geosynthetics which is tied with ropes. The soil properties are shown in Table 3 and the fine aggregate without fine-grained fraction of the wet density of γ_t =18kN/m^3 which tends to occur a collapse easily. The static failure of the reinforced embankment model is planned to be occurred by removing the panels and sandbags.

(a) View of embankment model

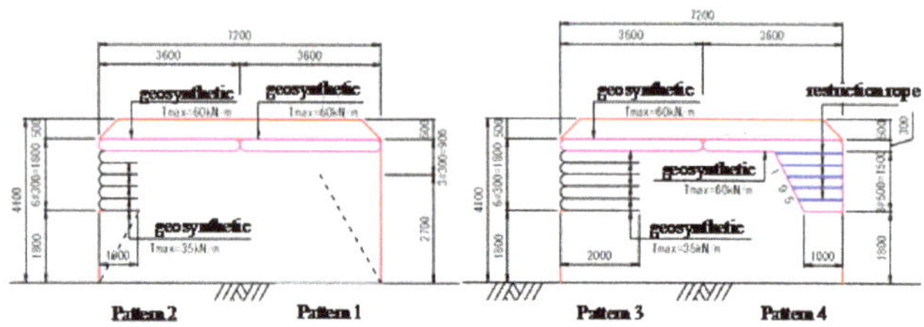

(b) 4 patterns of reinforced embankment models (unit:nm)

Fig. 9. Large-scalled embankment model in field test

The collapse mode of each reinforced embankment model is illustrated in Fig. 10(a). In case of Pattern 1, the large sliding failure occurred at the whole area of the embankment body beneath the surface layers. In case of Patterns 2 and 3, each deformation of the embankment body is smaller than Pattern 1. Moreover, the deformation of the embankment model in case of Pattern 3 with a longer reinforcement than that of Pattern 2 became smaller than that of Pattern 2. The difference of each damage level can be known in Fig. 10(b) which shows the side plane conditions and the surface condition of the crest of the embankment models. In case of Pattern 4, the whole reinforced part slid down severely because of the weight of the reinforced body.

Geosynthetics	Tensile strength : 35(kN/m), 60 (kN/m) Elongation : 5.0(%)	Core : aramid fiber Covering : polyethylene
Metal fittings	ϕ 3mm, 6mm	Ordinary iron wire
Restriction rope	Tensile strength : 22(kN) Elongation : 40%	Polyester spinning rope ϕ 12mm
Embankment	γ =18kN/m^3, c=0kN/m^2, ϕ =40°	

Table 3. Properties of test material

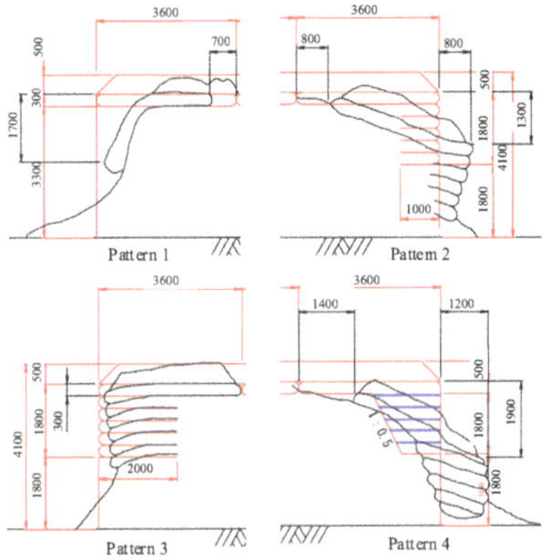

(a) Deformation of models (unit: mm)

(b) Difference of Settlement Level

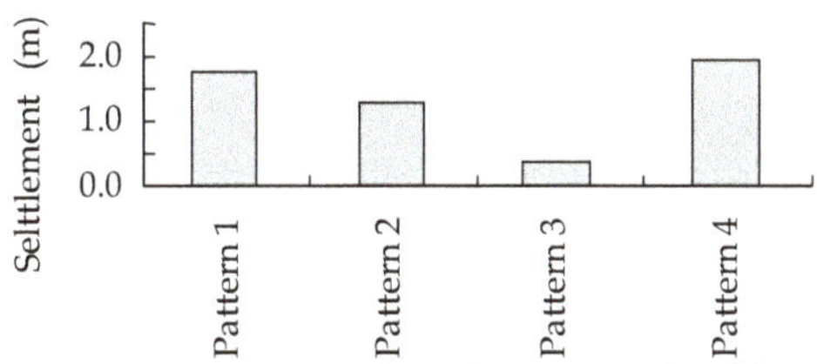

(c) Comparison of settlement at shoulder

Fig. 10. Comparison of damage level between different reinforcement patterns

The test results on the settlements at the shoulders of each embankment model can be summarized in Fig. 10(c) where the settlements of Pattern 1, Pattern 2, Pattern 3 and Pattern 4 are 1.7 m, 1.3 m, 0.3m and 1.9m, respectively. Therefore, it can be indicated that Pattern 3 where the horizontal length of the geosynthetics is set deeper than the estimated sliding plane is most effective to keep the shape of the embankment during an earthquake.

6.1.2 Dynamic centrifuge model test

Applying the steel container which is 900mm wide, 300mm thick and 280mm high as shown in Fig. 11, several centrifuge tests are carried out under the acceleration of 30g. A half-sized embankment on the solid base ground is made using the DL clay and silicon oil to adjust the water content to 5% in height of 290 mm which corresponds to the actual height of 8.7m. The density of total fill is 1.52g/cm and the shear wave velocity of embankment model measured is 158m/sec in 1g and 185m/sec in 30g. As shown in Fig. 11, the horizontal and vertical accelerometers are installed in three sites to measure the input acceleration on the bottom of soil container and response acceleration of embankment model near the toe of slope in height of 90mm and under the edge of slope in height of 230mm from the bottom. To measure the final residual deformation of model, several colored sand columns are installed into the central cross section of model from the surface in the interval of 60mm and several markers put at the surface of the embankment in the interval of 50mm. Furthermore, two thin plates with the thickness of 0.3mm on which the strain gauges are pasted on both sides are placed in the upper and central parts of embankment to investigate the failure mechanism.

Fig. 12(a) shows the cross section cut at the center of the embankment model which is not reinforced and slid largely by the inputted motion with the maximum acceleration of 488gal. Fig. 12(b) compares the initial and slid shapes of the embankment model drawn by markers located at the surface and the colored sands columns. Comparing the positions of markers before and after dynamic motions in Fig. 12(b), it can be seen that however the embankment body around the shoulder is disturbed according to the sliding, the sliding body is driven over all because the displacement of markers at the shoulder of 89mm and the one along the slope of 87mm are almost same. On the other hand, the vertical settlement at the shoulder is 65mm which is converted to 1.95m at the field. As mentioned above, it can be clarified that it's possible to reproduce the dynamic sliding plane artificially by the centrifuge test.

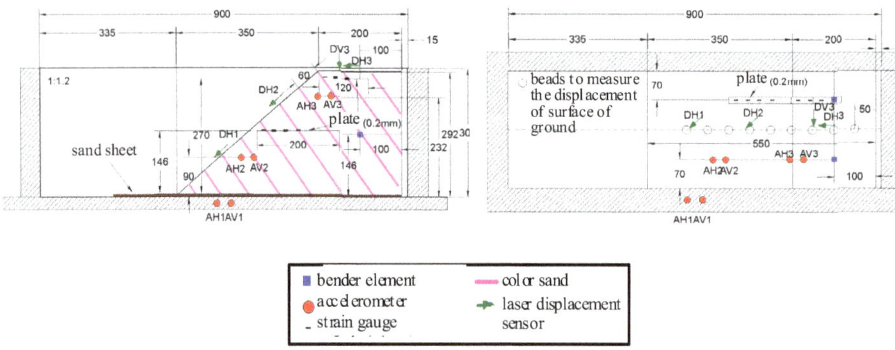

(a) Cross sectrion (unit: mm) (b) Plane section (unit:mm)

Fig. 11. Embankment model and measurement for dynamic centrifuge test

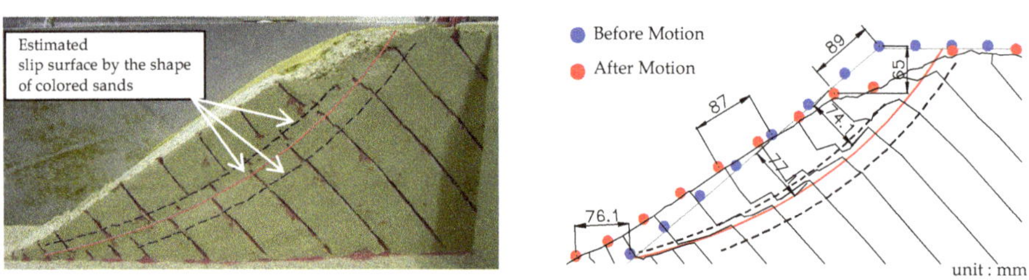

(a) Sliding plane occurred (b) Before and after dynamic motion

Fig. 12. Sliding failure in case of non-reinforced model

As shown in Fig. 13, four types of embankment models which are aimed the Crest Reinforcement Structure reinforced with changing the numbers of reinforced layers and the installed length of geosynthetics. The embankment modei in Case 1 is reinforced at the whole surface of crest with two layers by three long geosynthetics sheets which are fixed with the side wall. The one in Case 2 is reinforced with one layer by two long geosynthetics sheets and two layers with two short one which are set more deeply than the estimated sliding plane occurred in the case of nonreinforced embankment model shown in Fig. 12.

These two cases are supposed for the newly constructed embankment or the existing one to reinforce wholly. The ones in Case 3 and Case 4 are reinforced with three layers by four short geosynthetics sheets and two layers with three short ones, respectively. These two cases are supposed to apply to the existing embankment not to disturb the traffic activity under construction works.

In the above four cases, the thickness of both long and short reinforced layer is set to be 30mm which scale is 0.9m in the field. To ensure the total performance of reinforced zone, geosynthetics sheets are connected with each other at the surface of slope.

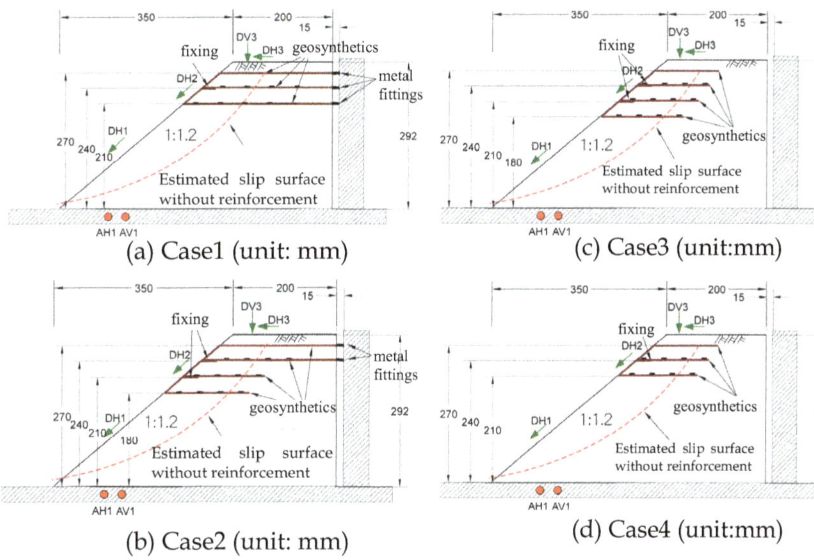

Fig. 13. Embankment Model Reinforced with Geosynthetics

Fig. 14 compares the slid conditions between four cases measured at the central cross section after inputting the sinusoidal motion with the maximum acceleration of about 500gal. It can be indicated qualitatively as follows based on the results of Fig. 14.

Case 1: The deformation on the crest of embankment is small and the sliding occurs at the area lower than the reinforced zone including the lower side of reinforced zone. The volume of sliding area is larger than that of Case 2.

Case 2: The deformation on the crest of embankment is small extremely and the sliding occurs at only the area lower than the reinforced zone which is not affected by the sliding. The effectiveness of the reinforcement can be known clearly in Photo. 10

Case 3: The deformation on the crest of embankment is very small and the sliding occurs at only the area lower than the reinforced zone which is not affected by the sliding.

Case 4: The deformation on the crest of embankment is occurred a little and the sliding occurs not only at the area lower than the reinforced zone but also at the boundary of reinforced zone which is reached to the surface of the crest.

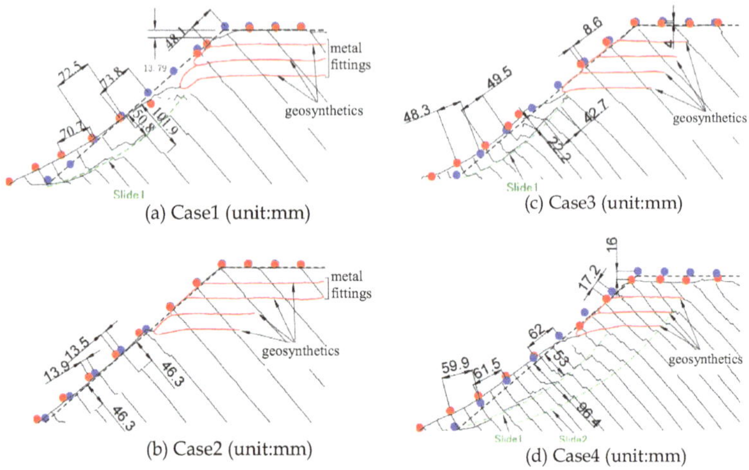

Fig. 14. Sliding failure in case of reinforced model with geosynthetics

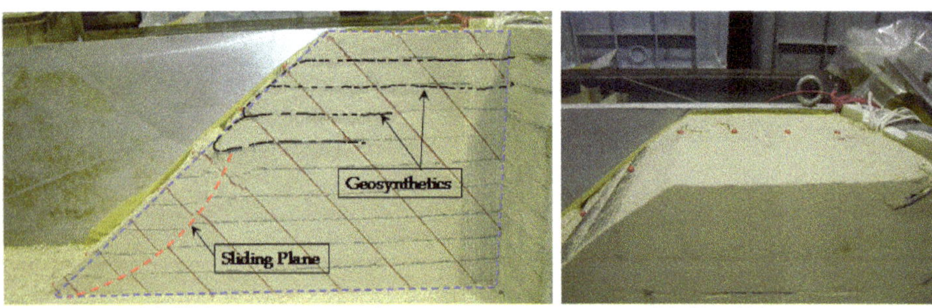

(a) Sliding at cross section (b) Condition at crest

Photo 10. Overview of deformation of reinforced embankment model in Case 2

The residual displacements measured at the shoulder and upper and lower slope of the four reinforced models comparing with the non-reinforced one in Fig. 12 are summarized in Fig. 15. Compared the reinforced models with the one without reinforcement, both horizontal and vertical residual displacements at the shoulder of the embankment model are decreased greatly in the whole reinforced models as shown in Fig. 15(a). Furthermore, the residual displacements along the upper-side slope are also decreased greatly as shown in Fig. 15(b).

However the residual displacements along the lower-side slope are not so smaller than that without reinforcement, it is not an important problem because the slid zone below the reinforced zone doesn't affect directly on the traffic performance will be repaired easily later.

As mentioned above, it can be indicated that the Crest Reinforcement Method with geosynthetics sheets is very effective for both newly constructed embankment and existing ones to get minimum traffic performance just after earthquakes. Some lessons can be obtained that the number of reinforced layers is very important for both long sheets and short sheets and the short sheets should be set deeper than the estimated sliding plane and almost fixed at the inside end.

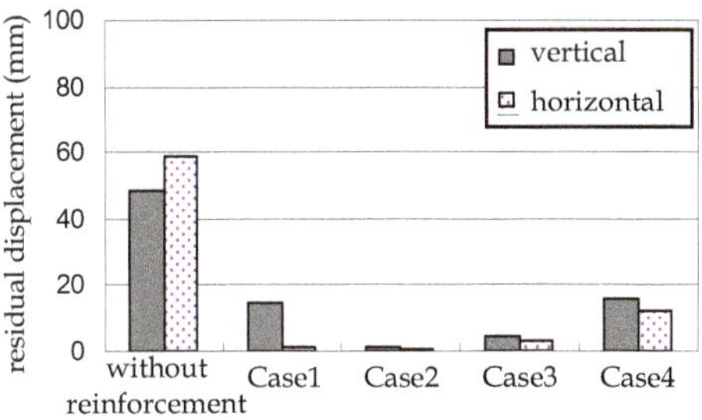

(a) At shoulder

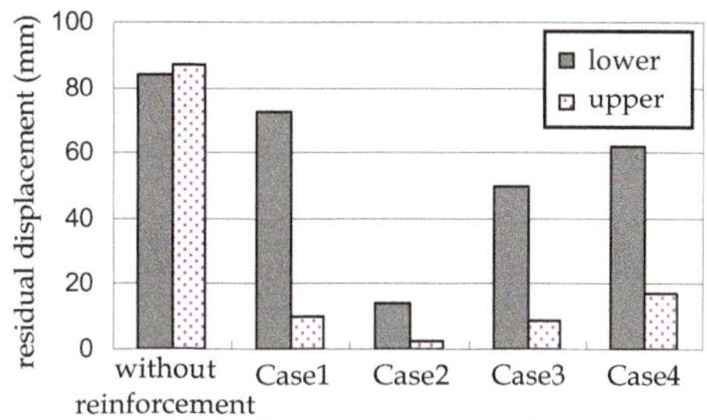

(b) At lower or upper slope

Fig. 15. Comparison of Residual Displacement

6.1.3 Analytical estimation

The example of analytical application of the Crest Reinforcement Structure with geosynthetics should be referred to Chapter 6.3.

6.1.4 Practical application in field

The Crest Reinforcement Structure has been applied at a certain site where the road embankment approaching to the other construction site is constructed temporally as shown in Fig. 16(a). The maximum height of the embankment is 20m and the slope ratio of that is 1:1.8. The soil materials used for the embankment body are classified into the silty clay with high natural water content of 50% and the average cohesion is 12.8kN/m^2 .as shown in Fig. 16(b). Therefore, because the embankment body is necessary to be improved to achieve the sliding safety factor of 1.2 in general conditions, the embankment body is improved to the shear strength of 74kN/m^2 with cements. Furthermore, the Crest Reinforcement Structure with geosynthetics as shown in Fig. 16(a) is designed additionally.

As shown in Fig. 16(a), one reinforced layer with two long sheets of 9m and two reinforced layers with two short sheets of 2m which are similar to the Case 2 in the centrifuge test in Fig. 13. The thickness of each layer is 0.5m. However the seismic design hasn't been applied in this case, it's very important that this case is the first trial to apply the Crest Reinforcement Structure in Japan as shown in Fig. 16(b) and can be obtained important lessons to construct the structures in the field.

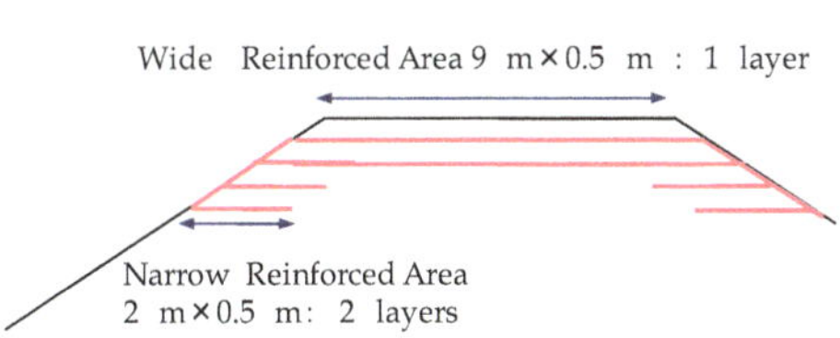

(a) Design condition for cross section

(b) Completed embankment

Fig. 16. Application of crest reinforcement structure to actual road

6.2 Toe reinforcement structure

In Japan, the groundwater in road embankments causes severe damage of road embankments due to earthquakes, and a lot of embankments to be damaged have been reported in recent years. A large-scale gabion is given as a typical structure of the Toe Reinforcement Structure in this Chapter. As for the large-scale gabion, the merits to apply it are low-cost, good construction workability, and there are a lot of actual results in the field as the disaster restoration method for the purpose of drainage as shown in Photo.11 (a).

(a) Disaster restoration (b) Moved gabions and occurred sliding

Photo 11. Application of large-scaled gabion at present

However the gabions were set only for drainage in advance, they could not work against the seismic motion as shown in Fig. 11(b). Then it is necessary and effective to apply the large-scale gabions for not only drainage but also reinforcement of the embankment.

For the reinforcing design of gabions, friction factors are necessary to be estimated. As for the friction factors between the base ground and the superstructure, Table 4 summarizes the static friction factors applied to the erosion cntrol dam, the retaining wall and the river wall in Japan based on Erosion Control & Land Slide Technical Center (2009), Japan Road Association (1999) and Japan River Association (1997), respectively. However Table 4 can be applied for the road embankment design, the drag force tests are tried as follows to investigate both static and dynamic friction factors for road embankments in detail.

Condition of Base Ground		Super-structure Condition on Base Ground				
		Erosion Control Dam		Retaining Wall		River Wall
		Gravel or Soil (φ_A)	Concrete	Concrete worked at Site	Pre-casted Concrete	Concrete Block or Gabion
Rock	(φ_B)	Smaller value of $\tan\varphi_A$ and $\tan\varphi_B$ Ex. $\varphi=30°$: $\tan\varphi=0.58$ $\varphi=35°$: $\tan\varphi=0.70$ $\varphi=40°$: $\tan\varphi=0.84$	0.7 ~ 1.2	$\tan\varphi_B$ or 0.7	$\tan\varphi_B$ or 0.6 $\leqq 0.6$	-
Blocked Rock, Stone			0.7	$\tan\varphi_B$ or 0.6		-
Gravel Layer			0.6			-
Sand Layer			0.55	$\tan\varphi_B$ or 0.6		0.65
Clay Layer			0.45	$\tan\varphi_B$ or 0.5	$\tan\varphi_B$ or 0.5 $\leqq 0.6$	

Table 4. Friction coefficients applied to several structures in Japan

6.2.1 Friction coefficient of large-scaled gabion

The field tests using full-scaled gabions were carried out and the effects on friction factors by the mesh-structure, the ground surface condition and the weight and/or load of gabions are discussed. Photo. 12 shows the three types of the conditions at the ground surface where the gabions are placed: a mat-placed ground, a sandy ground and a crushed-stone-layered ground, respectively. The sheet of a mat is permeable and possible to be set at the bottom of gabions to prevent fine soil particles to be flown out from the beneath the ground.

(a) Diamond-shaped (b) Grid-shaped (c) Ground surface condition to place gabions

Photo 12. Structure of gabion and ground surface condition

Structure of Gabion	Static Friction Coefficient : SFC Dynamic Friction Coefficient : DFC	Base Ground Condition		
		Crushed stone	Sandy soil	Mat- placed
Steel-made assembly mesh (diamond-shaped wire-mesh)	SFC	0.64	0.61	0.51
	DFC	0.59	0.54	0.48
Steel-made assembly mesh (grid-shaped wire-mesh)	SFC	0.79	0.77	0.63
	DFC	0.74	0.68	0.55

Table 5. Static and dynamic friction coefficient

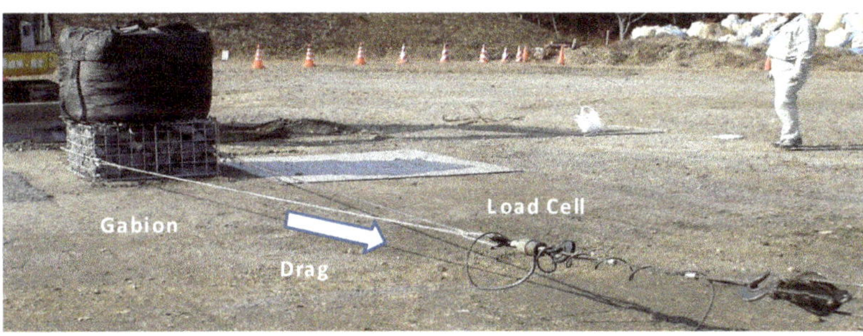

Photo 13. Typical method to drag gabion and measure drag-Load

Furthermore, three types of the weight and/or load condition are considered. The weight at Stage 1: 7.487kN are the same with that of the practical gabion and each weight at Stage 2: 15.249kN and at Stage 3: 21.619kN is about twice and three times the weight of the practical one, respectively. Photo.13 shows the typical method to drag a gabion and to measure the drag-load where the steel wire connected to the gabion is pulled by the winch which is set on the right side in Photo. 13.

Both the static friction coefficients and dynamic ones can be summarized in Table 5 from which the findings obtained are summarized as follows.

1. The static friction coefficients used in a present design differ depending on the structure of the large-scale gabion or the foundation ground condition. The static friction coefficients for all conditions range 0.51~0.79 and the ones of the diamond-shaped

wire-mesh and the grid-shaped one at the crushed stone and sandy soil ground range 0.61～0.74 and 0.77～0.79, respectively. Furthermore, it can be found that the ones at mat-placed ground are generally smaller than that at crushed-stone and sandy soil one. This indicates that it's necessary to pay attention to use a sheet of mat at the bottom of gabions.

2. The static friction coefficients obtained in these tests are same with or more than that in the cases of gravel layers and sand layers shown in Table 4.
3. On the other hand, the dynamic ones range 0.48～0.74. The relations of dynamic friction coefficients between mat-placed ground and crushed-stone ground or sandy soil ground are similar with that of static friction coefficients as mentioned above.

6.2.2 Seismic potential improvement with gabion

The centrifuge test is conducted on the effects by the Toe Reinforcement Structure with a rigid body shown in Fig. 17(a). The test conditions and measurement system are same with the former tests in Fig. 11. The toe structure set on the bottom of the container is not fixed. However the friction coefficient between the rigid body and the surface of the bottom is measured to be 0.78 in 30g, the rigid body isn't moved by dynamic motion with the maximum acceleration of 485 gal. Fig. 17(b) shows the displacements of the embankment model after shaking. It can be seen that the sliding plane is shallower and each displacement measured is smaller than that without reinforcement in Fig. 12. Furthermore, it can be indicated that it isn't necessary to fix the toe structure on the ground surface and it's enough to keep the high friction coefficient not to be slid. As the rigid body, the retaining wall and the large-scaled gabion etc. can be considered as the practical toe reinforcement structures.

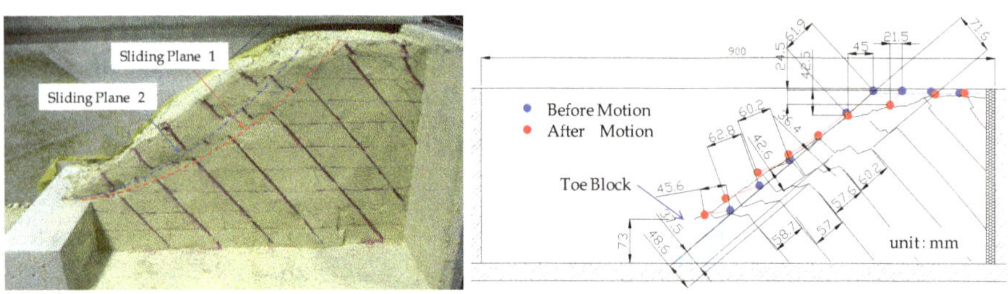

(a) Toe reinforced with rigid body (b) Residual displacement after shaking

Fig. 17. Centrifuge test on toe reinforcement structure

6.3 Artificial base-ground structure

The practical examples on the application of the Artificial Base-ground Structure and the Crest Reinforcement Structure with geosynthetics are introduced as follows. Fig. 18 shows a certain site same with Fig. 16 where the road embankment indicated by Layer 4 shown in Fig. 18(a) approaching to the other construction site is constructed temporally. Fig. 18(a) shows the soil conditions in each layer, i.e. Layer 1 and Layer 2 are natural rigid base grounds, and Layer 3 is a little soft clayey deposit. The groundwater table is shown as the blue line. The soil conditions of each layer are estimated in Fig. 18(b) based on the soil investigation.

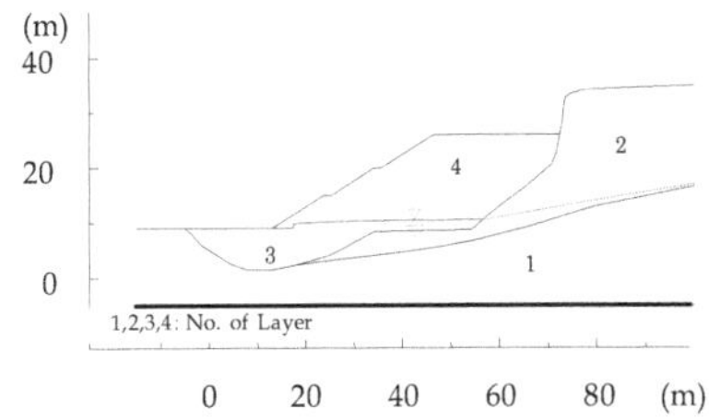

Layer No.	Saturated Unit Weight (kN/m³)	Wet Unit Weight (kN/m³)	Friction Angle (°)	Cohesion (kN/m²)
1	18.0	18.0	0.0	500.0
2	17.0	17.0	0.0	100.0
3	16.0	16.0	0.0	28.5
4	17.0	17.0	0.0	12.8

(a) Cross section (b) Conditions of soil profile

Fig. 18. Analytical model of in-situ field shown in Fig. 16

Four analytical cases of the reinforced embankment model are prepared as shown in Fig. 19. Table 6 shows the design conditions for the geosynthetics to reinforce the crest of the embankment and/or the improved strength of the embankment body as the artificial base-ground where the sliding plane don't pass through.

Case	Geosynthetics		Artificial Base-ground	
	Number of Sheet	Tensile Strengh (kN/m²)	Improved Area (m²)	Cohesion (kN/m²)
1	-	-	410	120
2	-	-	385	169
3	10	150	250	150
4	11	50	245	169

Table 6. Reinforcement conditions

The seismic ground motion measured at Togi City during the Noto Peninsula Earthquake in 2007 which is same with the one in Fig. 7(a) is applied with the maximum acceleration of 508gal converted at the base ground.

The analytical results on the sliding displacement and reach length identified as shown in Fig. 19(a) are shown in Fig. 19. These results can be summarized as follows.

Case 1: However the reach length of 12.8m is large, the sliding displacement of 0.16m can be reduced to be small. The sliding displacement of 0.16m can be judged to be Level 2 performance identified in Table 2.

Case 2: However the sliding displacement of 0.69m is very large, the reach length of 1.7m can be reduced to be small. Because the reach length of 1.7 is located around the shoulder of the embankmentt, Level 2 performance identified in Table 2 can be expected.

Case 3: However the reach length of 23.4m is very large, the sliding displacement of 0.08m can be reduced to be very small. The sliding displacement of 0.08m can be judged to be Level 2 performance identified in Table 2.

Case 4: The sliding displacement of 0.23m is small a little, the reach length of 2.4m can be also reduced to be small a little. Both the sliding displacement and the reach length can show Level 2 performance identified in Table 2.

The construction costs corresponding to the unit length in the longitudinal direction of the embankment among these four cases can be ordered in Case 4, Case 1, Case 2 and Case 3 from the low cost to the high one. Because the cost ratio of Case 3 against Case 4 is about 1.13, Case 4 can be indicated to be the best reinforcement plan based on the seismic performance-based design.

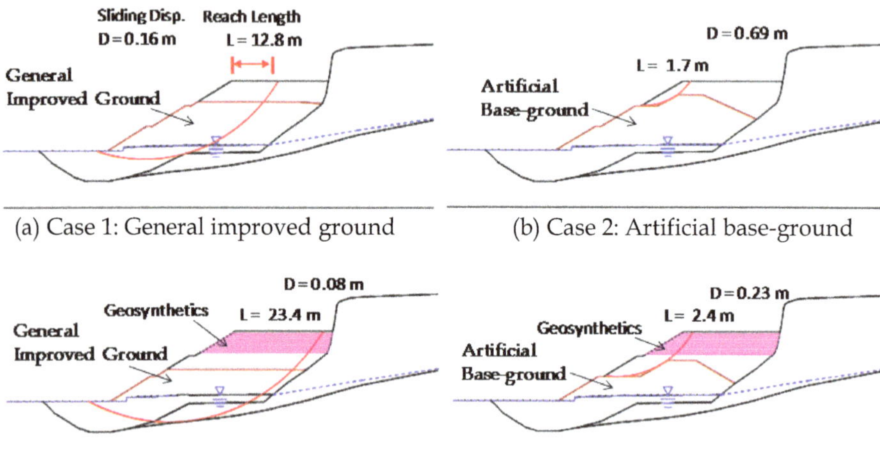

(a) Case 1: General improved ground

(b) Case 2: Artificial base-ground

(c) Case 3: General improved ground and geosynthetics

(d) Case 4: Artificial base-ground and geosynthetics

Fig. 19. Sliding characteristics according to reinforcement structures

6.4 Barrier reinforcement structure

The centrifuge test is conducted on the effects by Barrier Reinforcement Structure with a rigid plate shown in Photo. 14. The test conditions and measurement system are same with the former tests in Fig. 11. The under edge of the barrier structure is set to be rotated and not to be moved horizontally, i.e. assumed to be a hinge. The upper edge of the one is free and not to be connected with the side wall of the container.

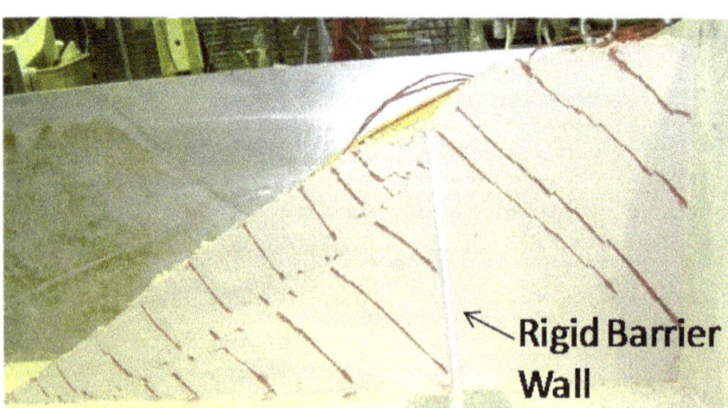

Photo 14. Centrifuge test on the Barrier Reinforcement Structure

It can be known that however the rigid plate is inclined a little toward the slope, the sliding plane can be controlled toward the shoulder of the crest where the traffic performance isn't affected severely. If the upper edge is fixed, the effectiveness of the barrier may become larger. The Barrier Reinforcement Structure can be assumed to be effective to decrease the residual displacement at the crest of road embankments. As the rigid plate, the sheet piles and the reinforced concrete wall etc. can be considered as the practical barrier reinforcement structures.

7. Conclusion

Lessons obtained in this chapter can be summarized as follows.

1. Based on the damages of road embankments induced by such typical earthquakes as the 2004 Mid-Niigata Prefecture Earthquake, the 2007 Noto Peninsula Earthquake, the 2009 Suruga Bay Earthquake and the 2011 off Pacific coast of Tohoku Earthquake, several important lessons for the performance-based design of road embankments can be obtained from the engineering view points.
2. The macroscopic seismic assessment method proposed can be applied to estimate the seismic risk potential of the existing road embankments for the future detail investigation and seismic reinforcement plan.
3. The performance level proposed for the performance-based design of road embankments can be applied to set the objective performance level of the sliding failure of road embankments in the design of both new embankments and existing ones in the future.
4. The practical performance-based design concepts such as "Crest Reinforcement Structure", "Barrier Reinforcement Structure", "Artificial Reinforcement Structure" and "Toe Reinforcement Structure" can be proposed and their applications are discussed and clarified qualitatively and/or quantitatively.

8. Acknowledgment

The research and development on the economical seismic reinforcement technology considering the performance of the road embankments conducted in 2007 to 2009 have been supported by the Committee of New Highway Technology presided by Ministry of Land, Infrastructure and Transport. The author would like to express grateful acknowledgement to the Committee members and the cooperative researchers: Dr. Akinori Nakahira of CTI Engineering Co. Ltd., Dr. Kenji Hayashi of Forest Engineering Inc., Dr. Naoki Tatta of Maedakosen Co. Ltd., Dr. J. Jang of Shimizu Corporation and Mr. Yusuke Egawa of CTI Engineering Co. Ltd..

9. References

Erosion Control & Land Slide Technical Center (2009). *Design Guideline for Erosion Control Steel Structure*, pp.32 (in Japanese) Japan Meteorological Agency (JMA). http://jma.go.jp/jma/press/1103/11g/kaisetsu201103112200_pdf (in Japanese)

Japan River Association (1997). *Technical Code -Design II- for River & Erosion Control Structure*, pp.40 (in Japanese)

Japan Road Association (1999). *Design Criteria for Retaining Wall*, p.21 (in Japanese)

National Research Institute for Earth Science and Disaster Prevention (NIED). http://www.bosai.go.jp/kyoshin/topics/html20110311144626/main_20110311144626.html (in Japanese)

Newmark, N. M. (1965). Effects of earthquakes on dams and embankments, *Geotechnique*, Vo.15, No.2, pp.139-159

Tateyama, M., Tatsuoka, F., Koseki, J. & Horii, K. (1998). Research on Seismic Design Method of Embankment, *Report of Railway research Institute*, Vo.12, No.4, pp.7-12 (in Japanese)

Tatta, N., Jang, J., Tokida, K., Oda, K. & Nakahira, A. (2009). Experimental Study on Seismic Reinforcement Method at Crest of Road Embankment by Geosynthetics, *Proceedings of The International Symposium on Earthquake Geotechnical Case Histories for Performance-Based Design*, Is-Tokyo, pp.963-970, Tsukuba, Japan, June 15-18

Tatta, N., Jan, J., Tokida, K., Oda, K., Nakahira, A. (2010). Seismic Reinforcement Method at Crest of Road Embankment by Geosynthetics, *Proceedings of The 9th International Conference on Geosynthetics*, IGC2010, pp.1611-1618, Guaruj&aacute, Brazil, May 23-27

Tokida, K., Oda, K., Nabeshima, Y. & Egawa, Y. (2007). Damage Level of Road Infra-structure and Road Traffic Performance in The Mid Niigata Prefecture Earthquake of 2004, *Structural Engineering/Earthquake Engineering*, JSCE, Vol. 24, No. 1, pp.51-61.

Tokida, K., Oda, K. & Nakahira, A. (2008). Research and Development on Economical Seismic reinforcing Technology for Road Embankment Based on Performance-based Concepts, *Report of Research and Development for Improving Highway Policy*, Committee of New Highway Technology of Ministry of Land, Infrastructure and Transport, No.17-4, pp.1-47 (in Japanese)

Tokida, K., Oda, K., Hayashi, K. & Nakahira, A. (2009). Macroscopic Aseismic Assessment Method of Road Embankment, *Proceedings of The International Symposium on Earthquake Geotechnical Case Histories for Performance-Based Design*, Is-Tokyo, pp.1633-1638, Tsukuba, Japan, June 15-18

Tokida, K., Chen, W.Z, Nakahira, A., Echigawa, H., Minamimoto, M., Hayashi, K. (2010). Field Test on Friction Factor between Gabion and Ground, *Proceedings of The 4th Japan-China Geotechnical Symposium*, pp.141-147, Okinawa, Japan, April 12-14

13

Seismic Protection of Monolithic Objects of Art Using a Constrained Oscillating Base

Alessandro Contento and Angelo Di Egidio
Department of Structural, Hydraulic and Geotechnical Engineering, University of L'Aquila Italy

1. Introduction

The model of rigid block is well known in literature. In the past, several papers analyzed the behaviour of rigid blocks under different kind of excitations because many monolithic objects of art, such as statues, obelisks and fountains, subject to earthquake excitation, can be modelled as rigid blocks. In [Shenton & Jones, 1991] a general bi-dimensional formulation of the rigid block has been obtained and rocking and slide-rock approximated conditions have been written. More recently this model has been used to describe the behaviour of monolithic bodies subject to base excitations as a one-sine pulse excitation in [Zhang & Makris, 2001; Makris & Black, 2004, Kounadis 2010] and earthquake excitation in [Agbabian et al., 1988; Pompei et al, 1998; Taniguchi, 2002]. Almost all the papers on rigid blocks subject to base excitation focus their attention on symmetric rigid bodies. Only a few papers concern non-symmetric rigid bodies that, usually, represent objects of art better than symmetric rigid blocks. In [Boroscheck & Romo, 2004] the influence of the eccentricity of the centre of mass on the motion of the system has been studied. In [Purvance, 2005; Purvance et al., 2008] an analytical and experimental estimation of overturning events under seismic excitations has been carried out, both for symmetric and non-symmetric rigid bodies. In particular, in [Zhang & Makris, 2001] for a one-sine pulse excitation and in [Purvance, 2005; Purvance et al., 2008] for seismic excitation, the existence of survival regions that lie above the *PGA* (Peak Ground Acceleration) associated with the first overturning occurrence have been shown. In recent years, methods to reduce the effects of seismic excitation on art objects have been studied in some papers. In [Fujita et al, 2008] a critical excitation problem for a rigid block subjected to horizontal and vertical simultaneous base inputs is considered. In [Vestroni & Di Cinto, 2000] a base isolation system has been used to protect statues from seismic effects. The work of art has been modeled through an equivalent elastic beam. In [Caliò & Marletta, 2003] the same problem has been analyzed, but the art object has been modeled as a symmetric rigid block simply supported on an oscillating base connected to the ground by a visco-elastic device. The sliding of the body is prevented by special seismic restraints. These analyses have shown the effectiveness of the isolation system and the role of many parameters. To make things more realistic, in [Contento & Di Egidio, 2009], the model presented in [Caliò & Marletta, 2003] has been enriched considering also the eccentricity of the centre of mass of the rigid body and the presence of security stops, able to prevent the breaking of the isolation device by limiting the displacement of the oscillating base to a maximum safety value. More recently, in [Di Egidio & Contento, 2009; Di Egidio & Contento, 2010], they have introduced sliding effects, that make the model able to carry out more complicated motions like slide-rocking, and sliding constrains, that prevent the rigid body from falling off the base. Alternatively, the possibility for the rigid block to be partially removed from the oscillating base has been considered. Security stops to avoid damaging the base isolation system are also considered. The behaviour of the whole system is studied under two types of excitations: impulsive and seismic. Exact nonlinear equations of motion are written for the different phases of motion: full-contact, sliding, rocking, slide-rocking; transition phase conditions are obtained by generalizing to the case under analysis those obtained in [Taniguchi, 2002; Caliò & Marletta, 2003; Contento & Di Egidio, 2009]. To describe the motion when the body is outside the oscillating support, original equations of motion, describing the rocking and the slide-rocking motions of the rigid block around a different point from one of the corners of its base, are obtained. The influence of the friction coefficient, of the eccentricity, of the security stops and of other parameters are analyzed by performing an extensive parametric analysis via a direct numerical integration of the equations of motion. Comparison between results obtained for isolated rigid body and non-isolated rigid body are also carried out to show the effectiveness of base isolation with respect to the absence of this passive control system. Another possibility, not considered here, is the use of three-dimensional models of rigid blocks, mostly circular based. They are used to study the sloshing in circular shaped tanks [Taniguchi, 2004], the wobbling motion [Stefanou et al., 2011] and the motion of a disk of finite thickness on a planar surface [Koh & Mustafa, 1990; Batista 2006].

2. Description of the object considered and model hypotheses

The two-dimensional model presented is that of an isolated rigid body, where the isolation system is an oscillating base connected to the ground by a linear visco-elastic device. The geometric dimensions of the rigid body are taken with ratios similar to those of real works of art and its characteristics are reported in Fig.1(a). With respect to the traditional models presented in literature, here, also the eccentricity of the centre of mass of the rigid body is considered: being C and O the centres of mass of the body and of the base respectively and M the middle point of the base, the eccentricity $e = b_1 - b$ is positive when the vertical projection of the centre of mass C is shifted on the right side of M. In Fig.1(b) parameters characterising the base isolation are shown, where c and k are the damping and the stiffness of the linear visco-elastic device representing the base isolation system.

The rigid body here considered can undergo different kinds of motion: (a) full-contact motion where the rigid body remains in contact with the isolated base when it oscillates, (b) sliding motion where the rigid body slides on the oscillating base, (c) rocking motion where the rigid body rocks around one of the two bottom corners without sliding and (d) slide-rocking motion where a combination of rocking and sliding motion occurs (Fig.2). Three Lagrangian parameters are used to describe the different phases of motion (Fig.2): $u(t)$ that is the translation of the oscillating base, $x(t)$ that is the translation of the body due to slide with respect to the isolated base and $\vartheta(t)$ that is the rotation of the body around one of the bottom corners and can be positive or negative if the body rocks around the left or the right corner respectively. Quantities u_g and v_g, shown in Fig.1(b), are the horizontal and vertical ground displacements respectively. Since there are no constraints the rigid block can come partially away from the base so that the rocking motion can occur not only around one of the corners of the rigid body but also around one of the corners of the isolating base. For the same reason one of the possible collapse condition of motion is the body falling off the base. The other is the overturning of the body. In the model, frictional forces are expressed by using the Coulomb description and the friction coefficient has been varied in the different analyses carried out.

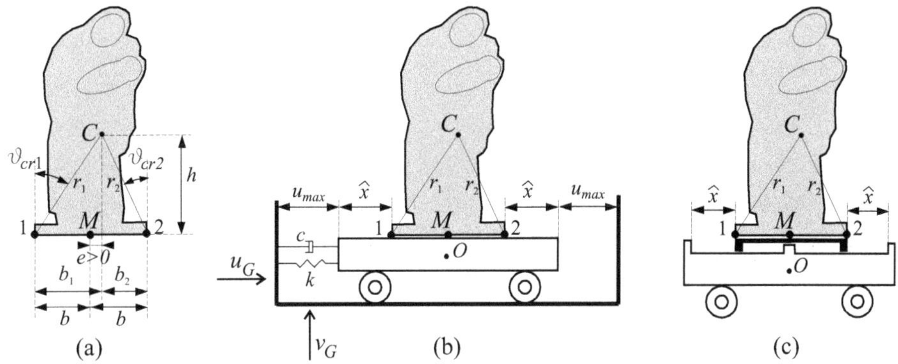

Fig. 1. Mechanical system: (a) geometrical parameters of the rigid body; (b) parameters of the base isolation system without security stops for the sliding motion; (c) security stops for the sliding motion.

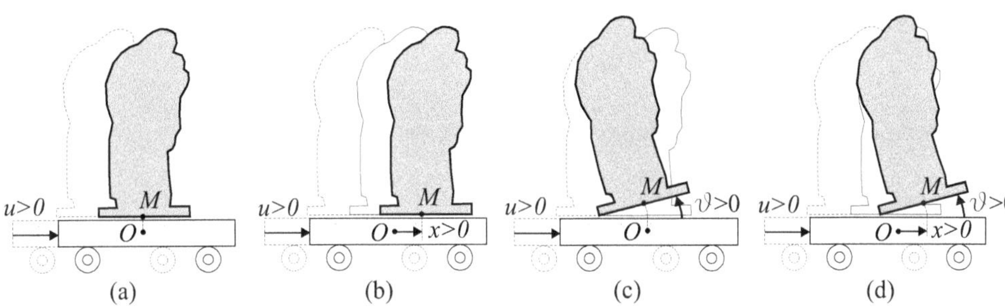

Fig. 2. Phases of motion: (a) full-contact; (b) sliding; (c) rocking; (d) slide-rocking.

3. General formulation

In this section the equations of motion for the different phases of motion, the transition conditions among them and the impact conditions are derived.

3.1 Equations of motion

Exact nonlinear equations of motion are written by using a Lagrangian approach for the different phases of motion.

3.1.1 Full-contact motion

The full-contact motion equation is the linear equation of a single degree of freedom system that represents the equilibrium of all forces acting on the system in the horizontal direction:

$$(m+m_b)(\ddot{u}+\ddot{u}_G)+c\dot{u}+ku=0 \tag{1}$$

where m and m_b are the mass of the rigid block and the mass of the isolated base, respectively.

3.1.2 Sliding motion

During a pure sliding motion kinetic energy T and potential energy V of the system can be written respectively:

$$T=\frac{1}{2}m\dot{u}_C^2+\frac{1}{2}m_B\dot{u}_O^2; \qquad V=\frac{1}{2}k(u_O-u_G)^2 \tag{2}$$

where u_c and u_o are the positions of the centres of the body C and of the base O expressed as functions of two Lagrangian parameters (Fig.3(a)):

$$u_C(t)=u_G(t)+u(t)+x(t), \qquad u_O(t)=u_G(t)+u(t) \tag{3}$$

Lagrange's equations of motion are derived as follows:

$$\begin{gathered} m(\ddot{u}+\ddot{u}_g+\ddot{x})+F=0 \\ (m+m_b)(\ddot{u}+\ddot{u}_g)+m\ddot{x}+c\dot{u}+ku=0 \end{gathered} \tag{4}$$

where F is the Coulomb kinetic friction force given by:

$$F=\mu_k\, sign(\dot{x})m(g+\ddot{v}_g) \tag{5}$$

and where μ_k is the kinetic friction coefficient and g the gravity acceleration.

3.1.3 Rocking motion

For the rocking motion, due to the fact that the model is non-symmetric, two different sets of equations of motion are necessary, depending on which corner the motion occurs. When the relative displacement $x\in[-\hat{x},\hat{x}]$, the base of the body is inside the oscillating support (see Fig.1(b)). In this case the positions of the centres of the rigid body C and of the base O, when a rocking around the corner 1 occurs, can be obtained by referring to Fig.3(b):

$$\begin{gathered} u_C(t)=u_G(t)+u(t)+(b_1'-\hat{b}_1) \\ v_C(t)=v_G(t)+(h_1'-h) \\ u_O(t)=u_G(t)+u(t) \end{gathered} \tag{6}$$

Kinetic energy, potential energy and vitual work δW of the non-conservative forces then read:

$$\begin{gathered} T=\frac{1}{2}m(\dot{u}_C^2+\dot{v}_C^2)+\frac{1}{2}m_b\dot{u}_O^2+\frac{1}{2}I_C\dot{\vartheta}^2; \\ V=\frac{1}{2}ku^2+mg(h_1'-h); \\ \delta W=-c\dot{u}(t)\delta u. \end{gathered} \tag{7}$$

Lagrange's equations of motion read:

$$\begin{aligned}(m+m_b)(\ddot{u}+\ddot{u}_G)+c\dot{u}+ku-m\ddot{\vartheta}h_1'-m\dot{\vartheta}^2b_1'=0\\ I_1\ddot{\vartheta}-m(\ddot{u}+\ddot{u}_G)h_1'+m(g+\ddot{v}_G)b_1'=0\end{aligned} \tag{8}$$

where

$$b_1'=\hat{b}_1\cos(\vartheta)-h\sin(\vartheta);\ h_1'=\hat{b}_1\sin(\vartheta)+h\cos(\vartheta) \tag{9}$$

are the horizontal and vertical distances between centre of mass and corner 1 (that is the centre of rotation) in the actual position, after rotation (see Fig.3(b)), and where $\hat{b}_1=b_1$ (see Fig.1(a)). In Eqs.(8) $I_1:=I_C+mr_1^2$ is the polar inertia around corner 1; I_C is the polar inertia with respect to centre C of the body. The first equation of motion of Eqs.(8) represents the equilibrium of the forces acting on the body together with the base in the horizontal direction, while the second represents the equilibrium of the moments around the corner on which the rotation occurs.

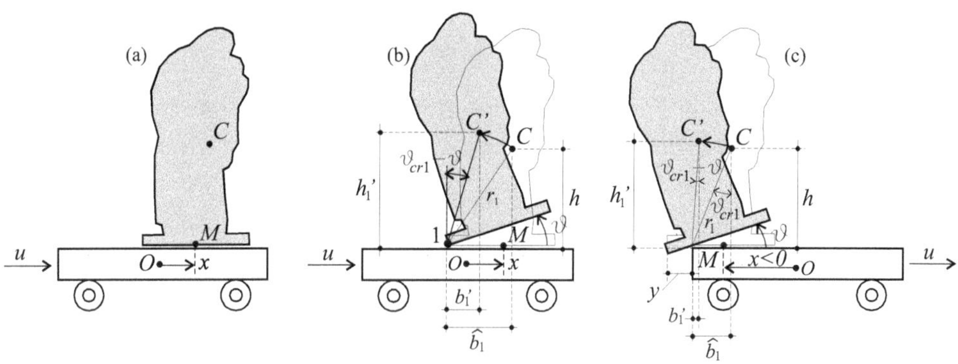

Fig. 3. Geometrical description of the different phases of motion: (a) sliding; (b) (slide-) rocking around corner 1; (c) (slide-)rocking where the left corner of the body is outside the oscillating base.

When the variable $x\notin[-\hat{x},\hat{x}]$, the base of the rigid body is outside the oscillating support. In this case the centre of rotation of the body is located in one of the corners of the oscillating support (see Figs. 3(c)). The positions of the centres of the rigid body C and the base O, when a rocking around the left corner of the support occurs ($-\hat{x}>x$), are equal to Eqs.(6) where, however, $\hat{b}_1=b_1+y$ with $y=x+\hat{x}$ ($x<0$ not depending on time, see Fig.3(c)). By using Eqs.(7) it is possible to obtain Lagrange's equations of motion equal to Eqs.(8). In this case I_1 is the polar inertia around the left corner of the support; b_1' and h_1' represent the horizontal and vertical distances between the centre of mass and the left corner of the oscillating support (and not the left corner of the rigid body as in the previous case, see Fig.3(c)). They are still given by Eqs.(9). The equations of motion of rocking around the right corner of the base or of the support can be obtained in a similar way (cases $x\in[-\hat{x},\hat{x}]$ and $x>\hat{x}$ respectively).

3.1.4 Slide-rocking motion

Also for the slide-rocking motion, due to the non-symmetry of the rigid body, two different sets of equations of motion are necessary, depending on which corner the rotation occurs. When during the motion the relative displacement $x \in [-\hat{x}, \hat{x}]$, by evaluating kinetic and potential energies and taking into account the virtual work of the generalized forces given by Eq.(7), Lagrange's equations of motion around the left corner can be found and read:

$$\begin{aligned} m(\ddot{u} + \ddot{u}_G + \ddot{x}) - m\ddot{\vartheta}h_1' - m\dot{\vartheta}^2 b_1' + F_1 = 0 \\ (m + m_b)(\ddot{u} + \ddot{u}_G) + m\ddot{x} + c\dot{u} + ku - m\ddot{\vartheta}h_1' - m\dot{\vartheta}^2 b_1' = 0 \\ I_1\ddot{\vartheta} - m(\ddot{u} + \ddot{u}_G + \ddot{x})h_1' + m(g + \ddot{v}_G)b_1' = 0 \end{aligned} \tag{10}$$

where b_1' and h_1', given by Eqs.(9), are the horizontal and vertical distances between the centre of mass and corner 1 (that is the centre of rotation) in the actual position, after rotation (see Fig.3(b)), and where $\hat{b}_1 = b_1$. Eq.(10_1) represents the equilibrium in the horizontal directions of all the forces acting on the rigid body (Fig.4(b)), while Eqs.(10_2) and (10_3) have the same meaning of Eqs.(8). The Coulomb kinetic friction force F_1 is given by:

$$F_1 = \mu_k \, sign(\dot{x}) Y_1 \tag{11}$$

where the vertical reaction Y_1 can be obtained by evaluating the vertical component of the total force acting on the rigid body (Fig.4(b)):

$$Y_1 = m(g + \ddot{v}_G) + mb_1'\ddot{\vartheta} - mh_1'\dot{\vartheta}^2 \tag{12}$$

The equations of motion of rocking around corner 2, in the case in which the relative displacement $x \in [-\hat{x}, \hat{x}]$ can be obtained similarly.

When the variable $x \notin [-\hat{x}, \hat{x}]$, the centre of rotation of the body is located in one of the corners of the oscillating support (see Fig.3(c)). It is possible to obtain Lagrange's equations of motion for the left corner:

$$\begin{aligned} m(\ddot{u} + \ddot{u}_G)\cos(\vartheta) + m\ddot{x} + m(g + \ddot{v}_G)\sin(\vartheta) - m\ddot{\vartheta}h - m\dot{\vartheta}^2\hat{b}_1 + F_1 = 0 \\ (m + m_b)(\ddot{u} + \ddot{u}_G) + m\ddot{x}\cos(\vartheta) + c\dot{u} + ku - m\ddot{\vartheta}h_1' - m\dot{\vartheta}^2 b_1' - 2m\dot{x}\dot{\vartheta}\sin(\vartheta) = 0 \\ I_1\ddot{\vartheta} - m(\ddot{u} + \ddot{u}_G)h_1' - m\ddot{x}h + m(g + \ddot{v}_G)b_1' + 2m\dot{x}\dot{\vartheta}\hat{b}_1 = 0 \end{aligned} \tag{13}$$

In this case I_1 is the polar inertia around the left corner of the support; b_1' and h_1' represent the horizontal and vertical distances between the centre of mass C and the left corner of the oscillating support (and not the left corner of the rigid body as in the previous case, see Fig.3 (c)). They are still given by Eqs.(9). It is interesting to observe that the first of Eqs.(13) represents the equilibrium of the forces acting during the slide-rocking motion along the slide direction parallel to the base of the body, when it is in a rotated configuration. The interpretation of the other two equations of motion do not change with respect to the previous case. A representation of the forces acting on the system during the slide-rocking motion around the left corner of the support is reported in Fig.4(c). The kinetic friction force

F_1 is still given by Eq.(11) where Y_1 represents the total force orthogonal to the inclined slide direction and it can be obtained by the forces acting on the rigid body (Fig.4(c)):

$$Y_1 = m(g+\ddot{v}_G)\cos(\vartheta) - m(\ddot{u}+\ddot{u}_G)\sin(\vartheta) + 2m\dot{x}\dot{\vartheta} + m\hat{b}_1\ddot{\vartheta} - mh\dot{\vartheta}^2 \quad (14)$$

It is interesting to observe that in this case the Coriolis force $(-2m\dot{x}\dot{\vartheta})$ appears in the system due to the fact that the centre of rotation of the rigid body changes its position with respect the body during the motion.

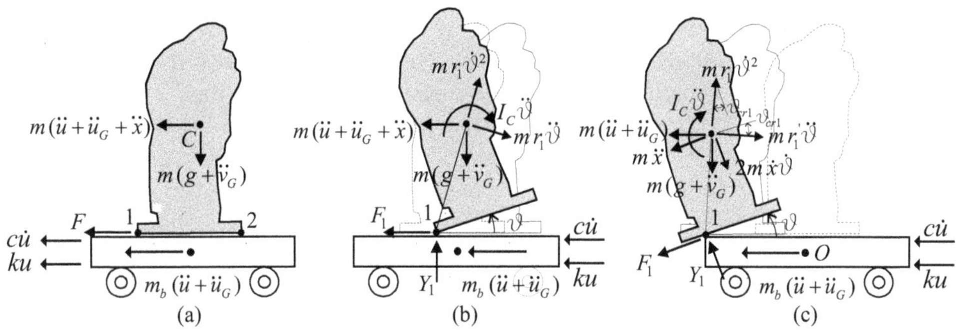

Fig. 4. Forces acting on rigid body: (a) during a sliding motion; (b) during (slide-)rocking around corner 1; (c) during (slide-)rocking where the left corner of the body is outside the oscillating base.

3.2 Transition and termination conditions

Transition phase conditions are obtained by the equilibrium among forces or moments acting on the rigid body during the motion. In the following only the case related to motions around the left corner is presented while transition and termination conditions evaluated for a motion occurring around the right corner can be found similarly.

3.2.1 Sliding motion

A pure sliding phase can take place starting from the rest or from the full-contact phase. By referring to the forces acting on the system during a full-contact phase shown in Fig.4(a) with $\ddot{x}=0$, a sliding phase can occur when the static frictional force $F_s = \mu_s m(g+\ddot{v}_G)$ is lesser than, or equal to, the absolute value of the horizontal inertial force $m(\ddot{u}+\ddot{u}_G)$. This condition reads:

$$\mu_s(g+\ddot{v}_G) \le |\ddot{u}+\ddot{u}_G| \quad (15)$$

where μ_s is the static friction coefficient. Equation (15) refers to starting condition from the rest when $\ddot{u}=0$ at $t=0$.

It is assumed that the termination condition of a sliding phase corresponds to the vanishing of the sliding velocity $|\dot{x}|=0$.

3.2.2 Rocking motion

The pure rocking phase can take place from the rest or the full-contact phase. A rocking phase can occur when the resisting moment M_R due to the vertical accelerations $m(g+\ddot{v}_G)$ of the body is smaller than the overturning moment M_O due to inertial force $m(\ddot{u}+\ddot{u}_G)$ (see Fig.4(a) with $\ddot{x}=0$). Considering the moments around the left corner (of the body or of the support), the following equation can be obtained:

$$\ddot{u}+\ddot{u}_G \geq \frac{\hat{b}_1}{h}(g+\ddot{v}_G) \tag{16}$$

where $\hat{b}_1 = b_1$ when the relative displacement $x \in [-\hat{x},\hat{x}]$ during the full-contact motion, (see Fig.3(b)). On the contrary if the relative displacement $-\hat{x} > x$, then $\hat{b}_1 = b_1 + y$ with $y = x + \hat{x}$ ($x < 0$, see Fig.3(c)). Equation (16) refers to a starting condition from the rest when $\ddot{u}=0$ at $t=0$ (and $x=0$). There is a favoured direction for the beginning of the rocking phase associated with the sign of the total acceleration $\ddot{u}+\ddot{u}_G$. In some cases when $\hat{b}_1 = 0$ or $\hat{b}_2 = 0$, which are the cases where the centre of mass is placed on the vertical projection of the centre of rotation of the body, an unstable equilibrium position manifests itself.

No particular conditions are assumed to describe the termination condition of a rocking phase. This means that the rocking phase finishes when the energy associated to this phase is completely dissipated.

3.2.3 Slide-rocking motion

A slide-rocking motion can take place following the rest condition, the full-contact phase, the pure sliding phase or the pure rocking phase. Depending on which phase it starts, different starting conditions must be considered. First the starting condition of a rocking phase during a sliding motion will be analyzed. By referring to forces acting on the system during a sliding phase shown in Fig.4(a), a rocking phase can occur when the resisting moment M_R due to the vertical accelerations $m(g+\ddot{v}_G)$ of the body is smaller than the overturning moment M_O due to inertial force $m(\ddot{u}+\ddot{u}_G+\ddot{x})$. Considering the moments with respect to the left corner (of the body or of the support), the following equation can be obtained:

$$\ddot{u}+\ddot{u}_G+\ddot{x} \geq \frac{\hat{b}_1}{h}(g+\ddot{v}_G) \tag{17}$$

where, if the displacement $x \in [-\hat{x},\hat{x}]$, then $\hat{b}_1 = b_1$ (see Fig.3(b)); if the relative displacement $-\hat{x} > x$, then $\hat{b}_1 = b_1 + y$ with $y = x + \hat{x}$ ($x < 0$, see Fig.3(c)). There is a favoured direction for the beginning of the rocking phase also in this case. The starting condition of a sliding phase during a rocking motion can be evaluated by referring to the forces acting on the rigid body during a rocking motion shown in Fig.4(b) with $\ddot{x}=0$. When the body is rocking around the left corner (of the body or of the support), the starting condition for a slide can be easily written as:

$$\mu_s Y_1 \leq |X_1| \tag{18}$$

where X_1 and Y_1 are the reactions of the base as in Fig.4(b),(c). If the displacement $x \in [-\hat{x}, \hat{x}]$, force Y_1 is given by Eq.(12), while X_1 can be evaluated by computing the total horizontal force acting on the rigid body. By referring to forces in Fig.4(b) (with $\ddot{x} = 0$), it reads:

$$X_1 = m(\ddot{u} + \ddot{u}_G) - mh_1'\ddot{\vartheta} - mb_1'\dot{\vartheta}^2 \tag{19}$$

When $-\hat{x} > x$, the force Y_1 is given by Eq.(14) with $\dot{x} = 0$, while X_1 can be evaluated by computing the total force acting on the rigid body along the sliding direction. By referring to forces in Fig.4(c) (with $\dot{x} = \ddot{x} = 0$), it reads:

$$X_1 = m(\ddot{u} + \ddot{u}_G)\cos(\vartheta) + m(g + \ddot{v}_G)\sin(\vartheta) - mh\ddot{\vartheta} - m\hat{b}_1\dot{\vartheta}^2 \tag{20}$$

Finally the starting condition of a slide-rocking phase from the full-contact phase is analyzed. A slide-rocking motion can take place from this phase if pure sliding and rocking conditions occur simultaneously. In particular a slide-rocking around the left corner (of the body or of the support) occurs when Eq.(15) and Eq.(16) are simultaneously proven. A slide-rocking phase takes place directly from the rest if the same previous equations, with $\ddot{u} = 0$ at $t = 0$, are simultaneously proven ($\hat{b}_1 = b_1$).

Also in this case, the termination condition of the sliding motion is taken to be $|\dot{x}| = 0$. When this condition is satisfied only the rocking motion remains active. The termination of the rocking motion is associated with the total dissipation of the energy associated to this phase; in this case only the sliding motion remains.

3.3 Impact conditions

Both during pure rocking motion and during slide-rocking motion an impact among the rigid block and the isolated base occurs when the angle ϑ approaches zero. No bouncing phenomenon is taken into account. Post impact quantities can be found, assuming that the impact happens instantly and the body position remains unchanged, imposing the conservation of the angular momentum and the conservation of the linear momentum along the horizontal direction. Referring to an impact that occurs when the object approaches the base by rocking around the left corner (of the body or of the support), the angular momentum after the impact $I_2\dot{\vartheta}^+ + mhu^+$ has to be equal to the angular momentum before the impact $I_1\dot{\vartheta}^- - 2mr_1b_2 sin(\alpha_2)\dot{\vartheta}^- + mhu^-$. This condition reads:

$$I_1\dot{\vartheta}^- - 2mr_1 b sin(\alpha_1)\dot{\vartheta}^- + mh\dot{u}^- = I_2\dot{\vartheta}^+ + mh\dot{u}^+ \tag{21}$$

where $r_1^2 = \hat{b}_1^2 + h^2$. The conservation of the linear momentum is needed to relate u^+ and u^- under the assumption that the impact (that happens instantly by hypothesis) does not affect the sliding motion:

$$mh\dot{\vartheta}^- + m(\dot{u}^- + \dot{x}) + m_b\dot{u}^- = mh\dot{\vartheta}^+ + m(\dot{u}^+ + \dot{x}) + m_b\dot{u}^+ \tag{22}$$

If the relative displacement $x \in [-\hat{x}, \hat{x}]$, then $\hat{b}_1 = b_1$ (see Fig.3(b)); if the relative displacement $-\hat{x} > x$, then $\hat{b}_1 = b_1 + y$ with $y = x + \hat{x}$ ($x < 0$, see Fig.3(c)). Taking Eq.(21) and

Eq.(22) into account, which means considering an impact that happens when the body is approaching the base by rocking around the left corner and successively rocking around the right corner, for a rectangular body, the following post-impact velocities are obtained:

$$\dot{\vartheta}^{+} = \frac{m^2h^2 - I_1(m+m_b) + 2mbr_1\sin(\alpha_1)(m+m_b)}{m^2h^2 - I_2(m+m_b)}\dot{\vartheta}^{-} \tag{23}$$

and

$$\dot{u}^{+} = \dot{u}^{-} + \frac{hm(I_1 - I_2 - 2bmr_1\sin(\alpha_1)}{h^2m^2 - I_2(m+m_b)}\dot{\vartheta}^{-} \tag{24}$$

The maximum value of the coefficient of restitution that allows rocking motion of a block on an isolated base then reads:

$$r_1 = \left(\frac{\dot{\vartheta}^{+}}{\dot{\vartheta}^{-}}\right)^2 = \left(\frac{m^2h^2 - I_1(m+m_b) + 2mbr_1\sin(\alpha_1)(m+m_b)}{m^2h^2 - I_2(m+m_b)}\right)^2 \tag{25}$$

In Fig.5 some values of the restitution coefficient are shown. The dashed and dotted lines represent r_1 obtained for $e=0.3b$ and $e=-0.3b$ respectively, while the solid line represents $r_1 = r$ $(e=0)$ as in [Vassiliou & Makris, 2011]. In the analyses the restitution coefficient has been taken accordingly with Eq.(25).

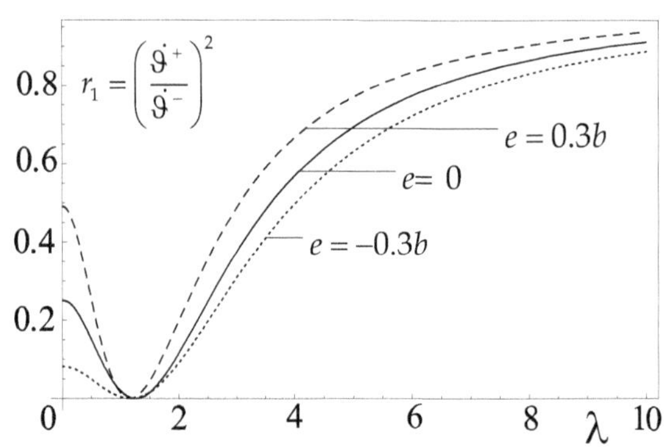

Fig. 5. Restitution coefficient for different values of the eccentricity.

4. Description of the excitations

Two types of base excitations are considered in the following: impulsive excitation and seismic excitation. Results are obtained in an analytical or in a numerical way depending on the excitation considered. Several geometrical and mechanical parameters are introduced to characterise the behaviour of the system. In particular:

$$\omega = \sqrt{\frac{k}{m+m_b}}\ ;\ T = \frac{2\pi}{\omega};\ \xi = \frac{c}{2(m+m_b)\omega} \tag{26}$$

where ω is the frequency, T is the period and ξ is the damping ratio of the isolated system in the full contact phase, that are well known quantities. Other quantities are introduced to characterize the system:

$$\varepsilon = \frac{e}{b} \ ; \ \lambda = \frac{h}{b} ; \ \alpha = \frac{m_b}{m} \tag{27}$$

where ε is the eccentricity ratio with respect to the base of the rigid body, λ is the slenderness of the body and α is the mass ratio.

4.1 Impulsive excitation

An horizontal impulsive ground excitation has been considered. The solution of the full-contact, Eq.(1), under an impulsive ground acceleration I, can be found in closed form and is well known, so, in this case, most of the results have been found analytically. Since the system exhibits a symmetry in the rocking motion, conditions where there is a rocking motion have been evaluated only referring to a rocking around corner 1 and for and eccentricity $\varepsilon < 0$. Results are exposed trough maps describing the criteria for the different phases of motion. These maps, firstly obtained for non isolated, symmetric rigid body subjects to a horizontal ground acceleration in [Shenton, 1996], were successively extended to vertical ground acceleration in [Tung, 2007]. Here they are extended to base isolated non-symmetric rigid body subjects to a horizontal ground impulsive excitation and comparison between not isolated and isolated systems is carried out.

4.2 Seismic excitation

In this case the system has been excited with two different Italian registered seismic ground motions. Also for this kind of excitation only horizontal components of the seismic source

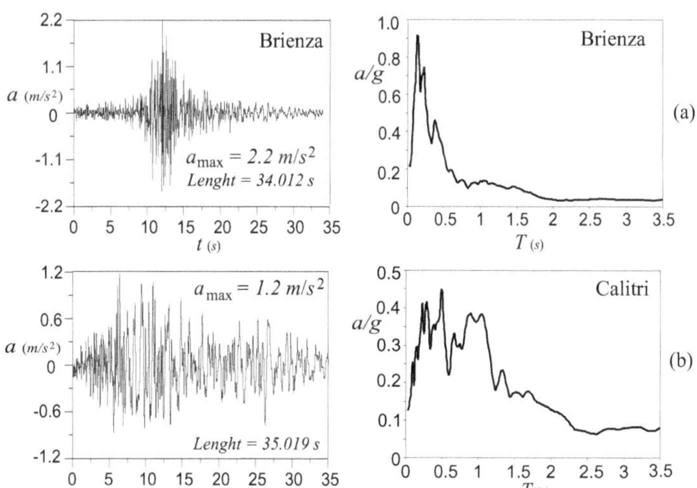

Fig. 6. Time-history and response spectrum of the Italian seismic accelerations used in the analysis: (a) Brienza; (b) Calitri.

$\ddot{u}_g(t) = \gamma f(t)$ are considered, where $f(t)$ is the recorded seismic acceleration and γ is a variable coefficient used to scale the maximum amplitude of the seismic accelerations (*PGA*: Peak Ground Acceleration). In Fig.6 the seismic inputs used and the time-histories with their elastic response spectrums are shown. The Brienza seismic source (Fig.6(a)) is used to analyze the behaviour of the isolated rigid block and its dependence on several parameters, especially on the friction coefficient; the Calitri seismic source (Fig.6(b)) is used to evaluate the influence of the spectral characteristic of the earthquake in the dynamics of the isolated system.

4.3 Description of the simulation

Results are obtained via a direct numerical integration of the equations of motion using a Runge-Kutta fourth order scheme. Special care is devoted to the choice of the integration time step and appropriate procedures are followed to iteratively identify the transitions among the different phases. Extensive parametric analyses are conducted to evaluate the influence of the friction coefficient, the eccentricity and the slenderness as well as the period of the isolated base on the behavior of the system. Analyses are performed by increasing the *PGA* or the impulsive excitation to find the first occurrence of rocking, sliding and falling or overturning of the body. Maps of behavior are produced to compare the results obtained with similar systems.

5. Description of the results

Results have been obtained for two different cases, the first one where only the rocking motion is allowed and a second one where the body is able to slip as well.

5.1 Introduction of safety devices

Two types of security stops have been introduced in the model considered and their effects on the system investigated. The first is able to prevent isolation device breakage by limiting the displacement of the oscillating base to a maximum safety value, the others are introduced to prevent the rigid body from falling off the base.

5.1.1 Security stops for the base isolation

This kind of security stops is introduced to prevent the breaking of isolation devices that generally cannot support displacements greater than a boundary value $u_{\max}$ (see Fig.1(b)). During the motion, if the velocity $\dot{u}$ is different from zero, an impact among oscillating base and security stops occurs when:

$$|u| = u_{\max} \tag{28}$$

To evaluate the condition after an impact, it is assumed that: the impact happens instantly and the body position remains unchanged; during the impact, a fraction of the horizontal momentum is dissipated and after the impact there is an inversion in the direction of motion. The horizontal momentum after the impact $Q_H^+ = (m + m_b)\dot{u}^+$ becomes a fraction η_i of the horizontal momentum before the impact $Q_H^- = (m + m_b)\dot{u}^-$. By letting $|Q_H^+| = \eta |Q_H^-|$, it is possible to obtain the post-impact velocity of the body:

$$\dot{u}^{+} = -\eta_i \, \dot{u}^{-} \tag{29}$$

where η_i is the restitution coefficient that can assume values in the range [0, 1].

5.1.2 Security stops for the slip of the body

Special sliding constrains that prevent the rigid body from falling off the base have been introduced in the model as an alternative to the possibility of the rigid block coming partially away from the base (Fig.1(c)). An impact among the body and the security stops, during a slide or a slide-rocking motion occurs when:

$$|x| = \hat{x} \tag{30}$$

if the velocity $\dot{x}$ is different from zero. After the impact it is supposed that the horizontal momentum due to slide motion $m\dot{x}^{-}$ completely vanishes since it is partially dissipated and partially transferred to the horizontal momentum of the oscillating base. In particular:

$$(m + m_b)\dot{u}^{-} + \eta_s \, m\dot{x}^{-} = (m + m_b)\dot{u}^{+} \tag{31}$$

from which it is possible to obtain the post-impact velocity of the base:

$$\dot{u}^{+} = \dot{u}^{-} + \eta_s \frac{m}{m + m_b} \dot{x}^{-} \tag{32}$$

where η_s is the restitution coefficient that can assume values in the range [0, 1], under the assumptions that the impact happens instantly.

5.2 Pure rocking motion

Analyses are performed to evaluate the first time at which rocking and overturning of the body occur. This evaluation is done for several values of parameters ε, T and λ. The analyses here performed do not permit to obtain the so-called survival regions that lies above the first overturning occurrence as in [Zhang & Makris, 2001; Purvance, 2005; Purvance et al., 2008].

5.2.1 Impulsive excitation

The rocking condition around corner 1 can be obtained by equating the maximum overturning moment $mh|\ddot{u}_{\max}|$, due to the inertia forces, with the resisting moment $mg(b+e)$, due to gravity, in which $|\ddot{u}_{\max}|$ is the absolute maximum acceleration generated by an impulse. This equation can be written, taking into account Eq.(27):

$$|\ddot{u}_{\max}| = \frac{g}{\lambda}(1 + \varepsilon) \tag{33}$$

The solution of the full-contact equation (1), under an impulsive ground acceleration I, is well known and reads:

$$u(t)=\frac{I}{\omega}\frac{1}{\sqrt{1-\xi^2}}e^{-\xi\omega t}\sin(\omega t\sqrt{1-\xi^2}) \tag{34}$$

By solving a maximum problem, it is possible to obtain the absolute maximum acceleration:

$$\left|\ddot{u}_{\max}\right|=\left|I\right|\omega a(\xi)=\frac{2\pi\left|I\right| a(\xi)}{T} \tag{35}$$

where the function $a(\xi)$, which is always positive, reads:

$$a(\xi)=1 \qquad \text{for } \xi=0$$

$$a(\xi)=e^{-\frac{\xi}{\sqrt{1-\xi^2}}\mathrm{Arctan}\left(\frac{\sqrt{1-\xi^2}}{\xi}\frac{1-4\xi^2}{3-4\xi^2}\right)} \qquad \text{for } 0<\xi\le\frac{1}{2} \tag{36}$$

Starting rocking condition can be obtained by Eq.(16) (with $\ddot{u}_g=\ddot{v}_g=0$), taking into account Eq.(35):

$$\mathcal{C}_R:\left|I\right|=\frac{T\,g}{2\pi\,a(\xi)}\frac{1}{\lambda}(1+\varepsilon) \tag{37}$$

In the following, with solid lines overturning curves are indicated, while with dashed lines rocking curves are indicated; moreover thin lines refer to an isolated rigid block, while thick lines refer to a non-isolated rigid block. In Fig.7(a) minimum rocking and overturning impulses versus the period of the isolated base T are plotted for a fixed damping ξ. In grey regions a rocking phase occurs. Below these regions only a full-contact phase is possible, while above an overturning of the rigid block occurs. It is observed that when T is increased the object rocks or overturns with higher impulse amplitude, however, the distance between the rocking and the overturning curves decreases. It is also possible to observe that for increased eccentricities the behaviour of the system worsens since smaller impulses are able to cause the rocking or the overturning. Also the amplitude of the rocking regions becomes smaller when the eccentricity is increased. In Fig.7(b) rocking and overturning impulses versus the eccentricity are shown both for isolated and non-isolated systems. Referring to the isolated system, also in this case, the grey area indicates the rocking phase region. The best behaviour of the system is obtained when $\varepsilon=0$ (symmetric body); the presence of an eccentricity reduces the performances since a smaller impulse is required to cause rocking or overturning. When $\varepsilon=\pm1$ (the centre of mass is located on a vertical side of the body), since no resisting moment is present, the system is in an unstable equilibrium position. By comparing rocking and overturning curves for isolated and non-isolated systems, it is possible to observe the efficiency of the base isolation. It has to be highlighted that the rocking curve for non-isolated system is coincident with the horizontal axis (the minimum rocking impulse is always zero) because, given that an impulse corresponds to an initial velocity, a non-isolated body has always to start its motion with an angular velocity $\dot{\vartheta}(0)\neq 0$. Finally in Fig.7(c) rocking and overturning impulses versus the slenderness λ are shown both for isolated and non-isolated systems. Curves are plotted starting by the value of the slenderness $\lambda = 1/\sqrt{2}$ above which no bouncing phenomena occur. For a fixed eccentricity an increase of the slenderness causes an evident worsening of the behaviour. For higher values of the eccentricity both isolated and non-isolated systems reduce their performances. Grey region refers to rocking phase for $\varepsilon = \pm 0.5$, both for isolated and non isolated structure; an increasing of the eccentricity also causes a reduction in the dimensions of the rocking region.

5.2.2 Seismic excitation

In Fig.8 comparison between the behaviour of an isolated rigid body with and without security stops for the oscillating base is shown under Buia earthquake (security stops for the sliding are not considered here). In the vertical left side of the graphs the scale factor γ is reported, while in the vertical right side the *PGA* is reported.

In particular in Fig.8(a) rocking and overturning curves versus eccentricity ε are drawn in the case of absence of security stops (thick lines) and the case in which these safety devices are present (thin lines). Dotted thick line refers to impact events: it gives the scaling factor γ (or *PGA*) at which impacts on security stops occur for each value of eccentricity ε. It is possible to observe that after an impact rocking and overturning happens for a lower value of the scaling factor with respect to the case in which security stops are not considered. Time-histories of $u(t)$ and $\vartheta(t)$ show better what happens when an impact occurs (Fig.8(b),(c)). Also in these figures thick lines denote the absence of security stops while thin lines denote the presence of this safety device. For the case labelled with *H* in Fig.8(a), when the displacement of the oscillating base $u(t)$ reaches security stops ($|u(t)| = u_{\max}$) one or more successive impacts can occur, as shown in Fig.8(b). The angle of rocking $\vartheta(t)$ (Fig.8(c)) in general increases with respect to the case in which security stops are not considered and causes the overturning of the rigid body for lower values of the scaling factor. Quantity t_r refers to the time at which rocking phase starts, that is the instant at which Eq.(16) is satisfied.

In the following, when safety devices are introduced, these characteristics are always considered: $\eta = 0.70$, $u_{\max} = 0.20$. In Fig.9 comparison between isolated and non isolated system is shown to better evaluate the efficiency of the isolation system under Buia earthquake for a fixed damping and in presence of security stops. In particular in Fig.9(a) rocking and overturning curves versus the period *T* are plotted for several eccentricity ε. Dots on γ-axis refer to the behaviour of the non isolated system. It is possible to observe that increasing the level of the protection (that is increasing period *T*) scaling factors at which rocking occurs increase with respect to the non isolated system, instead overturning scaling factors do not monotonically grow with *T* due to the high nonlinearity of the system. For example the minimum of the overturning curve that occurs in the range $T \in (0.5, 1.0)$ is related to the high contents of spectral energy of the Buia earthquake in the same range. Out of this range, the isolated system requires higher scaling factors to cause the overturning of the rigid body with respect the non isolated system. The behaviour of the system significantly decays when the rigid body has an eccentricity different to zero. The grey area in the graph refers to the region in which a rocking phase occurs for $\varepsilon = 0.5$. Below this

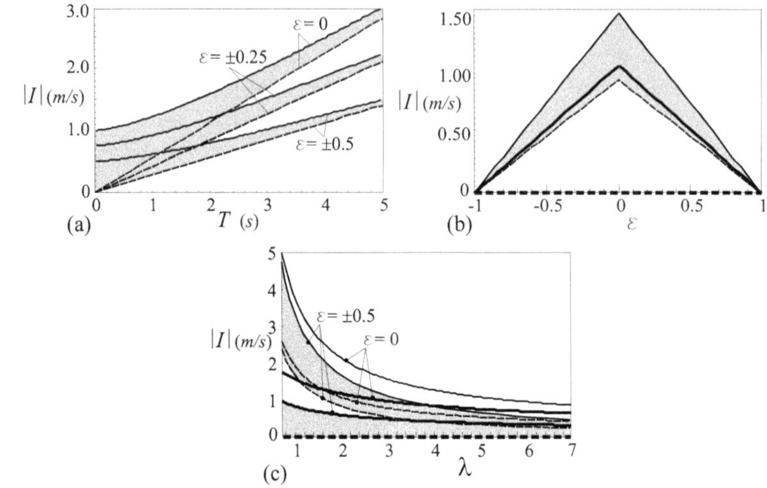

Fig. 7. Dependence on eccentricity ε : (a) Minimum absolute impulse versus period T ($\lambda = 3$); (b) Minimum absolute impulse versus eccentricity ε ($\lambda = 3,\ T = 2s$); (c) Minimum absolute impulse versus slenderness λ ($T = 2s$); ($b = 0.3m,\ \rho = 2000 kg / m^3,\ \xi = 0.05$).

region a full-contact phase takes place; above this region an overturning of the rigid body occurs. In Fig.9(b) rocking and overturning curves versus the eccentricity ε are plotted. Thick lines refer to the non isolated system, while thin lines to isolated system. Rocking and overturning curves for isolated system are the same shown in Fig.8(a). It is possible to note that the best behaviour of the system is obtained when $\varepsilon = 0$ (symmetric body); the presence of an eccentricity reduces the performances since a smaller impulse is required to cause rocking or overturning. By comparing rocking and overturning curves for isolated and non-isolated systems, it is possible to observe the efficiency of the base isolation. Also in this case grey areas indicate the rocking phase regions. Below these areas there is no rocking, above them there is a critical region in which overturning is not guaranteed everywhere because of the existence of survival regions that lie above the curve associated with the first overturning occurrence. Finally in Fig.9(c) rocking and overturning curves versus the slenderness λ, for a fixed eccentricity ($\varepsilon = 0.5$), are plotted. Also in this figure thick lines refer to the non isolated system, while thin lines to isolated system. Also in this case curves are plotted starting by the value of the slenderness $\lambda = 1/\sqrt{2}$. By increasing the slenderness λ both the non isolated and the isolated systems generally show a decay of the performances, but it is possible to observe also in this case the greater efficiency of the base isolated system.

The thick dotted line is the impact curve. Also in this case grey areas indicate the rocking phase regions. It is useful to summarize the effects due to the presence of security stops. In Fig.8(a) it is possible to observe that a range of eccentricity ($|\varepsilon| < 0.75$), in which impact events have the greater effects, exists. Into this range, when an impact occurs before the rocking in the system without security stops, the impact and the rocking curves become practically coincident. This means that the impact is the principal cause of rocking. Instead, when an impact occurs after the rocking, the impact and the overturning curves become coincident. This means that the impact is the main cause of overturning. This limit value of the absolute eccentricity seems to be possessed by every systems and it depends on geometrical and mechanical characteristics of the system itself. Also above a slenderness limit value $(\lambda > 3)$ it happens that impact curve and overturning curve coincide, as shown in Fig.9(c). This slenderness limit value also depends on geometrical and mechanical characteristics of the system.

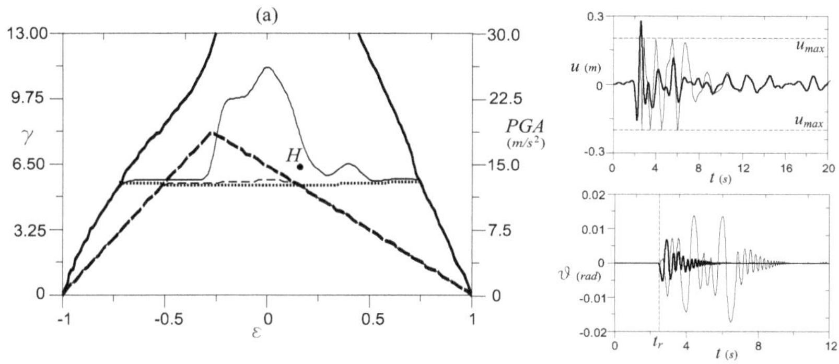

Fig. 8. Effects of the security stops under Buia earthquake: (a) Seismic scale factor versus eccentricity ε ($b = 0.3m,\ \lambda = 3,\ \rho = 2000\,kg/m^3,\ \xi = 0.20,\ \eta = 0.70,\ u_{\max} = 0.20,\ T = 2s$); (b) Time history of displacements $u(t)$ for the case labelled with H in Fig.8a ($\varepsilon = 0.3, \gamma = 6.5$); (c) Time history of rotation $\vartheta(t)$ for the case labelled with H.

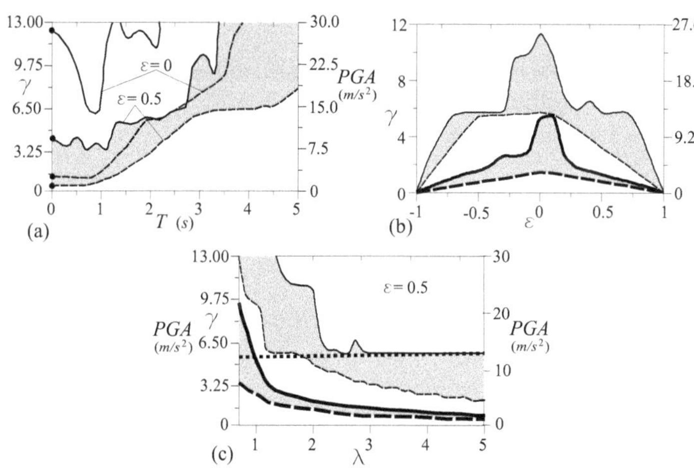

Fig. 9. Seismic analysis under Buia earthquake: (a) Seismic scale factor versus period T ($\lambda = 3$); (b) Seismic scale factor versus eccentricity ε ($\lambda = 3$, T = 2 s); (c) Seismic scale factor versus slenderness λ (T = 2 s); ($b = 0.3m,\ \rho = 2000\,kg/m^3, \xi = 0.20$).

The last analysis reported here points out the effects of the spectral characteristics of the earthquake. A simplified analysis is conducted by using Brienza and Calitri Italian registered earthquakes (Fig.6(b),(c)) with comparable lengths both normalized by assuming an acceleration peak equal to g. In Fig.10(a) rocking and overturning curves, obtained by evaluating for which slenderness λ the body rocks or overturns by varying the eccentricity ε, are plotted. It is clear that under Brienza earthquake the system shows a better behaviour since rocking and overturning happen for higher slenderness with respect to the Calitri earthquake. Grey areas indicate the rocking phase regions. Below these areas there is full-contact, above them there is overturning. In Fig.10(b) rocking (dashed lines) and overturning (solid lines) curves, obtained by evaluating the level of protection (that is the period *T*) to prevent rocking or overturning of the body by varying the eccentricity ε, are plotted. It is evident that the Brienza earthquake requires a lower level of protection with respect to the Calitri earthquake, to prevent rocking or overturning. For Brienza input in the range $\varepsilon \in [\varepsilon_A, \varepsilon_B]$, base isolation ($T = 0$) is not necessary to prevent overturning. Also in this case grey areas indicate the rocking phase regions but on the contrary, over these areas there is full-contact, below them there is a critical region in which overturning is not guaranteed everywhere because of the existence of survival regions. The isolated system shows a better behaviour for earthquake with a more narrow spectrum as found in [Purvance et al., 2008].

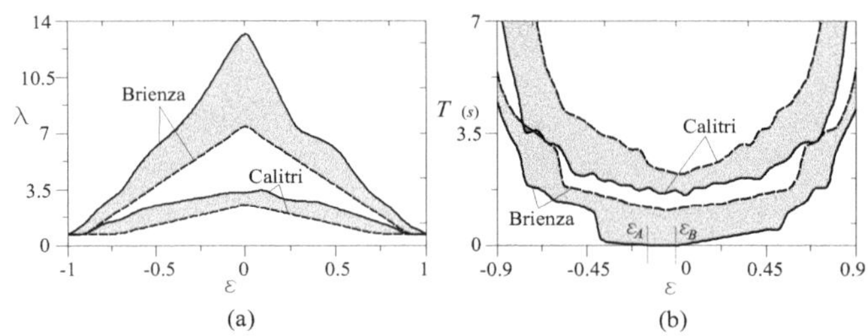

Fig. 10. Comparison between normalized Brienza and Calitri earthquakes (PGA=g): (a) slenderness λ versus eccentricity ε ($b = 0.3m,\ \lambda = 3,\ T = 2s$) and (b) period *T* versus eccentricity ε ($b = 0.3m,\ \lambda = 3,\ \rho = 2000 kg/m^3,\ \xi = 0.20$).

5.3 Slide-rocking motion

As for the rocking motion, analyses are performed to evaluate the first time at which slide, rocking or slide-rocking and overturning or falling of the body occur. Results are found for several values of parameters ε, T and λ.

5.3.1 Impulsive excitation

A horizontal impulsive ground excitation is considered individually in the following. Maps describing the criteria for the different phases of motion are obtained for base isolated non-symmetric rigid body subjects to a horizontal ground impulsive excitation and comparison between not isolated and isolated systems is carried out. These criteria maps are behaviour maps that divides the parameter plane ($I - \mu_s$) in different regions where the motion starts with a different phase of motion. Conditions where there is a rocking motion will be evaluated only referring to a rocking around corner 1 and $\varepsilon < 0$. This is possible since, as already pointed out, the system exhibits a symmetry in the rocking motion. Considering sliding condition given by Eq.(15) (with $\ddot{u}_g = \ddot{v}_g = 0$) and taking into account Eq.(35) and Eq.(37), it is possible to obtain the starting condition of the sliding motion:

$$\mathcal{C}_S : |I| = \mu_s \frac{T g}{2\pi\, a(\xi)} \tag{38}$$

Starting rocking condition is still expressed by Eq.(37). By equating Eqs.(37, 38) it is possible to obtain the curve where, during rocking motion, the starting condition of sliding motion also becomes true:

$$\mathcal{C}_{SR} : \mu_s = \frac{1}{\lambda}(1+\varepsilon) \tag{39}$$

When $\mu_s > (1+\varepsilon)/\lambda$ only the rocking motion is allowed (case considered in section 5.2). In order to find the condition at which a slide-rocking motion occurs starting from a pure rocking phase Eq.(18) must be taken into account. At the beginning of a rocking motion it can be considered that $\vartheta = \dot{\vartheta} = 0$ and, by referring to Fig.3(a), $b_1' = b_1, h_1' = h$. From Eq.(8$_2$) (with $\ddot{u}_g = \ddot{v}_g = 0$) it is possible to obtain the maximum angular acceleration $\ddot{\vartheta}_{\max}$:

$$\ddot{\vartheta}_{\max} = |I| \frac{2\pi m\, a(\xi) h}{T I_1} - \frac{m g b_1}{I_1} \tag{40}$$

where use of Eq.(35) is done. The curve along which a sliding motion occurs starting to a rocking phase can be obtained by Eq.(18), by taking into account Eq.(12), Eq.(19) and Eq.(40). It reads:

$$\mathcal{C}_{RS} : \mu_s = \frac{|I|(3+6\varepsilon+3\varepsilon^2+\lambda^2)\omega\, a(\xi) + 3g(1+\varepsilon)\lambda}{\lambda[3|I|(1+\varepsilon)\omega\, a(\xi) + 4g\lambda]} \tag{41}$$

where use of Eq.(27) is done. Previous curves divide the parameter plane in several regions as shown in Fig.11(b)-(f). It is possible to observe that, when a base isolation is applied to the rigid body, a full-contact region appears.

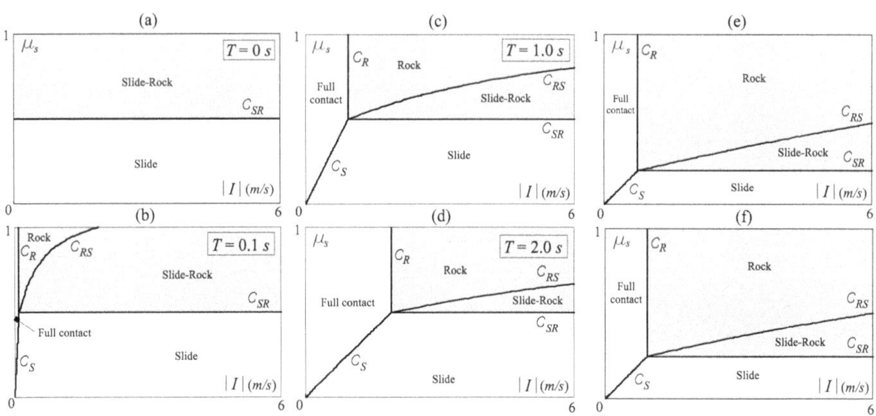

Fig. 11. Criteria maps: (a) not isolated system ($\lambda = 3,\ \varepsilon = 0,\ T = 2s$); (b) isolated system ($\lambda = 3,\ \varepsilon = 0,\ T = 0.1s$); (c) isolated system ($\lambda = 3,\ \varepsilon = 0,\ T = 1s$); (d) isolated system ($\lambda = 3,\ \varepsilon = 0,\ T = 2s$); (e) isolated system ($\lambda = 5,\ \varepsilon = 0,\ T = 2s$); (f) isolated system ($\lambda = 3,\ \varepsilon = -0.5,\ T = 2s$); ($\xi = 0.20$).

By considering the rocking motion particularly dangerous for the body, it is possible to observe that by increasing the period T of the oscillating base, the regions where there is rocking (grey regions) decrease (Fig.11(b)-(d)). By referring to the case where $T = 2s$ (Fig.11(d)), both an increase of the slenderness λ (Fig.11(e)) and of the eccentricity ε (Fig.11(f)) cause a deterioration in the performance to the system, since a reduction of the full-contact region and an increase of the rocking regions are observed.

5.3.2 Seismic excitation

In the numerical simulations performed here, $\mu_k = 0.8\mu_s$ is always used as in [Shenton & Jones, 1991]; according to results found in [Caliò & Marletta, 2003] and [Contento & Di Egidio, 2009], in this paper, it is always assumed $\xi = 0.2$. Where the security stop collision is concerned, $u_{\max} = 0.2m$, $\hat{x} = 0.1m$ and the restitution coefficients $\eta_{i,s} = 0.7$ are always chosen. In the following figures, these graphic conventions are always used: dashed curves are related to sliding motion and in particular heavy dashed curves are the curves at which a sliding motion begins, while thick dashed curves are the collapse curves due to the precipitation of the rigid block from the oscillating base; solid curves are related to rocking motion and in particular heavy solid curves are the curves at which rocking begins, while thick solid curves are the collapse curves due to overturning of the rigid body. Finally heavy dotted curves refer to the reaching of a slide $x = \hat{x}$ above which the body is partially out of the oscillating base, thick dotted curves refer to a collision with the slide security stops and thick dash-dot curves refer to a collision with the base isolation security stops when they are considered.

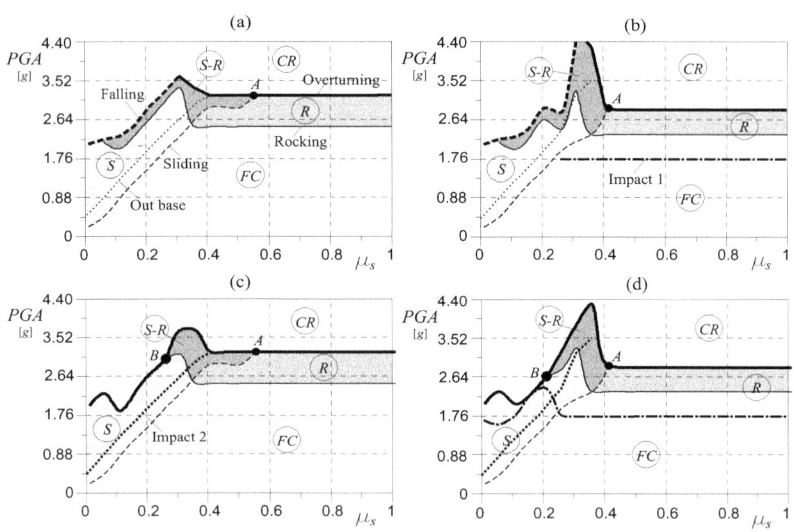

Fig. 12. Seismic analysis under Brienza earthquake: (a) *PGA* versus static friction coefficient μ_s - no security stops; (b) *PGA* versus static friction coefficient μ_s - only base stops; (c) *PGA* versus static friction coefficient μ_s - only sliding stops; (d) *PGA* versus static friction coefficient μ_s - both base and sliding stops; ($\lambda = 3$, $\varepsilon = 0$, $T = 2s$).

Results of the first analysis are shown in Fig.12 where the *PGA* associated with the first occurrence of sliding, rocking, falling or overturning is plotted versus the static friction coefficient μ_s, under Brienza earthquake for a body with slenderness $\lambda = 3$ and eccentricity $\varepsilon = 0$. In each graph of Fig.13 it is possible to observe the existence of five different regions: a region *S* where only sliding motion can occur, a region *S-R* where rocking or slide-rocking motion can occur, a region *R* of pure rocking motions, a region *FC* where during the motion the body and the base remain in full contact and a critical region *CR*. Here, as for the rocking case, the survival regions could exist. At the labelled point *A* a collision between the sliding curve and the overturning one manifests itself. While the sliding curve and the critical falling curve strongly depend on the friction coefficient μ_s, it is very interesting to note that the rocking and the overturning curve are practically independent from μ_s above the point *A*. As observed in [Di Egidio & Contento, 2009] in this region, where the behaviour of the system does not depend on the friction, only a pure rocking motion is possible. Differently from the results obtained in [Di Egidio & Contento, 2009] where the body could not exit from the oscillating base, here, also for small values of the static friction coefficient, a rocking motion manifests itself. This is possible since, when the body is out the base (this happens over the curve labelled Out base), it can rock easily because the resisting moment becomes smaller and smaller as the body comes out the base (see Figs. 12(a),(b)). The effects of the presence of the security stops on the behaviour of the system are very interesting. Figure 12(a) refers to the absence of these stops. When security stops on oscillating base are considered (Fig.12(b)) a worsening of the performance of the system in the *R* region can be observed, because rocking and overturning can occur for smaller values of the *PGA*. On the contrary, around the conjunction between the falling and the overturning curves it is possible to observe an improvement of the performance of the system. It is useful to note for next comments, that the impact on the security stops (above the curve labelled Impact 1) happens before a rocking motion manifests itself. The introduction of security stops on the sliding motion (Fig.12(c), curve labelled Impact 2) change the falling collapse events in overturning collapse events. Since below the point *B* rocking and overturing curves coincide, rocking and overturning events happen for the same *PGA* and no *S-R* region exists. However the presence of these kind of security stops, do not change significantly the values of *PGA* at which a collapse can occur. Finally in Fig.12(d) results obtained from the contemporary presence of the two kind of security stops are shown.

The effects of the eccentricity and specially of the slenderness on the performance of the system are shown in Fig.13. In Figs. 13(a)-(d) the results obtained for a body with slenderness $\lambda = 3$ and eccentricity $\varepsilon = 0.35$ are reported. First of all the worsening of the behaviour of the system due to an increasing of the eccentricity with respect to the results shown in Fig.12 must be observed, since smaller *PGA* are now able to cause falling and overturning collapse events. Another interesting aspect is related to the presence of security stops on the oscillating base (Fig.13(b)). Due to the increased eccentricity, the rocking starts before a collapse on security stops happens. This fact makes the system less sensitive to the presence of this kind of security stops compared to the case analyzed previously and shown in Fig.12. The presence of the security stops on the sliding motion (Fig.13(c)) or the contemporary presence of the two kind of security stops (Fig.13(d)) have, in this case, a small impact on the performance of the system. On the contrary, in Figs. 13(e)-(h), results obtained for a lesser slenderness $\lambda = 2$ (and the same eccentricity $\varepsilon = 0.35$) are shown. First of all it must be observed that a decreasing of the slenderness causes a reduction of the region R where a pure rocking motion manifest itself, since the sliding motion becomes possible for higher values of the friction coefficient. By comparing results obtained without security stops (Fig.13(e)) and with only the security stops on the oscillating base (Fig.13(f)), it is possible to observe the great positive influence of this type of stops specially on the overturning collapse condition. The great sensitivity to the presence of these security stops is related to the fact that the impact on them happens before the rocking motion starts as observed also in the previous case.

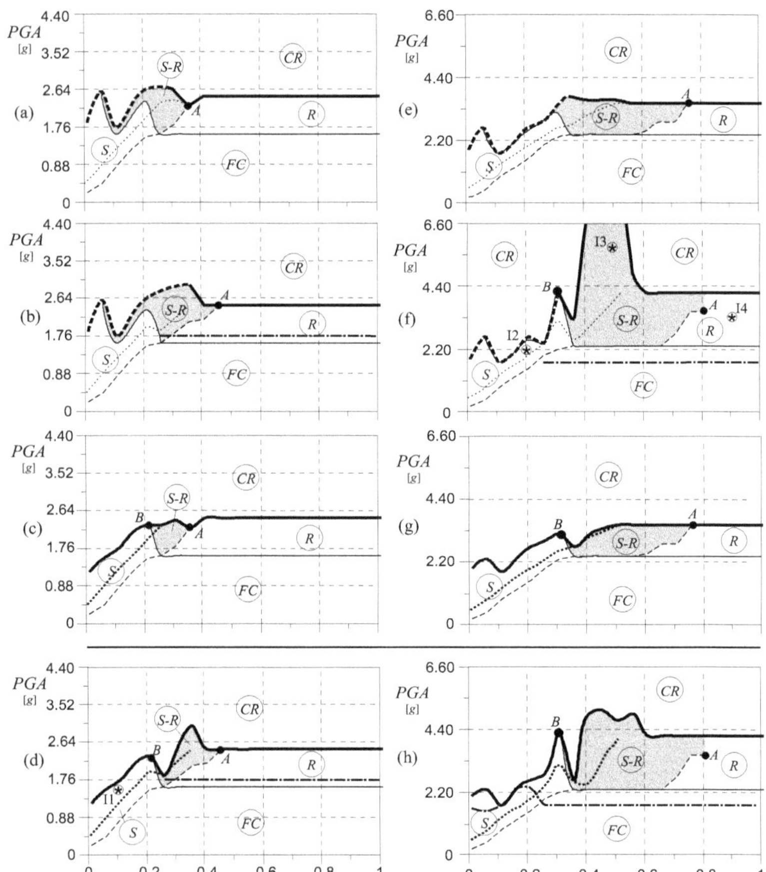

Fig. 13. Seismic analysis under Brienza earthquake: (a)-(d) *PGA* versus static friction coefficient μ_s ($\lambda = 3$); (a) no security stops; (b) only base stops; (c) only sliding stops; (d) both base and sliding stops. (e)-(h) *PGA* versus static friction coefficient μ_s ($\lambda = 2$); (e) no security stops; (f) only base stops; (g) only sliding stops; (h) both base and sliding stops; ($\varepsilon = 0.35$, $T = 2s$).

Unlike the case with $\lambda = 3$ (shown in Fig.12(b)), where security stops on an oscillating base cause a worsening of the performance of the system, for smaller slenderness they improve the behaviour of the system. The presence of security stops on sliding motion have a small impact on the behaviour of the system as shown in Fig.13(g). Finally the contemporary presence of the two types of security stops causes a general improvement of the performance of the system compared to their total absence. To conclude, the comparison between the case with $\lambda = 3$ (Figs. 13(a)-(d)) and $\lambda = 2$ (Figs. 13(e)-(h)) shows that the performances of the two system are similar as to the falling collapse conditions (when no security stops on sliding are considered), while in the overturning conditions the behaviour of the system with lesser slenderness is much better than the other one.

In Fig.14 some time-histories of different cases contained in different regions are shown. A case of pure slide motion in the presence of security stops is shown in Fig.14(a) for a point labelled with I1 in Fig.13(d). It is possible to observe the impossibility of the system to overcome the threshold value $|\hat{x}|$. When no sliding security stops are considered, the body can exit from the base as shown in Fig.14(b) for a point labelled I2 in Fig.13(f). In Fig.14(c) a case contained in the *S-R* region in presence of only security stops on the oscillation base is shown (point labelled I3 in Fig.13(f)). It is possible to observe the impacts of the oscillating base on the stops from the time-history of the displacement u that never exceed the threshold value $|u_{max}|$. From the time-histories of x and ϑ two different slide-rocking motions can be observed. Below the time t_{SR}, as the body is inside the support, rocking or slide-rocking motion around the corners of the of the body manifest themselves; above the time t_{SR}, as the body is outside the support, rocking and slide-rocking motion happen around the corners of the support. In particular, in this last case the rocking motion happens around the left corner of the support since $-\hat{x} > x$ and the right corner of the body. In Fig.14(d) the time-histories of a pure rocking motion are finally shown (point labelled I4 in Fig.13(f)).

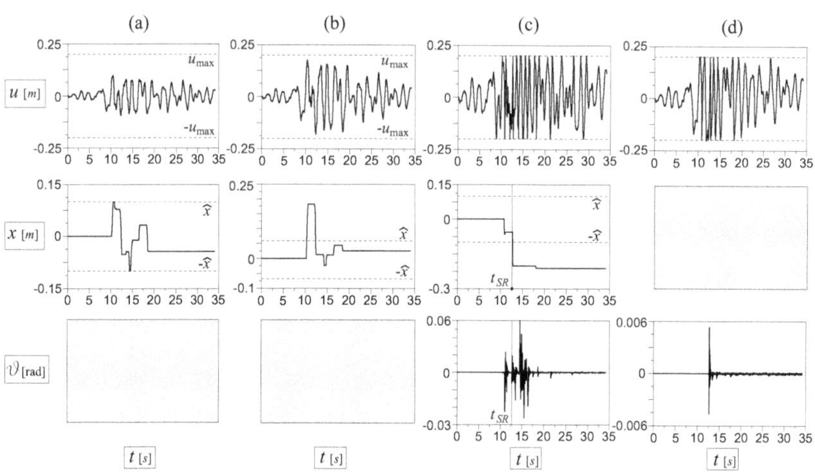

Fig. 14. Time-histories under Brienza earthquake: (a) system marked with I1 in Fig.13(d); (b) system marked with I2 in Fig.13(f); (c) system marked with I3 in Fig.13(f); (d) system marked with I4 in Fig.13(f).

In Figs. 15 results obtained for a body with slenderness $\lambda = 1$ and eccentricity $\varepsilon = 0.35$ are shown. It is very interesting to observe that in this case, the collapse of the system, also when no security stops are considered (Fig.15(a)), is related only to the overturning of the body. This fact can be explained by observing the great attitude of a body with less slenderness to slide instead of rocking. Since these kinds of bodies easily exit from the limits of the support by sliding and can easily reach large sliding displacement x, when a rocking motion starts, a overturning condition follows due to the small value of the resisting moment. When the two different security stops are considered a great improvement of the performance of the system is observed. This fact confirms the positive effects of the stops to system with a smaller slenderness. Above the points labelled with C rocking and sliding curves coincide. Finally it can be noted that no pure rocking region exists.

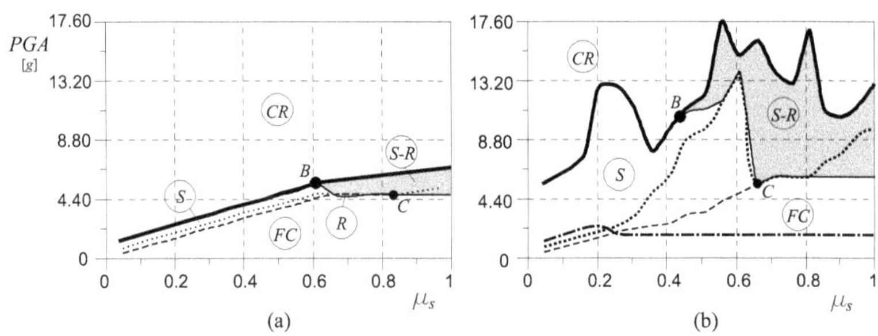

Fig. 15. Seismic analysis under Brienza earthquake, PGA vs static friction coefficient μ_s :(a) no security stops; (b) base and sliding stops; ($\lambda = 1$, $\varepsilon = 0.35$, $T = 2s$).

6. Conclusion

A model of a non-symmetric rigid body, where the centre of gravity is not equally distant from the two base corners can be used to investigate the behaviour of a real base isolated art object. The base under the rigid body is connected to the ground by a linear visco-elastic device representing the passive control system. Exact nonlinear equations of motion have been written by using a Lagrangian approach and transition conditions have also been derived. In the case in which no sliding security stops are considered original equations of motion describing the slide-rocking motion of the rigid block when it is partially outside the support have been obtained. For the impulsive excitation exact and approximated (for damped systems) results have been obtained in closed form while seismic excitation has been considered by using two Italian registered earthquakes. Two different kinds of collapse condition have been considered: the fall from the oscillating base of the rigid block and the overturning of the body. The analysis has been conducted with the aim of pointing out the effects of the friction coefficient, the influence of the slenderness and of the eccentricity of the body and the analysis confirms that base isolation can be more effective for rigid bodies with geometrical parameters similar to those of real works of art. Security stops have been considered one kind able to prevent isolation device breakage by limiting the displacement of the oscillating base to a maximum safety value, another introduced to prevent the rigid body from falling off the base. For wider bodies security stops turn out to have a positive influence on the performance of the system while above a certain value of slenderness they can worsen the behaviour of the system.

7. References

Agbabian M.S., Masri F.S., Nigbor R.L. & Ginel W.S. (1988). Seismic damage mitigation concepts for art objects in museum, *Proceeding of the 9th World Conference on Earthquake Engineering*, No.7, pp. 235-240, Tokyo-Kyoto, Japan, 1988

Batista M. (2006). Steady motion of a rigid disc of finite thickness on a horizontal plane. Journal of Nonlinear Mechanics, Vol.41, pp.850 - 859

Boroscheck R.L. & Romo D. (2004). Overturning criteria for non-anchored non-symmetric rigid bodies, *Proceeding of the 13th World Conference on Earthquake Engineering*, Vancouver, B.C., Canada, August 1-6, 2004

Caliò I. & Marletta M. Passive control of the seismic response of art objects. *Engineering Structures*, Vol.25, pp. 1009-1018

Contento A. & Di Egidio A. (2009). Investigations into Benefits of Base Isolation for Non-Symmetric Rigid Blocks. *Earthquake Engineering and Structural Dynamics*, Vol.38, pp. 849-866

Di Egidio A. & Contento A. (2009). Base Isolation of Sliding-Rocking Non-Symmetyric Rigid Blocks Subjected to Impulsive and Seismic Excitations. *Engineering Structures*, Vol.31, pp. 2723-2734

Di Egidio A. & Contento A. (2010). Seismic response of a non-symmetric rigid block on a constrained oscillating base. *Engineering Structures*, Vol.32, pp. 3028-3039

Fujita K., Yoshitomi S., Tsuji M. & Takewaki I. (2008). Critical cross-correlation function of horizontal and vertical ground motions for uplift of rigid block. *Engineering Structures*, Vol.30, No.5, pp.1199-1213

Koh A.S. & Mustafa G. (1990). Free rocking of cylindrical structures. *Journal of Engineering Mechanics*, Vol.116, pp.34 - 54

Kounadis A.N. (2010). On the overturning instability of a rectangular rigid block under ground excitation. *The Open Mechanics Journal*, Vol.4, pp. 43 - 57

Makris N. & Black C.J. (2004). Dimensional analysis of bilinear oscillators under pulse-type Excitations. *Journal of Engineering Mechanics* (ASCE), Vol.130, No.9, pp. 1019-1031

Pompei A., Scalia A. & Sumbatyan M. (1998). Dynamics of rigid block due to horizontal ground motion. *Journal of Engineering Mechanics*, Vol.124, No.7, pp. 713-717

Purvance M.D. (2005). Overturning of slender blocks: numerical investigation and application to precariously balanced rocks in southern California. Ph.D. Dissertation, University of Nevada, Reno, Available from http://www.seismo.unr.edu/PrecRock/gradresearch.html

Purvance M.D., Anooshehpoor A. & Brune J.N. (2008). Freestanding block overturning fragilities: Numerical simulation and experimental validation. *Earthquake Engineering and Structural Dynamics*, Vol.37, No.5, pp. 791-808

Shenton H.W. III. (1996). Criteria for initiation of slide, rock, and slide-rock rigid-body modes. *Journal of Engineering Mechanics* ASCE, Vol.122, No.7, pp. 690-693

Shenton H.W. & Jones N.P. (1991). Base excitation of rigid bodies. I: Formulation. *Journal of Engineering Mechanics*, Vol.117, No.10, pp. 2286-306

Stefanou I., Vardoulakis I. & Mavraganis A. (2011). Dynamic motion of a conical frustum over a rough horizontal plane. *International Journal of Non Linear Mechanics*, Vol.46, pp.114 - 124

Taniguchi T. (2002). Non-linear response analyses of rectangular rigid bodies subjected to horizontal and vertical ground motion. *Earthquake Engineering and Structural Dynamics*, Vol.31, pp. 1481-1500

Taniguchi T. (2004). Rocking behavior of unanchored flat-bottom cylindrical shell tanks under action of horizontal base excitation. *Engineering Structures*, Vol.26, pp.415 - 426

Tung CC. (2007). Initiation of motion of a free-standing body to base excitation. *Earthquake Engineering and Structural Dynamics*, Vol.36, pp. 1431-1439

Vassiliou M.F. & Makris N. (2011). Analysis of the rocking response of rigid blocks standing free on a seismically isolated base. *Earthquake Engineering and Structural Dynamics*, DOI: 10.1002/eqe.1124

Vestroni F. & Di Cinto S. (2000). Base isolation for seismic protection of statues, *Proceeding of the 12th World Conference on Earthquake Engineering*, New Zeland, 2000.

Zhang J. & Makris N. (2001). Rocking Response of Free-Standing Blocks Under Cycloidal Pulses. *Journal of Engineering Mechanics*, Vol.127, No.5, pp. 473-483

14

Numerical Prediction of Fire Whirlwind Outbreak and Scale Effect of Whirlwind Behavior

Seigo Sakai
Yokohama National University
Japan

1. Introduction

Our Japanese, especially the residents in east area of Japan, have experienced a large earthquake on March 11, 2011, i.e. East Japan great earthquake disaster (Takewaki et al., 2011). There were a lot of fires in the northeast area of Japan, for example in Kesen-numa City. Despite of a number of town area fires, a fire whirlwind was never observed in this disaster. However, fire whirlwind is still one of the concerned accidents in the earthquake (Hough & Bilham; 2005).

When a large-scale wide area fire such as a town area fire or a forest fire occurs, there can be a strong rotating flow to be called fire whirlwind. A fire whirlwind is a tornado that includes flames, hot winds and sparks. The fire whirlwind is regarded as the worst case which we should avoid at the time of a large-scale fire, because the whirlwind itself is critical and scatters sparks widely to promote spread of a fire.

As a fire occurs, a flame makes an ascending current of air and uses up neighboring oxygen. Furthermore, to collect oxygen from a wide area, there is a current of air against the flame, resulting in big natural convection in the fire current of air. When the wind from a certain specified direction blows in this fire current of air, homogeneity of suction of air with an ascending current of air collapses. Then, a vortex is easy to come to occur, the fire current of air becomes a fire whirlwind that is an ascending current of air accompanied with rotating. Fire whirlwind may be pushed away by wind downstream, or may move in search of oxygen.

Aiming at a property and elucidation of an outbreak factor of fire whirlwind as examples of the pasts for a lesson, investigation and a reproduction experiment of the outbreak situation (Graham, 1955; Emmons & Ying, 1967; Byram & Martin, 1970; Haines & Updike, 1971; Martin et al., 1976; Muraszew et al., 1979; Emori & Saito, 1982; Satoh & Yang, 1996; Hayashi et al., 2003; Liu, 2005; Kuwana et al., 2007; Liu et al., 2007; Kuwana et al., 2008; Chuah et al., 2011), numerical analysis are performed till now (Satoh & Yang, 1997; Battaglia et al., 2000a; Battaglia et al., 2000b; Snegirev et al., 2004; Hassan et al., 2005; Chuah et al., 2007; Grishin, 2007; Grishin et al., 2009). Though various factors are thought about outbreak of a fire whirlwind, such as climatic condition or existence of underground flammable gas, it is hard to say that property and outbreak mechanism of a fire whirlwind are to be elucidated enough.

Therefore, in this study, it is firstly aimed to get basic knowledge to elucidate outbreak mechanism of fire whirlwind and the property, by evaluating influence between a natural convection generated by the fire and the wind blowing to fire whirlwind on the ground utilizing numerical analysis. It is secondly examined whether a relationship exists between a real phenomenon and the phenomenon in reduction models, performing the numerical analysis of a fire whirlwind with respect to scale effect.

Three dimensional analyses are performed to investigate the thermal and flow fields by using an analytical software FLUENT 6.3. Natural convection is caused from a plane source of constant heat flux or constant temperature in the flat ground large enough. It is observed and evaluated whether a fire whirlwind occurs or not in a constant wind of air. It is also observed and evaluated how the whirlwind behaves in case that the whirlwind occurs. Then, it is analyzed that those swirling flow in original scale, 1/10 scale, 1/50 scale, 1/100 scale from the original brake out to vanish. As an analytical condition, parameter calculation is repeated to get the velocity of a parallel flow which is the easiest to occur the swirling flow for each reduction model, and then scale effect is discussed by comparing the velocity of the natural convection, the velocity of the parallel flow, the center pressure of the whirlwind and the continuance time of the swirling flow. For making of analysis models, a representative example of the fire whirlwind that occurred at Tokyo in the Great Kanto Earthquake (1923) is referred, and three types of heat source model (L-character model, random model and C-character model) are constructed.

2. Nomenclature

a	Absorption coefficient of air
$C_{1\varepsilon}, C_{2\varepsilon}, C_{3\varepsilon}$	Constants for standard k- ε turbulent model
$c_{p,j}$	Specific heat of species of j at constant pressure
E	Energy
$\vec{F}$	External force vector
G	Irradiance
G_b	Turbulent kinetic energy production due to buoyancy force
G_k	Turbulent kinetic energy production due to gradient of time-averaged velocity
$\vec{g}$	Gravitational acceleration vector
h	Sensible heat enthalpy
h_j	Sensible heat enthalpy of chemical species j
I	Unit tensor
$\vec{J}_j$	Diffusive flux vector of chemical species j
k	Turbulent kinetic energy
k_{eff}	Effective thermal conductivity
p	Pressure

q_r	Radiative heat flux
S_h	Heat source including volumetric heat generation
S_k, S_ε	Source terms for k and ε
T	Temperature
T_{ref}	Reference temperature
t	Time
$\vec{v}$	Velocity vector
Y_j	Mass fraction of chemical species j
Y_M	Expansion dissipation term for k

Greek symbols

ε	Turbulent energy dissipation rate
μ	Viscosity of air
μ_t	Turbulent viscosity of air
ρ	Density of air
σ	Stefan-Boltzmann constant
$\sigma_k, \sigma_\varepsilon$	Turbulent Prandtl numbers for k and ε
$\overline{\overline{\tau}}$	Shear stress tensor
$\overline{\overline{\tau}}_{eff}$	Effective shear stress

3. Governing equations

In this analysis, chemical reaction and combustion phenomenon are not dealt with, and driving force for convective heat transfer is due to density change of air from the heat source. Therefore, governing equations of the phenomenon are as follows (Fluent Inc/Flient Asia Pacific, 2006).

Mass conservation equation

$$\frac{\partial \rho}{\partial t} + \nabla \cdot (\rho \vec{v}) = 0 . \tag{1}$$

Navier-Stokes equation

$$\frac{\partial}{\partial t}(\rho \vec{v}) + \nabla \cdot (\rho \vec{v} \vec{v}) = -\nabla p + \nabla \cdot \left(\overline{\overline{\tau}}\right) + \rho \vec{g} + \vec{F} , \tag{2}$$

where

$$\overline{\overline{\tau}} = \mu \left[(\nabla \vec{v} + \nabla \vec{v}^T) - \frac{2}{3} \nabla \cdot \vec{v} I \right] , \tag{3}$$

and the second term of right side shows effect of volume expansion.

Energy equation

$$\frac{\partial}{\partial t}(\rho E)+\nabla\cdot\{\vec{v}(\rho E+p)\}=\nabla\cdot\left\{k_{eff}\nabla T-\sum h_j\vec{J}_j+\left(\bar{\bar{\tau}}_{eff}\cdot\vec{v}\right)\right\}+S_h\,. \tag{4}$$

The first, second and third term of right side show energy transfers by conduction, diffusion of chemical species, and viscous dissipation, respectively. In equation (4),

$$E=h-\frac{p}{\rho}+\frac{v^2}{2}, \tag{5}$$

and given for ideal gas as

$$h=\sum_j Y_j h_j\,. \tag{6}$$

h_j is given as follows.

$$h_j=\int_{T_{ref}}^{T} c_{p.j}dT\,. \tag{7}$$

where T_{ref} is set to 300K in this analysis.

Equation of Turbulent model

Turbulent flow is treated in the analysis, and the standard *k-ε* model is employed.

$$\frac{\partial}{\partial t}(\rho k)+\frac{\partial}{\partial x_i}(\rho k u_i)=\frac{\partial}{\partial x_j}\left[\left(\mu+\frac{\mu_t}{\sigma_k}\right)\frac{\partial k}{\partial x_j}\right]+G_k+G_b-\rho\varepsilon-Y_M+S_k\,, \tag{8}$$

$$\frac{\partial}{\partial t}(\rho\varepsilon)+\frac{\partial}{\partial x_i}(\rho\varepsilon u_i)=\frac{\partial}{\partial x_j}\left[\left(\mu+\frac{\mu_t}{\sigma_\varepsilon}\right)\frac{\partial\varepsilon}{\partial x_j}\right]+C_{1\varepsilon}\frac{\varepsilon}{k}(G_k+C_{3\varepsilon}G_b)-C_{2\varepsilon}\rho\frac{\varepsilon^2}{k}+S_\varepsilon\,. \tag{9}$$

Radiation model

P-1 radiative heat exchange model (Siegel & Howell, 2002) is also employed to simulate the radiative heat transfer from the heat source at high temperature. The equation of P-1 model is as follows.

$$-\nabla q_r=aG-4a\sigma T^4\,. \tag{10}$$

Numerical analysis is performed three-dimensionally with FLUENT6.3.

4. Analysis model

A calculation domain is taken enough widely to avoid the influence of natural convection due to heat source. Depth and width are 2000m and height is 500m. Heat source is modeled by a combination of 16 squares (200m×200m) at the center of bottom in the domain. Each square is given a constant value of heat flux individually, and various profiles of large-scale fires are easily constructed. While heat flux of house fire is about 1.0 MW/m^2, simulation is performed at 0.03MW/m^2 of heat flux. Fig. 1 shows the calculation domain for analysis.

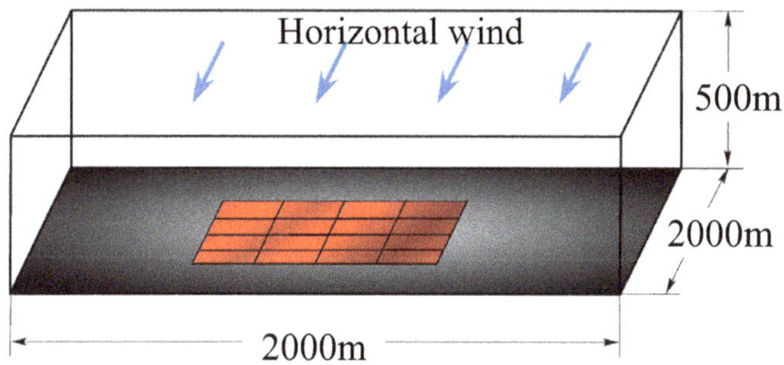

Fig. 1. Calculation domain for analysis

Three models are considered in the analysis. One is based on the large-scale fire at the Great Kanto Earthquake (1923 in Japan), and the profile of the heat source at the bottom is "L-character" type (Model 1). The second one is the profile in which the heat source is randomly dispersed at the bottom (Model 2). The third one is the profile of the heat source at the bottom is "C-character" type (Model 3). Fig. 2 shows the profiles of the heat source.

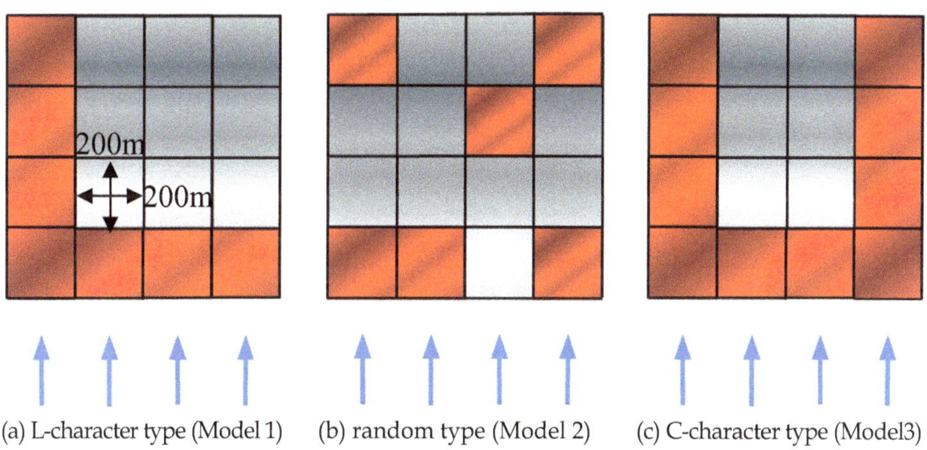

Fig. 2. Profiles of heat source at the bottom

Initially, air is uniformly enclosed in the domain at atmospheric pressure and temperature of 300K. Bottom of the domain assumes adiabatic except 16 squares of modeled heat source. Four sides and top of the domain is free for inflow and outflow.

5. Results and discussions

5.1 Outbreak of fire whirlwind

In this section, it is aimed to get basic knowledge to elucidate outbreak mechanism of fire whirlwind and the property, by evaluating influence between a natural convection generated by the fire and the wind blowing to fire whirlwind on the ground utilizing numerical analysis (Sakai & Watanabe, 2007).

For 5 minutes from calculation start, natural convection is developed in the domain. Then, horizontal air flow is introduced from one side of the domain at 4, 6 and 8m/s of the velocity. Wind on the ground is modeled by the air flow. Calculation is continued for several minutes from the introduction of the air flow.

5.1.1 Model 1

Figs. 3 and 4 are the sample of streamline distribution after 5 minutes and that after 30 minutes for the wind of 4m/s. These streamlines are drawn from the pressure surface whose pressure is lower by 30 Pa from the atmospheric pressure, and are colored by velocity magnitude. Number of generated fire whirlwind and behavior of the whirlwind (movement and extinction) are altered dependent on the velocity of wind. It is summarized in the Fig. 5.

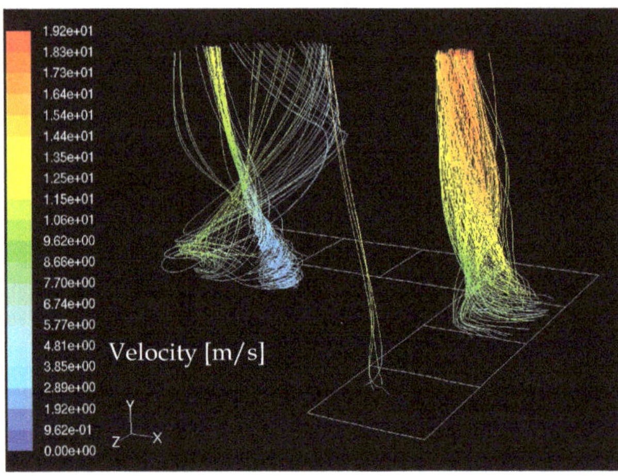

Fig. 3. Distribution of streamlines after 5 minutes for the wind of 4m/s

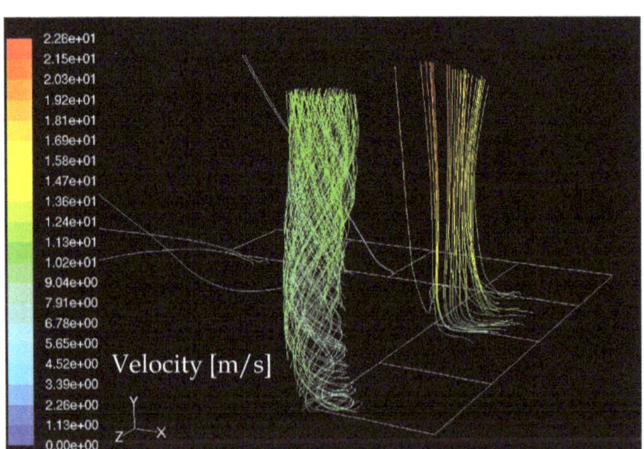

Fig. 4. Distribution of streamlines after 30 minutes for the wind of 4m/s

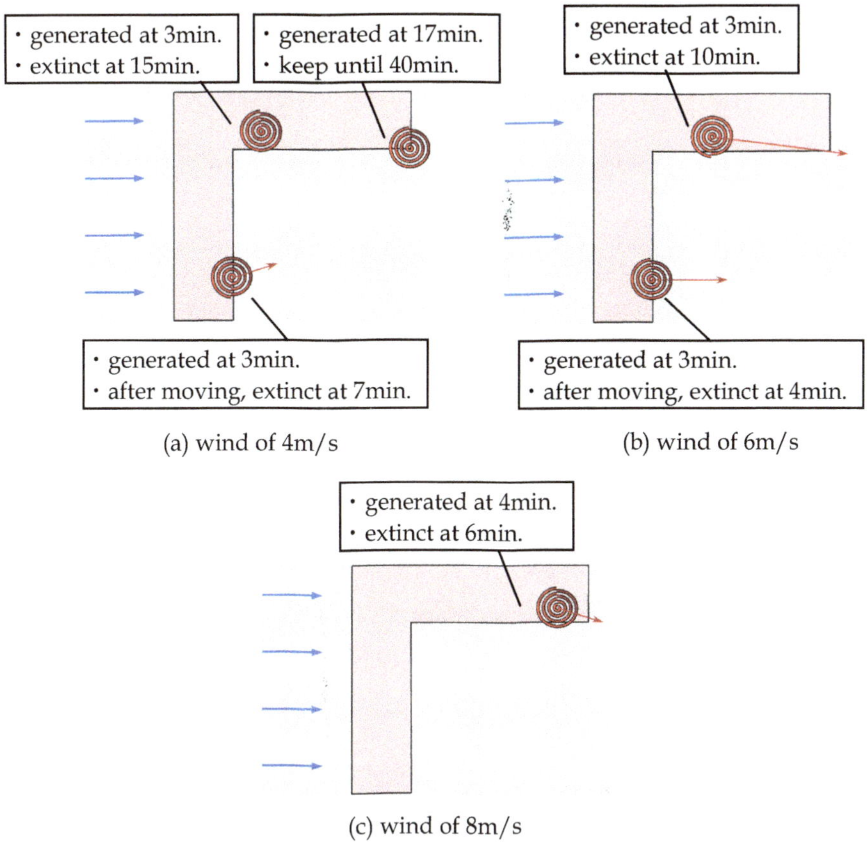

Fig. 5. Summary of generation and extinction of fire whirlwind in Model 1

Red arrows in the figure show the distance from the generation point for the duration time of the whirlwind. In the case of wind of 4m/s, three whirlwinds are generated, and duration time is longer than the other 2 cases. In the case of wind of 6m/s, two whirlwinds are generated, and duration time is shorter than the case of 4m/s. In the case of wind of 8m/s, just one whirlwind is generated, and duration time is the shortest in 3 cases.

Generation points of the whirlwind are located at the tips and corner of "L" character. Fig. 6 shows a numerical result of streamlines in natural convective heat transfer at 5 minutes (no wind). The points where the natural convective flow develops become origins of the fire whirlwind.

The number of the generated whirlwind is decreased as the velocity of wind is increased. This is because too fast wind blows the ascending current of air by natural convection to the downstream. Figs. 7 and 8 are side views of streamlines for the wind of 4m/s in Model 1. The former is streamlines after 18.6 minutes, and the latter is that after 26.6 minutes. As time passes, it is shown that the ascending current of air by natural convection is blown to the downstream by the wind.

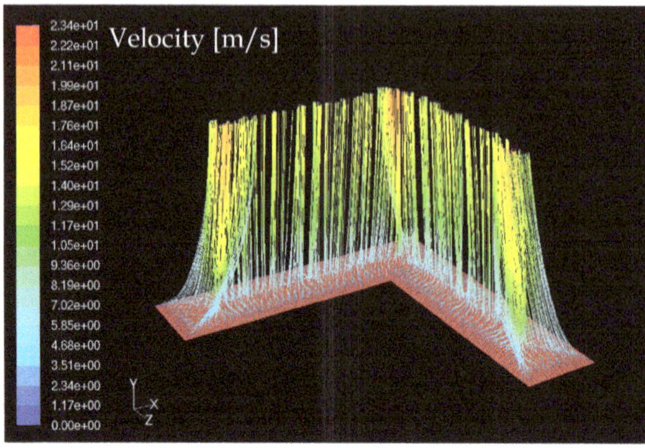

Fig. 6. Distribution of streamlines after 5 minutes in natural convective heat transfer

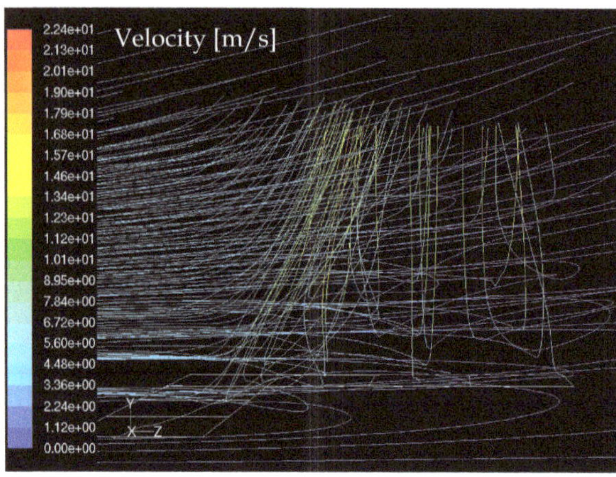

Fig. 7. Side view of streamlines after 18.6 minutes for the wind of 4m/s in Model 1

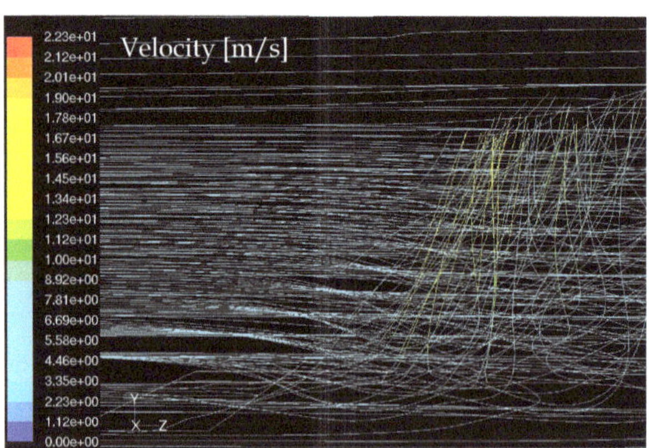

Fig. 8. Side view of streamlines after 26.6 minutes for the wind of 4m/s in Model 1

5.1.2 Model 2

Figs. 9 and 10 are the samples of streamline distribution after 15 minutes and that after 35 minutes for the wind of 4m/s. These streamlines are also drawn from the pressure surface whose pressure is lower by 30 Pa from the atmospheric pressure, and are colored by velocity magnitude.

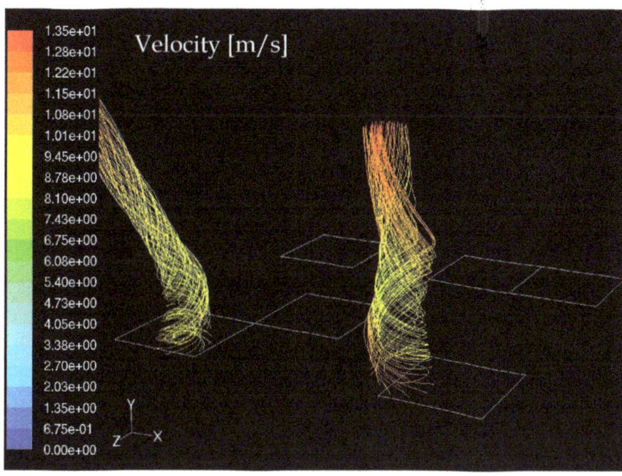

Fig. 9. Distribution of streamlines after 15 minutes for the wind of 4m/s

In contrast with the results in the "L" character type, number of generated fire whirlwind is almost the same, i.e. two, even if the velocity of wind is altered. Behavior of the whirlwind (movement and extinction) is also almost the same in the both cases of wind velocity 6 and 8m/s. In the case of wind velocity 4m/s, a whirlwind generated after 12 minutes around two heat sources surfaces moves and once weakened because it is apart from the heat source. Then, it is strengthened again as it returns on the heat source. The point that we should pay attention to is that two fire whirlwinds in the case of 4m/s (one is generated after 12 minutes and remains until the end of simulation, and the other is generated at 35 minutes and extinct at 39 minutes) move to the upstream.

Generation points of the whirlwind are located at the downstream heat source, and the number of the generated whirlwind is almost the same though the velocity of wind is varied. In the random type of heat source surfaces, there is a lack of heat source in the upstream. Therefore, interaction between the ascending current of air by natural convection and wind is disturbed (See Fig. 11). Then, the interacted flow arrives at the downstream heat source, and easily generates the fire whirlwind because heat is supplied.

It is summarized in the Fig. 12. Duration time of the whirlwind becomes shorter as the velocity of wind is increased.

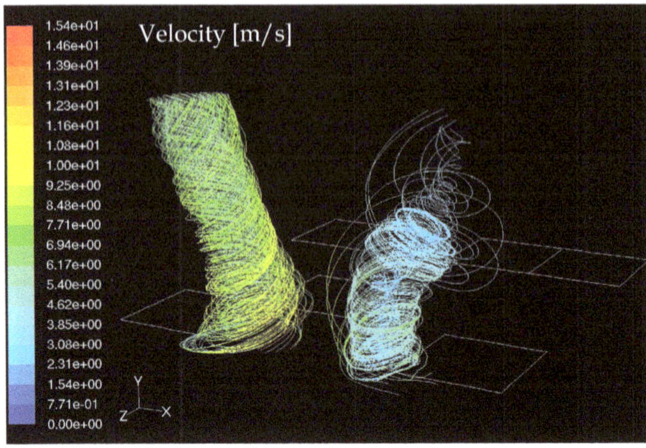

Fig. 10. Distribution of streamlines after 35 minutes for the wind of 4m/s

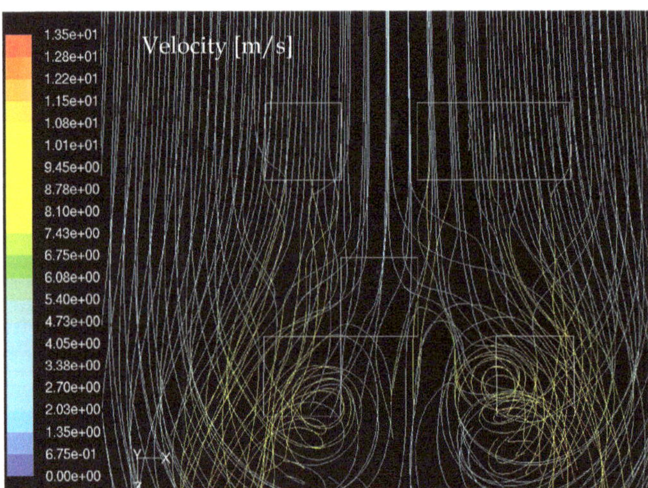

Fig. 11. Upper view of streamlines after 18 minutes for the wind of 4m/s in Model 2.

5.1.3 Summary

As a result of numerical analysis, interaction between the ascending current of air by natural convection and wind results in a fire whirlwind. The whirlwind sometimes moves and extinct. Generation points of the whirlwind are located at the tips and corner in the model of "L" character type. The points where the natural convective flow develops becomes origins of the fire whirlwind. Generation points of the whirlwind in the model of random type are located at the downstream heat source, and the number of the generated whirlwind is almost the same though the velocity of wind is varied. This is because the ascending current of air by natural convection is partially disturbed by the wind.

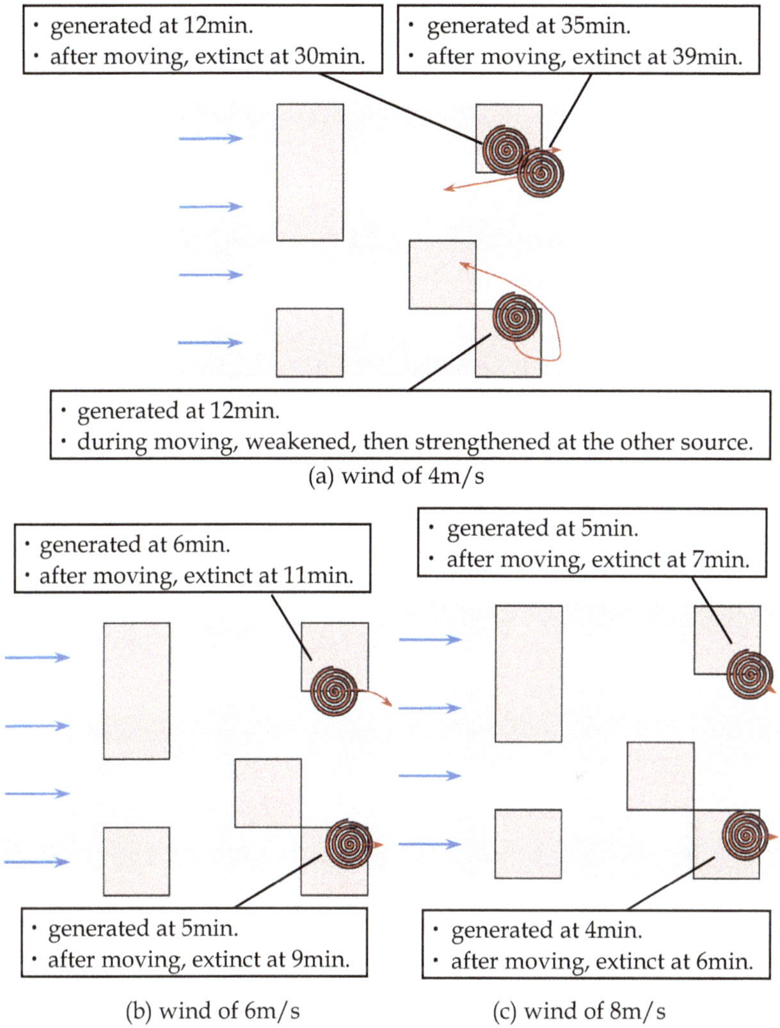

Fig. 12. Summary of generation and extinction of fire whirlwind in Model 2

5.2 Scale effect analysis

In this section, performing the numerical analysis of fire whirlwind with respect to scale effect, it is examined whether a relationship exists between a real phenomenon and the phenomenon in the reduction model with taking into account radiative heat transfer (Sakai & Miyagi, 2010).

Preceding the parametric calculation study, grid sensitivity for numerical result is evaluated. Residuals of continuity, velocity components, energy, turbulent energy and dissipation ratio of turbulent energy are derived and those absolute criteria are set to 10^{-3}

except 10^{-6} for energy to judge the convergence. For all scale ratios, grid number is altered and grid sensitivity is checked. Even if the coarse mesh number of 570700 cells and 1442164 faces is utilized for calculation, those residuals are well converged within 1000 iterations for each time step. Therefore, mesh number of 570700 cells and 1442164 faces is employed to attain the precision and calculation speed.

5.2.1 Effect of radiative heat exchange

In our previous study, radiative heat exchange was not considered to simulate the whirlwind, because we had to simplify the real phenomenon of the whirlwind into the fundamental thermal and flow fields as much as possible (Sakai and Watanabe, 2007). However, radiative heat exchange plays an important role in the heat transfer at the higher temperature field. Therefore, the P-1 model is utilized to simulate the radiative heat transfer from the heat source at high temperature.

Fig. 13 shows numerical results of streamlines in natural convective heat transfer at 5 minutes (no horizontal wind) for the L-character heat source in the original scale; (a) is the case of no radiative heat exchange considered, and (b) is the case of radiative heat exchange considered. Colors of streamlines show the temperature of the upward flow.

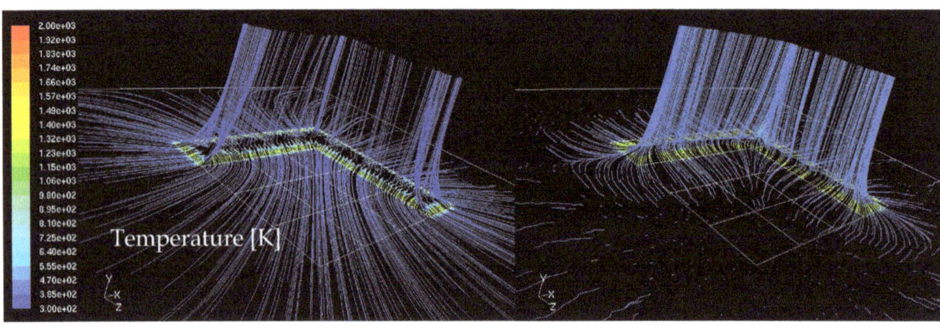

(a) No radiative heat exchange considered (b) Radiative heat exchange considered

Fig. 13. Distribution of streamlines after 5 minutes in natural convective heat transfer

In the case of no radiative heat exchange considered, temperature decreases within a few decade meters from the heat source. Therefore, air above the heat source plays an adiabatic role to prevent the heat transfer to the upper air layer. On the other hand, in the case of radiative exchange considered, heat from the heat source flows more to the upper air layer, resulting in more developed upward flow. In the case of no radiative exchange considered, the maximum upward velocity of natural convection and the appropriate velocity of horizontal wind to generate swirling flow are 23.5 m/s and 4 m/s, respectively. In the case of radiative exchange considered, the maximum upward velocity of natural convection and the appropriate velocity of horizontal wind are increased to 65.2 m/s and 11 m/s, respectively, due to the effective heat transfer. It is clarified that the radiative heat exchange must be considered to precisely simulate the fire whirlwind. The following numerical results are taking into account the radiative heat exchange.

5.2.2 Generation and extinction of swirling flow in Model 1

Fig. 14 shows samples of pressure field at the bottom surface and streamline distributions for the original scale simulation; (a) is at 2 minutes since the introduction of horizontal wind, (b) is at 5.5 minutes since the introduction of horizontal wind, and (c) is at 7.5 minutes since the introduction of horizontal wind. These streamlines are drawn from the bottom surface, and are colored by velocity magnitude.

From Fig. 14 (a), two whirlwinds are generated and generation points of the whirlwind are located at the tip and the corner of "L" character. At 5.5 minutes since the introduction of horizontal wind, a whirlwind generated at the tip is extinguished, and the other whirlwind is flown to the downstream (See Fig. 14 (b)). The remaining whirlwind is the strongest at this time, and the maximum center pressure of the whirlwind is lower than the atmospheric pressure by 1,061Pa. At 7.5 minutes since the introduction of horizontal wind, the remaining whirlwind is divided into two whirlwinds (See Fig. 14 (c)), and is going to be extinguished at 10 minutes since the introduction of horizontal wind.

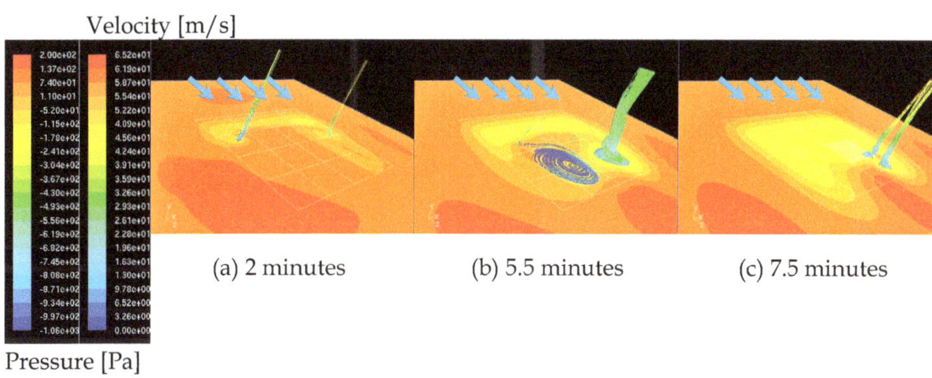

Fig. 14. Pressure field at the bottom surface and streamline distributions since the introduction of horizontal wind in Model 1

Numerical calculation is performed not only for the case of the original size, but also for 1/10 scale, 1/50 scale and 1/100 scale. In the original size calculation, natural convection is developed in the domain after 5 minutes. The smaller the scale is, the shorter the time necessary to develop the natural convection is. The times necessary to develop the natural convection are 90 seconds for 1/10 scale model, 45 seconds for 1/50 scale model, and 30 seconds for 1/100 scale model. This is because dynamics between the original size and scaled models are said to be similar if the rate of the scale ratio to the square of velocity (or time) are the same, in which Froude numbers are the same.

After defining the appropriate velocity of horizontal wind to generate swirling flow, behavior of the whirlwind (movement and extinction) is observed for all the cases. Number of generated fire whirlwind and behavior of the whirlwind are almost the same for all the cases. Therefore, characteristic physical quantities of whirlwind are discussed by comparing with scale ratio and square root of scale ratio in the followings.

5.2.2.1 Pressure of whirlwind

Fig. 15 shows relation among the scale ratio, the square root of scale ratio, and pressure at the bottom of natural convection after each period necessary to develop the natural convection. Taken the original size as a standard, the pressure at the bottom of natural convection is more coincident with the scale ratio rather than the square root of scale ratio.

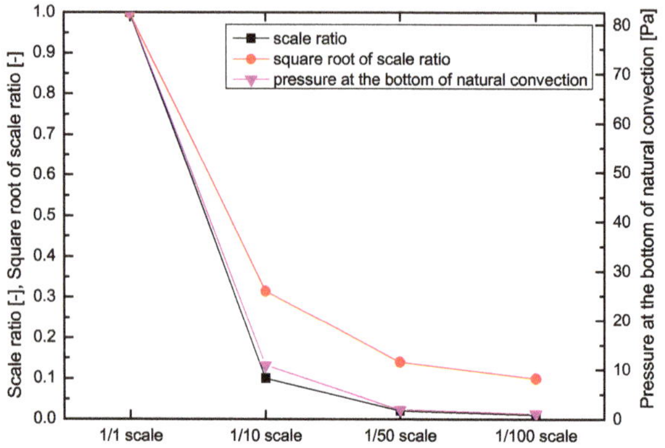

Fig. 15. Relation among scale ratio, square root of scale ratio, and pressure at the bottom of natural convection in Model 1

Fig. 16 shows relation among the scale ratio, the square root of scale ratio, and the maximum center pressure of whirlwind, which is lower than the atmospheric pressure. (The times of the strongest whirlwind generation after the introduction of the horizontal wind, i.e. the times when the center pressure of whirlwind is maximum, are different for all cases, and discussed in the next subsubsection.) The maximum center pressure of whirlwind is also more coincident with the scale ratio rather than the square root of scale ratio. Therefore, scale effect is recognized, in which pressure of whirlwind is coincident with the scale ratio.

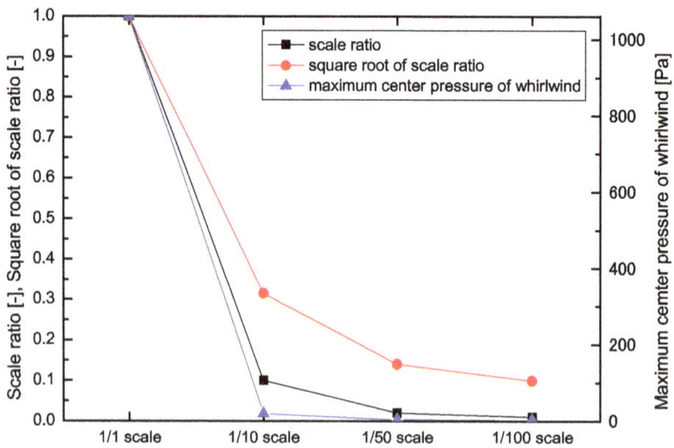

Fig. 16. Relation among scale ratio, square root of scale ratio, and maximum center pressure of whirlwind in Model 1

5.2.2.2 Time variation of whirlwind

Figs. 17 and 18 show relation among the scale ratio, the square root of scale ratio, and times of whirlwind generation and extinction since introduction of horizontal wind. Taken the original size as a standard, the times of whirlwind generation and extinction since introduction of horizontal wind are more coincident with the square root of scale ratio rather than the scale ratio. Duration time of whirlwind is derived from the times of whirlwind generation and extinction since introduction of horizontal wind. As a matter of course, the duration time is more coincident with the square root of scale ratio rather than the scale ratio.

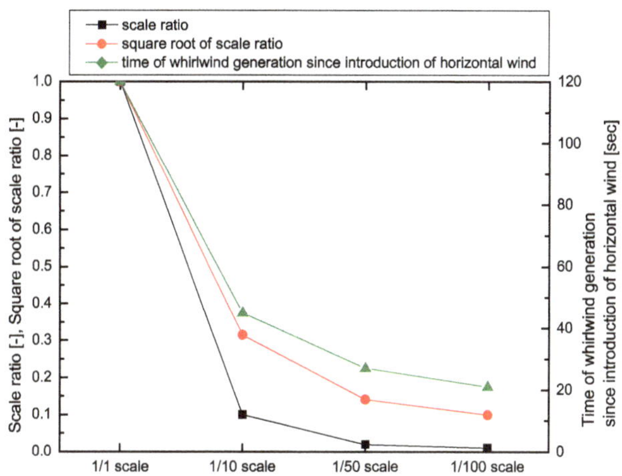

Fig. 17. Relation among scale ratio, square root of scale ratio, and time of whirlwind generation since introduction of horizontal wind in Model 1

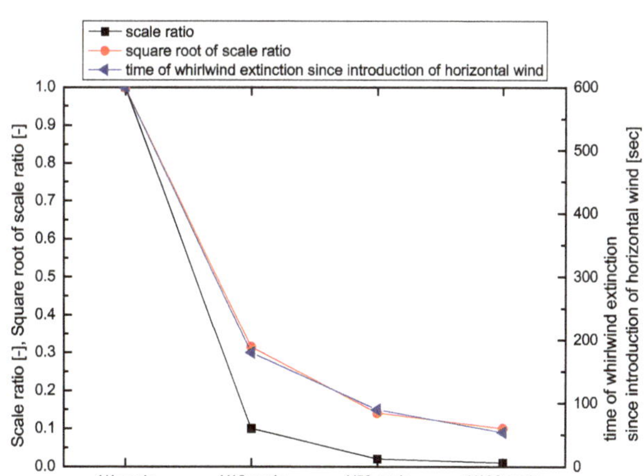

Fig. 18. Relation among scale ratio, square root of scale ratio, and time of whirlwind extinction since introduction of horizontal wind in Model 1

Fig. 19 shows relation among the scale ratio, the square root of scale ratio, and time of the strongest whirlwind generation since introduction of horizontal wind. The time of the strongest whirlwind generation is also more coincident with the square root of scale ratio rather than the scale ratio. Therefore, scale effect is recognized, in which time variation of whirlwind is coincident with the square root of scale ratio.

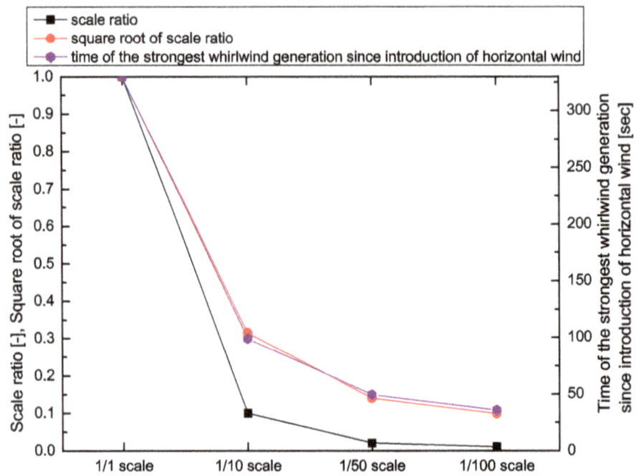

Fig. 19. Relation among scale ratio, square root of scale ratio, and time of the strongest whirlwind generation since introduction of horizontal wind in Model 1

5.2.2.3 Velocity of horizontal wind

Fig. 20 shows relation among the scale ratio, the square root of scale ratio, and appropriate velocity of horizontal wind to generate swirling flow. The appropriate velocity of horizontal wind to generate swirling flow is also more coincident with the square root of scale ratio rather than the scale ratio. Therefore, scale effect is recognized, in which velocity of horizontal wind is coincident with the square root of scale ratio.

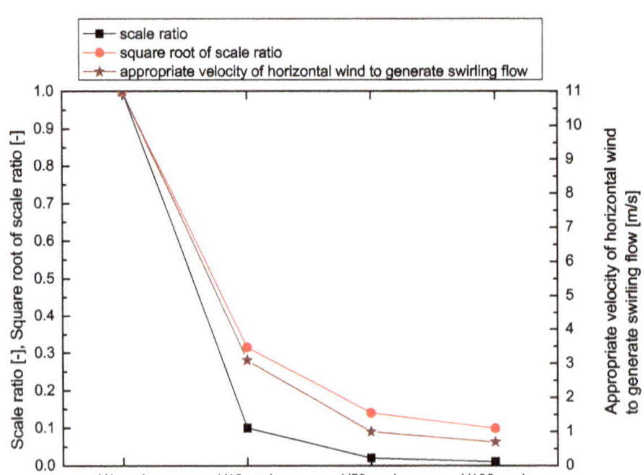

Fig. 20. Relation among scale ratio, square root of scale ratio, and appropriate velocity of horizontal wind to generate swirling flow in Model 1

5.2.3 Generation and extinction of swirling flow in Model 3

Fig. 21 shows samples of pressure field at the bottom surface and streamline distributions for the original scale simulation; (a) is at 2 minutes since the introduction of horizontal wind, (b) is at 4.5 minutes since the introduction of horizontal wind, and (c) is at 5.5 minutes since the introduction of horizontal wind. These streamlines are drawn from the bottom surface, and are colored by velocity magnitude.

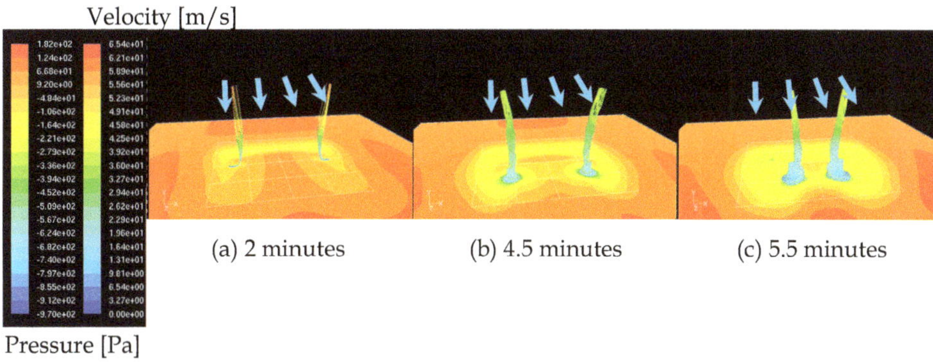

Fig. 21. Pressure field at the bottom surface and streamline distributions since the introduction of horizontal wind in Model 3

From Fig. 21 (a), two whirlwinds are generated and generation points of the whirlwind are located at the corners of "C" character. The whirlwinds are flown to the downstream, and are the strongest at this time (See Fig. 21 (b)). The maximum center pressure of the whirlwinds is lower than the atmospheric pressure by 970Pa. Then, the whirlwinds approach each other regardless of horizontal wind direction (See Fig. 21 (c)), and is remaining at 10 minutes since the introduction of horizontal wind.

Numerical calculation is performed not only for the case of the original size, but also for 1/10 scale, 1/50 scale and 1/100 scale. In the original size calculation, natural convection is developed in the domain after 5 minutes. The smaller the scale is, the shorter the time necessary to develop the natural convection is. The times necessary to develop the natural convection are 30 seconds for 1/10 scale model, 15 seconds for 1/50 scale model, and 9 seconds for 1/100 scale model. These values are smaller than those of L-character heat source. This is because the wider the area of heat source is, the more natural convection develops.

Introducing the appropriate velocity of horizontal wind to generate swirling flow in Model 1 as a velocity of horizontal wind, behavior of the whirlwind (movement and extinction) is observed for all the cases. While number of generated fire whirlwind is the same for all the cases, whirlwinds are not extinguished, and are still remained in the cases of the original size and 1/50 scale model. This is because the appropriate velocity of horizontal wind to generate swirling flow in Model 1 is larger or smaller compared with the appropriate velocity of horizontal wind to generate swirling flow in Model 3. Relation between configuration of heat source and horizontal wind velocity is one of the important roles to simulate generation and extinction of fire whirlwind. Despite of the difference in behavior of the whirlwind, characteristic physical quantities of whirlwind are discussed by comparing with scale ratio and square root of scale ratio as well as in the case of Model 1.

5.2.3.1 Pressure of whirlwind

Fig. 22 shows relation among the scale ratio, the square root of scale ratio, and pressure at the bottom of natural convection after each period necessary to develop the natural convection.

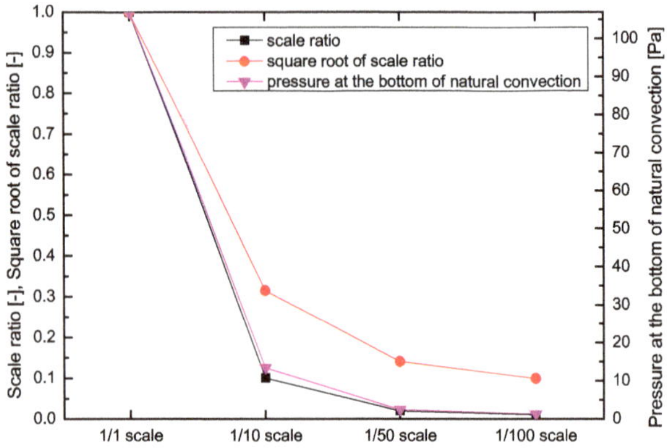

Fig. 22. Relation among scale ratio, square root of scale ratio, and pressure at the bottom of natural convection in Model 3

Taken the original size as a standard, the pressure at the bottom of natural convection is more coincident with the scale ratio rather than the square root of scale ratio.

Fig. 23 shows relation among the scale ratio, the square root of scale ratio, and the maximum center pressure of whirlwind, which is lower than the atmospheric pressure. The maximum center pressure of whirlwind is also more coincident with the scale ratio rather than the square root of scale ratio. Therefore, scale effect may be recognized, in which pressure of whirlwind is coincident with the scale ratio.

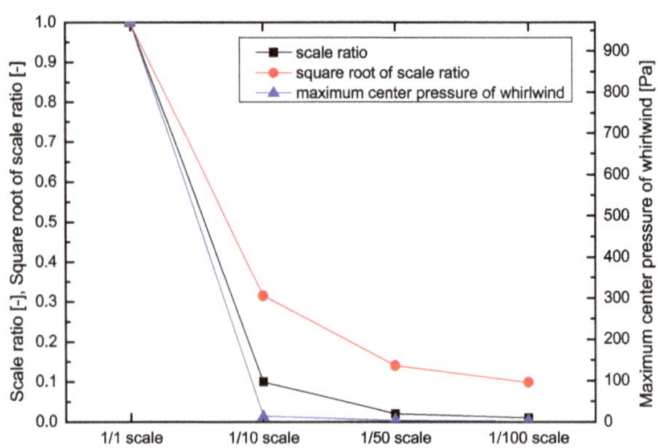

Fig. 23. Relation among scale ratio, square root of scale ratio, and maximum center pressure of whirlwind in Model 3

5.2.3.2 Time variation of whirlwind

Figs. 24 and 25 show relation among the scale ratio, the square root of scale ratio, and times of whirlwind generation and the strongest whirlwind generation since introduction of horizontal wind. Here is no discussion for time of whirlwind extinction, because whirlwinds are not extinguished, and are still remained in the cases of the original size and 1/50 scale model. The times of whirlwind generation and extinction since introduction of horizontal wind are more coincident with the square root of scale ratio rather than the scale ratio. Therefore, scale effect may be recognized, in which time variation of whirlwind is coincident with the square root of scale ratio.

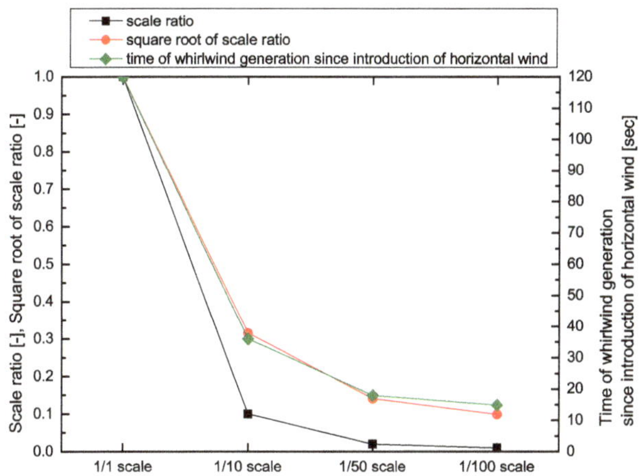

Fig. 24. Relation among scale ratio, square root of scale ratio, and time of whirlwind generation since introduction of horizontal wind in Model 3

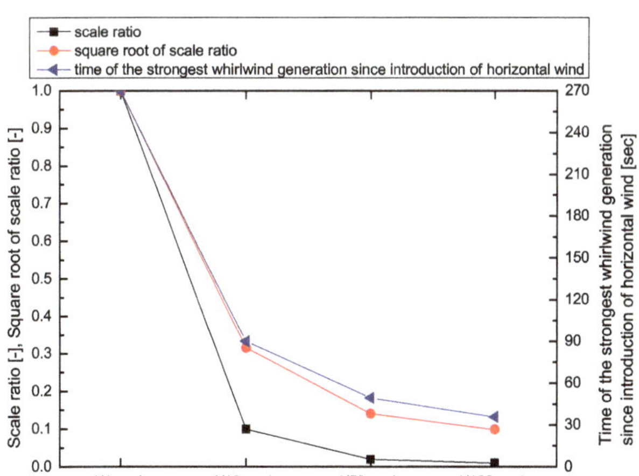

Fig. 25. Relation among scale ratio, square root of scale ratio, and time of whirlwind extinction since introduction of horizontal wind in Model 1

5.2.4 Summary

In this section, performing the numerical analysis of fire whirlwind with respect to scale effect, it is examined whether a relationship exists between a real phenomenon and the phenomenon in the reduction model with taking into account radiative heat transfer. It is clarified that the radiative heat exchange must be considered to precisely simulate the fire whirlwind, because the radiative heat transfer encourages the maximum upward velocity of natural convection and the appropriate velocity of horizontal wind due to the effective heat transfer. Scale effect is recognized in the L-character heat source model, in which pressure of whirlwind is coincident with the scale ratio, time variation of whirlwind and velocity of horizontal wind are coincident with the square root of scale ratio.

6. Conclusions

In this study, it is firstly aimed to get basic knowledge to elucidate outbreak mechanism of fire whirlwind and the property, by evaluating influence between a natural convection generated by the fire and the wind blowing to fire whirlwind on the ground utilizing numerical analysis.

- Interaction between the ascending current of air by natural convection and wind results in a fire whirlwind. The whirlwind sometimes moves and extinct. Generation points of the whirlwind are located at the tips and corner in the model of "L" character type. The points where the natural convective flow develops becomes origins of the fire whirlwind.
- Generation points of the whirlwind in the model of random type are located at the downstream heat source, and the number of the generated whirlwind is almost the same though the velocity of wind is varied.

It is secondly examined whether a relationship exists between a real phenomenon and the phenomenon in reduction models, performing the numerical analysis of a fire whirlwind with respect to scale effect.

- It is clarified that the radiative heat exchange must be considered to precisely simulate the fire whirlwind, because the radiative heat transfer encourages the maximum upward velocity of natural convection and the appropriate velocity of horizontal wind due to the effective heat transfer.
- Scale effect is recognized in the L-character heat source model, in which pressure of whirlwind is coincident with the scale ratio, time variation of whirlwind and velocity of horizontal wind are coincident with the square root of scale ratio.

7. References

Battaglia, F.; McGrattan, K. B.; Rehm, R. G. & Baum, H. R. (2000b). Simulating Fire Whirls, *Combustion Theory and Modelling*, Vol.4, No.3, pp.123-138

Battaglia, F.; Rehm, R. G. & Baum, H. R. (2000a). Fluid Mechanics of Fire Whirls: An Inviscid Model, *Physics of Fluids*, Vol.12, No.11, pp.2859-2867

Byram, G. M. & Martin, R. E. (1970). The Modeling of Fire Whirlwinds, *Forest Science*, Vol.16, No.4, pp.386-399

Chuah, K. H. & Kushida, G. (2007). The prediction of flame heights and flame shapes of small fire whirls, *Proceedings of the Combustion Institute*, Vol. 31, pp.2599–2606

Chuah, K. H.; Kuwana, K.; Saito, K. & Williams, F. A. (2011). Inclined fire whirls, *Proceedings of the Combuation Institute*, Vol.33, pp.2417-2424

Emmons, H. W. & Ying, S. J. (1967) The fire whirl, *Proceedings of the Combustion Institute*, Vol.11, pp. 475–488

Emori, R. I. & Saito, K. (1982). Model experiment of hazardous forest fire whirl, *Fire Technology*, Vol.18, No.4, pp.319-327

Fluent Inc./Fluent Asia Pacific (2006). *FLUENT 6.3 User's Guide*, Fluent Incorporated

Graham, H. E. (1955). Fire whirlwinds, *Bulletin American Meteorological Society*, Vol.36, pp. 99–103

Grishin, A. M. (2007). Effect of the Interaction between Fire Tornadoes on Their Propagation, *Doklady Physics*, Vol.52, No.10, pp. 521-522

Grishin, A. M.; Matvienko, O. V. & Rudi, Y. A. (2009). Mathematical Modeling of Gas Combustion in A Twisted Jet and of The Formation of A Fiery Whirlwind, *Journal of Engineering Physics and Thermophysics*, Vol.82, No.5, pp.906-913

Haines, D. A. & Updike, G. H. (1971). Fire Whirlwind Formation over Flat Terrain, *USDA Forest Service Research Paper*, NC-71

Hassan, M. I.; Kuwana, K.; Saito, K. & Wang, F. (2005). Flow Structure Of A Fixed-frame Type Firewhirl, *Proceedings of the Eighth International Symposium*, International Association for Fire Safety Science, Beijing, China, September 18-23, pp.951-962

Hayashi, Y.; Ohmiya, Y.; Iwami, T. & Saga, T. (2003). Experimental Study on Fire and Plume Properties Using BRI's Fire Wind Tunnel Facility, *International Journal for Fire Science and Technology*, Vol.22, pp.17-35

Hough, S. E. & Bilham, R. G. (2005). *After the Earth Quakes: Elastic Rebound on an Urban Planet*, Oxford University Press USA

Kuwana, K.; Sekimoto, K.; Saito, K. & Williams, F. A. (2007). Can We Predict the Occurrence of Extreme Fire Whirls?, *AIAA JOURNAL*, Vol.45, pp.16-19

Kuwana, K.; Sekimoto, K.; Saito, K. & Williams, F. A. (2008). Scaling fire whirls, *Fire Safety Journal*, Vol.43, No.4, pp.252-257

Liu, N. A. (2005). Experimental and Theoretical Investigation on Fire Interactions and the Induced Firewhirls in Square Fire Arrays, *Proceedings of Fifth NRIFD International Symposium on Forest Fires*, Tokyo, pp.293-301

Liu, N. A.; Liu, Q.; Deng, Z. H.; Satoh, K. & Zhu, J. P. (2007). Burn-out Time Data Analysis on Interaction Effects among Multiple Fires in Fire Arrays, *Proceedings of the Combustion Institute*, Vol.31, pp.2589-2597

Martin, R. E.; Pendleton, D. W. & Burgess, W. (1976). Effect of Fire Whirlwind Formation on Solid Fuel Burning Rates, *Fire Technology*, Vol.12, No.1, pp.33-40

Muraszew, A.; Fedele, J. B. & Kuby W.C. (1979). The fire whirl phenomenon, *Combustion and Flame*, Vol. 34, pp.29-45

Sakai, S. & Miyagi, N. (2010). Numerical Study of Fire Whirlwind Taking into Account Radiative Heat Transfer, *IOP Conference Series: Materials Science and Engineering*, Vol.10, 012031

Sakai, S. & Watanabe, Y. (2007). Numerical Study of Interaction between Natural Convection Flow and horizontal wind, *Proceedings of the FEDSM2007, 5th Joint*

ASME/JSME Fluids Engineering Conference, FEDSM2007-37212, San Diego, California, USA, July 30-August 2, 2007

Satoh, K. & Yang, K. T. (1996). Experimental Observations of Swirling Fires, *Proceedings of ASME Heat Transfer Division,* HTD-Vol.335, No.4, pp.393-400

Satoh, K. & Yang, K. T. (1997). Simulations of swirling fires controlled by channeled self-generated entrainment flows [A], *Fire Safety Sci., Proceedings of the 5th International Symposium[C],* pp.201-212

Siegel, R. & Howell, J. (2002). *Thermal Radiation Heat Transfer, 4th Edition,* Taylor & Fransis

Snegirev, A. Y.; Marsden, J. A.; Francis, J. & Makhviladze, G. M. (2004). Numerical studies and experimental observations of whirling flames, *International Journal of Heat and Mass Transfer,* Vol.47, pp.2523–2539

Takewaki, I.; Murakami, S.; Fujita, K.; Yoshitomi, S. & Tsuji, M. (2011). The 2011 off the Pacific coast of Tohoku earthquake and response of high-rise buildings under long-period ground motions, *Soil Dynamics and Earthquake Engineering,* in press

Application of a Highly Reduced One-Dimensional Spring-Dashpot System to Inelastic SSI Systems Subjected to Earthquake Ground Motions

Masato Saitoh
Saitama University, Shimo-Okubo Sakura-Ku Saitama, Saitama
Japan

1. Introduction

Recent devastating earthquakes in many countries, particularly the 2010 Haiti earthquake (e.g., Eberhard et al., 2010) and the 2011 off the Pacific coast of Tohoku earthquake (e.g., Takewaki et al., 2011) in Japan, have caused severe damages to buildings and structures. The past and the recent seismic events have led us to attempt to improve technologies used in practical applications for evaluating the dynamic behavior of structural systems subjected to earthquake ground motions. To accomplish this, it is considered that an appropriate representation of soil-structure interaction (SSI) effects may be of great importance in earthquake engineering.

SSI has been studied since the late 19th century. Since then, developments in SSI science over the years have resulted in the latest SSI technologies (Kausel, 2010). In recent years, the effects of SSI have been taken into consideration in various practical numerical computations. For analyzing the dynamic response of SSI systems, a substructure method is often used for performing more efficient computations with lesser degrees of freedom (DOFs) rather than more comprehensive models with extremely large number of DOFs. In a substructure method, impedance functions (IFs) are generally used to represent the dynamic stiffness and damping of soil-foundation systems. Most IFs of soil-foundation systems exhibit various frequency-dependent characteristics; they usually occur as a result of the reflection and refraction of traveling waves originating from the foundations. Numerous studies associated with the frequency-dependent characteristics of IFs have been conducted over the past several years. IFs exhibit the following typical frequency-dependent characteristics: (a) slight oscillation shown in soil reaction and surface rigid foundations or embedded rigid foundations (Baranov, 1967; Beredugo & Novak, 1972; Novak, 1974; Novak et al., 1978; Veletsos & Dotson, 1988; Gazetas, 1991; Saitoh & Watanabe, 2004; Tileylioglu et al., 2011); (b) multiple oscillations typically exhibited in pile groups (Kaynia & Kausel, 1982; Dobry & Gazetas, 1988; Makris & Gazetas, 1993; Mylonakis & Gazetas, 1998); and (c) a cut-off frequency below which damping is negligible and above which damping increases rapidly (Novak & Nogami, 1977; Nogami & Novak, 1977; Kausel & Roesset, 1975; Elsabee &

Morray, 1977; Takemiya & Yamada, 1981). These studies have indicated that considering frequency dependency in the response analysis of structures can yield more accurate calculations in SSI systems.

In contrast to the findings of many of the latest studies of various foundations under diverse conditions and in recent numerical computations, frequency dependency in SSI systems has been recognized as a hindrance. In accordance with the recent performance-based seismic design of structures, nonlinearity in structural members, such as cracking, yielding, and collapse, needs to be considered in computations in order to predict the inelastic response of structural systems subjected to the expected earthquake waves. In calculating the response, step-by-step numerical procedures in the time domain are usually used with constitutive models, because the nonlinearity of structural members strongly depends upon the stress pass being integrated stepwise. Therefore, a conventional method for considering frequency dependency in the frequency domain cannot be applied. This has been one of the most important problems in structural engineering and geotechnical engineering over the past few decades.

Recently, various methods that are ready to use in practice have been proposed to overcome this issue, when frequency dependency in IFs and nonlinearity in structures are simultaneously taken into account. Basically, there are two main streams of thought in dealing with this problem. One is to use a method for transforming IFs into impulse responses in the time domain. The origins of this method date back to Wolf & Obernhuber, 1985, who developed a numerical method in the time domain in which the impulse response obtained from dynamic soil stiffness by using inverse Fourier transform was applied to the response analysis. Since this method was first developed in the 1980s, many transform methods in the time domain have been proposed and improved to overcome difficulties in various frequency dependencies in IFs (Wolf & Motosaka, 1989; Meek, 1990; Motosaka & Nagano, 1992; Hayashi & Katsukura, 1990). Nakamura (2006a; 2006b; 2008a; 2008b) has developed various sophisticated transform methods that can deal with strong frequency dependency in IFs, non-causal impedance with large hysteretic damping, and soil nonlinearity.

The other method for dealing with the SSI problem is the use of a lumped parameter model (LPM), which is considered to be a very powerful tool for solving this problem. In general, an LPM consists of springs, dashpots, and masses having frequency-independent coefficients. A particular combination of these elements can simulate a frequency-dependent impedance characteristic. The advantage of LPMs is that they can be easily incorporated into a conventional numerical analysis in the time domain, even under nonlinear conditions of superstructures. Time stepping methods that have conventionally been applied to structural analyses, such as the central difference method, Newmark's method, and Wilson's method—employed as conventional integration schemes—can be used with LPMs, while time-domain transform methods usually need a specific scheme to incorporate the impulse response into the response analysis.

From the viewpoint of construction schemes in LPMs, the existing LPMs can be categorized into three types: a) semi-empirical LPMs, b) systematic LPMs, and c) modal LPMs.

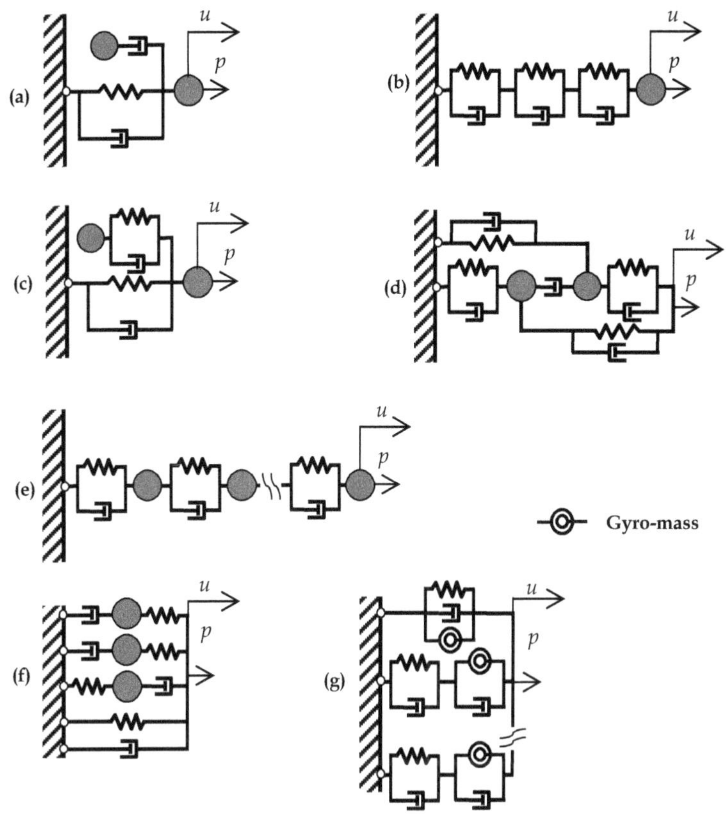

Fig. 1. Semi-empirical lumped parameter models proposed in previous studies [(a) Meek & Veletsos, 1974; Wolf & Somaini, 1986; (b) Nogami & Konagai, 1986; 1988; (c) de Barros & Luco, 1990; (d) Jean et al., 1990; (e) Wu & Chen, 2001; (f) Taherzadeh et al., 2009; and (g) Saitoh, 2011a].

In semi-empirical LPMs, mechanical elements such as springs, dashpots, and masses, which have frequency-independent coefficients, are arranged appropriately, depending on how significantly an optimal fit of the IFs obtained using LPMs with the corresponding target (exact) IFs can be achieved using specific arrangements. The values of the elements are usually determined by minimizing the errors from the target IFs. Meek & Veletsos, 1974, initially presented an LPM that represents the lateral and rocking impedance characteristics for a truncated semi-infinite cone that can be considered as an analog of an elastic half space. In an LPM, a mass has an additional DOF that is not directly attached to the foundation node but rather is connected to it through a dashpot, as shown in Fig. 1a. Wolf & Somaini, 1986, extended the above model to a rigid disk, an embedded cylinder, a rectangle, and a strip, and specified the coefficients of the elements comprising the foundations. Nogami & Konagai, 1986; 1988, represented the subgrade reaction of soil surrounding single piles by using three Voigt models connected in series (in the axial direction) and those with a mass (in the flexural direction), as shown in Fig. 1b. Alternative LPMs (de Barros & Luco, 1990

(Fig. 1c); Jean et al., 1990 (Fig.1d)) were proposed for a better fit to the impedance functions than the precedent LPMs. For a more concise usage of LPMs in practical applications, the coefficients of simple LPMs have been specified in tables for rigid foundations on the surface and with embedment in all translational and rotational motions (Wolf & Paronesso, 1992; Wolf, 1997). Wu & Chen, 2001, proposed an LPM consisting of a set of units in which each unit consists of a mass connected to a spring and a dashpot arranged in parallel, as shown in Fig. 1e. Wu & Chen, 2002, adopted the LPM to a simple SSI analysis for seismic excitations. Recently, other forms of LPMs have been proposed by researchers (Taherzadeh et al., 2009, (Fig. 1f); Khodabakhshi et al., 2011).

As described in Wolf, 1994; Wu & Chen, 2002; and Wu & Lee, 2002, however, the inclusion of any mass in an LPM becomes a drawback when an LPM is straightforwardly used in seismic excitations, that is, the driving forces at the foundation node induced by a foundation input motion differ from the corresponding forces by applying the same input motion at the bottom end (the far end) of an LPM because of the mass. This problem does not occur when an LPM consists of springs and dashpots. However, the lack of mass in an LPM tends to show difficulty in an appropriate fit to the exact impedance functions, and thus, a substantial increase in the number of elements and DOFs is necessary.

To overcome this issue, Saitoh, 2007, proposed an LPM that uses a gyro-mass element instead of an ordinary mass. The gyro-mass element generates a reaction force proportional to the relative acceleration of the nodes between which it is placed. The use of the gyro-mass element does not influence the driving forces acting at the foundation because the gyro-mass element does not generate any inertia force. An LPM with a gyro-mass element (GLPM) accomplishes a rapid change in frequency in IFs with a small number of elements without an ordinary mass. In Saitoh's study, two types of LPMs—*Type I* and *Type II* models—are presented for simulating IFs that have cut-off frequencies and frequency-dependent oscillations, respectively. In Saitoh, 2011a, the accuracy in the GLPMs was verified using an example of 2 × 4 pile groups embedded in a layered soil medium, supporting a 1DOF system having inelasticity in the structural member when subjected to ground motions. In his study, a more generalized GLPM, called "the *Type III* model" was proposed for a more appropriate fit (Fig. 1g).

In the development of semi-empirical LPMs, attempts have been made to obtain an easier and more systematic determination of the values for the coefficients in LPMs. A systematic procedure was first proposed by Wolf, 1991a; 1991b. In the procedure, the dynamic-stiffness coefficient is approximated as a ratio of two polynominals, which is then formulated as a partial-fraction expansion whose each term is represented by a discrete model comprising parallel-form LPMs, as shown in Fig. 2a. Wu & Lee, 2002, proposed series-form LPMs (Fig. 2b) formulated with a ratio of two polynominals approximating the flexibility functions (compliance functions) instead of using IFs. Moreover, Wu & Lee, 2004, alternatively proposed nested LPMs (Fig. 2c) based on a continued-fraction expansion instead of a partial-fraction expansion. The advantage of nested LPMs is that the configuration is independent of the soil-foundation systems being dealt with, whereas the previous series-form LPMs, which use the partial-fraction expansion, depend upon them. Another new systematic LPM (Fig. 2d) based on the continued-fraction expansion was proposed by Zhao & Du, 2008.

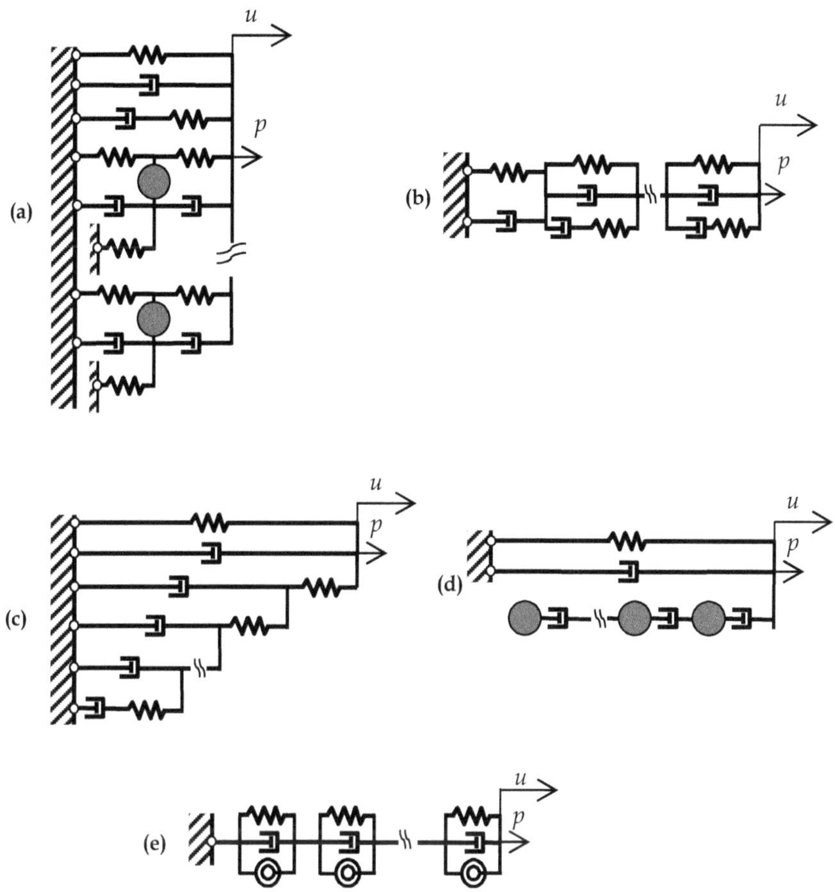

Fig. 2. Systematic lumped parameter models [(a) Wolf, 1991a; 1991b; (b) Wu & Lee, 2002; (c) Wu & Lee, 2004; and (d) Zhao & Du, 2008] and a modal LPM [(e) Saitoh, 2010a] proposed in previous studies.

The aforementioned LPMs need to approximate the target IFs by using specific functions such as the ratio of two polynominals in the case of systematic LPMs. Recently, two new transform methods for constructing an exact LPM from the original systems have been developed in the field of computational mechanics (Saitoh, 2010a; 2010b). In Saitoh, 2010a, an LPM consisting of units arranged in series—in which each unit consists of a spring, a dashpot, and a gyro-mass element arranged in parallel (Fig. 2e)—is formulated, i.e., calculated from a closed-form solution based on a modal expansion. In Saitoh's study, an LPM that represents the impedance characteristics at the extremity of a uniform, isotropic, and homogeneous rod supported by continuously distributed springs and dashpots (the Kelvin-Voigt model is assumed as a viscoelastic medium) was proposed. This method can be applied to systems where the conventional modal expansion is available for solving the differential equation of motion. In a later study (Saitoh, 2010b), a transformation procedure based on a conventional complex modal analysis was proposed in which the impedance function in general linearly elastic systems with non-classical damping is transformed into an exact one-dimensional spring-dashpot system (1DSD) comprising units arranged in series. Each unit, which is directly related to each vibrating mode of the original system, is a parallel system consisting of a spring, a dashpot, and a unit having a spring and a dashpot arranged in series. The properties of the elements comprising the 1DSDs are automatically determined through the proposed procedure by using complex modal quantities. The advantage of 1DSDs is that 1DSD transformation offers compatibility with the merit of complex modal analysis, that is, a large number of units associated with high modes beyond the target frequency region can be removed from the 1DSDs as an approximate expression of impedance functions. Accordingly, a marked decrease in the computational domain size and time with the use of the 1DSDs can be achieved. Extremely complicated frequency dependency in impedance functions over a wide frequency range tends to appear in diverse technological applications, as exhibited in Saitoh, 2010b. In such cases, the approximation of the target IFs by using specific functions in the previous LPMs may encounter a certain limitation in terms of accuracy and may not be accomplished sufficiently. On the other hand, the 1DSDs transform procedure provides an exact LPM at the initial step, that is, we can adjust the number of DOFs (the number of units) in the reduced LPM by achieving a balance with the accuracy from the exact LPM. A transformation procedure for general linearly elastic systems with classical damping was also proposed in his recent study (Saitoh, 2011b).

Therefore, the main aim of this study is to verify the applicability of the transformation method of 1DSDs to SSI problems influenced by seismic excitations. This study deals with an application example of a four-story shear building supported by a shallow foundation embedded in layered soil resting on a rigid bedrock. The soil-foundation system is modeled using two-dimensional isoparametric finite elements, as shown in Fig. 4. In most previous studies, nonlinearity in structural members was not considered when verifying the performance of the proposed LPMs. In this study, therefore, the Clough model (Clough & Johnson, 1966), which has a bilinear skeleton curve, is applied to each inter-story in the building to compare the relationship between force and inter-storey drift obtained with 1DSDs with that obtained with the original finite element (FE) model.

2. Methodology for transforming structural systems with non-classical damping into reduced 1DSDs

This section presents an overview of the method for transforming structural systems into a 1DSD. The configuration of a 1DSD is shown in Fig. 3. In general, the impedance function $S_{IJ}(\omega)$ is defined as the ratio of the dynamic force P_J applied at an arbitrarily selected DOF (denoted as "*J*-th DOF") and the response displacement u_I at the same or another DOF (denoted as "*I*-th DOF"). Here ω is the excitation frequency. The properties of the elements comprising a 1DSD are derived from a proposed procedure based on complex modal analysis. In accordance with the detailed description in Saitoh, 2010b, the properties of the elements are evaluated using the stiffness matrix, the damping matrix, the mass matrix, and complex eigenvalues and eigenvectors. Once these values are obtained by a one-time complex modal calculation, the properties are automatically determined from the mathematical formula presented in Saitoh's study.

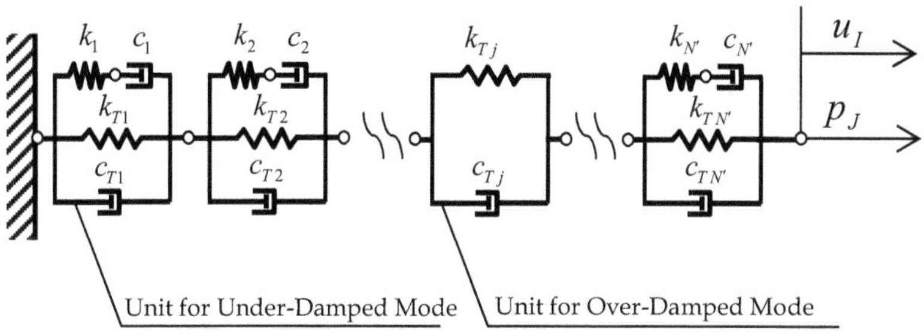

Fig. 3. One-dimensional lumped parameter model with spring and dashpot elements for simulating the impedance function $S_{IJ}(\omega) = p_J / u_I$ in general structural systems [Saitoh, 2010b].

In this method, the equations of motion of general linearly elastic structural systems comprising N DOFs are dealt with and are expressed by the following form:

$$[M]\{\ddot{u}\} + [C]\{\dot{u}\} + [K]\{u\} = \{p\} \tag{1}$$

where $[M]$, $[C]$, and $[K]$ are the mass matrix, damping matrix, and stiffness matrix, respectively, of the original structural systems. Each matrix has order $N \times N$. $\{u\}$ and $\{p\}$ are the response displacements and the external forces at the nodes, respectively, and each vector has order N. The dots denote partial derivatives with respect to time t. In this study, the damping matrix $[C]$ is assumed to be based on non-classical damping.

In complex modal analysis, the following $2N$ first-order equations are considered instead of the N second-order equations of Eq. 1:

$$[R]\{\dot{z}\} + [S]\{z\} = \{f\} \tag{2}$$

where

$$[R] = \begin{bmatrix} [C] & [M] \\ [M] & [0] \end{bmatrix}, \; [S] = \begin{bmatrix} [K] & [0] \\ [0] & -[M] \end{bmatrix}, \; \{z\} = \begin{Bmatrix} \{u\} \\ \{\dot{u}\} \end{Bmatrix}, \; \{f\} = \begin{Bmatrix} \{p\} \\ \{0\} \end{Bmatrix}.$$

The complex eigenvalues and eigenvectors can be obtained according to the conventional complex modal procedure. Each complex eigenvalue λ_n is known to have an eigenvalue $\bar{\lambda}_n$, which is the complex conjugate of λ_n; the corresponding vector $\{\varphi_n\}$ has a vector $\{\bar{\varphi}_n\}$, whose components are complex conjugates of those of $\{\varphi_n\}$. The eigenvectors are assembled compactly into a matrix using diagonal matrices $[\Omega]$ and $[\bar{\Omega}]$ comprising the eigenvalues λ_n and $\bar{\lambda}_n$, respectively, as

$$[\Psi] = \begin{bmatrix} [\varphi] & [\bar{\varphi}] \\ [\varphi][\Omega] & [\bar{\varphi}][\bar{\Omega}] \end{bmatrix} \tag{3a}$$

where

$$[\varphi]=\left[\{\varphi_1\} \quad \{\varphi_2\} \quad \cdots \quad \{\varphi_N\}\right] \tag{3b}$$

$$[\overline{\varphi}]=\left[\{\overline{\varphi}_1\} \quad \{\overline{\varphi}_2\} \quad \cdots \quad \{\overline{\varphi}_N\}\right] \tag{3c}$$

$$[\Omega]=\left[diag\,\lambda_n\right],\ n=1,2,\cdots,N \tag{3d}$$

$$\left[\overline{\Omega}\right]=\left[diag\,\overline{\lambda}_n\right],\ n=1,2,\cdots,N \tag{3e}$$

The matrix $[\Psi]$ is called the modal matrix. In general, $[\Psi]^T[R][\Psi]$ becomes a diagonal matrix owing to the orthogonality relationships. Here the upper N components of the matrix are denoted as α_n, while the lower N components are the complex conjugates of α_n, denoted as $\overline{\alpha}_n$.

At the end of the mathematical derivation in Saitoh's study, the properties of the elements can be determined using the following formula:

$$k_{Tn}=\frac{\sigma_n^2+\omega_{dn}^2}{2\left(G_n\sigma_n-R_n\omega_{dn}\right)} \tag{4}$$

$$c_{Tn}=\frac{1}{2G_n} \tag{5}$$

$$k_n=\frac{-\left(G_n^2+R_n^2\right)\omega_{dn}^2}{2G_n^2\left(G_n\sigma_n-R_n\omega_{dn}\right)} \tag{6}$$

$$c_n=\frac{-\left(G_n^2+R_n^2\right)\omega_{dn}^2}{2G_n\left(G_n\sigma_n-R_n\omega_{dn}\right)^2} \tag{7}$$

where $G_n+iR_n=\dfrac{\varphi_{nI}\varphi_{nJ}}{\alpha_n}$ and $G_n-iR_n=\dfrac{\overline{\varphi}_{nI}\overline{\varphi}_{nJ}}{\overline{\alpha}_n}$.

Here φ_{nI} and φ_{nJ} are the components of the n-th eigenvector at the I-th and J-th DOFs, respectively; $\overline{\varphi}_{nI}$ and $\overline{\varphi}_{nJ}$ are the complex conjugates of the components φ_{nI} and φ_{nJ}, respectively. σ_n is the n-th modal decay rate and ω_{dn} is the n-th damped natural circular frequency defined as

$$\lambda_n=-\sigma_n+i\omega_{dn} \tag{8}$$

$$\overline{\lambda}_n=-\sigma_n-i\omega_{dn} \tag{9}$$

In practice, over-damped modes often appear. In this case, eigenvalues λ_n are real and negative. In Saitoh, 2010b, it was mathematically derived that the impedance function associated with over-damped modes is expressed as a Kelvin–Voigt unit comprising spring k_{Tn} and dashpot c_{Tn}, as shown in Fig. 3.

$$k_{Tn} = \frac{\sigma_n}{G_n} \tag{10}$$

$$c_{Tn} = \frac{1}{G_n} \tag{11}$$

Note that over-damped modes generally appear with even numbers $2m$ in $2N$ modes, so the total unit number N changes to $N'(=N+m)$ when over-damped modes exist.

3. Application to soil-shallow foundation-structure system

3.1 FE model studied

This is the first application of a 1DSD to SSI problems. In principle, the dynamic response of the original structural systems, which have even inelasticity in their superstructures, can be correctly simulated using 1DSDs. However, the accuracy of the transfer functions and the time histories of the dynamic response of the structural system comprising 1DSDs have never been verified. In this section, therefore, a soil-shallow foundation system interacting with a multiple-DOF system with inelasticity is used for verification.

The overall system is shown in Fig. 4. A shallow foundation 10 m wide, 50 m long, and 2 m deep is embedded in layered soil up to the middle of the foundation. The elastic modulus of the foundation is assumed to be rigid, imposing unique displacements u_f and θ_f at the center of gravity in the horizontal and rotational directions, respectively. The mass and the mass moment of inertia of the foundation are $m_f = 1000$ t and $J_f = 8500$ tm², respectively. A four-story building supported by the foundation is represented by a 4-DOF system. The properties of the system are shown in the figure. In this study, the inelasticity in each story

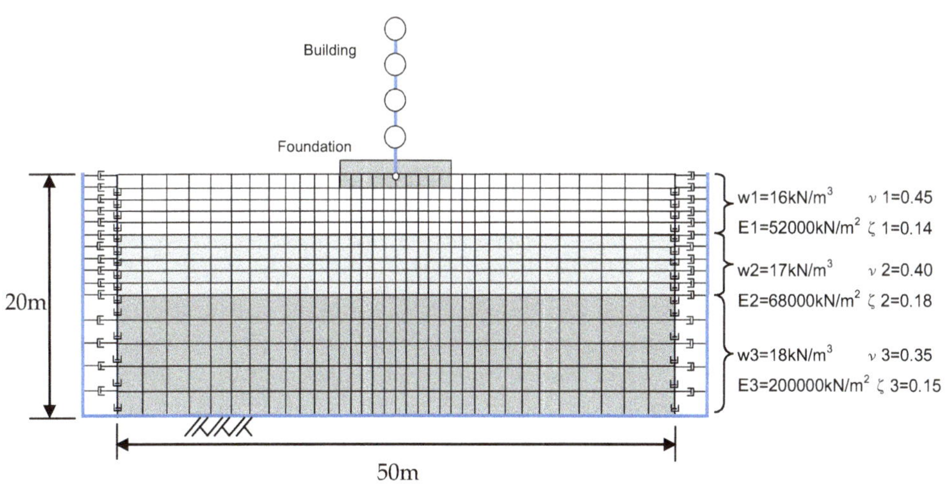

Fig. 4. Two-dimensional FE model for a soil-shallow foundation system supporting a four-story building. The unit weight, modulus of elasticity, Poisson's ratio, and damping ratio of the *i*-th soil layer are denoted as w_i, E_i,v_i, and ζ_i, respectively.

is taken into account. The soil-foundation system is modeled using conventional two-dimensional rectangular isoparametric elements (Weaver et al., 1990), in which each element has eight DOFs. The soil strata are composed of three soil layers resting on a rigid bedrock. The bottom of the layered soil is fixed in the vertical and lateral directions, whereas a viscous boundary proposed by Lysmer and Kuhlemeyer, 1969, is applied to the sidewalls of the soil as a fictitious boundary that dissipates energy toward an infinite region of soil. The moduli of elasticity and the damping ratios of the soil shown in the figure are assumed to approximately account for the appreciable levels of strain during ground shaking. The total number of nodes comprising the isoparametric elements is 560, whereas the degrees of freedom minus the fixed degrees of freedom are 1,050. The thickness of the elements is the same as the length of the foundation (50 m) under a plane-strain condition.

3.2 Equations of motion of FE model with structures

In this model, the soil-foundation system consists of conventional isotropic elements, whereas the structural system comprising the superstructure and the mass of the foundation is discretized by springs, dashpots, and masses. Therefore, the global mass matrix, stiffness matrix, and damping matrix in the equations of motion of the total system are obtained by superimposing local matrices in both equilibrium equations. The details of the equations are described as follows.

The equations of motion of the soil-foundation system are expressed as

$$[M_G]\{\ddot{u}_G\}+[C_G]\{\dot{u}_G\}+[K_G]\{u_G\}=\{f_G\} \tag{12}$$

where $[M_G]$, $[C_G]$, and $[K_G]$ are the mass matrix, damping matrix, and stiffness matrix, respectively, of the soil-foundation system. $\{u_G\}$ and $\{f_G\}$ are the response displacements and the forces at the DOFs in the system, respectively. In this study, the damping matrix for each element is constructed on the basis of

$$[C_e]=\beta_l[K_e] \tag{13}$$

where

$$\beta_l=\frac{2\zeta_l}{\omega_1} \tag{14}$$

where $[C_e]$ and $[K_e]$ are the local damping matrix and stiffness matrix, respectively, of the isoparametric element; the parameter ζ_l is the damping ratio given in each soil layer at the fundamental natural frequency of the soil-foundation system; and ω_1 is the fundamental natural circular frequency of the undamped system. The damping ratio of each soil layer used in this model is shown in Fig. 4. In the FE model, the elastic modulus of the elements in the foundation is appreciably higher than that of the soil for realizing rigid body movement of the foundation. In addition, the mass density of the elements is negligible because the mass and the mass moment of inertia of the foundation are taken into account in the model of the building.

The equilibrium equation of the 4-DOF system connected with the rigid mass of the foundation is expressed as follows:

$$[M_I]\begin{Bmatrix}\{\ddot{u}_s\}\\ \ddot{u}_f\\ \ddot{\theta}_f\end{Bmatrix}+[C_I]\begin{Bmatrix}\{\dot{u}_s\}\\ \dot{u}_f\\ \dot{\theta}_f\end{Bmatrix}+[K_I]\begin{Bmatrix}\{u_s\}\\ u_f\\ \theta_f\end{Bmatrix}=\begin{Bmatrix}\{0\}\\ 0\\ 0\end{Bmatrix} \tag{15}$$

where

$$[M_I]=\begin{bmatrix} m_1 & 0 & 0 & 0 & 0 & m_1H_1\\ 0 & m_2 & 0 & 0 & 0 & m_2H_2\\ 0 & 0 & m_3 & 0 & 0 & m_3H_3\\ 0 & 0 & 0 & m_4 & 0 & m_4H_4\\ 0 & 0 & 0 & 0 & m_f & 0\\ m_1H_1 & m_2H_2 & m_3H_3 & m_4H_4 & 0 & J_f+\sum_{i=1}^{4}m_iH_i^2 \end{bmatrix} \tag{16}$$

$$[C_I]=\begin{bmatrix} \overline{c}_1 & -\overline{c}_1 & 0 & 0 & 0 & 0\\ -\overline{c}_1 & \overline{c}_1+\overline{c}_2 & -\overline{c}_2 & 0 & 0 & 0\\ 0 & -\overline{c}_2 & \overline{c}_2+\overline{c}_3 & -\overline{c}_3 & 0 & 0\\ 0 & 0 & -\overline{c}_3 & \overline{c}_3+\overline{c}_4 & -\overline{c}_4 & 0\\ 0 & 0 & 0 & -\overline{c}_4 & \overline{c}_4 & 0\\ 0 & 0 & 0 & 0 & 0 & 0 \end{bmatrix} \tag{17}$$

$$[K_I]=\begin{bmatrix} \overline{k}_1 & -\overline{k}_1 & 0 & 0 & 0 & 0\\ -\overline{k}_1 & \overline{k}_1+\overline{k}_2 & -\overline{k}_2 & 0 & 0 & 0\\ 0 & -\overline{k}_2 & \overline{k}_2+\overline{k}_3 & -\overline{k}_3 & 0 & 0\\ 0 & 0 & -\overline{k}_3 & \overline{k}_3+\overline{k}_4 & -\overline{k}_4 & 0\\ 0 & 0 & 0 & -\overline{k}_4 & \overline{k}_4 & 0\\ 0 & 0 & 0 & 0 & 0 & 0 \end{bmatrix} \tag{18}$$

where $\{u_s\}\left(=\{u_1\ \ u_2\ \ u_3\ \ u_4\}^T\right)$ are the absolute displacements of the DOFs in the superstructure, excluding the displacements related to the rotational response of the foundation, and $\overline{c}_i$ and $\overline{k}_i$ are the damping coefficient and the initial stiffness of each story in the superstructure, respectively. Here the suffix i is numbered from top to bottom. The damping matrix is constructed as Rayleigh damping by using the mass matrix (Eq. 16) and the stiffness matrix (Eq. 18) with a first and second modal damping constant of 0.05.

In fact, the DOF of the foundation in the rotational direction is incompatible with the DOFs in the FE model as no rotational DOF is explicitly considered at each node in the FE model. Therefore, the rotational displacement θ_f in the rigid mass of the foundation is transformed in the FE model based on the geometrical relationship using the relative displacement at the center of the foundation to the displacement of the successive node located at the bottom of the foundation. In addition, the rotational moment is transmitted to the corresponding nodes in the FE model using an equivalent couple.

3.3 Transforming the original system into a 1DSD

According to the procedure shown in Section 3.1, the FE model is to be transformed into an equivalent 1DSD. Complex modal analysis is performed to obtain the fundamental quantities by which the properties of the elements in the 1DSD are determined. Complex modal analysis has commonly been used for estimating modal components. In this study, a conventional modal analysis (Foss, 1958) is used. First, the impedance function in the horizontal direction is considered. In this case, although movement of the foundation in the rotational direction may be restrained more accurately to obtain a net horizontal impedance function (by extracting the interactive components), it was not restrained in this study because the foundation is embedded in the soil only a little in the model. Thus, no difference was observed in either of the impedance functions.

The results of complex modal analysis show that the number of over-damped modes is 2,026, whereas that of under-damped modes is 74 (the total number of modes is 2,100). Therefore, the 1DSD consists of 2,026 Kelvin–Voigt units and 37 standard units. The results show that the lowest natural frequency is 2.11 Hz, whereas the highest frequency is 4,760,000 Hz. As described above, a significant advantage of a 1DSD is that the units comprising a 1DSD are associated with the vibration modes of the original structural system. Therefore, a small set of units associated with modes from the lowest order can appropriately express the dynamic characteristics of structural systems without using all units. Recently, Saitoh, 2011a, studied the influence of frequency dependency in pile-group impedance functions upon the elastic and inelastic responses of superstructures. The results indicate that the important frequency range is the dominant frequency of foundation input motions that excite the inertial structural systems. Figure 5 shows the time-history response acceleration at the ground surface calculated using the conventional one-dimensional wave propagation theory with the soil properties shown in Fig. 4. An observed earthquake record, 2004 Ojiya EW, is applied to the bottom soil layer. This response acceleration is used in the following calculations as the foundation input motion in this study, which indicates that, for simplicity, no adjustment for the kinematic interaction effects is conducted. The figure shows that the foundation input motion contains a wide range of frequency components showing the dominant frequency at around 1.4 Hz. The amplitude of acceleration ranges from 0 to 10 Hz. Therefore, this frequency range is considered to be the target frequency range in this study.

The modal analysis results show that 1,110 vibrating modes appear in this frequency range. In fact, many units associated with these modes contain a relatively large spring constant k_{Tn} than other units. These units can appreciably be removed provided the impedance functions of the reduced 1DSD are in sufficient agreement with those of the original system. In this study, only 35 units (12 under-damped modes and 23 over-damped modes) remain after applying a threshold to the spring constant k_{Tn} of 1.0×10^{10} kN/m, above which the corresponding unit is removed. This threshold is adjusted on the basis of both the number of units remaining suitable for effective computation and the accuracy in the approximation of impedance functions. The properties of the reduced 1DSD are shown in Table 1. Figure 6 shows a comparison of the impedance functions K_{hh} obtained from the reduced 1DSD with those obtained with the FE model. The results indicate that the impedance functions obtained from the 1DSD agree closely with those obtained with the FE model within the target frequency range.

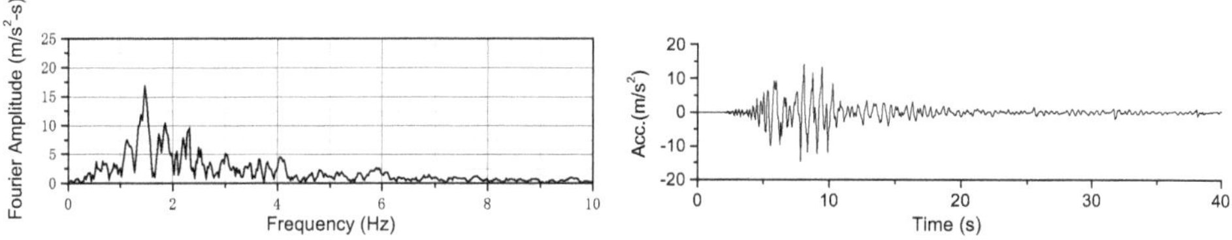

Fig. 5. Fourier amplitude and time history of the foundation input motion.

The impedance functions K_{rr} in the rotational direction are obtained along similar lines to the impedance functions in the horizontal direction. To obtain the relationship between rotational moment and displacement in the FE model, the center of the foundation is fixed in the horizontal and vertical directions and lateral force is applied to the successive node at the bottom of the foundation. The results of the complex modal analysis show that the number of over-damped modes is 2,018, while that of under-damped modes is 78 (total number of modes is 2,096). Therefore, the 1DSD consists of 2,018 Kelvin–Voigt units and 39 standard units. In this direction, 44 units (27 under-damped modes and 17 over-damped modes) comprise the reduced 1DSD after applying a threshold to the spring constant k_{Tn} in the rotational direction of 7.0×10^{10} kNm/rad. The properties of the reduced 1DSD are shown in Table 2. The resultant impedance functions in the rotational direction at the center of the foundation are shown in Fig. 6. Although a slight difference in the impedance functions appear in a high-frequency region when compared with those with the FE model, fairly close agreement can be observed over the target frequency range, on the whole.

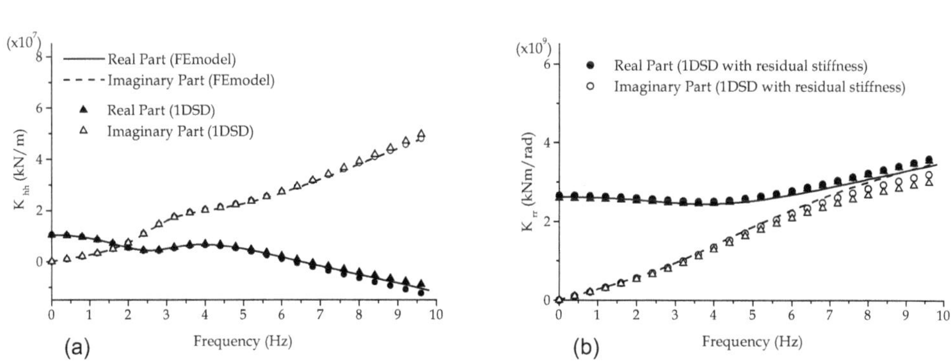

Fig. 6. Impedance functions of a soil-shallow foundation system using 1DSDs and 1DSDs with residual stiffness [(a) K_{hh} in the horizontal direction and (b) K_{rr} in the rotational direction]. Results obtained from the original FE model are shown for comparison.

In Saitoh, 2010b, a mechanical element associated with residual stiffness R_{IJ} was proposed to improve the accuracy of the reduced 1DSD. Residual stiffness has often been applied to approximate expressions of structural systems in conventional modal analysis. The residual stiffness representing the stiffness of high-frequency modes is expressed as

$$\frac{1}{R_{IJ}} = \sum_{n=l+1}^{N'} \frac{1}{\overline{K}_{nIJ}} = \frac{1}{\overline{K}_{l+1IJ}} + \frac{1}{\overline{K}_{l+2IJ}} + \cdots + \frac{1}{\overline{K}_{N'IJ}} \tag{19}$$

where

$$\text{For under-damped modes: } \overline{K}_{nIJ} = \frac{\sigma_n^2 + \omega_{dn}^2}{2\left(\sigma_n G_n - \omega_{dn} R_n\right)} \tag{20}$$

$$\text{For over-damped modes: } \overline{K}_{nIJ} = \frac{\sigma_n}{G_n} \tag{21}$$

where l is the maximum mode number considered in the 1DSDs without residual stiffness.

The residual stiffness R_{IJ} can be incorporated into the 1DSDs as a mechanical element arranged in series with the 1DSDs. The residual stiffnesses in both horizontal and rotational

Mode (n)	1	2	3	4	5	6
k_n	-4.5692E+08	3.7591E+08	1.0618E+09	-5.1042E+08	-1.6506E+10	0.0000E+00
c_n	-6.6604E+06	2.1711E+07	7.9324E+07	-3.9891E+07	2.6594E+07	0.0000E+00
k_{Tn}	1.9143E+07	-4.2683E+08	-1.4162E+09	6.8531E+08	4.7634E+07	8.2805E+09
c_{Tn}	7.4465E+06	-1.1586E+07	-2.0610E+07	7.6590E+06	-2.5143E+07	2.0113E+08
	7	8	9	10	11	12
k_n	0.0000E+00	0.0000E+00	0.0000E+00	0.0000E+00	0.0000E+00	0.0000E+00
c_n	0.0000E+00	0.0000E+00	0.0000E+00	0.0000E+00	0.0000E+00	0.0000E+00
k_{Tn}	1.9166E+09	2.9441E+09	6.5445E+09	8.8819E+09	1.9163E+09	6.5327E+09
c_{Tn}	4.6386E+07	7.1043E+07	1.5783E+08	2.1356E+08	4.5797E+07	1.5576E+08
	13	14	15	16	17	18
k_n	0.0000E+00	0.0000E+00	0.0000E+00	0.0000E+00	0.0000E+00	0.0000E+00
c_n	0.0000E+00	0.0000E+00	0.0000E+00	0.0000E+00	0.0000E+00	0.0000E+00
k_{Tn}	1.1882E+09	5.3841E+09	4.9079E+09	1.7577E+09	8.5567E+09	4.9203E+09
c_{Tn}	2.8224E+07	1.2766E+08	1.1612E+08	4.1497E+07	2.0073E+08	1.1436E+08
	19	20	21	22	23	24
k_n	0.0000E+00	0.0000E+00	0.0000E+00	0.0000E+00	0.0000E+00	0.0000E+00
c_n	0.0000E+00	0.0000E+00	0.0000E+00	0.0000E+00	0.0000E+00	0.0000E+00
k_{Tn}	7.8980E+09	7.1101E+08	6.3409E+09	3.0631E+09	2.2748E+09	1.7152E+09
c_{Tn}	1.7967E+08	1.6147E+07	1.4262E+08	6.8550E+07	5.0782E+07	3.7956E+07
	25	26	27	28	29	30
k_n	0.0000E+00	0.0000E+00	0.0000E+00	-1.8100E+10	0.0000E+00	-3.0252E+08
c_n	0.0000E+00	0.0000E+00	0.0000E+00	3.9047E+08	0.0000E+00	-3.5556E+06
k_{Tn}	5.6713E+09	2.0277E+08	5.6121E+09	5.6507E+09	4.3924E+09	2.4506E+08
c_{Tn}	1.2154E+08	4.2003E+06	1.1373E+08	-1.0463E+08	8.4557E+07	7.5807E+06
	31	32	33	34	35	Resi. Stif.
k_n	-9.4744E+09	-3.4928E+09	-4.8323E+09	-6.2970E+09	-1.1486E+10	0.0000E+00
c_n	5.7840E+07	4.4528E+07	-2.1066E+07	3.2994E+07	-7.5059E+07	0.0000E+00
k_{Tn}	6.6702E+08	6.8297E+08	6.2645E+08	4.3951E+08	7.6216E+09	-6.6866E+08
c_{Tn}	-3.5218E+07	-1.7234E+07	3.5920E+07	-1.8667E+07	2.2991E+08	0.0000E+00

*units: kN/m for k_n and k_{Tn}; and kNsec/m for c_n and c_{Tn}

Table 1. Properties of mechanical elements in reduced 1DSDs in the horizontal direction directions are presented in Table 1 and Table 2, respectively. In Fig. 6, the impedance functions of the residual 1DSDs with the residual stiffness are plotted. An appreciable improvement can be seen in the high-frequency region when incorporating the residual stiffness in both directions.

3.4 Dynamic response of structural system in frequency domain

In this section, the dynamic response of the structural system computed by using the reduced 1DSDs in the frequency domain is verified by comparing it with the dynamic response obtained with the original FE model. The complete structural system using the 1DSDs in both horizontal and rotational directions is shown in Fig. 7. The properties of the superstructure are summarized in Table 3. The reduced 1DSDs obtained above are connected with each DOF in the foundation, as shown in the figure. The equations of motion of the structural system can be easily constructed using Eq. 15 with conventional spring-dashpot matrices (details are described in Saitoh, 2010b) expressing the reduced 1DSDs. The resultant equilibrium equations of the total system can be formulated as

Mode (n)	1	2	3	4	5	6
k_n	-8.7367E+10	-3.2511E+10	5.9885E+11	-8.9724E+10	-2.0509E+12	-8.2557E+10
c_n	-8.6002E+08	-3.2912E+08	-5.3876E+09	7.6112E+08	-2.2854E+09	8.3561E+08
k_{Tn}	6.3069E+09	3.8708E+09	-3.4692E+10	5.3955E+09	4.5267E+09	8.6083E+09
c_{Tn}	9.9889E+08	4.1663E+08	3.9650E+09	-5.4771E+08	2.4152E+09	-4.9286E+08
	7	8	9	10	11	12
k_n	0.0000E+00	-6.2547E+09	0.0000E+00	0.0000E+00	0.0000E+00	0.0000E+00
c_n	0.0000E+00	-2.1604E+08	0.0000E+00	0.0000E+00	0.0000E+00	0.0000E+00
k_{Tn}	2.9016E+10	1.0142E+10	3.3727E+10	4.2553E+10	1.7739E+10	4.1057E+10
c_{Tn}	6.8559E+08	1.5672E+08	7.7893E+08	9.5936E+08	3.7922E+08	8.7675E+08
	13	14	15	16	17	18
k_n	0.0000E+00	-5.5850E+10	0.0000E+00	0.0000E+00	-1.1840E+12	3.6129E+09
c_n	0.0000E+00	9.3275E+08	0.0000E+00	0.0000E+00	3.3843E+09	1.5195E+08
k_{Tn}	1.9758E+10	1.3439E+10	5.1907E+10	1.5146E+10	2.1723E+10	-6.6498E+09
c_{Tn}	4.0089E+08	-3.1988E+08	1.0150E+09	2.9275E+08	-2.7039E+09	-5.0781E+07
	19	20	21	22	23	24
k_n	-1.0096E+09	-8.9520E+09	-3.2909E+09	-1.1530E+12	3.8877E+10	-7.9236E+08
c_n	-2.2987E+07	-1.0163E+08	-3.8159E+07	-3.6097E+09	6.5584E+09	-7.6230E+06
k_{Tn}	2.4822E+09	1.2331E+10	4.6047E+09	6.8295E+10	-4.5729E+10	2.1613E+09
c_{Tn}	3.2532E+07	3.0091E+08	1.0301E+08	5.3063E+09	-5.3463E+07	3.8351E+07
	25	26	27	28	29	30
k_n	0.0000E+00	0.0000E+00	0.0000E+00	-3.5501E+09	0.0000E+00	-1.8681E+09
c_n	0.0000E+00	0.0000E+00	0.0000E+00	-6.3417E+06	0.0000E+00	-2.5699E+06
k_{Tn}	-4.6466E+10	-3.5291E+10	-4.6258E+10	2.0957E+09	-4.9346E+10	2.2527E+09
c_{Tn}	-4.6411E+08	-2.1706E+08	-1.9382E+08	1.5135E+07	-1.6920E+08	9.2640E+06
	31	32	33	34	35	36
k_n	0.0000E+00	0.0000E+00	-1.4155E+09	-1.7451E+09	8.1243E+06	-2.3990E+06
c_n	0.0000E+00	0.0000E+00	-3.9743E+06	-2.7057E+06	1.5958E+04	-4.7181E+03
k_{Tn}	-2.9367E+10	-1.0747E+10	1.3208E+10	9.9997E+09	-1.9771E+10	5.1191E+09
c_{Tn}	-6.6288E+07	-2.3854E+07	2.3000E+07	2.5153E+07	-3.9248E+07	9.6705E+06
	37	38	39	40	41	42
k_n	0.0000E+00	-3.4634E+08	-2.6621E+09	-7.3699E+06	-1.6447E+07	-3.2997E+08
c_n	0.0000E+00	-5.8881E+05	-3.8641E+06	-1.3444E+04	-2.8127E+04	-5.2243E+05
k_{Tn}	1.2511E+10	4.5378E+10	3.0996E+10	6.1565E+10	3.8787E+09	3.1127E+10
c_{Tn}	2.3189E+07	8.9426E+07	7.0953E+07	1.1015E+08	7.0887E+06	5.8484E+07
	43	44	Resi. Stif.			
k_n	0.0000E+00	0.0000E+00	0.0000E+00			
c_n	0.0000E+00	0.0000E+00	0.0000E+00			
k_{Tn}	-5.8039E+10	-3.1792E+10	-1.0587E+10			
c_{Tn}	-9.5838E+07	-5.0192E+07	0.0000E+00			

*units: kNm/rad for k_n and k_{Tn}; and kNm sec/rad for c_n and c_{Tn}

Table 2. Properties of mechanical elements in reduced 1DSDs in the rotational direction

$$[M_T]\{\ddot{u}\}+[C_T]\{\dot{u}\}+[K_T]\{u\}=\{0\} \tag{22}$$

where

$$\{u\}=\left[u_s\ u_f\ u_1\ u_2\cdots u_i\cdots u_m\ \theta_f\ \theta_1\ \theta_2\cdots\theta_i\cdots\theta_n\right]^T \tag{23}$$

where the mass matrix $[M_T]$, the damping matrix $[C_T]$, and the stiffness matrix $[K_T]$ are the resultant matrices formed by superimposing the partial matrices. u_i and θ_i are the displacements at the DOFs in the reduced 1DSDs. The maximum DOFs in both directions are represented by m $(=47)$ and n $(=71)$, respectively. In order to estimate the response of the structural system, the absolute displacements are expressed by the sum of the displacements of the inertial response and the input motion, as follows:

$$\{u\}=\{U\}+\{U_g\} \tag{24}$$

where

$$\{U\}=\left[U_s\ U_f\ U_1\ U_2\cdots U_i\cdots 0\ \Theta_f\ \Theta_1\ \Theta_2\cdots\Theta_i\cdots 0\right]^T \tag{25}$$

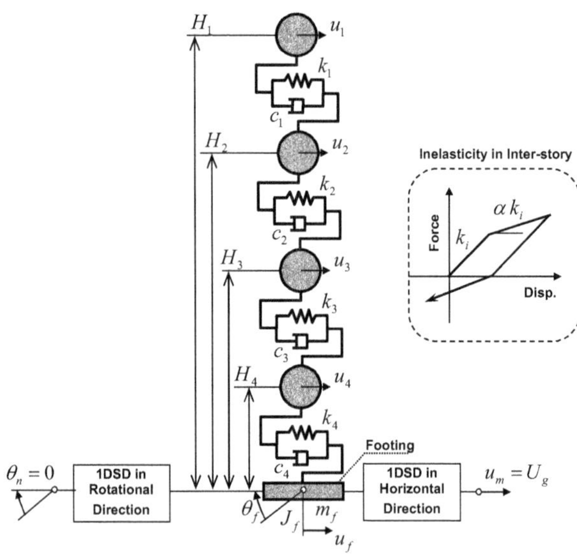

Fig. 7. Numerical model representing complete structural system.

Story No.	Units	1	2	3	4
Mass m_i	t	750	750	750	750
Stiffness k_i	kN/m	2000000	2000000	2000000	2000000
Height H_i	m	12	9	6	3
Yield Strength p_i	kN	15000	30000	33000	44000

Table 3. Properties of four-story building

$$\{U_g\} = \left[U_g\, U_g\, U_g\, U_g \cdots U_g \cdots U_g\, \Theta_g\, \Theta_g\, \Theta_g \cdots \Theta_g \cdots \Theta_g\right]^T \tag{26}$$

where U_g and Θ_g are the foundation input motions to the inertial system in the horizontal and rotational directions, respectively. In this study, the foundation input motion in the rotational direction Θ_g is neglected as the amplitude is negligible because of the shallow embedment of the foundation in the soil. Substituting Eq. 24 into Eq. 22 leads to

$$[M_T]\{\ddot{U}\} + [C_T]\{\dot{U}\} + [K_T]\{U\} = -[M_T]\{\ddot{U}_g\} \tag{27}$$

The comparison is performed using the transfer functions (TFs) of the dynamic responses of the superstructure with respect to the foundation input motion defined as $T_{sa} = \left(\ddot{U}_1 + \ddot{U}_g + H_1\ddot{\Theta}_f\right) / \ddot{U}_g$ for absolute acceleration. The TFs of the footing are also computed. They are defined as $T_{ha} = \left(\ddot{U}_f + \ddot{U}_g\right) / \ddot{U}_g$ and $T_{ra} = H_1\ddot{\Theta}_f / \ddot{U}_g$ for the absolute acceleration associated with the horizontal and rotational motions, respectively.

Figure 8 shows the real part and the imaginary part of the transfer functions of the structural systems defined above. The figure indicates that the transfer functions obtained with the reduced 1DSDs are compatible with those obtained with the original FE model. This implies that 1DSDs with almost one-tenth of the DOFs in the original system can correctly represent the impedance functions in the target frequency region. Here the transfer functions obtained using the 1DSDs with residual stiffness are not presented as negligibly small differences that appeared in the impedance functions.

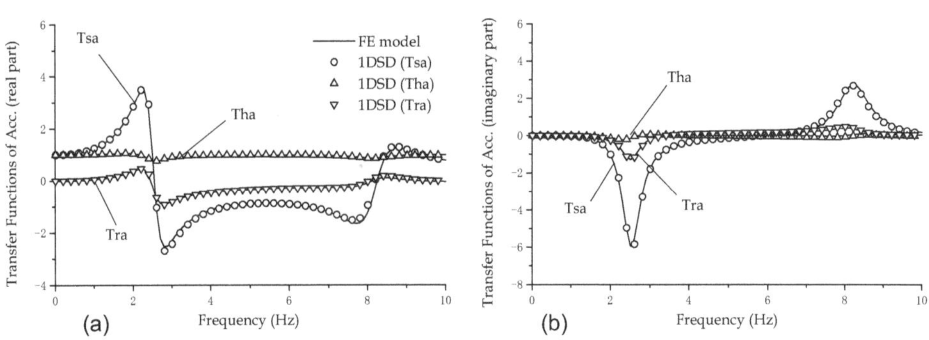

Fig. 8. Comparisons of the transfer functions of the structural system computed using 1DSDs without residual stiffness and the original FE model [(a) real part and (b) imaginary part].

3.5 Dynamic response of structural system in time domain

3.5.1 Elastic response of structural system

In the previous section, it was shown that a structural system with reduced 1DSDs can correctly simulate the transfer functions in the real part and the imaginary part in the frequency domain. This implies that the time-history response of a structural system having linearly elastic members can be appropriately calculated using 1DSDs. In this section, the time-history response of a structural system with 1DSDs when subjected to foundation input motion is computed and compared with that obtained with the original FE model. An attempt is then made to compute the time-history response of the structural system with inelasticity in the superstructure.

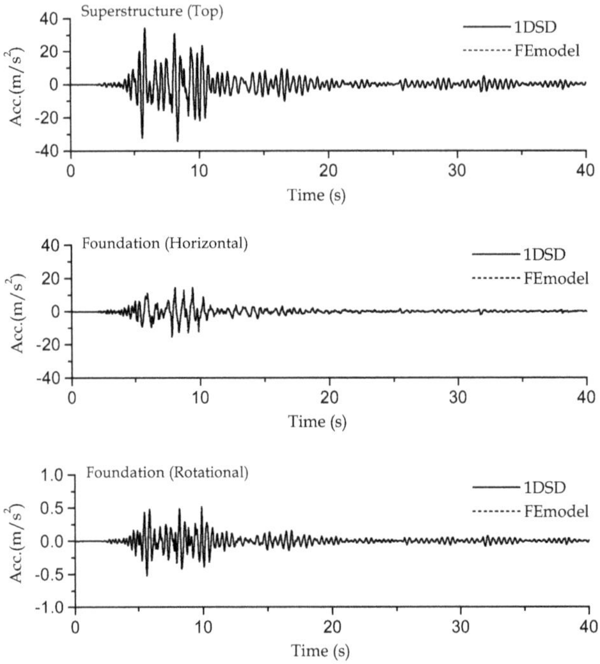

Fig. 9. Time-history response accelerations of superstructure and footing in the horizontal and rotational directions when subjected to ground motion associated with 2004 Ojiya EW by using 1DSDs and the original FE model.

The time-history analysis is performed on the basis of Eq. 27 by using Newmark's β method ($\beta = 1/4$) as a numerical integration scheme in which the time interval Δt is 0.001 s. The acceleration response at the ground surface shown in Fig. 5 is applied as the foundation input motion. Here the response accelerations defined in the previous section (the same definitions as in the transfer functions) are calculated. Figure 9 shows the time-history responses of the acceleration of the structural systems using reduced 1DSDs without residual stiffness. The figure shows that the time histories of the 1DSDs are compatible with those of the original FE model.

Note that a structural system using 1DSDs with residual stiffness cannot be obtained correctly because the dynamic response of the system is not converged (the response increases oscillatory with time). This implies that the structural system becomes unstable, which is attributed to the residual stiffness components. The unit for residual stiffness in the horizontal and rotational directions comprises only a negative stiffness, as shown in Table 1 and Table 2. This negative stiffness could be a drawback to stabilizing the dynamic response of structures, although the units can achieve better agreement with the impedance functions of the original systems in the frequency domain.

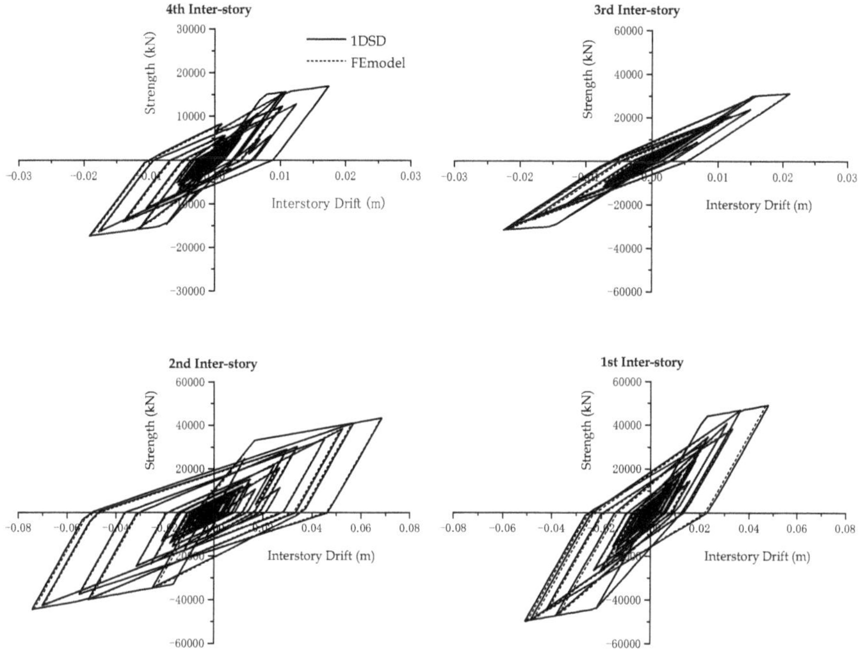

Fig. 10. Relationship between shear force and inter-story drift in each story when subjected to ground motion associated with 2004 Ojiya EW by using 1DSDs and the original FE model.

3.5.2 Inelastic response of structural system

Comparisons are extended to the structural system to allow inelasticity in the superstructure. The inelasticity of the superstructure is represented by the Clough model (Clough & Johnson, 1966), which is generally used to model reinforced concrete members. The spring of the superstructure has a bilinear skeleton curve in which the ratio of tangent stiffness to initial stiffness is assumed to be 0.1, as shown in Fig. 7. The yield strength p_i in each story is presented in Table 3. In this study, the modified Newton–Raphson method is applied to calculate the nonlinear response of the system.

Figure 10 shows the relationship between shear force and inter-story drift in each story. Figure 11 shows the time-history responses of the displacement of the foundation u_f and $H_1\theta_f$ in the horizontal and rotational directions, respectively. The results indicate that although a slight difference appears in the relationship between shear force and inter-story drift in the first inter-storey, the inelastic responses obtained with the 1DSDs show sufficiently close agreement with those obtained from the original FE model. From a practical viewpoint, a dominant advantage is that the computational time can be considerably reduced by transforming the original system into 1DSDs. According to a rough measurement using the author's PC (CPU 3.40 GHz, RAM 3.00 GB), the inelastic responses shown above were obtained in about 11,965 s with the original FE model, while those with 1DSDs were obtained in about 68 s.

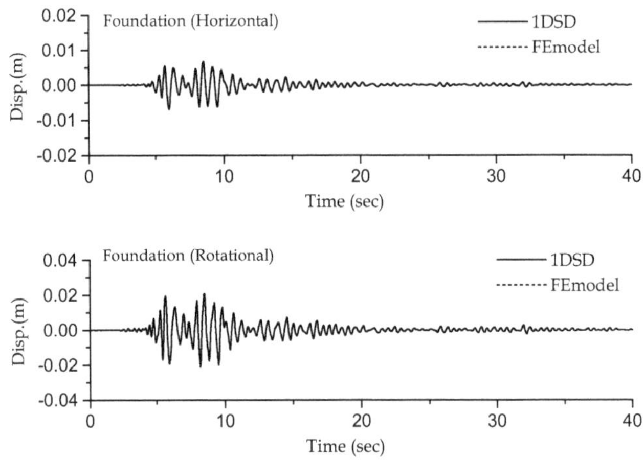

Fig. 11. Time-history response displacements of the foundation in the horizontal and rotational directions when subjected to ground motion associated with 2004 Ojiya EW by using 1DSDs and the original FE model.

4. Conclusion

This study demonstrates the transformation procedure using a 1DSD for a multi-story building supported on a shallow foundation embedded in layered soil. In accordance with the conventional modal concept, the reduced 1DSDs in both horizontal and rotational directions are constructed with a small number of units associated with important modes in the target frequency region in this application example. The impedance functions obtained with the 1DSDs correctly simulate the impedance functions of the original FE model. The transfer functions of the structural systems in the frequency domain using the 1DSD show fairly good agreement with those obtained with the FE model. The time-history responses of structures in both linearly elastic and inelastic cases can be simulated using the 1DSDs. The results indicate that a significant decrease in the structural system with the 1DSD lead to a marked decrease in the computational time taken for the results. Therefore, it may be concluded that 1DSD transformation is effective and efficient for numerical computations in SSI problems influenced by ground motions.

5. References

Baranov, V. A. (1967). On the calculation of excited vibrations of an embedded foundation (*in Russian*), *Voprosy Dynamiki I Prochnocti*, Polytechnical Institute of Riga, No. 14, pp. 195-209

Beredugo, Y. O. & Novak, M. (1972). Coupled horizontal and rocking vibration of embedded footings, *Can. Geotech. J.*, 9 (4), pp. 477-497

de Barros, F. C. P. & Luco, J. E. (1990). Discrete models for vertical vibrations of surface and embedded foundations, *Earthquake Engineering and Structural Dynamics*, 19, pp. 289-303

Clough, R. W. & Johnson, S. B. (1966). Effect of stiffness degradation on earthquake ductility requirements, *The second Earthquake Engrg. Sym., Proc.*, pp. 227-232

Dobry, R. & Gazetas, G. (1988). Simple method for dynamic stiffness and damping of floating pile groups, *Geotechnique*, 38 (4), pp. 557-674

Elsabee, F. & Morray, J. P. (1977). Dynamic behavior of embedded foundation, *Research Report*, R77-33, Dep. of Civil Engrg., MIT. Cambridge, Mass.

Foss, K. A. (1958). Co-ordinates which uncouple the equations of motion of damped linear dynamic systems, *Journal of Applied Mechanics*, 57, pp. 361-364

Gazetas, G. (1991). Formulas and charts for impedances of surface and embedded foundations, *J. Geotech. Engrg.* ASCE, 117(9), pp. 1363-1381

Hayashi, Y. & Katsukura, Y. (1990). Effective time domain soil-structure interaction analysis based on FFT algorithm with causality condition, *Earthquake Engineering and Structural Dynamics*, 19, pp. 693-708

Jean, W. Y., Lin, T. W. & Penzien, J. (1990). System parameters of soil foundations for time domain dynamic analysis. *Earthquake Engineering and Structural Dynamics*, 19, pp. 541-553

Kausel, E. (2010). Early history of soil-structure interaction, *Soil Dynamics and Earthquake Engineering*, 30, pp. 822-832

Kausel, E. & Roesset, J. M. (1975). Dynamic Stiffness of Circular Foundations, *J. Engrg. Mech. Div.*, ASCE, Vol. 101, No. EM6, pp. 770-1731

Kaynia, A. & Kausel, E. (1982). Dynamic stiffness and seismic response of pile groups, *Research Report*, R82-03, Dept. of Civil Engrg., MIT, Cambridge, Mass.

Khodabakhshi, P., Jahankhah, H., & Ghannad, M. A. (2011). A discrete model for response estimation of soil-structure systems with embedded foundations, *Earthquake Engineering and Engineering Vibration*, 10 (2), pp. 263-276

Lysmer, J. & Kuhlemeyer, R. L. (1969). Finite dynamic model for infinite media, *J. Eng. Mech. Div.*, ASCE, 95, pp. 859-877

Makris, N. & Gazetas, G. (1993). Displacement phase differences in a harmonically oscillating pile, *Geotechnique*, 43 (1), pp. 135-150

Meek, J. W. & Veletsos, A. S. (1974). Simple models for foundations in lateral and rocking motion, *Proc., 5th World Conf. on Earthquake Engrg.*, 2, Rome, Italy, pp. 2610-2613

Meek, J. W. (1990). Recursive analysis of dynamic phenomena in civil engineering, *Bautechnik*, 67, pp. 205-210, (in German)

Motosaka, M. & Nagano, M. (1992). Recursive evaluation of convolution integral in nonlinear soil-structure interaction analysis and its applications, *Journal of Structural and Construction Engineering (AIJ)*, 436, pp. 71-80, (in Japanese)

Mylonakis, G. & Gazetas, G. (1998). Vertical vibration and additional distress of grouped piles in layered soil, *Soils and Foundations*, JGS, 38 (1), pp. 1-14

Nakamura, N. (2006a). A Practical method to transform frequency dependent impedance to time domain, *Earthquake Engineering and Structural Dynamics*, 35(2), pp. 217-234

Nakamura, N. (2006b). Improved methods to transform frequency dependent complex stiffness to time domain, *Earthquake Engineering and Structural Dynamics*, 35(8), pp. 1037-1050

Nakamura, N. (2008a). Transform Methods for Frequency Dependent Complex Stiffness to Time Domain Using Real or Imaginary Data Only, *Earthquake Engineering and Structural Dynamics*, 37(4), pp. 495-515

Nakamura, N. (2008b). Time history Response Analysis Considering Dynamic Stiffness with Both Frequency and Strain Dependencies, *J. of Eng. Mechanics*, ASCE, 134(4), pp. 530-541

Nogami, T. & Novak, M. (1977). Resistance of soil to a horizontally vibrating pile, *Earthquake Engineering and Structural Dynamics*, 5, pp. 249-261

Nogami, T. & Konagai, K. (1986). Time domain axial response of dynamically loaded single piles, *J. Engrg. Mech.*, ASCE, 112(11), pp. 1241-1252

Nogami, T. & Konagai, K. (1988). Time domain flexural response of dynamically loaded single piles, *J. Engrg. Mech.*, ASCE, 114(9), pp. 1512-1525

Novak, M. (1974). Dynamic stiffness and damping of piles, *Can. Geotech. J.*, 11, pp. 574-598.

Novak, M. & Nogami, T. (1977). Soil-pile interaction in horizontal vibration, *Earthquake Engineering and Structural Dynamics*, 5, pp. 263-281

Novak, M., Nogami, T., Konagai, K., & Aboul-Ella, F. (1978). Dynamic soil reactions for plane strain case, *J. Engrg. Mech.*, ASCE, 104(4), pp. 953-959

Saitoh, M. & Watanabe, H. (2004). Effects of flexibility on rocking impedance of deeply embedded foundation, *Journal of Geotechnical and Geoenvironmental Engineering*, ASCE, 130(4), pp. 438-445

Saitoh, M. (2007). Simple model of frequency-dependent impedance functions in soil-structure interaction using frequency-independent elements, *Journal of Engineering Mechanics*, ASCE, 133(10), pp. 1101-1114

Saitoh, M. (2010a). Lumped parameter models representing impedance functions at the interface of a rod on a viscoelastic medium, *Journal of Sound and Vibration*, 330, pp. 2062-2072

Saitoh, M. (2010b). Equivalent One-Dimensional Spring-Dashpot System Representing Impedance Functions of Structural Systems with Non-Classical Damping, *CMES: Computer Modeling in Engineering & Sciences*, 67(3), pp. 211-238

Saitoh, M. (2011a). On the performance of lumped parameter models with gyro-mass elements for the impedance function of a pile-group supporting a single-degree-of-freedom system, *Earthquake Engineering and Structural Dynamics*, DOI: 10.1002/eqe.1147

Saitoh, M. (2011b). A one-dimensional lumped parameter model representing impedance functions in general structural systems with proportional damping, International Journal for Numerical Methods in Engineering, DOI: 10.1002/nme.3323

Taherzadef, R., Clouteau, D., & Cottereau, R. (2009). Simple formulas for the dynamic stiffness of pile groups, *Earthquake Engineering and Structural Dynamics*, 38(15), pp. 1665-1685

Takemiya, H. & Yamada, Y. (1981). Layered soil-pile-structure interaction, *Earthquake Engineering and Structural Dynamics*, 9, pp. 437-452

Takewaki, I., Murakami, S., Fujita, K., Yoshitomi, S., Tsuji, M. (2011). The 2011 off the Pacific coast of Tohoku earthquake and response of high-rise buildings under long-period ground motions, *Soil Dynamics and Earthquake Engineering*, doi:10.1016/j. soildyn. 2011. 06. 001

Tileylioglu, S., Stewart, J. P., & Nigbor, R. L. (2011). Dynamic stiffness of a shallow foundation from forced vibration of a field test structure, *Journal of Geotechnical and Geoenvironmental Engineering*, Vol. 137 (4), pp. 344-353

Eberhard, M. O., Baldridge, S., Marshall, J., Mooney, W., G. J., Rix (2010). *The Mw 7.0 Haiti Earthquake of January 12, 2010*, USGS/EERI Advance Reconnaissance Team: Team Report V.1.1

Veletsos, A. S. & Dotson, K. W. (1988). Horizontal impedances for radially inhomogeneous viscoelastic soil layers, *Earthquake Engineering and Structural Dynamics*, 16, pp. 947-966

Weaver, J. W., Timoshenko, S. P., & Young, D. H. (1990). Vibration Problems in Engineering, Fifth Edition, Wiley Interscience

Wolf, J. P. (1994). Foundation Vibration Analysis Using Simple Physical Models. Englewood Cliffs, NJ, Prentice-Hall

Wolf, J. P. & Obernhuber, P. (1985). Nonlinear soil-structure-interaction analysis using dynamic stiffness or flexibility of soil in the time domain, *Earthquake Engineering and Structural Dynamics*, 13, pp. 195-212

Wolf, J. P. & Motosaka, M. (1989). Recursive evaluation of interaction force of unbounded soil in the time domain, *Earthquake Engineering and Structural Dynamics*, 18, pp. 345-363

Wolf, J. P. & Somaini, D. R. (1986). Approximate dynamic model of embedded foundation in time domain, *Earthquake Engineering and Structural Dynamics*, 14, pp. 683-703

Wolf, J. P. & Paronesso, A. (1992). Lumped-parameter model for a rigid cylindrical foundation embedded in a soil layer on rigid rock, *Earthquake Engineering and Structural Dynamics*, 21, pp. 1021-1038

Wolf, J. P. (1997). Spring-dashpot-mass models for foundation vibrations. *Earthquake Engineering and Structural Dynamics*, 26, pp. 931-949

Wolf, J. P. (1991a). Consistent lumped-parameter models for unbounded soil: physical representation, *Earthquake Engineering and Structural Dynamics*, 20(1), pp. 11-32

Wolf, J. P. (1991b). Consistent lumped-parameter models for unbounded soil: frequency-independent stiffness, damping and mass matrices, *Earthquake Engineering and Structural Dynamics*, 20(1), pp. 33-41

Wu, W. H. & Chen, C. Y. (2001). Simple lumped-parameter models of foundation using mass-spring-dashpot oscillators, *J Chin Inst Eng*, 24(6), pp. 681-97

Wu, H. W. & Chen, C. Y. (2002). Simplified Soil-structure Interaction Analysis Using Efficient Lumped-parameter Models for Soils, *Soils and Foundations*, 42, pp. 41-52

Wu, W. H. & Lee, W. H. (2002). Systematic lumped-parameter models for foundations based on polynomial-fraction approximation. *Earthquake Engineering and Structural Dynamics*, 31, pp. 1383-1412

Wu, H. W. & Lee, H. W. (2004). Nested Lumped-parameter Models for Foundation Vibrations, *Earthquake Engineering and Structural Dynamics*, 33, pp. 1051-1058

Zhao, M. & Du, X. (2008). High-order lumped-parameter model for foundation based on continued fraction, Beijing, 14WCEE

16

The Vibration of a Layered Rotating Planet and Bryan's Effect

Michael Y. Shatalov[1], Stephan V. Joubert[2] and Charlotta E. Coetzee[2]
[1]*CSIR (Pretoria) and Tshwane University of Technology*
[2]*Tshwane University of Technology*
South Africa

1. Introduction

Because, among other seismological observations, it is important to be able to predict the location of the vibrating pattern of an earthquake, in this chapter we take the first tentative steps towards including "Bryan's effect" in a mathematical model for the seismic vibration pattern of a layered slowly rotating vibrating planet or moon.

Historically, G.H. Bryan observed (in his article (Bryan, 1890)) that when a vibrating structure is subjected to an inertial rotation, the vibrating pattern rotates within the structure at a rate proportional to (but in general not equal to) the inertial angular rate. This effect has come to be known as " Bryan's effect". It is interesting to note that (Rayleigh, 1894) mentions Bryan's effect in § 233, but thereafter investigations of Bryan's effect appear to have lain dormant for about 75 years, reappearing in connection with a resonator gyroscope in 1965 in the small Delco Wakefield, MA, USA R&D facility, according to (Rozelle, 2009). Rozelle states that the resonator gyroscope "has been utilized in many applications over its developmental lifetime: aircraft navigation, strategic missile navigation, underground borehole navigation, communication satellite stabilization, precision pointing, and in deep space missions". The effect may be useful in understanding the dynamics of pulsating stars in astrophysics and this was mentioned in the paper (Shatalov et al., 2009) in which the theoretical background on this chapter was discussed in general terms. For the constant of proportionality, Bryan made the following calculation for a body consisting of a ring or cylinder:

$$BF = \frac{\text{Angular rate of the vibration pattern}}{\text{Inertial angular rate of the vibrating body}}. \qquad (1)$$

This constant of proportionality BF has come to be known as "Bryan's factor". The authors (Zhuravlev & Klimov, 1988) investigated Bryan's effect for an elastic, isotropic, spherically symmetric body, rotating in three-dimensional space. Among other results, they demonstrated that Bryan's factor depends on the vibration mode. Their concise formulation is given in general terms without computational detail or assumptions on the magnitude of rotation or illustrative examples. In (Shatalov et al., 2009) "slow rotation" (explained below) was assumed for spherical bodies consisting of concentric layers of elastic and/or acoustic media. In that paper, some detail was supplied for computations and an illustrative example was presented. They did not assume a "thin shell theory", as in (Loveday & Rogers, 1998), where Bryan's effect is considered in a thin cylindrical shell for both high and low rotational rates. Consequently the model described by (Shatalov et al., 2009) is readily adaptable to the structure of an ideal planet or moon that is rotating slowly.

In this chapter we study Bryan's effect applied to a rotating planet or moon that may have several solid and liquid layers that are assumed to be rotating at the same rate. The rotation rate is assumed to be small when compared to the lowest eigenvalue of vibration $\omega = \frac{2\pi}{T}$ where T is the period of oscillation of the vibration pattern for a natural mode of vibration. As a first tentative approach, we discuss isotropic solid spherical layers that may be isotropically damped as well as isotropic inviscid spherical liquid layers. If damping is present in the media we expect to encounter, as a first approach, we assume that it will be isotropic and be light in the sense that the "damping factor" will be substantially smaller than the lowest eigenvalue of the system. Using Rayleigh's dissipation function (Goldstein, et al.), we demonstrate that light, isotropic, viscous damping does not influence Bryan's effect or Bryan's factor (Equation (1)). Hence, in the sequel, we assume that the body is subjected to nondecaying vibrations in one of its natural modes. The introduction of "impurities" such as prestress, mass-stiffness imperfections and anisotropic damping effects into the calculations is important for real-life situations (as opposed to ideal situations with at most isotropic, viscous damping) and has been earmarked by us for further study. Indeed, we have recently published a spherical model with varying mass density (Shatalov et al., 2011) that indicates that the rotation rate of the vibrating pattern is nonlinear (as opposed to the linear rate for the ideal state discussed below) and that this nonlinearity manifests itself as a "capture effect" whereby the rotation angle of the vibration pattern varies periodically.

2. Preliminaries

Consider a coordinate system $Oxyz$ and a composite spherical body, with its centre at the origin O, consisting of concentric solid and or acoustic layers. Let N be the number of concentric spherical media in the system and a_i and a_{i-1} the inner and outer radii of the ith layer respectively, $i = 1, \cdots, N$. We convert to spherical coordinates $Or\theta\phi$ as depicted in Figure ??, where we have adopted the notation of (Spiegel, 1967).

Consider the position of rest $P(r, \theta, \phi)$ of a vibrating particle in the ith layer where $a_i \leq r \leq a_{i-1}$. Let $\hat{\boldsymbol{r}}$ be the unit vector in the direction of increasing r. Hence the position vector of the point $P(r, \theta, \phi)$ is $\boldsymbol{r} = r\hat{\boldsymbol{r}}$. Consider the usual unit vectors $\hat{\boldsymbol{\phi}} = \dfrac{\partial \boldsymbol{r}}{\partial \phi} / \left|\dfrac{\partial \boldsymbol{r}}{\partial \phi}\right|$ (in the direction of increasing ϕ) and $\hat{\boldsymbol{\theta}} = \dfrac{\partial \boldsymbol{r}}{\partial \theta} / \left|\dfrac{\partial \boldsymbol{r}}{\partial \theta}\right|$ (in the direction of increasing θ). Let $\boldsymbol{w}_i + \boldsymbol{u}_i + \boldsymbol{v}_i$ (where $\boldsymbol{w}_i = w_i\hat{\boldsymbol{r}}$, $\boldsymbol{u}_i = u_i\hat{\boldsymbol{\theta}}$ and $\boldsymbol{v}_i = v_i\hat{\boldsymbol{\phi}}$) represent the displacement from the position of rest of the vibrating particle in the ith layer. For the sake of simplicity, we suppress subscripts i when no confusion is expected. The position vector of the vibrating particle is thus

$$\mathbf{R} = (r + w)\hat{\boldsymbol{r}} + u\hat{\boldsymbol{\theta}} + v\hat{\boldsymbol{\phi}}. \tag{2}$$

Now consider an inertial coordinate system $OXYZ$ with origin O, where initially the X, Y, Z-axes correspond to the x, y, z-axes respectively. Let the spherical body (the $Or\theta\phi \equiv Oxyz$ system) rotate about the z-axis with respect to inertial space $OXYZ$ with a small constant angular rate. If $\hat{\boldsymbol{k}}$ is the unit vector in the direction of increasing z, then let the angular velocity $\boldsymbol{\Omega}$ of the body be

$$\boldsymbol{\Omega} = \varepsilon\Omega\hat{\boldsymbol{k}} = \varepsilon\Omega(\hat{\boldsymbol{r}}\cos\theta - \hat{\boldsymbol{\theta}}\sin\theta), \tag{3}$$

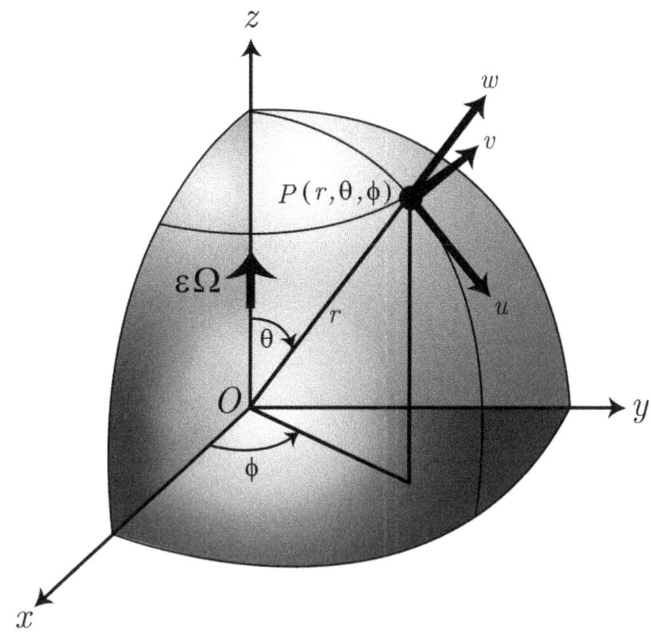

Fig. 1. The spherical coordinate system $Or\theta\phi$ and its relation to the reference frame $Oxyz$ that is rotating slowly about the $z-$axis at rate $\varepsilon\Omega$, showing a particle in the ith layer with position of rest $P(r, \theta, \phi)$ and displacement from the position of rest u in the direction of increasing angle θ, v in the direction of increasing angle ϕ and w in the radial direction.

where the dimensionless parameter ε is a measure of smallness. By the word "smallness" of the angular rate of rotation $\varepsilon\Omega$ we mean (as mentioned above) that this rate is substantially smaller than the lowest eigenvalue of the system.

3. Gyroscopic effects in distributed bodies

The mathematical formulation given below (in spherical coordinates) is presented within the framework of the linearised, three-dimensional theory of elasticity (Redwood, 1960). With Lagrange's equations (Spiegel, 1967) in mind, we formulate expressions for the (approximate) kinetic and potential energies of the system of concentric spherical bodies. The absolute linear velocity of the vibrating particle is

$$\begin{gathered} \boldsymbol{V} = \frac{d\boldsymbol{R}}{dt} + \boldsymbol{\Omega} \times \boldsymbol{R} = \\ (\dot{w} - \varepsilon\Omega v \sin\theta)\hat{\boldsymbol{r}} + \\ (\dot{u} - \varepsilon\Omega v \cos\theta)\,\hat{\boldsymbol{\theta}} + \\ (\dot{v} + \varepsilon\Omega\,(u\cos\theta + (r+w)\sin\theta))\,\hat{\boldsymbol{\phi}}. \end{gathered} \tag{4}$$

Neglecting terms of $O(\varepsilon^2)$, the approximate kinetic energy of the system of concentric spherical bodies is given by

$$K = \frac{1}{2}\sum_{i=1}^{N}\rho_i \int_0^{2\pi}\int_0^{\pi}\int_{a_i}^{a_{i-1}} \left\{\left(\dot{u}_i^2 + \dot{v}_i^2 + \dot{w}_i^2\right) + 2\varepsilon\Omega\left[(u_i\dot{v}_i - \dot{u}_i v_i)\cos\theta + \right.\right.$$
$$\left.\left.(\dot{v}_i(r+w_i) - v_i\dot{w}_i)\sin\theta\right]\right\} r^2 \sin\theta dr d\theta d\phi. \tag{5}$$

The potential energy of the system of concentric spheres is

$$P = \frac{1}{2}\sum_{i=1}^{N}\int_0^{2\pi}\int_0^{\pi}\int_{a_i}^{a_{i-1}} \left\{\sigma_{i,rr}\epsilon_{i,rr} + \sigma_{i,\theta\theta}\epsilon_{i,\theta\theta} + \sigma_{i,\phi\phi}\epsilon_{i,\phi\phi} + \right.$$
$$\left.\sigma_{i,r\theta}\epsilon_{i,r\theta} + \sigma_{i,\theta\phi}\epsilon_{i,\theta\phi} + \sigma_{i,r\phi}\epsilon_{i,r\phi}\right\} r^2 \sin\theta dr d\theta d\phi \tag{6}$$

where we use "ρ" for mass density, "σ" for stress and "ϵ" for strain. We indicate Lamé's constants (from the theory of elasticity) by λ_i and μ_i for the ith layer. In a spherical coordinate system, with the subscript i suppressed for the sake of brevity (i.e. we write σ_{rr} for $\sigma_{i,rr}$ etc.), a standard reference such as (Redwood, 1960) yields stresses

$$\sigma_{rr} = \lambda(\epsilon_{rr} + \epsilon_{\theta\theta} + \epsilon_{\phi\phi}) + 2\mu\epsilon_{rr}; \quad \sigma_{\theta\theta} = \lambda(\epsilon_{rr} + \epsilon_{\theta\theta} + \epsilon_{\phi\phi}) + 2\mu\epsilon_{\theta\theta};$$
$$\sigma_{\phi\phi} = \lambda(\epsilon_{rr} + \epsilon_{\theta\theta} + \epsilon_{\phi\phi}) + 2\mu\epsilon_{\phi\phi};$$
$$\sigma_{r\theta} = 2\mu\epsilon_{r\theta}; \quad \sigma_{\theta\phi} = 2\mu\epsilon_{\theta\phi}; \quad \sigma_{r\phi} = 2\mu\epsilon_{r\phi}; \tag{7}$$

and strains

$$\epsilon_{rr} = \frac{\partial w}{\partial r}; \quad \epsilon_{\theta\theta} = \frac{1}{r}\left(\frac{\partial u}{\partial \theta} + w\right); \quad \epsilon_{\phi\phi} = \frac{1}{r}\left(u\cot\theta + \frac{1}{\sin\theta}\frac{\partial v}{\partial \phi} + w\right);$$
$$\epsilon_{r\theta} = \frac{\partial u}{\partial r} + \frac{1}{r}\left(\frac{\partial w}{\partial \theta} - u\right); \quad \epsilon_{\theta\phi} = \frac{1}{r}\left(\frac{1}{\sin\theta}\frac{\partial u}{\partial \phi} + \frac{\partial v}{\partial \theta} - v\cot\theta\right);$$
$$\epsilon_{r\phi} = \frac{\partial v}{\partial r} + \frac{1}{r}\left(\frac{1}{\sin\theta}\frac{\partial w}{\partial \phi} - v\right). \tag{8}$$

Assume that we can express the magnitude of the displacements u_i, v_i and w_i of the vibrating particle P in the ith body as follows:

$$u_i(r,\theta,\phi,t) = U_i(r,\theta)\left[C(t)\cos m\phi + S(t)\sin m\phi\right], \tag{9a}$$
$$v_i(r,\theta,\phi,t) = V_i(r,\theta)\left[C(t)\sin m\phi - S(t)\cos m\phi\right], \tag{9b}$$
$$w_i(r,\theta,\phi,t) = W_i(r,\theta)\left[C(t)\cos m\phi + S(t)\sin m\phi\right], \tag{9c}$$

where the nature of the functions $C(t)$ and $S(t)$ is still to be determined, $U_i(r,\theta)$, $V_i(r,\theta)$ and $W_i(r,\theta)$ are unknown eigenfunctions of the system and m is the circumferential wave number. Substituting Equations (9) into Equations (8), (7), (6) and (5) involves a long algebraic calculation. A computer algebra system is handy for checking the calculation that yields:

$$K = \pi\left[I_0\left(\dot{C}^2 + \dot{S}^2\right) + 2\Omega I_1\left(\dot{C}S - C\dot{S}\right)\right] \tag{10}$$

and

$$P = \pi I_2 \left(C^2 + S^2 \right). \tag{11}$$

Here

$$I_0 = \frac{1}{2} \sum_{i=1}^{N} \rho_i \int_0^{\pi} \int_{a_i}^{a_{i-1}} \left(U_i^2 + V_i^2 + W_i^2 \right) r^2 \sin\theta dr d\theta, \tag{12}$$

$$I_1 = \sum_{i=1}^{N} \rho_i \int_0^{\pi} \int_{a_i}^{a_{i-1}} \left(U_i \cos\theta + W_i \sin\theta \right) V_i r^2 \sin\theta dr d\theta, \tag{13}$$

$$\begin{aligned} I_2 = \sum_{i=1}^{N} \rho_i \int_0^{\pi} \int_{a_i}^{a_{i-1}} & \left\{ 2\mu_i \left[\begin{array}{c} \left(m \csc\theta\, U_i + \cot\theta\, V_i - \dfrac{\partial V_i}{\partial \theta} \right)^2 + \\ \left(\dfrac{\partial W_i}{\partial \theta} + r \dfrac{\partial U_i}{\partial r} - U_i \right)^2 + \left(V_i + m \csc\theta\, W_i \right) - r \dfrac{\partial V_i}{\partial r} \right)^2 \end{array} \right] + \right. \\ & (\cot\theta\, U_i + m \csc\theta\, V_i + W_i) \times \\ & \left[(\lambda_i + 2\mu_i) \left[\begin{array}{c} \cot\theta\, U_i + \\ m \csc\theta\, V_i + 2(\lambda_i + \mu_i) W \end{array} \right] + \lambda_i \left(\frac{\partial U_i}{\partial \theta} + r \frac{\partial W_i}{\partial r} \right) \right] + \\ & \left(W_i + \frac{\partial U_i}{\partial \theta} \right) \left(\begin{array}{c} \lambda_i (\cot\theta\, U_i + m \csc\theta\, V_i) + \\ 2(\lambda_i + \mu_i) W_i + (\lambda_i + 2\mu_i) \dfrac{\partial U_i}{\partial \theta} + r \lambda_i \dfrac{\partial W_i}{\partial \theta} \end{array} \right) + \\ & \left. r \frac{\partial W_i}{\partial r} \left[\lambda_i \left(\begin{array}{c} \cot\theta\, U_i + m \csc\theta\, V_i + \\ 2 W_i + \dfrac{\partial U_i}{\partial \theta} \end{array} \right) + r (\lambda_i + 2\mu_i) \frac{\partial W_i}{\partial r} \right] \right\} \sin\theta dr d\theta. \end{aligned} \tag{14}$$

Because $K = K(C, S, \dot{C}, \dot{S})$ and $P = P(C, S)$, the Lagrangian

$$L(C, S, \dot{C}, \dot{S}) = K - P \tag{15}$$

yields two equations of motion from Lagrange's equations (Goldstein, et al.):

$$\frac{d}{dt} \frac{\partial L}{\partial \dot{C}} - \frac{\partial L}{\partial C} = - \frac{\partial \mathcal{F}}{\partial \dot{C}}, \tag{16a}$$

$$\frac{d}{dt} \frac{\partial L}{\partial \dot{S}} - \frac{\partial L}{\partial S} = - \frac{\partial \mathcal{F}}{\partial \dot{S}} \tag{16b}$$

where

$$\mathcal{F} = \frac{1}{2} (c\dot{C}^2 + s\dot{S}^2) \tag{17}$$

is Rayleigh's dissipation function, c and s are viscous damping constants. We assume isotropic damping, that is, say, $c = s = \pi D$. We further assume that for the media we will encounter, the "damping factor"

$$\varepsilon\delta = \frac{D}{2I_0} \tag{18}$$

is substantially smaller than the lowest eigenvalue of the vibrating system. Computation of Equations (16) yields a coupled system of second-order linear ordinary differential equations (ODE):

$$\ddot{C} + 2\eta\varepsilon\Omega\dot{S} + \omega^2 C + 2\varepsilon\delta\dot{C} = 0, \tag{19a}$$

$$\ddot{S} - 2\eta\varepsilon\Omega\dot{C} + \omega^2 S + 2\varepsilon\delta\dot{S} = 0, \tag{19b}$$

where

$$-1 \leq \eta = \frac{I_1}{I_0} \leq 1 \tag{20}$$

and

$$\omega^2 = \frac{I_2}{I_0}. \tag{21}$$

We now show that η in Equation (20) is Bryan's factor as given by Equation (1) and that ω in Equation (21) is an eigenvalue for the vibrating system.
In order to interpret what the system of ODE (19) represents, combine the two equations by considering the complex function $Z = C + iS$ to obtain the single equation

$$\ddot{Z} + 2\varepsilon(\delta - i\eta\Omega)\dot{Z} + \omega^2 Z = 0. \tag{22}$$

where, as is usual, $i^2 = -1$. Writing Z in polar form

$$Z(t) = Y(t)e^{i\beta(t)} \tag{23}$$

and assuming that $\beta(t)$ has the linear form

$$\beta(t) = at, \tag{24}$$

while $Y(t)$ decays according to

$$Y(t) = X(t)e^{-bt}, \tag{25}$$

then

$$Z = X(t)e^{(ia-b)t} \tag{26}$$

and substituting into Equation (22), we obtain the ODE

$$\begin{gathered} \ddot{X} + 2\left[(ia - b) + \varepsilon\left(\delta - i\eta\Omega\right)\right]\dot{X} + \\ \left[2\varepsilon\left(\delta - i\eta\Omega\right)(ia - b) + (ia - b)^2 + \omega^2\right] X = 0. \end{gathered} \tag{27}$$

If we choose $a = \eta\varepsilon\Omega$ and $b = \varepsilon\delta$, then the coefficient of $\dot{X}$ vanishes in Equation (27) and we obtain the ODE

$$\ddot{X} + \nu^2 X = 0 \tag{28}$$

where

$$\nu^2 = \omega^2 - \varepsilon^2\left(\delta - i\eta\Omega\right)^2 \tag{29}$$

Neglecting $O(\varepsilon^2)$, we obtain

$$\nu \approx \omega. \tag{30}$$

Consequently Equation (28) approximates the equation of motion of a harmonic oscillator. Equations (19) can now be viewed in the form

$$Z(t) = \left[e^{-\delta t} X(t)\right] e^{i\eta\varepsilon\Omega t}. \tag{31}$$

These equations represent a "vector" in the complex plane with its size varying like a damped harmonic oscillator and its position varying at a rate $\eta\varepsilon\Omega$ (in the rotating reference frame $Oxyz$). Thus, according to Equation (1), η is Bryan's factor for the system. Consequently, neither Bryan's effect nor the value of Bryan's factor η, depends on the inclusion of light isotropic viscous damping in the model and we will thus neglect damping in the sequel. The rotation of the vibrating pattern is in the direction of rotation of the system if $\eta > 0$ and in the opposite direction if $\eta < 0$. Equations (28) and (30) show that ω is an eigenvalue of the vibrating system.
Taking Equation (30) into account, Equation (28) has a general solution of the form

$$X(t) = A\cos(\omega t + \gamma) \tag{32}$$

where the amplitude A and phase angle γ are arbitrary constants. Consequently, neglecting damping, from Equation (31) we obtain

$$C + iS = A\cos(\omega t + \gamma)e^{i\eta\varepsilon\Omega t}. \tag{33}$$

Equating real and imaginary parts we have a good approximation

$$C(t) = A\cos\eta\varepsilon\Omega t\ \cos(\omega t + \gamma), \tag{34a}$$

$$S(t) = A\sin\eta\varepsilon\Omega t\ \cos(\omega t + \gamma) \tag{34b}$$

and so initially, that is, when $t \approx 0$, we find that

$$C(t) \approx A\cos(\omega t + \gamma), \tag{35}$$

$$S(t) \approx 0. \tag{36}$$

4. Eigenfunctions

In order to determine the nature of Bryan's constant η (Equation (20)) and the square of the corresponding eigenvalue ω (Equation (21)), we neglected $O(\varepsilon^2)$. We are going to demonstrate that it is also feasible to neglect the small rotation rate $\varepsilon\Omega$ when we derive expressions for the eigenfunctions $U_i(r,\theta)$, $V_i(r,\theta)$ and $W_i(r,\theta)$, $i = 1,\cdots,N$, as given in Equations (9). Indeed, assume that we have a system of partial differential equations (PDE) involving the eigenfunctions $U_i(r,\theta)$, $V_i(r,\theta)$ and $W_i(r,\theta)$. Using the small parameter ε and a perturbation method Nayfeh (1973), assume that we have an asymptotic expansion

$$U_i(r,\theta) = U_{i,0}(r,\theta) + \varepsilon U_{i,1}(r,\theta) + \varepsilon^2 U_{i,2}(r,\theta) + \cdots, \tag{37}$$

where the $U_{i,j}(r,\theta)$, $j = 0,1,2,\cdots$ are independent of ε and $U_{i,0}(r,\theta)$ is that part of the solution when $\varepsilon = 0$. A similar argument is valid for the eigenfunctions $V_i(r,\theta)$ and $W_i(r,\theta)$. Substituting these three expansions into Equations (12) and (13) and again neglecting all term of $O(\varepsilon^2)$, the formula for η given in Equation (20) yields

$$\eta = \frac{2\sum_{i=1}^{N}\rho_i\int_0^{\pi}\int_{a_i}^{a_{i-1}}\left[\left(U_{i,0}\cos\theta + W_{i,0}\sin\theta\right)V_{i,0} + O(\varepsilon)\right]r^2\sin\theta\, dr d\theta}{\sum_{i=1}^{N}\rho_i\int_0^{\pi}\int_{a_i}^{a_{i-1}}\left[U_{i,0}^2 + V_{i,0}^2 + W_{i,0}^2 + O(\varepsilon)\right]r^2\sin\theta\, dr d\theta}. \tag{38}$$

Now write Bryan's factor η as an asymptotic expansion

$$\eta = \eta_0 + \varepsilon\eta_1 + \varepsilon^2\eta_2 + \cdots \tag{39}$$

where the η_j, $j = 0, 1, 2, \cdots$ are independent of ε and η_0 is given by Equation (38) when $\varepsilon = 0$, that is

$$\eta_0 = \frac{2\sum_{i=1}^{N} \rho_i \int_0^{\pi} \int_{a_i}^{a_{i-1}} \left[\left(U_{i,0} \cos\theta + W_{i,0} \sin\theta \right) V_{i,0} \right] r^2 \sin\theta dr d\theta}{\sum_{i=1}^{N} \rho_i \int_0^{\pi} \int_{a_i}^{a_{i-1}} \left[U_{i,0}^2 + V_{i,0}^2 + W_{i,0}^2 \right] r^2 \sin\theta dr d\theta}. \tag{40}$$

Hence, because we neglect $O(\varepsilon^2)$, Equation (39) yields

$$\varepsilon\eta = \varepsilon\eta_0 + O(\varepsilon^2) \approx \varepsilon\eta_0 \tag{41}$$

and so

$$\eta \approx \eta_0. \tag{42}$$

A similar result is valid for the square of the eigenvalue ω determined by Equation (21). Consequently, because only the eigenfunctions of a non-rotating body need be considered for the purposes of numerically calculating η, in the sequel we will use the notation u_i for $u_{i,0}$, U_i for $U_{i,0}$, etc. Indeed, using (Redwood, 1960) and our notation for stresses, the equations of motion of an isotropic solid, nonrotating vibrating body in spherical coordinates are

$$\rho\frac{\partial^2 u}{\partial t^2} = \frac{\partial\sigma_{rr}}{\partial r} + \frac{1}{r}\frac{\partial\sigma_{r\theta}}{\partial\theta} + \frac{1}{r\sin\theta}\frac{\partial\sigma_{r\phi}}{\partial\phi} + \frac{2\sigma_{rr} - \sigma_{\theta\theta} - \sigma_{\phi\phi} + \cot\theta\,\sigma_{r\theta}}{r}, \tag{43a}$$

$$\rho\frac{\partial^2 v}{\partial t^2} = \frac{\partial\sigma_{r\theta}}{\partial r} + \frac{1}{r}\frac{\partial\sigma_{\theta\theta}}{\partial\theta} + \frac{1}{r\sin\theta}\frac{\partial\sigma_{\theta\phi}}{\partial\phi} + \frac{3\sigma_{r\theta} + \cot\theta\left(\sigma_{\theta\theta} - \sigma_{\phi\phi}\right)}{r}, \tag{43b}$$

$$\rho\frac{\partial^2 w}{\partial t^2} = \frac{\partial\sigma_{r\phi}}{\partial r} + \frac{1}{r}\frac{\partial\sigma_{\theta\phi}}{\partial\theta} + \frac{1}{r\sin\theta}\frac{\partial\sigma_{\phi\phi}}{\partial\phi} + \frac{3\sigma_{r\phi} + 2\cot\theta\,\sigma_{\theta\phi}}{r}. \tag{43c}$$

The stresses are given by Equation (7). The coupled system of second-order PDE given by Equations (43) for the three displacement components u, v and w can be uncoupled directly, but this leads to a sixth-order PDE. Hence, in a manner similar to that explained in (Eringen & Suhubi, 1975), we express the displacement components of Equations (43) in terms of derivatives of potentials $\Phi = \Phi(r,\theta,\phi)$, $\chi = \chi(r,\theta,\phi)$ and $\Psi = \Psi(r,\theta,\phi)$ as follows:

$$u = \left\{ \frac{1}{r}\frac{\partial}{\partial\theta}\left[\Phi + \frac{\partial(r\chi)}{\partial r}\right] + \frac{1}{a\sin\theta}\frac{\partial\Psi}{\partial\phi} \right\} \cos\left(\omega t + \gamma\right), \tag{44a}$$

$$v = \left\{ \frac{1}{r\sin\theta}\frac{\partial}{\partial\phi}\left[\Phi + \frac{\partial(r\chi)}{\partial r}\right] - \frac{1}{a}\frac{\partial\Psi}{\partial\theta} \right\} \cos\left(\omega t + \gamma\right), \tag{44b}$$

$$w = \left\{ \frac{\partial\Phi}{\partial r} + \frac{\partial^2(r\chi)}{\partial r^2} + r\nabla^2\chi \right\} \cos\left(\omega t + \gamma\right), \tag{44c}$$

where ∇^2 the Laplace operator in spherical coordinates, ω is the eigenvalue mentioned above, γ is some phase angle (see Equation (32)) and a is a nonzero constant with the dimension of length. When Equations (44) are substituted into Equations (43) and the resulting equations of motion are uncoupled, it is found that each potential Φ, χ and Ψ satisfies the Helmholtz equations

$$\nabla^2\Phi + k_1^2\Phi = 0, \quad \nabla^2\chi + k_2^2\chi = 0, \quad \nabla^2\Psi + k_2^2\Psi = 0 \tag{45}$$

with wave numbers

$$k_1 = k_1(\omega) = \omega\sqrt{\rho/(\lambda+2\mu)}, \tag{46a}$$

$$k_2 = k_2(\omega) = \omega\sqrt{\rho/\mu}. \tag{46b}$$

In order to obtain initial (at $t \approx 0$) agreement with our choice of Equations (9), the appropriate solutions to Equations (45) are

$$\Phi_{m,n}(r,\theta,\phi) = \left[B_1 j_n(k_1 r) + B_2 y_n(k_1 r)\right] P_n^m(\cos\theta)\cos(m\phi), \tag{47a}$$

$$\chi_{m,n}(r,\theta,\phi) = \left[B_3 j_n(k_2 r) + B_4 y_n(k_2 r)\right] P_n^m(\cos\theta)\cos(m\phi), \tag{47b}$$

$$\Psi_{m,n}(r,\theta,\phi) = \left[B_5 j_n(k_2 r) + B_6 y_n(k_2 r)\right] P_n^m(\cos\theta)\sin(m\phi), \tag{47c}$$

where, as usual, $j_n(kr) = \sqrt{\frac{\pi}{2kr}} J_{n+1/2}(kr)$ and $y_n(kr) = \sqrt{\frac{\pi}{2kr}} Y_{n+1/2}(kr)$ represent the spherical Bessel and Neumann functions respectively where $J_{n+1/2}$ and $Y_{n+1/2}$ represent the Bessel and Neumann functions respectively, while P_n^m is the associated Legendre polynomial. The symbols $B_1, B_2, \cdots, B_6$ are arbitrary constants (if the body contains the centre O, then the constants $B_2 = B_4 = B_6 = 0$). By substituting Equations (47) into Equations (44) we obtain the nature of the eigenfunctions $U(r,\theta)$, $V(r,\theta)$ and $W(r,\theta)$ as given in Equations (9).
The motion of a compressible, inviscid acoustic medium is represented by the following wave equation:

$$\nabla^2 p = \frac{\rho}{\kappa}\frac{\partial^2 p}{\partial t^2}, \tag{48}$$

where κ is the bulk modulus and ρ the mass density of the acoustic medium. With the wave number

$$k_3 = k_3(\omega) = \omega\sqrt{\rho/\kappa}, \tag{49}$$

an appropriate solution to PDE (48) is

$$p_{m,n}(r,\theta,\phi,t) = \left\{\left[B_7 j_n(k_3 r) + B_8 y_n(k_3 r)\right] P_n^m(\cos\theta)\cos(m\phi)\right\}\cos(\omega t+\gamma), \tag{50}$$

where $p = p_{m,n}(r,\theta,\phi,t)$ is the pressure in the acoustic medium and B_7, B_8 are arbitrary constants. Particle displacement of the acoustic medium in the radial direction is

$$w = \frac{1}{\rho\omega^2}\frac{\partial p}{\partial r}. \tag{51}$$

In the sequel, keep in mind that for $j = 1,2,3$ and layers $i = 1,\cdots,N$, we will write $k_{i,j}$ for k_j in layer i, etc.

5. Boundary and continuity conditions

Observing Equations (44) it is possible to distinguish between spheroidal and torsional modes. For the spheroidal mode we assume that $\Psi_i = 0$. In this case the stress components of the solids are

$$\sigma_{i,rr} = \left[2\mu_i \frac{\partial^2 \Phi_i}{\partial r^2} - \lambda_i k_{i,1}^2 \Phi_i\right] + 2\mu_i \left[\frac{\partial^3 (r\chi_i)}{\partial r^3} + k_{i,2}^2 \frac{\partial (r\chi_i)}{\partial r}\right], \tag{52a}$$

$$\sigma_{i,r\theta} = \frac{2\mu_i}{r}\frac{\partial}{\partial \theta}\left\{\left(\frac{\partial \Phi_i}{\partial r} - \frac{\Phi_i}{r}\right) + \left[r\frac{\partial^2 \chi_i}{\partial r^2} + \frac{\partial \chi_i}{\partial r} + \left(\frac{rk_{i,2}^2}{2} - \frac{1}{r}\right)\chi_i\right]\right\}, \tag{52b}$$

$$\sigma_{i,r\phi} = \frac{2\mu_i}{r\sin\theta}\frac{\partial}{\partial \phi}\left\{\left(\frac{\partial \Phi_i}{\partial r} - \frac{\Phi_i}{r}\right) + \left[r\frac{\partial^2 \chi_i}{\partial r^2} + \frac{\partial \chi_i}{\partial r} + \left(\frac{rk_{i,2}^2}{2} - \frac{1}{r}\right)\chi_i\right]\right\}, \tag{52c}$$

where $i = 1,2,3,4,6$ denotes a solid layer in the spherical structure. Indeed, here we assume $i = 1$ for the outer solid layer ("crust"), $i = 2$ for the next inner layer ("upper mantle-1"), $i = 3$ for the "upper mantle-2", $i = 4$ for the "lower mantle" and $i = 6$ for the "inner solid core". We do not use $i = 5$ in the list of the solids because this describes the "liquid outer core".
For the investigation of the torsional modes we assume that $\Phi_i = 0$ and $\chi_i = 0$ and hence the corresponding stress components are

$$\sigma_{i,r\theta} = \frac{\mu_i}{a\sin\theta}\frac{\partial}{\partial \phi}\left(\frac{\partial \Psi_i}{\partial r} - \frac{\Psi_i}{r}\right), \tag{53a}$$

$$\sigma_{i,r\phi} = -\frac{\mu_i}{a}\frac{\partial}{\partial \theta}\left(\frac{\partial \Psi_i}{\partial r} - \frac{\Psi_i}{r}\right). \tag{53b}$$

Let us model a spherical "planet" with solid "inner core", four solid layers and one liquid layer. Assume that the outer radius of the planet is $r = a$ and that the boundary between the "crust" and the "upper mantle-1" is located at radius $r = a_1$. Furthermore, assume that the boundary between the "upper mantle-1" and the "upper mantle-2" is situated at radius $r = a_2$, while at $r = a_3$ we find the boundary between the "upper mantle-2" and the "lower mantle". All these layers are considered as isotropic solid bodies. Assume that the boundary between the "lower mantle" and the "liquid outer core" is situated at radius $r = a_4$ and finally, radius $r = a_5$ is the boundary between the "liquid outer core" and the solid "inner core".
The elastic potentials and pressure that describe the spheroidal modes of the planet are solutions given by Equations (47a), (47b) and (50):

$$\Phi_i(r,\theta,\phi) = \left[A_{4i-3} j_n\left(k_{i,1}r\right) + A_{4i-2} y_n\left(k_{i,1}r\right)\right] P_n^m(\cos\theta)\cos m\phi, \tag{54a}$$

$$\chi_i(r,\theta,\phi) = \left[A_{4i-1} j_n\left(k_{i,2}r\right) + A_{4i} y_n\left(k_{i,2}r\right)\right] P_n^m(\cos\theta)\cos m\phi, \tag{54b}$$

$$p_5(r,\theta,\phi) = \left[A_{17} j_n\left(k_{5,3}r\right) + A_{18} y_n\left(k_{5,3}r\right)\right] P_n^m(\cos\theta)\cos m\phi, \tag{54c}$$

$$\Phi_6(r,\theta,\phi) = \left[A_{19} j_n\left(k_{6,1}r\right)\right] P_n^m(\cos\theta)\cos m\phi, \tag{54d}$$

$$\chi_6(r,\theta,\phi) = \left[A_{20} j_n\left(k_{6,2}r\right)\right] P_n^m(\cos\theta)\cos m\phi, \tag{54e}$$

where the symbols "A" with various subscripts represent arbitrary constants that will be determined from the boundary and continuity conditions described below and $i = 1,2,3,4$.
The "upper core" of the planet does not have a torsional mode and is considered to be an inviscid ideal fluid. Despite the fact that the "inner core" is solid and hence that it could support the torsional mode, we do not consider it when we analyse the torsional mode because the shear stress in the "inner core" cannot be transformed into the outer layers through the shear free liquid "outer core". The elastic potential describing the torsional modes of the planet is obtained from Equation (47c):

$$\Psi_i(r,\theta,\phi) = \left[B_{2i-1} j_n\left(k_{i,2}r\right) + A_{2i} y_n\left(k_{i,2}r\right)\right] P_n^m(\cos\theta)\cos m\phi, \tag{55}$$

where $i = 1,2,3,4$.

The boundary and continuity conditions below express the balance between the radial stresses and pressure and equality of the tangential stresses in the solid layers. They also express the equality of the radial and tangential components between the solid layers (correspondingly, only radial components between the solids and fluid layer). The outer layer is assumed to be free from radial and tangential stress components. Using these assumptions together with Equations (52) we obtain the following boundary and continuity conditions for the investigation of the spheroidal modes:

At $r = a_0 = a$

$$\left\{\left[2\mu_1 \frac{\partial^2 \Phi_1}{\partial r^2} - \lambda_1 k_{1,1}^2 \Phi_1\right] + 2\mu_1 \left[\frac{\partial^3 (r\chi_1)}{\partial r^3} + k_{1,2}^2 \frac{\partial (r\chi_1)}{\partial r}\right]\right\}\Bigg|_{r=a} = 0, \tag{56}$$

$$\frac{2\mu_1}{a}\left\{\left(\frac{\partial \Phi_1}{\partial r} - \frac{\Phi_1}{r}\right) + \left[r\frac{\partial^2 \chi_1}{\partial r^2} + \frac{\partial \chi_1}{\partial r} + \left(\frac{r k_{1,2}^2}{2} - \frac{1}{r}\right)\chi_1\right]\right\}\Bigg|_{r=a} = 0. \tag{57}$$

At $r = a_1$

$$\begin{aligned}&\left\{\left[2\mu_1 \frac{\partial^2 \Phi_1}{\partial r^2} - \lambda_1 k_{1,1}^2 \Phi_1\right] + 2\mu_1 \left[\frac{\partial^3 (r\chi_1)}{\partial r^3} + k_{1,2}^2 \frac{\partial (r\chi_1)}{\partial r}\right]\right\}\Bigg|_{r=a_1} = \\ &\left\{\left[2\mu_2 \frac{\partial^2 \Phi_2}{\partial r^2} - \lambda_2 k_{2,1}^2 \Phi_2\right] + 2\mu_2 \left[\frac{\partial^3 (r\chi_2)}{\partial r^3} + k_{2,2}^2 \frac{\partial (r\chi_2)}{\partial r}\right]\right\}\Bigg|_{r=a_1},\end{aligned} \tag{58}$$

$$\begin{aligned}&\frac{2\mu_1}{a_1}\left\{\left(\frac{\partial \Phi_1}{\partial r} - \frac{\Phi_1}{r}\right) + \left[r\frac{\partial^2 \chi_1}{\partial r^2} + \frac{\partial \chi_1}{\partial r} + \left(\frac{r k_{1,2}^2}{2} - \frac{1}{r}\right)\chi_1\right]\right\}\Bigg|_{r=a_1} = \\ &\frac{2\mu_2}{a_1}\left\{\left(\frac{\partial \Phi_2}{\partial r} - \frac{\Phi_2}{r}\right) + \left[r\frac{\partial^2 \chi_2}{\partial r^2} + \frac{\partial \chi_2}{\partial r} + \left(\frac{r k_{2,2}^2}{2} - \frac{1}{r}\right)\chi_2\right]\right\}\Bigg|_{r=a_1},\end{aligned} \tag{59}$$

$$\begin{aligned}&\left\{\frac{\partial \Phi_1}{\partial r} + \left[\frac{\partial^2 (r\chi_1)}{\partial r^2} + k_{1,2}^2 (r\chi_1)\right]\right\}\Bigg|_{r=a_1} = \\ &\left\{\frac{\partial \Phi_2}{\partial r} + \left[\frac{\partial^2 (r\chi_2)}{\partial r^2} + k_{2,2}^2 (r\chi_2)\right]\right\}\Bigg|_{r=a_1},\end{aligned} \tag{60}$$

$$\left\{\Phi_1 + \frac{\partial (r\chi_1)}{\partial r}\right\}\Bigg|_{r=a_1} = \left\{\Phi_2 + \frac{\partial (r\chi_2)}{\partial r}\right\}\Bigg|_{r=a_1}. \tag{61}$$

At $r = a_2$

$$\begin{aligned}&\left\{\left[2\mu_2 \frac{\partial^2 \Phi_2}{\partial r^2} - \lambda_2 k_{2,1}^2 \Phi_2\right] + 2\mu_2 \left[\frac{\partial^3 (r\chi_2)}{\partial r^3} + k_{2,2}^2 \frac{\partial (r\chi_2)}{\partial r}\right]\right\}\Bigg|_{r=a_2} = \\ &\left\{\left[2\mu_3 \frac{\partial^2 \Phi_3}{\partial r^2} - \lambda_3 k_{3,1}^2 \Phi_3\right] + 2\mu_3 \left[\frac{\partial^3 (r\chi_3)}{\partial r^3} + k_{3,2}^2 \frac{\partial (r\chi_3)}{\partial r}\right]\right\}\Bigg|_{r=a_2},\end{aligned} \tag{62}$$

$$\begin{aligned}&\frac{2\mu_2}{a_2}\left\{\left(\frac{\partial \Phi_2}{\partial r} - \frac{\Phi_2}{r}\right) + \left[r\frac{\partial^2 \chi_2}{\partial r^2} + \frac{\partial \chi_2}{\partial r} + \left(\frac{r k_{2,2}^2}{2} - \frac{1}{r}\right)\chi_2\right]\right\}\Bigg|_{r=a_2} = \\ &\frac{2\mu_3}{a_2}\left\{\left(\frac{\partial \Phi_3}{\partial r} - \frac{\Phi_3}{r}\right) + \left[r\frac{\partial^2 \chi_3}{\partial r^2} + \frac{\partial \chi_3}{\partial r} + \left(\frac{r k_{3,2}^2}{2} - \frac{1}{r}\right)\chi_3\right]\right\}\Bigg|_{r=a_2},\end{aligned} \tag{63}$$

$$\begin{aligned}&\left\{\frac{\partial \Phi_2}{\partial r} + \left[\frac{\partial^2 (r\chi_2)}{\partial r^2} + k_{2,2}^2 (r\chi_2)\right]\right\}\Bigg|_{r=a_2} = \\ &\left\{\frac{\partial \Phi_3}{\partial r} + \left[\frac{\partial^2 (r\chi_3)}{\partial r^2} + k_{3,2}^2 (r\chi_3)\right]\right\}\Bigg|_{r=a_2},\end{aligned} \tag{64}$$

$$\left\{\Phi_2 + \frac{\partial(r\chi_2)}{\partial r}\right\}\bigg|_{r=a_2} = \left\{\Phi_3 + \frac{\partial(r\chi_3)}{\partial r}\right\}\bigg|_{r=a_2}. \quad (65)$$

At $r = a_3$

$$\left\{\left[2\mu_3\frac{\partial^2\Phi_3}{\partial r^2} - \lambda_3 k_{3,1}^2\Phi_3\right] + 2\mu_3\left[\frac{\partial^3(r\chi_3)}{\partial r^3} + k_{3,2}^2\frac{\partial(r\chi_3)}{\partial r}\right]\right\}\bigg|_{r=a_3} = \left\{\left[2\mu_4\frac{\partial^2\Phi_4}{\partial r^2} - \lambda_4 k_{4,1}^2\Phi_4\right] + 2\mu_4\left[\frac{\partial^3(r\chi_4)}{\partial r^3} + k_{4,2}^2\frac{\partial(r\chi_4)}{\partial r}\right]\right\}\bigg|_{r=a_3}, \quad (66)$$

$$\frac{2\mu_3}{a_3}\left\{\left(\frac{\partial\Phi_3}{\partial r} - \frac{\Phi_3}{r}\right) + \left[r\frac{\partial^2\chi_3}{\partial r^2} + \frac{\partial\chi_3}{\partial r} + \left(\frac{rk_{3,2}^2}{2} - \frac{1}{r}\right)\chi_3\right]\right\}\bigg|_{r=a_3} = \frac{2\mu_4}{a_3}\left\{\left(\frac{\partial\Phi_4}{\partial r} - \frac{\Phi_4}{r}\right) + \left[r\frac{\partial^2\chi_4}{\partial r^2} + \frac{\partial\chi_4}{\partial r} + \left(\frac{rk_{4,2}^2}{2} - \frac{1}{r}\right)\chi_4\right]\right\}\bigg|_{r=a_3}, \quad (67)$$

$$\left\{\frac{\partial\Phi_3}{\partial r} + \left[\frac{\partial^2(r\chi_3)}{\partial r^2} + k_{3,2}^2(r\chi_3)\right]\right\}\bigg|_{r=a_3} = \left\{\frac{\partial\Phi_4}{\partial r} + \left[\frac{\partial^2(r\chi_4)}{\partial r^2} + k_{4,2}^2(r\chi_4)\right]\right\}\bigg|_{r=a_3}, \quad (68)$$

$$\left\{\Phi_3 + \frac{\partial(r\chi_3)}{\partial r}\right\}\bigg|_{r=a_3} = \left\{\Phi_4 + \frac{\partial(r\chi_4)}{\partial r}\right\}\bigg|_{r=a_3}. \quad (69)$$

At $r = a_4$

$$\left\{\left[2\mu_4\frac{\partial^2\Phi_4}{\partial r^2} - \lambda_4 k_{4,1}^2\Phi_4\right] + 2\mu_4\left[\frac{\partial^3(r\chi_4)}{\partial r^3} + k_{4,2}^2\frac{\partial(r\chi_4)}{\partial r}\right]\right\}\bigg|_{r=a_4} = -p_5|_{r=a_4}, \quad (70)$$

$$\frac{2\mu_4}{a_4}\left\{\left(\frac{\partial\Phi_4}{\partial r} - \frac{\Phi_4}{r}\right) + \left[r\frac{\partial^2\chi_4}{\partial r^2} + \frac{\partial\chi_4}{\partial r} + \left(\frac{rk_{4,2}^2}{2} - \frac{1}{r}\right)\chi_4\right]\right\}\bigg|_{r=a_4} = 0, \quad (71)$$

$$\frac{2\mu_4}{a_4}\left\{\left(\frac{\partial\Phi_4}{\partial r} - \frac{\Phi_4}{r}\right) + \left[r\frac{\partial^2\chi_4}{\partial r^2} + \frac{\partial\chi_4}{\partial r} + \left(\frac{rk_{4,2}^2}{2} - \frac{1}{r}\right)\chi_4\right]\right\}\bigg|_{r=a_4} = \frac{1}{\rho_5\omega^2}\frac{\partial p_5}{\partial r}\bigg|_{r=a_4}. \quad (72)$$

At $r = a_5$

$$\left\{\left[2\mu_6\frac{\partial^2\Phi_6}{\partial r^2} - \lambda_6 k_{6,1}^2\Phi_6\right] + 2\mu_6\left[\frac{\partial^3(r\chi_6)}{\partial r^3} + k_{6,2}^2\frac{\partial(r\chi_6)}{\partial r}\right]\right\}\bigg|_{r=a_5} = -p_5|_{r=a_5}, \quad (73)$$

$$\frac{2\mu_6}{a_5}\left\{\left(\frac{\partial\Phi_6}{\partial r} - \frac{\Phi_6}{r}\right) + \left[r\frac{\partial^2\chi_6}{\partial r^2} + \frac{\partial\chi_6}{\partial r} + \left(\frac{rk_{6,2}^2}{2} - \frac{1}{r}\right)\chi_6\right]\right\}\bigg|_{r=a_5} = 0, \quad (74)$$

$$\frac{2\mu_4}{a_5}\left\{\left(\frac{\partial\Phi_6}{\partial r} - \frac{\Phi_6}{r}\right) + \left[r\frac{\partial^2\chi_6}{\partial r^2} + \frac{\partial\chi_6}{\partial r} + \left(\frac{rk_{6,2}^2}{2} - \frac{1}{r}\right)\chi_6\right]\right\}\bigg|_{r=a_5} = \frac{1}{\rho_5\omega^2}\left(\frac{\partial p_5}{\partial r}\right)\bigg|_{r=a_5}. \quad (75)$$

The following eigenfunctions are obtained for the spheroidal modes by substituting Equations (54) into Equations (44):

$$U_i(r,\theta) = \left\{ \begin{array}{c} A_{4i-3}\left[\frac{1}{r} j_n\left(k_{i,1}r\right)\right] + A_{4i-2}\left[\frac{1}{r} y_n\left(k_{i,1}r\right)\right] + \\ A_{4i-1}\left[\frac{(n+1)}{r} j_n\left(k_{i,2}r\right) - k_{i,2} j_{n+1}\left(k_{i,2}r\right)\right] + \\ A_{4i}\left[\frac{(n+1)}{r} y_n\left(k_{i,2}r\right) - k_{i,2} y_{n+1}\left(k_{i,2}r\right)\right] \end{array} \right\} \times \left[-(n+1)\cot\theta P_n^m(\cos\theta) + \frac{n-m+1}{\sin\theta} P_{n+1}^m(\cos\theta)\right], \quad (76)$$

$$V_i(r,\theta) = \left\{ \begin{array}{c} A_{4i-3}\left[-\frac{m}{r\sin\theta} j_n\left(k_{i,1}r\right)\right] + A_{4i-2}\left[-\frac{m}{r\sin\theta} y_n\left(k_{i,1}r\right)\right] + \\ A_{4i-1}\left(-\frac{m}{r\sin\theta}\right)\left[(n+1) j_n\left(k_{i,2}r\right) - k_{i,2} r j_{n+1}\left(k_{i,2}r\right)\right] + \\ A_{4i}\left(-\frac{m}{r\sin\theta}\right)\left[(n+1) y_n\left(k_{i,2}r\right) - k_{i,2} r y_{n+1}\left(k_{i,2}r\right)\right] \end{array} \right\} \times P_n^m(\cos\theta), \quad (77)$$

$$W_i(r,\theta) = \left\{ \begin{array}{c} A_{4i-3}\left[\frac{n}{r} j_n\left(k_{i,1}r\right) - k_{i,1} j_{n+1}\left(k_{i,1}r\right)\right] + \\ A_{4i-2}\left[\frac{n}{r} y_n\left(k_{i,1}r\right) - k_{i,1} y_{n+1}\left(k_{i,1}r\right)\right] + \\ A_{4i-1}\left[\frac{n(n+1)}{r} j_n\left(k_{i,2}r\right)\right] + A_{4i}\left[\frac{n(n+1)}{r} y_n\left(k_{i,2}r\right)\right] \end{array} \right\} \times P_n^m(\cos\theta), \quad (78)$$

$$U_5(r,\theta) = \frac{1}{\rho_5\omega^2}\left\{A_{17}\left[\frac{1}{r} j_n\left(k_{5,3}r\right)\right] + A_{18}\left[\frac{1}{r} y_n\left(k_{5,3}r\right)\right]\right\} \times \left[-(n+1)\cot\theta P_n^m(\cos\theta) + \frac{n-m+1}{\sin\theta} P_{n+1}^m(\cos\theta)\right], \quad (79)$$

$$V_5(r,\theta) = \frac{1}{\rho_5\omega^2}\left\{A_{17}\left[-\frac{m}{r\sin\theta} j_n\left(k_{5,3}r\right)\right] + A_{18}\left[-\frac{m}{r\sin\theta} y_n\left(k_{5,3}r\right)\right]\right\} \times P_n^m(\cos\theta), \quad (80)$$

$$W_5(r,\theta) = \frac{1}{\rho_5\omega^2}\left\{ \begin{array}{c} A_{17}\left[\frac{n}{r} j_n\left(k_{5,3}r\right) - k_{5,3} j_{n+1}\left(k_{5,3}r\right)\right] + \\ A_{18}\left[\frac{n}{r} y_n\left(k_{5,3}r\right) - k_{5,3} y_{n+1}\left(k_{5,3}r\right)\right] \end{array} \right\} \times P_n^m(\cos\theta), \quad (81)$$

$$U_6(r,\theta) = \left\{A_{19}\left[\frac{1}{r} j_n\left(k_{6,1}r\right)\right] + A_{20}\left[\frac{(n+1)}{r} j_n\left(k_{6,2}r\right) - k_{6,2} j_{n+1}\left(k_{6,2}r\right)\right]\right\} \times \left[-(n+1)\cot\theta P_n^m(\cos\theta) + \frac{n-m+1}{\sin\theta} P_{n+1}^m(\cos\theta)\right], \quad (82)$$

$$V_6(r,\theta) = \left\{ \begin{array}{c} A_{19}\left[\left(-\frac{m}{r\sin\theta}\right) j_n(k_{6,1}r)\right] + A_{20}\left(-\frac{m}{r\sin\theta}\right) \times \\ \left[(n+1)\, j_n(k_{6,2}r) - k_{6,2} r j_{n+1}(k_{6,2}r)\right] \end{array} \right\} \times$$
$$P_n^m(\cos\theta), \quad (83)$$

$$W_6(r,\theta) = \left\{ \begin{array}{c} A_{19}\left[\frac{n}{r} j_n(k_{6,1}r) - k_{6,1} j_{n+1}(k_{6,1}r)\right] + \\ A_{20}\left[\frac{n(n+1)}{r} j_n(k_{6,2}r)\right] \end{array} \right\} \times$$
$$P_n^m(\cos\theta), \quad (84)$$

where $i = 1,2,3,4$.

For the investigation of torsional modes we formulate the boundary and continuity conditions which express the balance between the tangential stresses in the solid layers as well as the equality of the corresponding tangential displacements on junctions of the layers. The outer layer is assumed to be free from tangential stress components. Using these assumptions together with Equations (53) we obtain the following boundary and continuity conditions:

At $r = a$

$$\frac{\mu_1}{a}\left(\frac{\partial\Psi_1}{\partial r} - \frac{\Psi_1}{r}\right)\bigg|_{r=a} = 0. \quad (85)$$

At $r = a_1$

$$\frac{\mu_1}{a_1}\left(\frac{\partial\Psi_1}{\partial r} - \frac{\Psi_1}{r}\right)\bigg|_{r=a_1} = \frac{\mu_2}{a_1}\left(\frac{\partial\Psi_2}{\partial r} - \frac{\Psi_2}{r}\right)\bigg|_{r=a_1}, \quad (86)$$

$$\Psi_1\big|_{r=a_1} = \Psi_2\big|_{r=a_1}. \quad (87)$$

At $r = a_2$

$$\frac{\mu_2}{a_2}\left(\frac{\partial\Psi_2}{\partial r} - \frac{\Psi_2}{r}\right)\bigg|_{r=a_2} = \frac{\mu_3}{a_2}\left(\frac{\partial\Psi_3}{\partial r} - \frac{\Psi_3}{r}\right)\bigg|_{r=a_2}, \quad (88)$$

$$\Psi_2\big|_{r=a_2} = \Psi_3\big|_{r=a_2}. \quad (89)$$

At $r = a_3$

$$\frac{\mu_3}{a_3}\left(\frac{\partial\Psi_3}{\partial r} - \frac{\Psi_3}{r}\right)\bigg|_{r=a_3} = \frac{\mu_4}{a_3}\left(\frac{\partial\Psi_4}{\partial r} - \frac{\Psi_4}{r}\right)\bigg|_{r=a_3}, \quad (90)$$

$$\Psi_3\big|_{r=a_3} = \Psi_4\big|_{r=a_3}. \quad (91)$$

At $r = a_4$

$$\frac{\mu_4}{a_4}\left(\frac{\partial\Psi_4}{\partial r} - \frac{\Psi_4}{r}\right)\bigg|_{r=a_4} = 0. \quad (92)$$

For the torsional modes, the eigenfunctions are as follow for $i = 1,2,3,4$:

$$U_i(r,\theta) = \left\{B_{2i-1}\left[-\frac{m}{a\sin\theta} j_n(k_{i,2}r)\right] + B_{2i}\left[-\frac{m}{a\sin\theta} y_n(k_{i,2}r)\right]\right\} \times$$
$$P_n^m(\cos\theta), \quad (93)$$

$$V_i(r,\theta) = \left\{B_{2i-1}\left[-\frac{1}{a} j_n(k_{i,2}r)\right] + B_{2i}\left[-\frac{1}{a} y_n(k_{i,2}r)\right]\right\} \times$$
$$\left[-(n+1)\cot\theta P_n^m(\cos\theta) + \frac{n-m+1}{\sin\theta} P_{n+1}^m(\cos\theta)\right], \quad (94)$$

$$W_i(r,\theta) = 0. \quad (95)$$

	λ_i(Pa)	μ_i(Pa)	ρ_i (kg/m^3)	α_i (m/s)	β_i (m/s)
"Crust"	33.4×10^9	36.4×10^9	2.6×10^3	6.40×10^3	3.75×10^3
"Upper Mantle -1"	110×10^9	85.3×10^9	3.7×10^3	8.70×10^3	4.80×10^3
"Upper Mantle -2"	156×10^9	130×10^9	4.0×10^3	10.2×10^3	7.50×10^3
"Lower Mantle"	343×10^9	238×10^9	5.0×10^3	12.8×10^3	6.90×10^3
"Outer Core"	783×10^9	0	11.1×10^3	8.4×10^3	0
"Inner Core"	115×10^{10}	206×10^9	12.9×10^3	11.0×10^3	4.00×10^3

Table 1. Physical properties of the moon's layers and the phase velocities α_i and β_i of the P- and S- waves respectively.

5.1 Example

Consider the following fictional example: In the course of the "Avatar mission" it was found that the moon "Pandora" has a spherical layered structure with five concentric layers and an "inner core". It was determined that the outer radius of the moon is $a = 6371\,\text{km}$; the boundary between the upper layer ("crust") and the "upper mantle-1" is situated at radius $a_1 = 6336\,\text{km}$; the boundary between the "upper mantle-1" and "upper mantle-2" is at radius $a_2 = 5961\,\text{km}$; the boundary between the "upper mantle-2" and "lower mantle" is at radius $a_3 = 5711\,\text{km}$; the boundary between the "lower mantle" and the "outer core" is at radius $a_4 = 3482\,\text{km}$; and finally, the boundary between the "outer core" and "inner core" is at radius $a_5 = 1217\,\text{km}$. It is assumed that the outer layers ("crust", "upper mantle-1", "upper mantle-2", "lower mantle" and the "inner core") are solids and the "outer core" is liquid, which does not support shear stresses (i.e. it is simulated by the model of an inviscid ideal fluid). Table 1 gives the physical properties of the moon as well as the phase velocities of the P-wave

$$\alpha_i = \sqrt{\frac{\lambda_i + 2\mu_i}{\rho_i}} \tag{96}$$

and the S-wave

$$\beta_i = \sqrt{\frac{\mu_i}{\rho_i}}, \tag{97}$$

$i = 1, \cdots, 6$ where λ_i, μ_i are Lame coefficients (keeping in mind that $\lambda_5 = \kappa_5$ the bulk modulus with $\mu_5 = 0$) and ρ_i are mass densities of the layers.

For the spheroidal mode of vibration, for each pair of wave numbers n and m, by using Equations (54) and applying the boundary and continuity conditions Equations (56),$\cdots$,(75) we obtain a matrix equation

$$M \begin{pmatrix} A_1 \\ \vdots \\ A_{20} \end{pmatrix} = \begin{pmatrix} 0 \\ \vdots \\ 0 \end{pmatrix} \tag{98}$$

where the 20 × 20 matrix $M = M(\omega) = \left(a_{i,j}(\omega)\right)_{i,j=1,\cdots,20}$. Linear algebra tells us that Equation (98) has a nontrivial solution if the determinant

$$\det(M(\omega)) = 0. \tag{99}$$

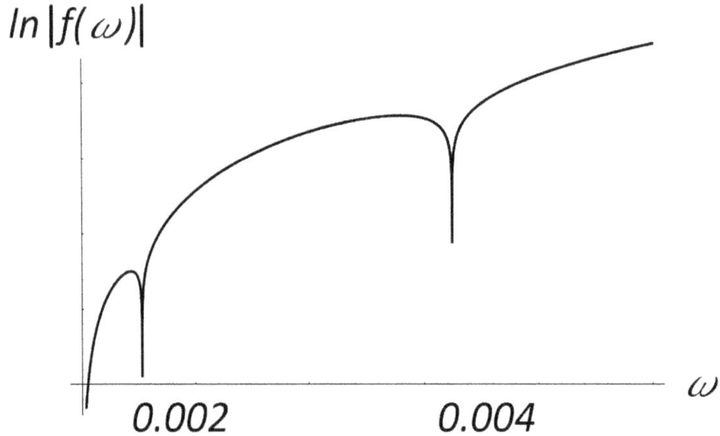

Fig. 2. A graph of the function $g(\omega) = \ln|\det(M(\omega))|$. The vertical "spikes" point towards roots of the function $f(\omega) = M(\omega)$. It appears that "good guesses" for the first and second eigenvalues are $\omega_1 \approx 0.002\,\text{rad}\,.\,\text{s}^{-1}$ and $\omega_2 \approx 0.004\,\text{rad}\,.\,\text{s}^{-1}$ respectively.

Determining the roots of the function

$$f(\omega) = \det(M(\omega)) \tag{100}$$

numerically yields a sequence of eigenvalues $\omega_1 < \omega_2 < \cdots$ (and hence periods of vibration $T_1 > T_2 > \cdots$ where $T_j = \frac{2\pi}{\omega_j}$). For instance, with $n = m = 2$, a graph of the function

$$g(\omega) = \ln|f(\omega)| \tag{101}$$

will be similar to that given in Figure 2. The vertical "spikes" indicate roots of the function $f(\omega)$. It appears that "good guesses" for the first and second eigenvalues are $\omega_1 \approx 0.002\,\text{rad.}\,\text{s}^{-1}$ and $\omega_2 \approx 0.004\,\text{rad.}\,\text{s}^{-1}$ respectively. These "good guesses" are then used in the numerical routine of a computer algebra system, producing ω_1 and ω_2 to, say, four significant figures of accuracy and hence yield the values $T_1 = \frac{2\pi}{\omega_1} = 68.7\,\text{min}$ and $T_2 = 24.3\,\text{min}$ as given in Table 2. Once a suitable number of eigenvalues have been determined, by choosing a fixed eigenvalue ω_k, the matrix $M(\omega_k)$ has known constant entries. Hence it is now possible to determine a non-trivial solution to

$$M(\omega_k)\begin{pmatrix} A_1 \\ \vdots \\ A_{20} \end{pmatrix} = \begin{pmatrix} 0 \\ \vdots \\ 0 \end{pmatrix}. \tag{102}$$

n	m	T_1 (min) η_1	T_2 (min) η_2	T_3 (min) η_3	T_4 (min) η_4	T_5 (min) η_5
2	2	68.7 −0.86	24.3 −0.63	15.7 −0.14	14.4 −0.24	9.2 −0.17
3	2	41.2 −0.47	17.7 −0.51	13.2 −0.05	10.3 −0.10	7.8 −0.13
3	3	41.2 −0.70	17.7 −0.76	13.2 −0.08	10.3 −0.14	7.8 −0.19
4	2	28.2 −0.25	14.4 −0.43	11.8 −0.03	7.9 −0.04	7.1 −0.12
4	3	28.2 −0.44	14.4 −0.64	11.8 −0.04	7.9 −0.06	7.1 −0.19
4	4	28.2 −0.58	14.4 −0.86	11.8 −0.05	7.9 −0.08	7.1 −0.25
5	2	21.0 −0.20	12.5 −0.32	10.5 −0.06	6.7 −0.07	6.5 −0.02
5	3	21.0 −0.30	12.5 −0.48	10.5 −0.09	6.7 −0.11	6.5 −0.03
5	4	21.0 −0.40	12.5 −0.64	10.5 −0.12	6.7 −0.14	6.5 −0.04
5	5	21.0 −0.50	12.5 −0.80	10.5 −0.15	6.7 −0.18	6.5 −0.05

Table 2. Period of oscillation T_j and Bryan's factor η_j of the spheroidal modes of vibration for wave numbers n and m.

Consequently the eigenfunctions given by Equations (76) to (84) are completely determined. Keeping Equations (40) and (42) in mind, Bryan's factor is calculated by using

$$\eta = \frac{2\int_0^{\pi}\left\{\sum_{i=1}^{6}\int_{a_i}^{a_{i-1}}\left[\rho_i\left(U_i\cos\theta + W_i\sin\theta\right)V_i\right]r^2dr\right\}\sin\theta\,d\theta}{\int_0^{\pi}\left\{\sum_{i=1}^{6}\int_{a_i}^{a_{i-1}}\rho_i\left(U_i^2+V_i^2+W_i^2\right)r^2dr\right\}\sin\theta\,d\theta}, \tag{103}$$

where $a_0 = a$ and $a_6 = 0$.

For the spheroidal modes of vibration, the periods of oscillation $T_j = \frac{2\pi}{\omega_j}$ as well as the corresponding Bryan's factors η_j for a pair of wave numbers m and n are given in Table 2. Doing a similar analysis (as explained above) for the torsional modes, the eigenfunctions given by Equations (93),$\cdots$, (95) as well as the eigenvalues are completely determined by applying the boundary and continuity conditions given by Equations (85),$\cdots$, (92) to Equations (55). Here Equation (103) for Bryan's factor may be simplified, because the radial displacements are zero $W_i = 0$. Bryan's factor for the torsional modes is calculated by using the formula

$$\eta_{TM} = \frac{2\int_0^{\pi}\left\{\sum_{i=1}^{6}\int_{a_i}^{a_{i-1}}U_iV_ir^2dr\right\}\sin\theta\cos\theta\,d\theta}{\int_0^{\pi}\left\{\sum_{i=1}^{6}\int_{a_i}^{a_{i-1}}\rho_i\left(U_i^2+V_i^2\right)r^2dr\right\}\sin\theta\,d\theta}. \tag{104}$$

Values of the periods of oscillation T_j and the corresponding Bryan's factors η_j for the n and m torsional modes are given in Table 3.

n	m	T_1 (min) η_1	T_2 (min) η_2	T_3 (min) η_3	T_4 (min) η_4	T_5 (min) η_5
2	2	42.4 0.33	11.9 0.33	7.2 0.33	5.0 0.33	3.7 0.33
3	2	27.2 0.17	11.0 0.17	7.0 0.17	4.9 0.17	3.7 0.17
3	3	27.2 0.25	11.0 0.25	7.0 0.25	4.9 0.25	3.7 0.25
4	2	20.7 0.10	10.0 0.10	6.8 0.10	4.8 0.10	3.6 0.10
4	3	20.7 0.15	10.0 0.15	6.8 0.15	4.8 0.15	3.6 0.15
4	4	20.7 0.20	10.0 0.20	6.8 0.20	4.8 0.20	3.6 0.20
5	2	16.9 0.07	9.1 0.07	6.5 0.07	4.7 0.07	3.5 0.07
5	3	16.9 0.10	9.1 0.10	6.5 0.10	4.7 0.10	3.5 0.10
5	4	16.9 0.13	9.1 0.13	6.5 0.13	4.7 0.13	3.5 0.13
5	5	16.9 0.17	9.1 0.17	6.5 0.17	4.7 0.17	3.5 0.17

Table 3. Period of oscillation T_j and Bryan's factor η_j of the torsional modes of vibration for wave numbers n and m.

6. Conclusions and discussions

After deriving expressions for Bryan's factor η in terms of eigenvalues and eigenfunctions of vibration we demonstrated that neither Bryan's effect nor the value of Bryan's factor η depend on the inclusion of light, isotropic, viscous damping in the model. Consequently we neglected damping in the model that we presented. However, we pointed out that preliminary studies indicate that the inclusion of "impurities" such as mass-stiffness, prestress and anisotropic damping into the model appears to produce "capture effects" such as the rotation angle of the vibration pattern varying periodically. Hence more realistic models (as opposed to ideal situations) that will include these "impurities" need to be developed.

Solutions were obtained for the dynamic equations of slowly rotating, vibrating planets or moons considered as ideal elastic solid and inviscid acoustic bodies composed of, for instance, four solid layers, one liquid layer and a solid core. Boundary conditions were formulated for calculating the eigenvalues and eigenfunctions of vibration and a brief discussion of how they are numerically calculated was given. It was pointed out that for each pair of wave numbers n and m, there is a sequence of eigenvalues $\omega_1 < \omega_2 < \cdots$ and hence periods of vibration $T_1 > T_2 > \cdots$ (where $T_j = \frac{2\pi}{\omega_j}$).

The model derived indicates that there are two modes of vibration, namely the "spheroidal" and "torsional" modes and Bryan's effect occurs for both modes.

Table 1 provides feasible physical properties for a fictional moon and gives possible phase velocities for both P- and S-waves.

Using Table 1, a numerical experiment produced Table 2, where it appears that for the spheroidal modes, for a given polar wave number n and a given non-negative integer j, the period of oscillation T_j does not depend on the circumferential wave number m. However, Bryan's factor η depends on both wave numbers n and m as well as the period of oscillation T_j for each pair n and m. Bryan's factor η appears to be negative in all cases and this means that an antinode of vibration will rotate through the moon's crusts in the **opposite direction** to the moon's rotation at a rate $\eta\varphi$, where φ is the moon's rotation rate.
Another numerical experiment produced Table 3, where it appears that for torsional modes, for a given wave number n and a given non-negative integer j, the period of oscillation T_j is independent of the circumferential wave number m and that it varies with n. However, Bryan's factor η appears to be invariant and independent of the period of oscillation for a given pair of wave numbers n and m, but appears to vary as each pair varies and it appears to be positive in all cases. Consequently Bryan's factors for the torsional modes depend only on angular components of the vibrating patterns, that is, on the corresponding wave numbers n and m, and not on radial vibrations because all radial displacements $W_i = 0$. These positive Bryan's factors η indicate that the antinodes of vibration will rotate through the moon's crusts in the **same direction** as the moon's rotation at a rate $\eta\gamma$, where γ is the moon's rotation rate.

7. Acknowledgement and disclaimer

This material is based upon work supported financially by the Tshwane University of Technology (TUT), the Council for Scientific and Industrial Research (CSIR) of South Africa and the National Research Foundation (NRF) of South Africa (NRF grant reference number EV2009011 400005). Any opinions, findings and conclusions or recommendations expressed in this material are those of the authors and therefore the TUT, the CSIR and the NRF do not accept any liability in regard thereto.

8. References

Bryan, G.H. (1890). On the beats in the vibrations of a revolving cylinder or bell, *Proceedings of the Cambridge Philosophical Society* 7, pp. 101-111.

Eringen, A.E. & Suhubi, E.S. (1975). *Elastodynamics, vol. I1.* Academic Press, New York.

Goldstein, H., Poole, C. & Safko, J. (2001). *Classical mechanics, 3rd ed*, Addison-Wesley, Reading, MA.

Loveday, P.W & Rogers, C.A. (1998). Free vibrations of elastically supported thin cylinders including gyroscopic effects, *Journal of Sound and Vibration* 217(3), pp. 547-562.

Nayfeh, A.H. (1973). *Perturbation methods*, John Wiley & Sons, Inc., New York.

Rayleigh, J.W.S. (1894). *The theory of sound, volume I, second edition*, MacMillan & Company, London.

Redwood, M. (1960). *Mechanical waveguides*, Pergamon Press, Oxford.

Rozelle, D.M. (2009). The hemispherical resonator gyro: From wineglass to the planets, *Proceedings of the 19th AAS/AIAA Space Flight Mechanics Meeting, February 8-12, Savannah, Georgia, USA*, pp. 1157-1178.

Shatalov, M.Y., Joubert, S.V., Coetzee, C.E. & Fedotov, I. (2009). Free vibration of rotating hollow spheres containing acoustic media, *Journal of Sound and Vibration*, 322(4-5), pp. 1038-1047.
Available online: doi:10.1016/j.jsv.2008.11.020, accessed July 2011.

Shatalov, M.Y., Joubert, S.V. & Coetzee, C.E. (2011). The influence of mass imperfections on the evolution of standing waves in slowly rotating spherical bodies, *Journal of Sound and Vibration*, 330, pp. 127–135.
Available online: doi:10.1016/j.jsv.2010.08.001, accessed July 2011.

Spiegel, M.R. (1967). *Theoretical mechanics.* McGraw-Hill, New York.

Zhuravlev, V. & Klimov, D. (1988). *Applied methods in the theory of oscillations.* Moscow: Nauka (in Russian).

Permissions

The contributors of this book come from diverse backgrounds, making this book a truly international effort. This book will bring forth new frontiers with its revolutionizing research information and detailed analysis of the nascent developments around the world.

We would like to thank Prof. Abbas Moustafa, for lending his expertise to make the book truly unique. He has played a crucial role in the development of this book. Without his invaluable contribution this book wouldn't have been possible. He has made vital efforts to compile up to date information on the varied aspects of this subject to make this book a valuable addition to the collection of many professionals and students.

This book was conceptualized with the vision of imparting up-to-date information and advanced data in this field. To ensure the same, a matchless editorial board was set up. Every individual on the board went through rigorous rounds of assessment to prove their worth. After which they invested a large part of their time researching and compiling the most relevant data for our readers. Conferences and sessions were held from time to time between the editorial board and the contributing authors to present the data in the most comprehensible form. The editorial team has worked tirelessly to provide valuable and valid information to help people across the globe.

Every chapter published in this book has been scrutinized by our experts. Their significance has been extensively debated. The topics covered herein carry significant findings which will fuel the growth of the discipline. They may even be implemented as practical applications or may be referred to as a beginning point for another development.

The editorial board has been involved in producing this book since its inception. They have spent rigorous hours researching and exploring the diverse topics which have resulted in the successful publishing of this book. They have passed on their knowledge of decades through this book. To expedite this challenging task, the publisher supported the team at every step. A small team of assistant editors was also appointed to further simplify the editing procedure and attain best results for the readers.

Our editorial team has been hand-picked from every corner of the world. Their multi-ethnicity adds dynamic inputs to the discussions which result in innovative outcomes. These outcomes are then further discussed with the researchers and contributors who give their valuable feedback and opinion regarding the same. The feedback is then collaborated with the researches and they are edited in a comprehensive manner to aid the understanding of the subject.

Apart from the editorial board, the designing team has also invested a significant amount of their time in understanding the subject and creating the most relevant covers. They scrutinized every image to scout for the most suitable representation of the subject and create an appropriate cover for the book.

The publishing team has been involved in this book since its early stages. They were actively engaged in every process, be it collecting the data, connecting with the contributors or procuring relevant information. The team has been an ardent support to the editorial, designing and production team. Their endless efforts to recruit the best for this project, has resulted in the accomplishment of this book. They are a veteran in the field of academics and their pool of knowledge is as vast as their experience in printing. Their expertise and guidance has proved useful at every step. Their uncompromising quality standards have made this book an exceptional effort. Their encouragement from time to time has been an inspiration for everyone.

The publisher and the editorial board hope that this book will prove to be a valuable piece of knowledge for researchers, students, practitioners and scholars across the globe.

List of Contributors

Neelima Satyam
Earthquake Engineering Research Centre, International Institute of Information Technology Hyderabad, India

Sarfraz Ali
NUST College of Civil Engineering, Pakistan

Robert C. Lo
Klohn Crippen Berger Ltd., Vancouver, B.C, Canada

Yumei Wang
Sustainable Living Solutions LLC, Portland, Oregon, U.S.A

Richard L. Handy
Iowa State University, U.S.A

Lin Bor-Shiun
Geotechnical Engineering Research Center, Sinotech Engineering Consultants, Taiwan

Zhenzhong Shen, Lei Gan, Juan Cui and Liqun Xu
State Key Laboratory of Hydrology-Water Resources and Hydraulic Engineering, Hohai University, China

Babak Ebrahimian
University of Tehran, Iran

Hasan Tosun
Eskisehir Osmangazi University, Civil Engineering Department, Eskisehir, Turkey

Evren Seyrek
Dumlupinar University, Civil Engineering Department, Kutahya, Turkey

Ahsan Sattar and Kazuo Konagai
Institute of Industrial Science, University of Tokyo, Japan

Xiao Shiyun
Faculty of Infrastructure Engineering, Dalian University of Technology, Dalian, China

Sayed Hemeda
Conservation Department, Faculty of Archaeology, Cairo University, Egypt

Ken-ichi Tokida
Osaka University, Japan

Alessandro Contento and Angelo Di Egidio
Department of Structural, Hydraulic and Geotechnical Engineering, University of L'Aquila, Italy

Seigo Sakai
Yokohama National University, Japan

Masato Saitoh
Saitama University, Shimo-Okubo Sakura-Ku Saitama, Saitama, Japan

Michael Y. Shatalov
CSIR (Pretoria) and Tshwane University of Technology, South Africa

Stephan V. Joubert and Charlotta E. Coetzee
Tshwane University of Technology, South Africa

www.ingramcontent.com/pod-product-compliance
Ingram Content Group UK Ltd.
Pitfield, Milton Keynes, MK11 3LW, UK
UKHW022206051026
15624UKWH00001B/15

9 781632 382146